# Plant Production: Natural Materials

Edited by Ben Davies

**SYRAWOOD**
PUBLISHING HOUSE

New York

Published by Syrawood Publishing House,
750 Third Avenue, 9th Floor,
New York, NY 10017, USA
www.syrawoodpublishinghouse.com

**Plant Production: Natural Materials**
Edited by Ben Davies

© 2018 Syrawood Publishing House

International Standard Book Number: 978-1-68286-501-9 (Hardback)

**Cataloging-in-Publication Data**

Plant production : natural materials / edited by Ben Davies.
    p. cm.
Includes bibliographical references and index.
ISBN 978-1-68286-501-9
1. Plants--Reproduction. 2. Plant propagation. 3. Plant cuttings.
4. Plants, Cultivated. I. Davies, Ben.
QK825 .P53 2018
575.6--dc23

# TABLE OF CONTENTS

Preface..................................................................................................................................IX

Chapter 1 **Genome-Wide Survey and Expression Analysis of Calcium-Dependent Protein Kinase in** *Gossypium raimondii*..................................................................1
Wei Liu, Wei Li, Qiuling He, Muhammad Khan Daud, Jinhong Chen, Shuijin Zhu

Chapter 2 **Effect of Late Planting and Shading on Cellulose Synthesis during Cotton Fiber Secondary Wall Development**..................................................................12
Ji Chen, Fengjuan Lv, Jingran Liu, Yina Ma, Youhua Wang, Binglin Chen, Yali Meng, Zhiguo Zhou, Derrick M. Oosterhuis

Chapter 3 **Gibberellin Overproduction Promotes Sucrose Synthase Expression and Secondary Cell Wall Deposition in Cotton Fibers**..................................................26
Wen-Qin Bai, Yue-Hua Xiao, Juan Zhao, Shui-Qing Song, Lin Hu, Jian-Yan Zeng, Xian-Bi Li, Lei Hou, Ming Luo, De-Mou Li, Yan Pei

Chapter 4 **Market Forces and Technological Substitutes Cause Fluctuations in the Value of Bat Pest-Control Services for Cotton**..................................................33
Laura López-Hoffman, Ruscena Wiederholt, Chris Sansone, Kenneth J. Bagstad, Paul Cryan, Jay E. Diffendorfer, Joshua Goldstein, Kelsie LaSharr, John Loomis, Gary McCracken, Rodrigo A. Medellín, Amy Russell, Darius Semmens

Chapter 5 **The *Li₂* Mutation Results in Reduced Subgenome Expression Bias in Elongating Fibers of Allotetraploid Cotton (*Gossypium hirsutum* L.)**..................................................40
Marina Naoumkina, Gregory Thyssen, David D. Fang, Doug J. Hinchliffe, Christopher Florane, Kathleen M. Yeater, Justin T. Page, Joshua A. Udall

Chapter 6 **Ecoinformatics Reveals Effects of Crop Rotational Histories on Cotton Yield**..................................................52
Matthew H. Meisner, Jay A. Rosenheim

Chapter 7 **Thirdhand Cigarette Smoke: Factors Affecting Exposure and Remediation**..................................................60
Vasundhra Bahl, Peyton Jacob III, Christopher Havel, Suzaynn F. Schick, Prue Talbot

Chapter 8 **The Effects of Fruiting Positions on Cellulose Synthesis and Sucrose Metabolism during Cotton (*Gossypium hirsutum* L.) Fiber Development**..................................................69
Yina Ma, Youhua Wang, Jingran Liu, Fengjuan Lv, Ji Chen, Zhiguo Zhou

Chapter 9 **Feeding and Dispersal Behavior of the Cotton Leafworm, *Alabama argillacea* (Hübner) (Lepidoptera: Noctuidae), on Bt and Non-Bt Cotton: Implications for Evolution and Resistance Management**..................................................78
Francisco S. Ramalho, Jéssica K. S. Pachú, Aline C. S. Lira, José B. Malaquias, José C. Zanuncio, Francisco S. Fernandes

Chapter 10 **Identification of Top-Down Forces Regulating Cotton Aphid Population Growth in Transgenic Bt Cotton in Central China**..................................................88
Peng Han, Chang-ying Niu, Nicolas Desneux

Chapter 11   **Competitive Ability and Fitness Differences between Two Introduced Populations of the Invasive Whitefly *Bemisia tabaci* Q in China**............97
Yi-Wei Fang, Ling-Yun Liu, Hua-Li Zhang, De-Feng Jiang, Dong Chu

Chapter 12   **A Meta-Analysis of the Impacts of Genetically Modified Crops**............106
Wilhelm Klümper, Matin Qaim

Chapter 13   **The Entomopathogenic Fungal Endophytes *Purpureocillium lilacinum* (Formerly *Paecilomyces lilacinus*) and *Beauveria bassiana* Negatively Affect Cotton Aphid Reproduction under both Greenhouse and Field Conditions**............113
Diana Castillo Lopez, Keyan Zhu-Salzman, Maria Julissa Ek-Ramos, Gregory A. Sword

Chapter 14   **Genetic Structure, Linkage Disequilibrium and Association Mapping of Verticillium Wilt Resistance in Elite Cotton (*Gossypium hirsutum* L.) Germplasm Population**............121
Yunlei Zhao, Hongmei Wang, Wei Chen, Yunhai Li

Chapter 15   **Quantification of the Pirimicarb Resistance Allele Frequency in Pooled Cotton Aphid (*Aphis gossypii* Glover) Samples by TaqMan SNP Genotyping Assay**............136
Yizhou Chen, Daniel R. Bogema, Idris M. Barchia, Grant A. Herron

Chapter 16   **Molecular Mapping and Validation of a Major QTL Conferring Resistance to a Defoliating Isolate of Verticillium Wilt in Cotton (*Gossypium hirsutum* L.)**............148
Xingju Zhang, Yanchao Yuan, Ze Wei, Xian Guo, Yuping Guo, Suqing Zhang, Junsheng Zhao, Guihua Zhang, Xianliang Song, Xuezhen Sun

Chapter 17   **Upland Cotton Gene *GhFPF1* Confers Promotion of Flowering Time and Shade-Avoidance Responses in *Arabidopsis thaliana***............156
Xiaoyan Wang, Shuli Fan, Meizhen Song, Chaoyou Pang, Hengling Wei, Jiwen Yu, Qifeng Ma, Shuxun Yu

Chapter 18   **Effects of Transgenic Cry1Ac + CpTI Cotton on Non-Target Mealybug Pest *Ferrisia virgata* and its Predator *Cryptolaemus montrouzieri***............167
Hongsheng Wu, Yuhong Zhang, Ping Liu, Jiaqin Xie, Yunyu He, Congshuang Deng, Patrick De Clercq, Hong Pang

Chapter 19   **Association Mapping for Epistasis and Environmental Interaction of Yield Traits in 323 Cotton Cultivars under 9 Different Environments**............177
Yinhua Jia, Xiwei Sun, Junling Sun, Zhaoe Pan, Xiwen Wang, Shoupu He, Songhua Xiao, Weijun Shi, Zhongli Zhou, Baoyin Pang, Liru Wang, Jianguang Liu, Jun Ma, Xiongming Du, Jun Zhu

Chapter 20   **Transcriptome Sequencing and *De Novo* Analysis of Cytoplasmic Male Sterility and Maintenance in JA-CMS Cotton**............185
Peng Yang, Jinfeng Han, Jinling Huang

Chapter 21   **Effects of Gibberellic Acid and N, N-Dimethyl Piperidinium Chloride on the Dose of and Physiological Responses to Prometryn in Black Nightshade (*Solanum nigrum* L.)**............198
Hailan Jiang, Xiaoxia Deng, Jungang Wang, Jing Wang, Jun Peng, Tingting Zhou

Chapter 22    **mRNA-seq Analysis of the *Gossypium* Arboreum Transcriptome Reveals Tissue
Selective Signaling in Response to Water Stress during Seedling Stage**................................................ 206
Xueyan Zhang, Dongxia Yao, Qianhua Wang, Wenying Xu, Qiang Wei,
Chunchao Wang, Chuanliang Liu, Chaojun Zhang, Hong Yan, Yi Ling, Zhen Su,
Fuguang Li

**Permissions**

**List of Contributors**

**Index**

# PREFACE

The products or physical matter that is extracted from plants are known as natural materials, such as cotton, vegetable, grains and timber. It studies sustainable methods of making plants more productive and resistant to diseases as well as maintaining soil nutrition and quality. This book contains some path-breaking studies in the field of plant production. Those with an interest in plant production would find this book helpful. Coherent flow of topics, student-friendly language and extensive use of examples make this book an invaluable source of knowledge. This book will serve as a reference to a broad spectrum of readers.

After months of intensive research and writing, this book is the end result of all who devoted their time and efforts in the initiation and progress of this book. It will surely be a source of reference in enhancing the required knowledge of the new developments in the area. During the course of developing this book, certain measures such as accuracy, authenticity and research focused analytical studies were given preference in order to produce a comprehensive book in the area of study.

This book would not have been possible without the efforts of the authors and the publisher. I extend my sincere thanks to them. Secondly, I express my gratitude to my family and well-wishers. And most importantly, I thank my students for constantly expressing their willingness and curiosity in enhancing their knowledge in the field, which encourages me to take up further research projects for the advancement of the area.

**Editor**

# Genome-Wide Survey and Expression Analysis of Calcium-Dependent Protein Kinase in *Gossypium raimondii*

Wei Liu[1⊕], Wei Li[1⊕], Qiuling He[1], Muhammad Khan Daud[2], Jinhong Chen[1]*, Shuijin Zhu[1]*

1 Department of Agronomy, Zhejiang University, Hangzhou, Zhejiang, China, 2 Department of Biotechnology and Genetic Engineering, Kohat University of Science and Technology, Kohat, Pakistan

## Abstract

Calcium-dependent protein kinases (CDPKs) are one of the largest protein kinases in plants and participate in different physiological processes through regulating downstream components of calcium signaling pathways. In this study, 41 CDPK genes, from *GrCPK1* to *GrCPK41*, were identified in the genome of the diploid cotton, *Gossypium raimondii*. The phylogenetic analysis indicated that all these genes were divided into four subgroups and members within the same subgroup shared conserved exon-intron structures. The expansion of *GrCPKs* family in *G. raimondii* was due to the segmental duplication events, and the analysis of Ka/Ks ratios implied that the duplicated *GrCPKs* had mainly undergone strong purifying selection pressure with limited functional divergence. The cold-responsive elements in promoter regions were detected in the majority of *GrCPKs*. The expression analysis of 11 selected genes showed that *GrCPKs* exhibited tissue-specific expression patterns and the expression of *GrCPKs* were induced or repressed by cold treatment. These observations would lay an important foundation for functional and evolutionary analysis of CDPK gene family in *Gossypium* species.

**Editor:** Baohong Zhang, East Carolina University, United States of America

**Funding:** The works are support by the National Basic Research Program (973 program, No: 2010CB126006) and the National High Technology Research and Development Program of China (2011AA10A102, 2013AA102601). The funders had no role in study design, data collection and analysis, decision to publish, or preparation of the manuscript.

**Competing Interests:** The authors have declared that no competing interests exist.

* E-mail: shjzhu@zju.edu.cn (SZ); jinhongchen@zju.edu.cn (JC)

⊕ These authors contributed equally to this work.

## Introduction

The plant growth and crop production are adversely affected by common stress conditions, such as drought, low temperature and high salinity [1]. Adaptation of plants to these environmental stresses includes the perception of stress signals and subsequent signal transduction, leading to the activation of various physiological and metabolic responses [2,3,4]. As a ubiquitous second messenger in cells, calcium ($Ca^{2+}$) plays an important role in the signal transduction pathways. The transient changes of cytoplasmic $Ca^{2+}$ concentration in response to various stresses were sensed and decoded by several $Ca^{2+}$ sensors or $Ca^{2+}$ binding proteins which relayed the signals into downstream response processes such as regulation of gene expression and phosphorylation cascades [5,6].

There are four major classes of $Ca^{2+}$ binding proteins characterized in plants, including calmodulins (CaM), calmodulin-like proteins (CaML), calcineurin B-like proteins (CBL) and calcium-dependent protein kinases (CDPK) [7,8,9]. Among them, CDPKs are the best characterized and are of particular interest, which constitute a large multigene family and are reported to be found throughout the plant kingdom from algae to angiosperms [10]. The CDPK protein possesses four characterized domains, a variable N-terminal domain, a catalytic Ser/Thr protein kinase domain, an autoinhibitory region, and a calmodulin-like domain [9,11,12,13]. The N-terminal domain is highly variable and contains myristoylation or palmotylation sites which may contribute to membrane localization [12,14]. And the calmodulin-like domain contains EF-hands structure for binding of $Ca^{2+}$ [11,12].

It was confirmed that CDPK genes were involved in regulating plant response to various stimuli, including abiotic and biotic stresses and hormones. Overexpression of the rice CDPK gene *OsCDPK7* enhanced the tolerance of the transgenic rice plants to cold, salt, and drought stress [15,16], and *OsCPK12* overexpression plants increased the tolerance to salt stress but to blast fungus [17]. The low temperatures induced *ZmCDPK1* expression in maize [18], and *ZmCPK11* participated in touch- and wound-induced pathways [19]. In tobacco, *NtCPK4* expression was increased in response to the treatment of gibberellin or NaCl [20]. In Arabidopsis, *AtCPK10* was reported to participate in ABA- and $Ca^{2+}$-mediated stomatal regulation in response to drought stress [21], and *AtCPK3* and *AtCPK6* were also shown as positive transducers of stomatal ABA signaling in guard cells [22]. *AtCPK23* was demonstrated to function in response to drought and salt stresses [23]. In addition, some CDPK genes were also involved in pollen tube growth [24], root development [25], cell division and differentiation, and cell death [26,27].

So far, many CDPK genes have been identified from numerous plant species. In Arabidopsis, 34 CDPK genes were revealed [9,12]. And similarly, 31 CDPK genes in rice genome has identified by a genome-wide analysis [28]. In wheat, 20 CDPK genes including 14 full-length cDNA sequences were comprehensively studied [29]. And in poplar, 30 CDPK genes were identified [30], and 35 CDPK genes were initially revealed in maize genome [31]. Other higher plants such as tomato [32], potato [33,34], tobacco [35,36] and grapevine [37] also have multiple CDPK genes characterized in recent years. However, compared to the extensive studies of CDPK genes in many other plant species, little is known about this gene family in cotton. Till now, only one CDPK gene, GhCPK1, has been identified from upland cotton. And its transcripts accumulated primarily in the elongating fiber, which suggested that GhCPK1 might play a vital role in the calcium signaling events associated with fiber elongation [38,39].

Cotton is the major source of natural fibers used in the textile industry and is cultivated worldwide. Cotton, which belongs to the genus of Gossypium, includes approximately 45 diploid and five tetraploid species and serves as an excellent model system for evolutionary studies of polyploidy plants [40]. With completion of the genome sequencing of the diploid cotton Gossypium raimondii [41], genome-wide analysis of all the genes belonging to specific gene families have been realized. Here, 41 CDPK genes from G. raimondii were identified by database searches and classified according to phylogenetic analysis. Furthermore, the expression profiles of CDPK genes in response to low temperature, which affects seed germination,plant development, and final yield in cotton, were investigated. The identification and comprehensive study for CDPK genes from G. raimondii will provide valuable information for breeding stress-resistant cotton and further studying of the biological function and evolutionary relationship of this family in cotton.

## Materials and Methods

### Gene retrieval and annotation

The G. raimondii genome database (release v2.1) [41] was downloaded from Phytozome (http://www.phytozome.net/cotton.php). The published CDPK protein information for Arabidopsis [12] and rice [28] were obtained from the Arabidopsis Information Resource (TAIR release 10, http://www.arabidopsis.org) and the Rice Genome Annotation Project Database (RGAP release 7, http://rice.plantbiology.msu.edu/index.shtml), respectively. To identify potential CDPK genes in G. raimondii, the Arabidopsis and rice CDPK proteins were used as queries by searching against the G. raimondii genome database using BlastP and tBlastN programs with default parameters. Subsequently, all hits were verified with the InterProScan program (http://www.ebi.ac.uk/Tools/pfa/iprscan/) [42] to confirm the presence of the protein kinase domain. Finally, the Pfam (http://pfam.sanger.ac.uk/) [43] and SMART (http://smart.embl-heidelberg.de/) [44] database were applied to manually determine each candidate member of the CDPK family. The EF hand was predicted by ScanProsite tool (http://prosite.expasy.org/scanprosite/) [45]. The N-myristoylation motif and the palmitoylation site were predicted by Myristoylator (http://web.expasy.org/myristoylator/) [46] and CSS-Plam program [47], respectively. The molecular weight (MW) of the full-length protein was calculated by Compute pI/Mw tool (http://web.expasy.org/cgi-bin/compute_pi/pi_tool) [48].

### Phylogenetic analysis and Gene structure prediction

Multiple alignments of the full-length protein sequences were performed using Clustal X version 2.0 program [49] with default parameters. Phylogenetic trees were constructed using the method of Neighbor Joining MEGA 5.2 [50] with pairwise deletion option and poisson correction model. For statistical reliability, bootstrap tests were carried out with 1000 replicates.

The gene structures were obtained through comparing the genomic sequences and their predicted coding sequences of GSDS(http://gsds.cbi.pku.edu.cn/) [51].

### Chromosomal locations and gene duplications

All CDPK genes were mapped on the G. raimondii chromosomes according to their starting positions given in the genome annotation document. The chromosome location image was generated by MapInspect software [52].

Gene duplication events of CDPK genes in G. raimondii were also investigated. The gene duplication was defined according to (1) the length of aligned sequence cover >80% of the longer gene, (2) the identity of the aligned regions >80%, and (3) only one duplication event was counted for tightly-linked genes [53,54]. With the chromosomal locations of CDPK genes, two types of gene duplications were recognized, i.e., tandem duplication and segmental duplication.

### Estimating Ka/Ks ratios for duplicated gene pairs

The CDPK duplicated gene pairs of G.raimondii were firstly aligned by Clustal X version 2.0 program [49]. Then Ks (synonymous substitution rate) and Ka (non-synonymous substitution rate) were calculated using the DnaSP v5.0 software (DNA polymorphism analysis) [55]. Finally, the Ka/Ks ratio was analyzed to assess the selection pressure for each gene pair.

### Cis-element analysis

For promoter analysis, 2000 bp genomic DNA sequences upstream of the initiation codon (ATG) were retrieved from the genome sequence. These sequences were subjected to search in the PLACE database (http://www.dna.affrc.go.jp/PLACE/signalscan.html) [56] to identify putative cis-elements in promoter regions.

### Plant materials and low temperature stress treatment

All the plants of G. raimondii were grown in a temperature-controlled chamber at 28°C with a photoperiod of 16 hours light and 8 hours dark. After ten days, the leaves, stems, roots, and cotyledons of some seedlings were collected to analyze tissue-specific expression, and the rest seedlings were used to examine the expression patterns of CDPK genes under low-temperature stress. Plant leaves of the seedlings grown in the temperature-controlled chamber treated at 10°C were harvested at 0, 3, 6, and 12 hours, which represented normal plants, slight stress, moderate stress, and severe stress, respectively. All collected samples were immediately frozen in liquid nitrogen and stored at −80°C.

### RNA isolation and quantitative real-time PCR (qRT-PCR)

Total RNA was extracted from all samples using EASYspin Plus RNAprep Kit, and the first-strand cDNAs were synthesized with PrimerScript 1st Strand cDNA synthesis kit (TaKaRa). All protocols followed the manufacturer's protocol. For quantitative real-time PCR (qRT-PCR) assay, gene-specific primers were designed for the 11 selected CDPK genes according to their CDSs (Table S1). The qRT-PCR was performed with SYBR premix Ex taq (TaKaRa) and CFX96 Realtime System (BioRad) by strictly following the manufacturer's instructions. Each reaction was done in a final volume of 20 μl containing 10 μl of SYBR premix Ex taq, 1.0 μl of cDNA sample, and 0.5 μl of each gene-specific

**Table 1.** The information of 41 CDPK genes from *G. raimondii*.

| Gene symbol | ID | Chromosome | Alternative splicing | CDS | Amino acid | MW (kDa) | No. of EF hands | N-terminal aa | N-Myristoylation | N-palmitoylation |
|---|---|---|---|---|---|---|---|---|---|---|
| GrCPK1 | Gorai.001G135000 | 1 | – | 1542 | 513 | 57.3 | 4 | MGNCCSRG | Yes | Yes |
| GrCPK2 | Gorai.001G138000 | 1 | 2 | 1590 | 529 | 59.5 | 2 | MGNCCATT | No | Yes |
| GrCPK3 | Gorai.002G088800 | 2 | 4 | 1614 | 537 | 60.5 | 4 | MGSCLTKS | Yes | Yes |
| GrCPK4 | Gorai.002G153600 | 2 | 2 | 1776 | 591 | 66.3 | 4 | MGNSCAKS | Yes | Yes |
| GrCPK5 | Gorai.003G009500 | 3 | 8 | 1626 | 541 | 61.3 | 4 | MGACLSAT | Yes | Yes |
| GrCPK6 | Gorai.003G084000 | 3 | – | 1527 | 508 | 57.5 | 3 | MGNCNGLP | No | Yes |
| GrCPK7 | Gorai.003G092900 | 3 | 3 | 1608 | 535 | 60.6 | 3 | MGNCCATP | Yes | Yes |
| GrCPK8 | Gorai.004G015100 | 4 | – | 1617 | 538 | 60.5 | 4 | MGNCCSRG | Yes | Yes |
| GrCPK9 | Gorai.005G019800 | 5 | – | 1599 | 532 | 60.7 | 4 | MGSCVARP | Yes | Yes |
| GrCPK10 | Gorai.005G074300 | 5 | – | 1527 | 508 | 56.9 | 4 | MNNQSSSI | No | No |
| GrCPK11 | Gorai.005G216500 | 5 | 6 | 1707 | 568 | 63.6 | 4 | MGNTCRGS | No | Yes |
| GrCPK12 | Gorai.006G124800 | 6 | – | 1575 | 524 | 58.6 | 4 | MGNCCSCG | Yes | Yes |
| GrCPK13 | Gorai.006G128200 | 6 | 3 | 1596 | 531 | 59.8 | 3 | MGNCCATP | No | Yes |
| GrCPK14 | Gorai.006G137800 | 6 | – | 1596 | 531 | 60.2 | 3 | MGNCCVTS | No | Yes |
| GrCPK15 | Gorai.006G147600 | 6 | – | 1836 | 611 | 68.4 | 4 | MGNNCFKT | No | Yes |
| GrCPK16 | Gorai.007G025000 | 7 | 5 | 1593 | 530 | 60.4 | 4 | MGNCNRPP | No | Yes |
| GrCPK17 | Gorai.007G035100 | 7 | 3 | 1653 | 550 | 62.3 | 4 | MGNCNACV | No | Yes |
| GrCPK18 | Gorai.007G194500 | 7 | 3 | 1665 | 554 | 62.8 | 4 | MGACLSTT | Yes | Yes |
| GrCPK19 | Gorai.007G378700 | 7 | – | 1584 | 527 | 59.2 | 3 | MGNCCRSP | No | Yes |
| GrCPK20 | Gorai.008G013700 | 8 | – | 1728 | 575 | 64.6 | 4 | MRLHYCMR | No | Yes |
| GrCPK21 | Gorai.008G251000 | 8 | 2 | 1605 | 534 | 60.2 | 4 | MGNCNSQP | Yes | Yes |
| GrCPK22 | Gorai.009G078000 | 9 | – | 1599 | 532 | 59.5 | 4 | MGNILCSRS | Yes | Yes |
| GrCPK23 | Gorai.009G191500 | 9 | – | 2622 | 873 | 97.1 | 4 | MGICQSLC | No | No |
| GrCPK24 | Gorai.009G290200 | 9 | 5 | 1617 | 538 | 60.4 | 4 | MNKKIAGS | No | Yes |
| GrCPK25 | Gorai.009G351200 | 9 | – | 1605 | 534 | 60.7 | 4 | MGSCISAP | Yes | Yes |
| GrCPK26 | Gorai.009G394700 | 9 | – | 1947 | 648 | 71.8 | 4 | MGNVCATL | No | Yes |
| GrCPK27 | Gorai.009G395400 | 9 | 5 | 1719 | 572 | 63.6 | 4 | MGNACAGP | No | Yes |
| GrCPK28 | Gorai.009G438300 | 9 | – | 1575 | 524 | 58.7 | 4 | MGGCLTKT | Yes | Yes |
| GrCPK29 | Gorai.010G001300 | 10 | – | 1581 | 526 | 59.1 | 4 | MGLCQSLG | Yes | Yes |
| GrCPK30 | Gorai.010G252400 | 10 | – | 1584 | 527 | 59.3 | 4 | MGCCSSKN | Yes | Yes |
| GrCPK31 | Gorai.011G014200 | 11 | – | 1656 | 551 | 62.0 | 4 | MGCFSSKH | Yes | Yes |
| GrCPK32 | Gorai.011G098300 | 11 | – | 1635 | 544 | 62.0 | 4 | MGICLSTT | Yes | Yes |
| GrCPK33 | Gorai.011G228500 | 11 | 5 | 1764 | 587 | 65.4 | 4 | MGNTCVGP | No | Yes |
| GrCPK34 | Gorai.012G045700 | 12 | 2 | 1494 | 497 | 55.9 | 4 | MSRTSSGT | No | No |
| GrCPK35 | Gorai.012G114600 | 12 | – | 1584 | 527 | 59.4 | 3 | MGNCCRSP | No | Yes |

**Table 1.** Cont.

| Gene symbol | ID | Chromosome | Alternative splicing | CDS | Amino acid | MW (kDa) | No. of EF hands | N-terminal aa | N-Myristoylation | N-palmitoylation |
|---|---|---|---|---|---|---|---|---|---|---|
| GrCPK36 | Gorai.012G138900 | 12 | 2 | 1659 | 552 | 62.0 | 4 | MGNTCRGP | No | Yes |
| GrCPK37 | Gorai.013G003100 | 13 | 2 | 1740 | 579 | 64.6 | 4 | MGNTCVGP | No | Yes |
| GrCPK38 | Gorai.013G064400 | 13 | 5 | 1464 | 487 | 54.6 | 4 | MRRAIDHQ | No | No |
| GrCPK39 | Gorai.013G064500 | 13 | 2 | 1572 | 523 | 58.3 | 3 | MGNTCLGS | No | Yes |
| GrCPK40 | Gorai.013G159300 | 13 | 3 | 1611 | 536 | 60.4 | 4 | MGGCLTKN | Yes | Yes |
| GrCPK41 | Gorai.013G253100 | 13 | – | 1584 | 527 | 58.8 | 4 | MGNCCTRG | No | Yes |

– represents no alternative splice variants.

primer. The qRT-PCR cycles were conducted with 40 cycles and an annealing temperature of 60°C, the amplification programs were as follows: 95°C for 30 seconds, 40 cycles of 94°C for 10 seconds, 60°C for 10 seconds, and 72°C for 15 seconds. The cotton *UBQ7* gene was used as internal reference for all the qRT–PCR analyses. Each cDNA sample was tested in three replicates. The relative expression levels were calculated according to the $2^{-\triangle\triangle Ct}$ method [57]. The expression profiles were clustered using the Cluster 3.0 software [58].

## Results and Discussion

### Identification and annotation of CDPK genes in *G. raimondii*

The availability of the *G. raimondii* genome sequences [41] makes it possible to identify all CDPK gene family members in the diploid cotton. BLASTP and TBLASTN programs were performed to search the candidate CDPK genes from the *G. raimondii* genome with the query sequences of Arabidopsis and rice CDPK genes. Then Pfam and SMART analyses were used to verify the retrieved sequences, and finally a total of 41 non-redundant CDPK genes were confirmed and described (Table1). According to the proposed nomenclature for CDPK genes [59], we designated these 41 CDPK genes as *GrCPKs*, from *GrCPK1* to *GrCPK41*. The numbering of these *GrCPKs* was based on their position from top to bottom on corresponding chromosomes, from chromosome 1 to chromosome 13. Through analyzing the transcriptome sequencing data downloaded from the NCBI Sequence Read Archive (SRA), it was found that at least 37 *GrCPKs* were actively expressed in *G. raimondii* (Figure S1). The total number of CDPK genes identified from *G. raimondii* was greater than that in Arabidopsis [12] and rice [28]. The length and molecular weights (MW) of 41 GrCPK proteins were deduced from their predicted protein sequences. All identified GrCPK genes encoded proteins with amino acid numbers varying from 487 to 873 and molecular weight range between 54.6 kDa and 97.1 kDa, which were comparable with CDPK genes from other plant species [12,28,30,31].

All 41 CDPKs identified in our study possessed the typical CDPK structure with four characterized domains, including a variable N-terminal domain, a catalytic Ser/Thr protein kinase domain, an autoinhibitory region, and a calmodulin-like domain. The N-terminal domain of several CDPK proteins contained myristoylation motif with a Gly residue at the second position, which was thought to be critical for mediating protein-protein and protein-membrane interactions [60]. It was reported that CDPK genes which had an N-myristoylation motif tended to localize in the plasma membranes [14,21,29,61,62], demonstrating that myristoylation was required for membrane binding. In addition, palmitoylation, as a second type of lipid modification, was also necessary to stabilize for membrane association [14]. Among 41 CDPKs in our study, 18 CDPKs were predicted to have myristoylation motifs at the N-terminus, and all of them possessed at least one palmitoylation site. However, the membrane association of CDPKs was complex and might be affected by other motifs. For example, *TaCPK3* and *TaCPK15* lacking myristoylation motifs could also be associated with the membranes [29], while *ZmCPK1* which was predicted to have an N-myristoylation motif was found to localize into the cytoplasm and nucleus [63]. Therefore, the subcellular localizations of these CDPKs were still needed to be further characterized experimentally.

The calmodulin-like domain contained $Ca^{2+}$ binding EF hand structure that allow CDPK proteins to function as a $Ca^{2+}$ sensor.

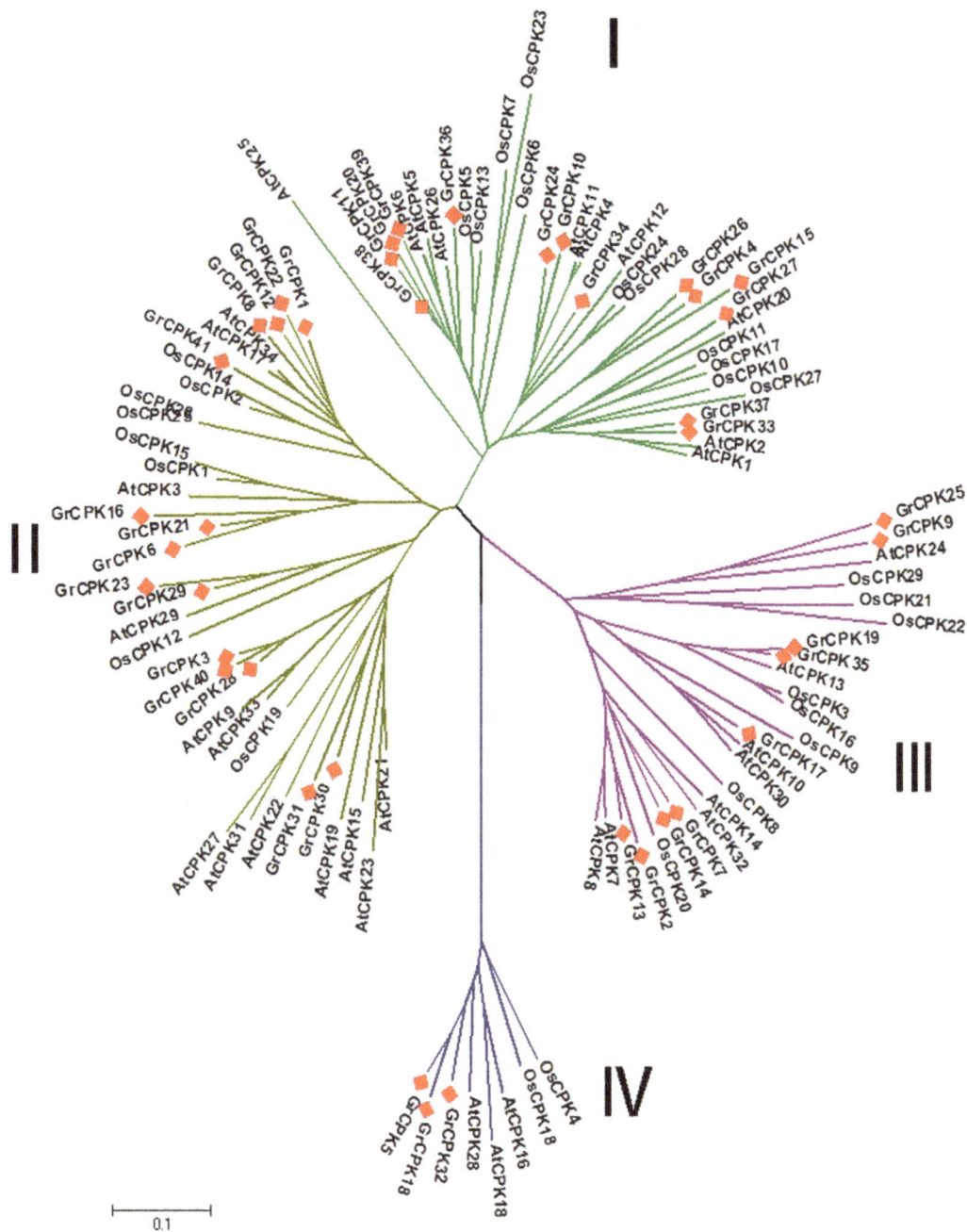

**Figure 1. Unrooted phylogenetic tree of CDPK genes from *G. raimondii*, Arabidopsis and rice.** Four subfamilies are labeled as I, II, III, and IV. And the branches of each subfamily are indicated in a specific color.

In *G. raimondii*, 33 CDPKs contained four $Ca^{2+}$ binding EF hands, seven CDPKs, including GrCPK6, GrCPK7, GrCPK13, GrCPK14, GrCPK19, GrCPK35, and GrCPK39, had three EF hand motifs each, whereas GrCPK2 had two EF hand motifs only. This difference in EF hand was also found in the CDPK family of other plants [9,30,31,64]. Studies have demonstrated that the number and position of EF hands might be important for determining the $Ca^{2+}$ regulation of CDPK activity [65,66], and compared to C-terminal EF3 and EF4 motifs, N-terminal EF1 and EF2 motifs with lower $Ca^{2+}$-binding affinities were more important for activating the kinases [67,68].

## Phylogenetic analysis of the CDPK gene family

To detect the evolutionary relationships of CDPK genes, an unrooted phylogenetic tree was generated from alignments of the full-length sequences of *G. raimondii*, Arabidopsis and rice CDPK proteins. All the CDPK proteins were divided into four subgroups (Figure 1): CDPK I, CDPK II, CDPK III and CDPK IV. CDPK I contained 14 CDPKs from *G. raimondii*, 11 from rice, and ten from Arabidopsis. CDPK II contained 15 CDPKs from *G. raimondii*, eight from rice, and 13 from Arabidopsis. The CDPKs numbers in CDPK I and CDPK II from *G. raimondii* were greater than that from Arabidopsis and rice, mainly due to more CDPK genes in *G. raimondii*. In CDPK III, there were nine CDPKs from *G. raimondii*,

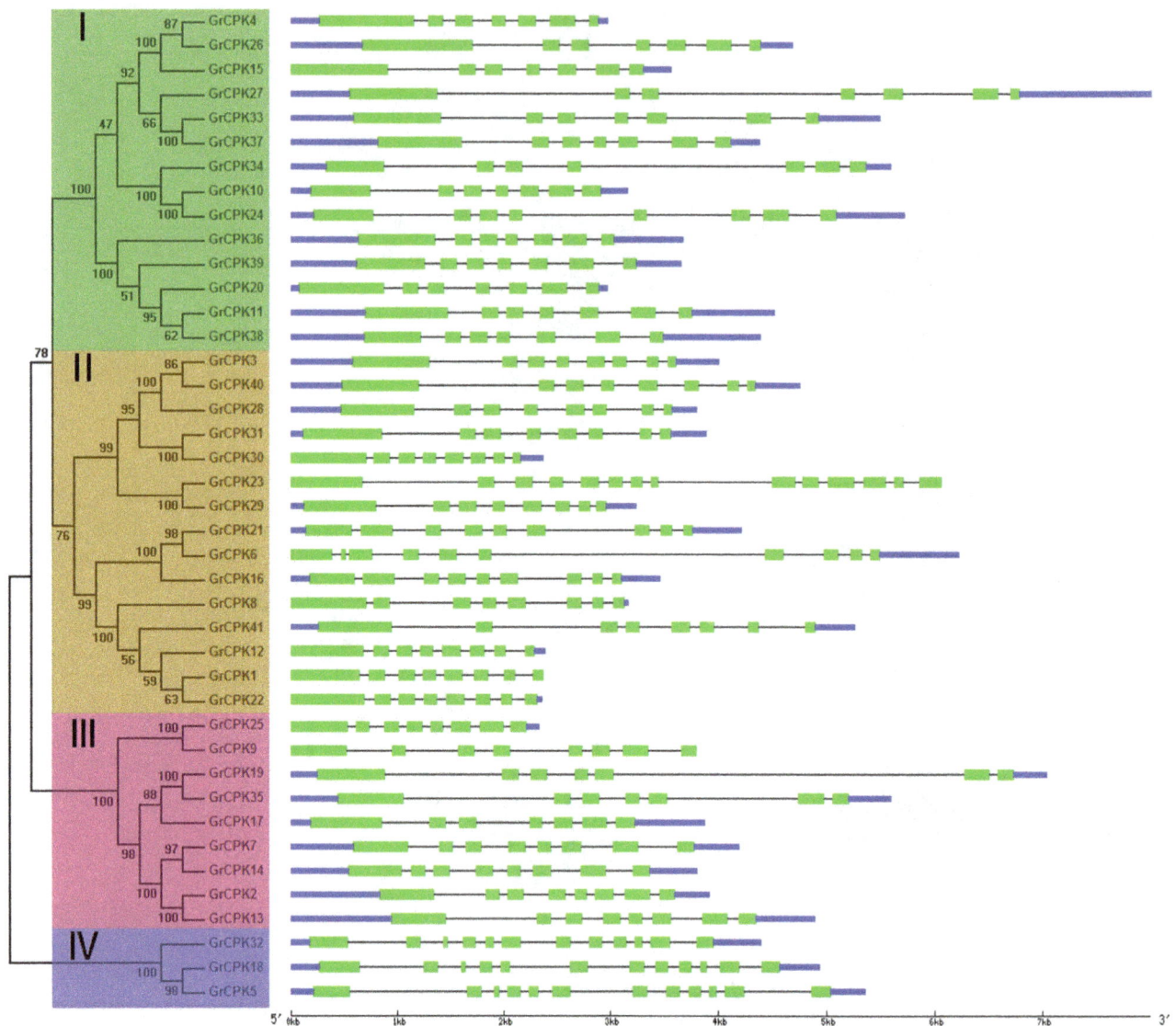

**Figure 2. Phylogenetic relationship and gene structure of CDPK genes from *G. raimondii*.** Four subfamilies labeled as I, II, III, and IV are marked with different color backgrounds. Exons are represented by green boxes and introns by black lines. The untranslated regions (UTRs) are indicated by thick blue lines.

eight from rice, and eight from Arabidopsis, while CDPK IV genes consisted of the smallest subfamilies in all of the three species, which contained three CDPKs from *G. raimondii*, two from rice, and three from Arabidopsis. And the number of genes in CDPK III and CDPK IV were approximately identical across these three species. Interestingly, among the four subfamilies, CDPK I in rice comprised the most members, while CDPK II in *G. raimondii* and Arabidopsis was the largest subfamily. Recent study indicated that poplar had total 30 CDPK genes with 11 genes in CDPK I, eight genes in CDPK II, nine genes in CDPK III and two genes in CDPK IV [30]. These results implied that the difference in the number of CDPK genes was mainly due to the occurrence of gene gain or loss in CDPK I or CDPK II subfamilies independently among the different organisms.

Phylogenetic analysis also showed that *GrCPKs* were more closely allied to *AtCPKs* than to *OsCPKs*, consistent with the evolutionary relationships among *G. raimondii*, Arabidopsis, and rice. Moreover, all the CDPK genes from the three species sorted

into four distinct clades, implying that these four subfamilies existed before the divergence of monocots and dicots, which also supported the hypothesis that CDPK genes radiated into four subfamilies before algae and land plants split [69].

## Structural analysis and chromosomal localization of GrCPKs

To get further insight into the possible structural evolution of *GrCPKs*, a separate unrooted phylogenetic tree was constructed using the protein sequences of all the CDPK genes from *G. raimondii*, and then the diverse exon-intron organizations of *GrCPKs* were compared. As shown in Figure 2, the *GrCPKs* clustered in the same subfamily shared very similar exon-intron structures. Most members in Group I possessed seven exons, except for *GrCPK24*, which contained eight exons. Most genes in Group II had eight exons, but *GrCPK16* and *GrCPK21* had nine exons each, *GrCPK6* contained ten exons, and *GrCPK23* contained 14 exons. Six genes in group III had eight exons, whereas three genes contained seven

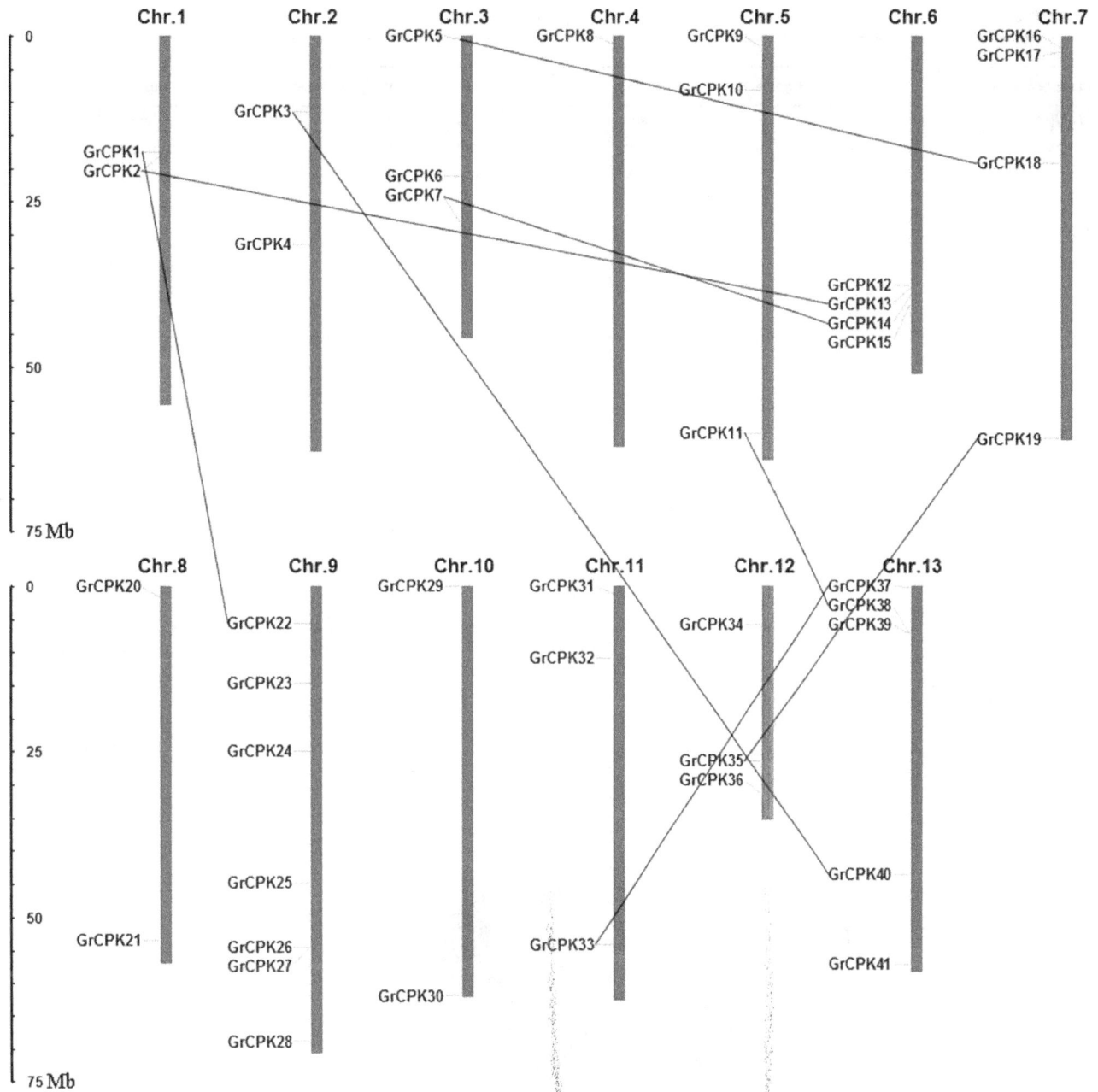

**Figure 3. Distributions of CDPK genes from *G. raimondii* in 13 chromosomes.** Chromosome numbers are indicated above each vertical bar. The scale represents megabases (Mb). The duplicated gene pairs are connected with black lines.

exons each. All genes in Group IV contained 12 exons. These conserved gene structure within each group supported their close evolutionary relationship.

The 41 nonredundant CDPK genes were mapped on the 13 *G. raimondii* chromosomes (Figure 3). They were found to be distributed unevenly among 13 chromosomes and the number of CDPK genes on each chromosome varied widely. Chromosome 9 which contained 7 CDPK genes was the highest one in gene numbers, followed by chromosome 13 on which five genes were found. Chromosome 6 and 7 both had four genes, and chromosome 3, 5, 11, and 12 contained three genes each. Chromosome 1, 2, 8, and 10 had two genes each only, whereas only single CDPK gene was localized on chromosome 4.

## CDPK gene duplications and functional divergence

To elucidate the expanded mechanism of the CDPK gene family in *G. raimondii*, gene duplication events, including tandem and segmental duplications, were investigated, which were thought to play a significant role in the amplification of gene family members in the genome [70,71]. A total of eight duplication events, *GrCPK2/GrCPK13*, *GrCPK7/GrCPK14*, *GrCPK18/GrCPK5*, *GrCPK19/GrCPK35*, *GrCPK22/GrCPK1*, *GrCPK37/GrCPK33*, *GrCPK38/GrCPK11*, and *GrCPK40/GrCPK3*, were found in the *G. raimondii* genome and all of them were segmental duplication events based on the chromosomal distribution of the CDPK genes (Figure 3). This result suggested that the CDPK gene family

**Table 2.** Ka/Ks analysis for the duplicated gene pairs.

| Duplicated gene 1 | Duplicated gene 2 | Ka | Ks | Ka/Ks | Purifying selection | Duplicate type |
|---|---|---|---|---|---|---|
| GrCPK2 | GrCPK13 | 0.068 | 0.422 | 0.161 | Yes | Segmental |
| GrCPK14 | GrCPK7 | 0.083 | 0.484 | 0.171 | Yes | Segmental |
| GrCPK18 | GrCPK5 | 0.057 | 0.589 | 0.096 | Yes | Segmental |
| GrCPK19 | GrCPK35 | 0.028 | 0.399 | 0.071 | Yes | Segmental |
| GrCPK22 | GrCPK1 | 0.080 | 0.683 | 0.118 | Yes | Segmental |
| GrCPK37 | GrCPK33 | 0.062 | 0.511 | 0.121 | Yes | Segmental |
| GrCPK38 | GrCPK11 | 0.034 | 0.515 | 0.065 | Yes | Segmental |
| GrCPK40 | GrCPK3 | 0.038 | 0.528 | 0.071 | Yes | Segmental |

expansion in *G. raimondii* was mainly attributed to segmental duplication events.

The duplicated gene pairs might experience three alternative outcomes during the process of evolution, i.e., (i) one copy may simply become silenced and lost original functions (nonfunctionalization); (ii) one copy may acquire a novel, beneficial function, with the other copy retaining the original function (neofunctionalization); or (iii) both copies may become partition of original functions (subfunctionalization) [72]. We subsequently calculated the non-synonymous to synonymous substitution ratio (Ka/Ks) for each pair of duplicated CDPK genes, which showed the selective force acting on the protein, to reveal whether Darwinian positive selection was associated with functional divergence after gene duplications. Generally, Ka/Ks >1 indicates positive selection, Ka/Ks = 1 indicates neutral selection, while Ka/Ks <1 indicates

negative or purifying selection. In this study, the Ka/Ks ratios for eight duplicated CDPK gene pairs were no larger than 0.2 (Table 2), which demonstrated that the CDPK genes from *G. raimondii* had mainly experienced strong purifying selection pressure with limited functional divergence after segmental duplications. These results suggested that functions of the duplicated CDPK genes did not diverge much during subsequent evolution.

## Expression analysis of the CDPK genes in *G. raimondii*

Cold stress is one of the serious environmental stresses that most land plants might encounter during the process of their growth. And so far, numerous of CDPK genes identified in various plant species have been proven to play crucial roles in cold stress response. Cotton is a subtropical crop and its cultivation has been

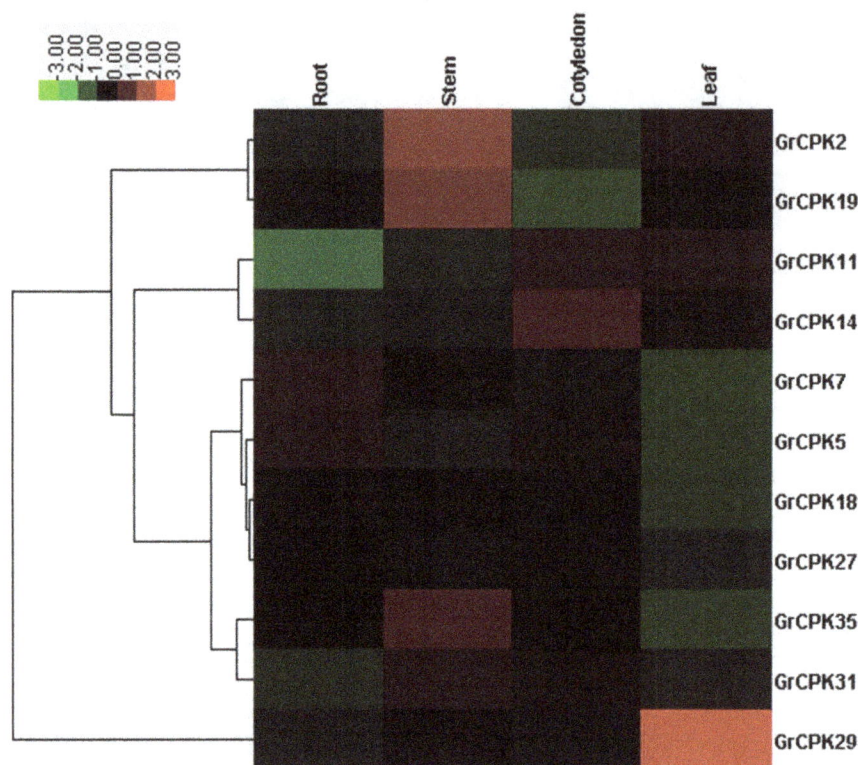

**Figure 4. Expression patterns of 11** *GrCPKs* **in four representative tissues of** *G. raimondii* **seedlings.** The color bar represents the relative signal intensity values.

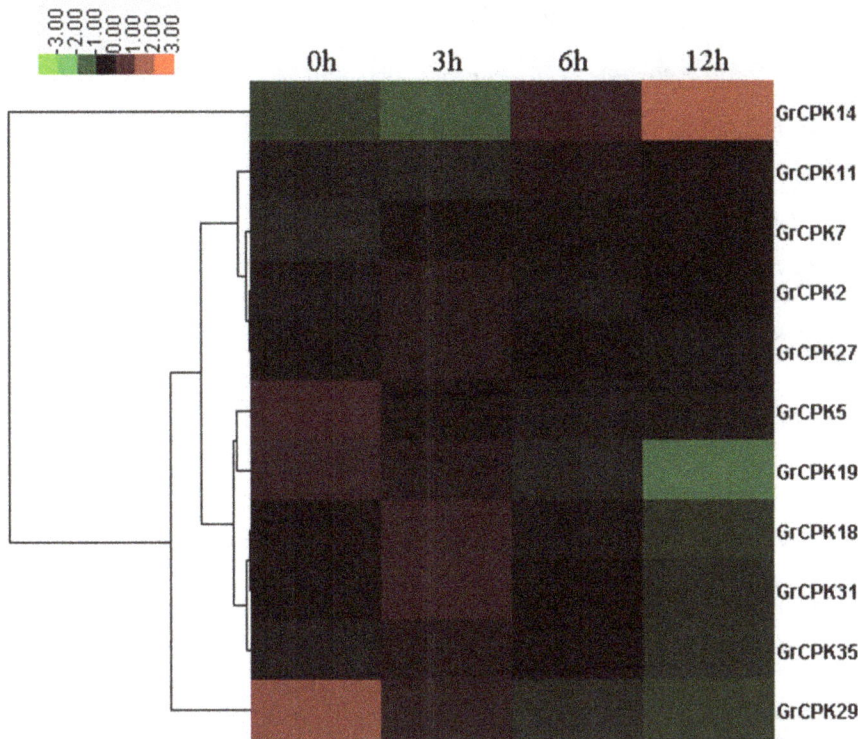

**Figure 5. Expression profiling of 11 *GrCPKs* in leaves under 10°C treatment.** The color bar represents the relative signal intensity values.

extended from tropical and subtropical to colder regions. Low temperature (<15°C ) can adversely affect plant development, resulting in poor germination, infection by fungi and other disease-causing organisms, and yield loss [73]. There is little information about the functions of CDPK genes in the cotton response to low temperature.

The *cis*-element analysis in promoter regions might provide some indirect evidence for the functional dissection of CDPK genes in stress responses [74]. To investigate the potential functions of *GrCPKs* in cotton under cold stress, the possible presence of low temperature-responsive element (LTRE) in the promoter regions of all the *GrCPKs* was detected by searching against the PLACE database (Table S2). The results revealed that the majority, 27 out of 41, of *GrCPKs* contained the LTRE in promoter sequences, indicating that these *GrCPKs* might partici-pate in the signal transduction of the plant response to cold stress. In order to verify the expression patterns of these genes, eight *GrCPKs* which had the LTRE in promoter regions were selected for qRT-PCR analysis. In addition, as the gene expression was a complex process, the existence of some *cis*-elements did not always correlate with the gene function [74], therefore, three *GrCPKs*, *GrCPK2*, *GrCPK11*, and *GrCPK27*, which had no LTRE in promoter regions were also selected in this expression analysis.

Firstly, qRT-PCR analysis was performed to examine expres-sion patterns of 11 selected *GrCPKs* in roots, stems, cotyledons, and leaves of 10-day-old seedlings. As shown in Figure 4, *GrCPK5* and *GrCPK7* were expressed at high levels in roots, while *GrCPK2*, *GrCPK19*, and *GrCPK35* shared high expressions in young stems. In cotyledons, *GrCPK14* exhibited relatively high transcript abundance, whereas *GrCPK29* were predominantly expressed in leaves, which showed about 6-fold higher expression than all other tissues. The results demonstrated that most of these 11 *GrCPKs* had

specific spatial expression patterns, which was similar with CDPK genes in maize [54] and rice [74] that also exhibited tissue-specific expression.

To further confirm the cold stress response of *GrCPKs*, the expression patterns of 11 *GrCPKs* in 10-day-old leaves under low temperature stress of 10°C were investigated. The expression levels of all the 11 CDPK genes responsive to slight cold stress (3 hours), moderate cold stress (6 hours), and severe cold stress (12 hours) were shown in Figure 5, comparing with normal plants. The result showed that the expression of all the 11 *GrCPKs* were induced or repressed by clod treatment. Among them, *GrCPK2*, *GrCPK7*, *GrCPK11*, *GrCPK14*, *GrCPK18*, *GrCPK27*, *GrCPK31*, and *GrCPK35* were positively regulated response to cold stress according to their transcription levels. For example, *GrCPK14* showed a significantly increase in its expression after long time cold stress treatment. Whereas, the transcription levels of the other three genes, *GrCPK5*, *GrCPK19*, and *GrCPK29*, were down-regulated in response to cold stress.

## Conclusions

A complete set of 41 CDPK genes were identified in the *G. raimondii* genome, which were categorized into four subgroups and mapped on the 13 chromosomes. Segmental duplication contrib-uted to the expansion of *GrCPKs*, and the duplicated genes mainly undergone strong purifying selection pressure with limited functional divergence. The majority of *GrCPKs* contained the low temperature-responsive elements in their promoter sequences. The expression of *GrCPKs* were also induced or repressed by the clod stress treatment. The results would be helpful for better understanding of the biological functions of the CDPK genes in cotton and the evolutionary relationship of this family in *Gossypium* species.

## Supporting Information

**Figure S1 Expression analysis of *GrCPKs* using the transcriptome sequencing data.** The transcriptome sequencing datasets of *G. raimondii* for three tissue samples (mature leaves, 0DPA ovules, and 3DPA ovules) were downloaded from the NCBI Sequence Read Archive (SRA) with accession numbers SRX111367, SRX111365 and SRX111366. Then sequenced reads of these three datasets were mapped to the sequences of *GrCPKs*, respectively. And matches were converted to RPKM to estimate gene expression levels. The expression profiles were clustered using the Cluster 3.0 software. The color bar represents the relative signal intensity values. DPA: Days Post Anthesis.

## Acknowledgments

We are grateful to Lei Mei, Ning Zhu, Cheng Li, Yue Chen, Yingchao Sun and Jingjing Pan (Zhejiang University, China) for their support in this study.

## Author Contributions

Conceived and designed the experiments: SJZ JHC. Performed the experiments: W. Liu W. Li MKD JHC. Analyzed the data: W. Liu W. Li JHC QLH. Contributed reagents/materials/analysis tools: SJZ. Wrote the paper: W. Liu W. Li SJZ MKD.

## References

1. Xiong L, Schumaker KS, Zhu JK (2002) Cell signaling during cold, drought, and salt stress. Plant Cell 14 Suppl: S165–S183.
2. Yamaguchi-Shinozaki K, Shinozaki K (2006) Transcriptional regulatory networks in cellular responses and tolerance to dehydration and cold stresses. Annu Rev Plant Biol 57: 781–803.
3. Valliyodan B, Nguyen HT (2006) Understanding regulatory networks and engineering for enhanced drought tolerance in plants. Curr Opin Plant Biol 9: 189–195.
4. Tran LS, Nakashima K, Shinozaki K, Yamaguchi-Shinozaki K (2007) Plant gene networks in osmotic stress response: from genes to regulatory networks. Methods Enzymol 428: 109–128.
5. Evans NH, McAinsh MR, Hetherington AM (2001) Calcium oscillations in higher plants. Curr Opin Plant Biol 4: 415–420.
6. Tuteja N, Mahajan S (2007) Calcium signaling network in plants: an overview. Plant Signal Behav 2: 79–85.
7. McCormack E, Braam J (2003) Calmodulins and related potential calcium sensors of Arabidopsis. New Phytologist 159: 585–598.
8. Kolukisaoglu U, Weinl S, Blazevic D, Batistic O, Kudla J (2004) Calcium sensors and their interacting protein kinases: genomics of the Arabidopsis and rice CBL-CIPK signaling networks. Plant Physiol 134: 43–58.
9. Cheng SH, Willmann MR, Chen HC, Sheen J (2002) Calcium signaling through protein kinases. The Arabidopsis calcium-dependent protein kinase gene family. Plant Physiol 129: 469–485.
10. Harmon AC, Gribskov M, Gubrium E, Harper JF (2001) The CDPK superfamily of protein kinases. New Phytologist 151: 175–183.
11. Harper JF, Sussman MR, Schaller GE, Putnam-Evans C, Charbonneau H, et al. (1991) A calcium-dependent protein kinase with a regulatory domain similar to calmodulin. Science 252: 951–954.
12. Hrabak EM, Chan CW, Gribskov M, Harper JF, Choi JH, et al. (2003) The Arabidopsis CDPK-SnRK superfamily of protein kinases. Plant Physiol 132: 666–680.
13. Klimecka M, Muszynska G (2007) Structure and functions of plant calcium-dependent protein kinases. Acta Biochim Pol 54: 219–233.
14. Martin ML, Busconi L (2000) Membrane localization of a rice calcium-dependent protein kinase (CDPK) is mediated by myristoylation and palmitoylation. Plant J 24: 429–435.
15. Saijo Y, Hata S, Kyozuka J, Shimamoto K, Izui K (2000) Over-expression of a single Ca$^{2+}$-dependent protein kinase confers both cold and salt/drought tolerance on rice plants. Plant J 23: 319–327.
16. Saijo Y, Kinoshita N, Ishiyama K, Hata S, Kyozuka J, et al. (2001) A Ca$^{2+}$-dependent protein kinase that endows rice plants with cold- and salt-stress tolerance functions in vascular bundles. Plant Cell Physiol 42: 1228–1233.
17. Asano T, Hayashi N, Kobayashi M, Aoki N, Miyao A, et al. (2012) A rice calcium-dependent protein kinase OsCPK12 oppositely modulates salt-stress tolerance and blast disease resistance. Plant J 69: 26–36.
18. Berberich T, Kusano T (1997) Cycloheximide induces a subset of low temperature-inducible genes in maize. Mol Gen Genet 254: 275–283.
19. Szczegielniak J, Borkiewicz L, Szurmak B, Lewandowska-Gnatowska E, Statkiewicz M, et al. (2012) Maize calcium-dependent protein kinase (ZmCPK11): local and systemic response to wounding, regulation by touch and components of jasmonate signaling. Physiol Plant 146: 1–14.
20. Zhang M, Liang S, Lu YT (2005) Cloning and functional characterization of NtCPK4, a new tobacco calcium-dependent protein kinase. Biochim Biophys Acta 1729: 174–185.
21. Zou JJ, Wei FJ, Wang C, Wu JJ, Ratnasekera D, et al. (2010) Arabidopsis calcium-dependent protein kinase CPK10 functions in abscisic acid- and Ca$^{2+}$-mediated stomatal regulation in response to drought stress. Plant Physiol 154: 1232–1243.
22. Mori IC, Murata Y, Yang Y, Munemasa S, Wang YF, et al. (2006) CDPKs CPK6 and CPK3 function in ABA regulation of guard cell S-type anion- and Ca$^{2+}$-permeable channels and stomatal closure. PLoS Biol 4: e327.
23. Ma SY, Wu WH (2007) AtCPK23 functions in Arabidopsis responses to drought and salt stresses. Plant Mol Biol 65: 511–518.
24. Estruch JJ, Kadwell S, Merlin E, Crossland L (1994) Cloning and characterization of a maize pollen-specific calcium-dependent calmodulin-independent protein kinase. Proc Natl Acad Sci U S A 91: 8837–8841.
25. Ivashuta S, Liu J, Liu J, Lohar DP, Haridas S, et al. (2005) RNA interference identifies a calcium-dependent protein kinase involved in Medicago truncatula root development. Plant Cell 17: 2911–2921.
26. Yoon GM, Cho HS, Ha HJ, Liu JR, Lee HS (1999) Characterization of NtCDPK1, a calcium-dependent protein kinase gene in Nicotiana tabacum, and the activity of its encoded protein. Plant Mol Biol 39: 991–1001.
27. Lee SS, Cho HS, Yoon GM, Ahn JW, Kim HH, et al. (2003) Interaction of NtCDPK1 calcium-dependent protein kinase with NtRpn3 regulatory subunit of the 26S proteasome in Nicotiana tabacum. Plant J 33: 825–840.
28. Ray S, Agarwal P, Arora R, Kapoor S, Tyagi AK (2007) Expression analysis of calcium-dependent protein kinase gene family during reproductive development and abiotic stress conditions in rice (Oryza sativa L. ssp. indica). Mol Genet Genomics 278: 493–505.
29. Li AL, Zhu YF, Tan XM, Wang X, Wei B, et al. (2008) Evolutionary and functional study of the CDPK gene family in wheat (Triticum aestivum L.). Plant Mol Biol 66: 429–443.
30. Zuo R, Hu R, Chai G, Xu M, Qi G, et al. (2013) Genome-wide identification, classification, and expression analysis of CDPK and its closely related gene families in poplar (Populus trichocarpa). Mol Biol Rep 40: 2645–2662.
31. Ma P, Liu J, Yang X, Ma R (2013) Genome-wide identification of the maize calcium-dependent protein kinase gene family. Appl Biochem Biotechnol 169: 2111–2125.
32. Chang WJ, Su HS, Li WJ, Zhang ZL (2009) Expression profiling of a novel calcium-dependent protein kinase gene, LeCPK2, from tomato (Solanum lycopersicum) under heat and pathogen-related hormones. Biosci Biotechnol Biochem 73: 2427–2431.
33. Gargantini PR, Giammaria V, Grandellis C, Feingold SE, Maldonado S, et al. (2009) Genomic and functional characterization of StCDPK1. Plant Mol Biol 70: 153–172.
34. Giammaria V, Grandellis C, Bachmann S, Gargantini PR, Feingold SE, et al. (2011) StCDPK2 expression and activity reveal a highly responsive potato calcium-dependent protein kinase involved in light signalling. Planta 233: 593–609.
35. Witte CP, Keinath N, Dubiella U, Demouliere R, Seal A, et al. (2010) Tobacco calcium-dependent protein kinases are differentially phosphorylated in vivo as part of a kinase cascade that regulates stress response. J Biol Chem 285: 9740–9748.
36. Yang DH, Hettenhausen C, Baldwin IT, Wu J (2012) Silencing Nicotiana attenuata calcium-dependent protein kinases, CDPK4 and CDPK5, strongly up-regulates wound- and herbivory-induced jasmonic acid accumulations. Plant Physiol 159: 1591–1607.
37. Dubrovina AS, Kiselev KV, Khristenko VS (2013) Expression of calcium-dependent protein kinase (CDPK) genes under abiotic stress conditions in wild-growing grapevine Vitis amurensis. J Plant Physiol 170: 1491–1500.
38. Huang QS, Wang HY, Gao P, Wang GY, Xia GX (2008) Cloning and characterization of a calcium dependent protein kinase gene associated with cotton fiber development. Plant Cell Rep 27: 1869–1875.
39. Wang H, Mei W, Qin Y, Zhu Y (2011) 1-Aminocyclopropane-1-carboxylic acid synthase 2 is phosphorylated by calcium-dependent protein kinase 1 during cotton fiber elongation. Acta Biochim Biophys Sin (Shanghai) 43: 654–661.

40. Chen ZJ, Scheffler BE, Dennis E, Triplett BA, Zhang T, et al. (2007) Toward sequencing cotton (*Gossypium*) genomes. Plant Physiol 145: 1303–1310.

41. Paterson AH, Wendel JF, Gundlach H, Guo H, Jenkins J, et al. (2012) Repeated polyploidization of *Gossypium* genomes and the evolution of spinnable cotton fibres. Nature 492: 423–427.

42. Quevillon E, Silventoinen V, Pillai S, Harte N, Mulder N, et al. (2005) InterProScan: protein domains identifier. Nucleic Acids Res 33: W116–W120.

43. Punta M, Coggill PC, Eberhardt RY, Mistry J, Tate J, et al. (2012) The Pfam protein families database. Nucleic Acids Res 40: D290–D301.

44. Letunic I, Doerks T, Bork P (2012) SMART 7: recent updates to the protein domain annotation resource. Nucleic Acids Res 40: D302–D305.

45. de Castro E, Sigrist CJ, Gattiker A, Bulliard V, Langendijk-Genevaux PS, et al. (2006) ScanProsite: detection of PROSITE signature matches and ProRule-associated functional and structural residues in proteins. Nucleic Acids Res 34: W362–W365.

46. Bologna G, Yvon C, Duvaud S, Veuthey AL (2004) N-Terminal myristoylation predictions by ensembles of neural networks. Proteomics 4: 1626–1632.

47. Ren J, Wen L, Gao X, Jin C, Xue Y, et al. (2008) CSS-Palm 2.0: an updated software for palmitoylation sites prediction. Protein Eng Des Sel 21: 639–644.

48. Bjellqvist B, Basse B, Olsen E, Celis JE (1994) Reference points for comparisons of two-dimensional maps of proteins from different human cell types defined in a pH scale where isoelectric points correlate with polypeptide compositions. Electrophoresis 15: 529–539.

49. Larkin MA, Blackshields G, Brown NP, Chenna R, McGettigan PA, et al. (2007) Clustal W and Clustal X version 2.0. Bioinformatics 23: 2947–2948.

50. Tamura K, Peterson D, Peterson N, Stecher G, Nei M, et al. (2011) MEGA5: molecular evolutionary genetics analysis using maximum likelihood, evolutionary distance, and maximum parsimony methods. Mol Biol Evol 28: 2731–2739.

51. Guo AY, Zhu QH, Chen X, Luo JC (2007) [GSDS: a gene structure display server]. Yi Chuan 29: 1023–1026.

52. Zhao Y, Zhou Y, Jiang H, Li X, Gan D, et al. (2011) Systematic analysis of sequences and expression patterns of drought-responsive members of the HD-Zip gene family in maize. PLoS One 6: e28488.

53. Wei H, Li W, Sun X, Zhu S, Zhu J (2013) Systematic analysis and comparison of nucleotide-binding site disease resistance genes in a diploid cotton *Gossypium raimondii*. PLoS One 8: e68435.

54. Kong X, Lv W, Jiang S, Zhang D, Cai G, et al. (2013) Genome-wide identification and expression analysis of calcium-dependent protein kinase in maize. BMC Genomics 14: 433.

55. Librado P, Rozas J (2009) DnaSP v5: a software for comprehensive analysis of DNA polymorphism data. Bioinformatics 25: 1451–1452.

56. Higo K, Ugawa Y, Iwamoto M, Korenaga T (1999) Plant cis-acting regulatory DNA elements (PLACE) database: 1999. Nucleic Acids Res 27: 297–300.

57. Livak KJ, Schmittgen TD (2001) Analysis of relative gene expression data using real-time quantitative PCR and the $2^{-\Delta\Delta CT}$ Method. Methods 25: 402–408.

58. de Hoon MJ, Imoto S, Nolan J, Miyano S (2004) Open source clustering software. Bioinformatics 20: 1453–1454.

59. Hrabak EM, Dickmann LJ, Satterlee JS, Sussman MR (1996) Characterization of eight new members of the calmodulin-like domain protein kinase gene family from *Arabidopsis thaliana*. Plant Mol Biol 31: 405–412.

60. Johnson DR, Bhatnagar RS, Knoll LJ, Gordon JI (1994) Genetic and biochemical studies of protein N-myristoylation. Annu Rev Biochem 63: 869–914.

61. Mehlmer N, Wurzinger B, Stael S, Hofmann-Rodrigues D, Csaszar E, et al. (2010) The Ca²⁺-dependent protein kinase CPK3 is required for MAPK-independent salt-stress acclimation in Arabidopsis. Plant J.

62. Lu SX, Hrabak EM (2013) The myristoylated amino-terminus of an Arabidopsis calcium-dependent protein kinase mediates plasma membrane localization. Plant Mol Biol 82: 267–278.

63. Wang C, Shao J (2013) Characterization of the *ZmCK1* gene encoding a calcium-dependent protein kinase responsive to multiple abiotic stresses in maize. Plant Molecular Biology Reporter 31: 222–230.

64. Asano T, Tanaka N, Yang G, Hayashi N, Komatsu S (2005) Genome-wide identification of the rice calcium-dependent protein kinase and its closely related kinase gene families: comprehensive analysis of the *CDPKs* gene family in rice. Plant Cell Physiol 46: 356–366.

65. Hong Y, Takano M, Liu C, Gasch A, Chye M, et al. (1996) Expression of three members of the calcium-dependent protein kinase gene family in *Arabidopsis thaliana*. Plant molecular biology 30: 1259–1275.

66. Zhao Y, Pokutta S, Maurer P, Lindt M, Franklin RM, et al. (1994) Calcium-binding properties of a calcium-dependent protein kinase from *Plasmodium falciparum* and the significance of individual calcium-binding sites for kinase activation. Biochemistry 33: 3714–3721.

67. Christodoulou J, Malmendal A, Harper JF, Chazin WJ (2004) Evidence for differing roles for each lobe of the calmodulin-like domain in a calcium-dependent protein kinase. J Biol Chem 279: 29092–29100.

68. Franz S, Ehlert B, Liese A, Kurth J, Cazale AC, et al. (2011) Calcium-dependent protein kinase CPK21 functions in abiotic stress response in *Arabidopsis thaliana*. Mol Plant 4: 83–96.

69. Chen F, Fasoli M, Tornielli GB, Dal Santo S, Pezzotti M, et al. (2013) The evolutionary history and diverse physiological roles of the grapevine calcium-dependent protein kinase gene family. PLoS One 8: e80818.

70. Cannon SB, Mitra A, Baumgarten A, Young ND, May G (2004) The roles of segmental and tandem gene duplication in the evolution of large gene families in *Arabidopsis thaliana*. BMC Plant Biol 4: 10.

71. Maere S, De Bodt S, Raes J, Casneuf T, Van Montagu M, et al. (2005) Modeling gene and genome duplications in eukaryotes. Proc Natl Acad Sci U S A 102: 5454–5459.

72. Lynch M, Conery JS (2000) The evolutionary fate and consequences of duplicate genes. Science 290: 1151–1155.

73. Kargiotidou A, Deli D, Galanopoulou D, Tsaftaris A, Farmaki T (2008) Low temperature and light regulate delta 12 fatty acid desaturases (FAD2) at a transcriptional level in cotton (*Gossypium hirsutum*). J Exp Bot 59: 2043–2056.

74. Wan B, Lin Y, Mou T (2007) Expression of rice Ca²⁺-dependent protein kinases (CDPKs) genes under different environmental stresses. FEBS Lett 581: 1179–1189.

# Effect of Late Planting and Shading on Cellulose Synthesis during Cotton Fiber Secondary Wall Development

Ji Chen[1], Fengjuan Lv[1], Jingran Liu[1], Yina Ma[1], Youhua Wang[1], Binglin Chen[1], Yali Meng[1], Zhiguo Zhou[1]*, Derrick M. Oosterhuis[2]

1 Key Laboratory of Crop Physiology & Ecology, Ministry of Agriculture, Nanjing Agricultural University, Nanjing, Jiangsu Province, PR China, 2 Department of Crop, Soil, and Environmental Sciences, University of Arkansas, Fayetteville, Arkansas, United States of America

## Abstract

Cotton-rapeseed or cotton-wheat double cropping systems are popular in the Yangtze River Valley and Yellow River Valley of China. Due to the competition of temperature and light resources during the growing season of double cropping system, cotton is generally late-germinating and late-maturing and has to suffer from the coupling of declining temperature and low light especially in the late growth stage. In this study, late planting (LP) and shading were used to fit the coupling stress, and the coupling effect on fiber cellulose synthesis was investigated. Two cotton (*Gossypium hirsutum* L.) cultivars were grown in the field in 2010 and 2011 at three planting dates (25 April, 25 May and 10 June) each with three shading levels (normal light, declined 20% and 40% PAR). Mean daily minimum temperature was the primary environmental factor affected by LP. The coupling of LP and shading (decreased cellulose content by 7.8%–25.5%) produced more severe impacts on cellulose synthesis than either stress alone, and the effect of LP (decreased cellulose content by 6.7%–20.9%) was greater than shading (decreased cellulose content by 0.7%–5.6%). The coupling of LP and shading hindered the flux from sucrose to cellulose by affecting the activities of related cellulose synthesis enzymes. Fiber cellulose synthase genes expression were delayed under not only LP but shading, and the coupling of LP and shading markedly postponed and even restrained its expression. The decline of sucrose-phosphate synthase activity and its peak delay may cause cellulose synthesis being more sensitive to the coupling stress during the later stage of fiber secondary wall development (38–45 days post-anthesis). The sensitive difference of cellulose synthesis between two cultivars in response to the coupling of LP and shading may be mainly determined by the sensitiveness of invertase, sucrose-phosphate synthase and cellulose synthase.

**Editor:** Xianlong Zhang, National Key Laboratory of Crop Genetic Improvement, China

**Funding:** This work was funded by the National Natural Science Foundation of China (30971735), China Agriculture Research System (CARS-18-20), and the Special Fund for Agro-scientific Research in the Public Interest (Impact of climate change on agriculture production of China, 200903003). The funders had no role in study design, data collection and analysis, decision to publish, or preparation of the manuscript.

**Competing Interests:** The authors have declared that no competing interests exist.

* Email: giscott@njau.edu.cn

## Introduction

Cotton fiber development is delineated into four stages: fiber initiation, elongation, secondary wall thickening and maturation [1]. Cotton fiber, which deposit almost pure cellulose into secondary cell walls, are referred to as a primary model system for cell wall biogenesis [2,3], and many of the textile properties of cotton fiber are directly dependent on the amount and property of cellulose, which is mainly formed during secondary wall development [4–7].

The deposition of fiber secondary wall cellulose begin at about 16 days post anthesis (DPA) (at least 5 days prior to elongation cessation) and last around 15–35 d [6,8], and the period would be prolonged by cool temperature [9]. Fiber cellulose synthesis is believed to be carried out by the plasma membrane-associated rosette structure [10]. In the rosette structure, sucrose synthase (SuSy) associated with the plasma membrane (M-SuSy) may form a complex with cellulose synthase (CesA) to channel carbon from sucrose into cellulose [2,11]. In the process, sucrose is degraded by SuSy to provide uridine diphosphate glucose (UDP-glucose) for

cellulose synthesis [2,5,12], and a portion of fructose maybe recycled to sucrose through sucrose-phosphate synthase (SPS) [5,13]. The energy and hexoses required for the maintenance of cell growth is provided by the soluble SuSy (S-SuSy) in the cytosol [11]. Sucrose can be converted at high rates to both cellulose and callose (β-1,3-glucan) [14]. In 20 DPA cotton boll, fiber callose is codistributed with abundantly present SuSy in the fiber cell wall region (CW-SuSy) [11,15]. The distribution of SuSy is consistent with its having a dual role in cellulose and callose synthesis in secondary-wall-stage cotton fiber [15]. In addition to SuSy, acidic invertase (either tightly bound to the cell wall or inside the vacuole) and alkaline invertase (a nonglycosylated cytosolic invertase) (INV) can also catalyze hydrolysis of sucrose [16,17].

Cotton fiber development is restricted by declining temperature or low light in many cotton-growing areas [18–21]. However, these two climatic factors often appear as a combined one. Multiple cropping cotton areas (such as cotton-rapeseed or cotton-wheat double cropping systems) are popular in the Yangtze River Valley and Yellow River Valley of China [22]. Due to the competition of temperature and light resources during the growing season of double cropping system, cotton is generally late-germinating and late-maturing and has to suffer from the coupling of declining temperature and low light especially in the late growth stage, e.g., in the Yangtze River Valley, cotton often suffers from rainy and overcast weather during the early stage of flowering and boll formation, as well as from declining temperature and overcast weather during the late stage of flowering and boll formation. These sub-optimal environmental condition during fiber development may hinder cellulose synthesis in fiber [5,23,24], and have a negative impact on fiber quality [25–27].

Declining temperature hinders cellulose synthesis within cotton fiber [5]. Sucrose synthesis is a particularly cool temperature-sensitive step in the partitioning of carbon to cellulose [23], and the activities of related enzymes in cellulose synthesis (SuSy, SPS, INV) are also affected by declining temperature, and leaded to restrain cellulose synthesis and sucrose metabolism [5,23,28]. The activities of SuSy, SPS and INV in various plants or organs are also affected by shading, and result in the decline of biomass and yield [29–31], cotton grown in reduced light environments produced inferior fiber with a lower quality [32,33].

Studies on the enzymological mechanism of carbon partitioning to cellulose synthesis have been carried out in plants under declining temperature or low light [5,33,34], but little is reported about the response of carbon partitioning to cellulose synthesis to the coupling of declining temperature and low light. Therefore, in the study, late planting (LP) and shading were used to fit the combined situation which cotton generally suffers from in the cotton-rapeseed or cotton-wheat double cropping systems, the impact of declining temperature and low light (formed by LP and shading) on sucrose metabolism, cellulose synthesis and related enzymes activity change during fiber secondary wall development (FSWD) were studied, and the physiological and biochemical mechanism of carbon partitioning to cellulose synthesis in response to adverse environmental conditions of declining temperature with low light would be elucidated.

## Materials and Methods

### Experimental Design

Field experiments were conducted at Pailou experimental station of Nanjing Agricultural University at Nanjing, China (32°02′N, 118°50′E), in the Yangtze River Valley in 2010 and 2011. The experimental soil was clay, mixed, thermic, Typic alfisols (udalfs; FAO luvisol) with 18.3 and 18.1 g kg$^{-1}$ organic matter, 1.1 and 1.0 g kg$^{-1}$ total N, 64.5 and 70.2 mg kg$^{-1}$ available N, 17.9 and 20.3 mg kg$^{-1}$ available P, and 102.3 and 111.1 mg kg$^{-1}$ available K contained in 20 cm depth of the soil profile before sowing cotton in 2010 and 2011, respectively.

Two cotton (*Gossypium hirsutum* L.) cultivars, Kemian 1 which was cool temperature-tolerant and Sumian 15 which was cool temperature-sensitive [6,35] were selected based on the categorization of cultivars widely grown in the Yangtze River Valley in its low temperature sensitivity. In the field, different environmental condition during fiber development were provided by planting cotton in different dates [6], 25 April, 25 May and 10 June in 2010 and 2011. Planting date of 25 April is comparatively appropriate to grow cotton in the Yangtze River Valley, and 25 May and 10 June are belong to late planting dates (LPD). Cotton seeds were sown in a nursery bed, and seedlings with three true leaves were transplanted to field at a spacing of 80 cm×25 cm.

When approximately 50% of flowers in the first fruiting node of the 6–7th sympodial branches of plants in each planting date bloomed, three shading treatments were imposed for the plots of each planting date, including an unshaded control (CRLR (crop relative light rate) 100%), mild shading (CRLR 80%), severe shading (CRLR 60%) achieved with white nylon cloth (12 m length, 7 m width, 2 m height, and two different kinds of cloth which reduced the incident light by 20% and 40%, respectively). Shading cloths were removed after cotton bolls in the first fruiting node of the twelfth synpodial branches opened. Experiments were arranged as a randomized complete block design in the field with three replications and each plot was 6 m wide and 11 m long. Furrow-irrigation was applied as needed during both seasons. Conventional insect and weed control methods were utilized as needed.

### Sampling and processing

Cotton flowers in the first or the second fruiting node of the 6–7th sympodial branches with the same anthesis date were tagged with small plastic tags listing the flowering date. About 6–8 cotton bolls in the similar size with the same anthesis date for each treatment were collected from once every 7 days starting from 10 DPA until boll opening. Cotton bolls were collected at 9:00–11:00 am, and cotton fiber were excised from bolls with a scalpel and were immediately put into liquid nitrogen for subsequent measurement.

### Weather data

Weather data were collected from the Nanjing weather station located about 6 km from the plot area. Table 1 shows the mean daily maximum temperature (MDTmax), mean daily temperature (MDT), mean daily minimum temperature (MDTmin) and mean daily radiation (MDR) during FSWD for different planting dates. FSWD was calculated from the initiation date of fiber biomass rapid-accumulation (data not shown) to the boll opening date [7,36].

### Field microclimate measurement

Field microclimate were measured at 15, 30 and 45 DPA (15, 30 and 45 days after initiation of shading). Air temperature, relative humidity and photosynthetically active radiation (PAR) were measured every two hours from 6:00am to 6:00pm, using a Hygro-Thermometer Psychrometer (DT-8892, CEM, Shenzhen, China) to measure air temperature and relative humidity at the position of 6–7th fruiting branches. PAR was measured at the position about 0.2 m above the canopy (PAR$_0$, below the shading cloth) by a Decagon AccuPAR LP-80 Ceptometer (Decagon Devices, Logan, Utah, USA). Measurements were taken only when the direct sunlight was not blocked by clouds.

### Cellulose content, sucrose content and callose content analyses

Cotton fiber was digested in an acetic-nitric reagent, and then the cellulose content was measured with anthrone according to Updegraff [37].

Sucrose was extracted and quantified by a modified method of Pettigrew [33]. About 0.3 g dry weight (DW) fiber samples were extracted with three successive 5 ml washes of 80% ethanol [5]. The ethanol samples were incubated in an 80°C water bath for 30 min.Then the samples were centrifuged at 10,000 g for 10 min, and three aliquots of supernatant were collected together for sucrose measurement [5]. The sucrose assay was conducted according to Hendrix [38].

Callose (β-1,3-glucan) was extracted and quantified by a modified method of Köhle [39]. Fiber was soaked for 2–3 h in 5 ml of ethanol to remove autofluorescent soluble material and

**Table 1.** Weather factors during cotton fiber secondary wall development (FSWD) under different planting dates.

| Years | Planting dates | Flowering dates | Starting date of FSWD | Boll opening date | Duration of FSWD | Weather factors | | | |
|---|---|---|---|---|---|---|---|---|---|
| | (dd–mm) | (dd–mm) | (dd–mm) | (dd–mm) | (d) | MDT (°C) | MDTmax (°C) | MDTmin (°C) | MDR (MJ m$^{-2}$) |
| 2010 | 25-Apr | 28-Jul | 11-Aug | 8-Sep | 29 | 28.3 | 32.4 | 25.3 | 15.8 |
| | 25-May | 19-Aug | 8-Sep | 18-Oct | 41 | 21.4 | 25.5 | 18.5 | 11.5 |
| | 10-Jun | 4-Sep | 28-Sep | 5-Nov | 39 | 16.7 | 21.3 | 13.2 | 11.1 |
| | | | | | CV(%) | 26.51 | 21.32 | 31.82 | 20.34 |
| 2011 | 25-Apr | 27-Jul | 12-Aug | 14-Sep | 34 | 25.9 | 29.5 | 23.0 | 13.0 |
| | 25-May | 25-Aug | 14-Sep | 24-Oct | 41 | 19.9 | 24.5 | 16.6 | 12.5 |
| | 10-Jun | 10-Sep | 30-Sep | 11-Nov | 44 | 17.4 | 21.4 | 14.4 | 10.2 |
| | | | | | CV(%) | 20.63 | 16.15 | 24.83 | 12.56 |

MDT, MDTmax, MDTmin and MDR stand for mean daily temperature, mean daily maximum temperature, mean daily minimum temperature and mean daily radiation.

then oven dried. The above sample (200 mg) was ground into a fine powder in liquid nitrogen followed by 5 ml of 1 N NaOH. The resulting suspension was incubated at 80°C for 30 min to solubilize the callose and centrifuged (15 min, 380 g). The supernatant was used for the callose assay. 0.6 ml of supernatant was mixed with 1.2 ml of 0.1% (w/v) aniline blue WS in water, resulting in a violet-red color. After addition of 0.63 ml of 1N HCl the color changes to deep blue, indicating neutral to acidic pH values [39]. The final pH value was adjusted by addition of 1.77 ml 1M glycine/NaOH buffer (pH 9.5) and the tubes were mixed vigorously. During the following incubation for 20 min at 50°C and further 30 min at room temperature, the aniline blue becomes almost completely decolorize [39]. Fluorescence of the assay was read in a Tecan Infinite M200 microplate reader (Tecan, Männedorf, Switzerland, excitation 400 nm, emission 510 nm, slit 5 nm). Calibration curves were established using a freshly prepared solution of the β-1,3-glucan in 1N NaOH [39].

## Enzymatic analyses

Enzyme extraction and assay were according to King [40] with minor modifications. Fiber cell samples, about 0.5 g fresh weight (FW), were ground into a fine powder in liquid nitrogen followed by grinding in cold extraction buffer (5:1, v/w), which contained 50 mM N-(2-hydroxyethyl) piperazine -N'-(2-ethanesulfonic acid)-NaOH (Hepes-NaOH) (pH 7.5), 10 mM MgCl$_2$, 1 mM ethylene-diamine tetraaceticacid (EDTA), 1 mM ethyleneglycol bis-(2-aminoethylether)-tetraacetic acid (EGTA), 0.5% (w/v) bovine serum albumin (BSA), 2% (w/v) polyvinylpyrrolidone (PVP), 0.1% (v/v) Triton X-100, 2 mM dithiothreitol (DTT), and 1 mM phenylmethylsulfonyl fluoride (PMSF) as described in the previous research [5]. The resulting homogenate was centrifuged at 15,000 g for 20 min, and the supernatant was stored at 4°C for analysis [5]. All extraction procedures were carried out at 0–4°C.

SuSy activity was assayed by measuring the cleavage of sucrose [40]. Each reaction contained 20 mM piperazine-N,N'-bis (2-ethanesulfonic acid)-KOH (Pipes-KOH) (pH 6.5), 100 mM sucrose, 2 mM UDP, and 200 μl of extract in a total volume of 650 μl as described previously [5]. Reactions were started by incubating at 30°C for 30 min. The reactions were stopped with 250 μl of 0.5 M N-tric-(hydroxy-methyl) methylglycine-KOH (Tricine-KOH) (pH 8.3), which were heated for 10 min in boiling water, and the amount of fructose in SuSy reactions was determined as described before [5].

Soluble acid and alkaline invertases' activities were measured by incubation of 100 ml of extract with 1M sucrose in 200 mM acetic acid-NaOH (pH 5.0) (acid invertase), or 100 mM sodium acetate-acetic acid (pH 7.5) (alkaline invertase), in a total volume of 2.5 ml [5]. Reactions were started by incubating at 30°C for 30 min and were stopped with 1 ml of 3,5-dinitro salicylic acid (DNS), and boiling for 5 min [5,40]. Glucose content was measured spectrophotometrically at 540 nm.

SPS activity was assayed by measuring the synthesis of sucrose-6-P from UDP-glucose and fructose-6-P [41]. Each reaction contained 14 mM UDP-glucose, 50 mM fructose-6-P, 50 mM extraction buffers, 50 mM MgCl$_2$ and 200 ml extract in a total volume of 650 ml as described in previous research [5]. Reactions were started by incubating the enzyme extracts at 30°C for 30 min and were stopped with 100 ml of 2N NaOH and 10 min of heating at 100°C to destroy unreacted hexoses and hexose phosphates [5]. After adding 1 ml of 0.1% (w/v) resorcin in 95% (v/v) ethanol, reactions were incubated for 30 min at 80°C. Sucrose-6-P content was calculated from a standard curve measured at 480 nm [5].

## Semi-quantitative RT-PCR analyses

Fiber sampling in 2011 was used in the semi-quantitative RT-PCR analyses. Total RNA was isolated from cotton fiber according to Jiang and Zhang [42]. For each reaction, 2 μg of RNA was reverse transcribed to cDNA with oligo(dT)$_{15}$. PCR was performed in a final volume of 25 μL containing 2U of *Taq* DNA polymerase. The gene-specific primers of two cellulose synthase catalytic subunit, CesA1 and CesA2 (*GhCesA1*, U58283 and *GhCesA2*, U58284),were designed to unique regions of both cDNA (Table 2), the cotton 18srRNA gene (*Gh18srR*,U42827) [43] was used as an internal control. PCR reaction was initially denatured at 94°C for 5 min and 30 cycles at 94°C for 30 s, proper annealing temperature for 30 s and72°C for extending 30–60 s (Table 2), a final extension of RT-PCR products at 72°C for 10 min. PCR products were size-separated by electrophoresis in a 1.8% agarose gel. All photographs were statistically analyzed with the software Quantity One.

## Data analysis

OriginPro 8.0 was adopted for data processing and drawing of figures. An analysis of variance was performed using SPSS statistic package Version 17.0. The means were separated using the least significant difference (LSD) test at 5% of probability level. The coefficient of variation (CV) was calculated as the ratio of the standard deviation to the mean. Changing amplitude ($\Delta\%$) = (Treatment-Control)/Control × 100%, control is the plot of 25 April plus *CRLR* 100%. The fiber growing period delineated by the two dotted lines in figures stood for the period of fiber secondary wall rapid-thickening (FSWR), which was calculated from the initiation date to the termination date of fiber biomass rapid-accumulation. The formation of cotton fiber biomass could be described by the logistic regression model and then the initiation and termination date of the fiber biomass rapid-formation were obtained (data not shown), which represented the initiation and termination date of FSWR, respectively, and the duration was the period of FSWR [7,36].

## Results

### Field environmental condition and field microclimate

Environmental condition of normal planting date (NPD) of 25 April was advantageous to develop cotton, and was the optimal planting date in the Yangtze River Valley [44], delaying the planting date would prolong FSWD from 29 to 41 days in 2010 and from 34 to 44 days in 2011 (Table 1). In two experimental years, MDT, MDTmax, MDTmin and MDR during FSWD decreased as planting date delayed. Fiber microclimate data in Table S1 were expressed as the mean of data measured from 6:00am to 6:00pm at 15, 30 and 45 DPA, air temperature and

PAR$_0$ in cotton field also decreased as planting dates delayed, but the CVs of MDT, MDTmax and MDTmin were higher than that of MDR (Table 1), it was indicated that effect of LP on temperature factors was greater than sunshine factors during FSWD. MDTmin was the primary environmental factor affected by LP, and reduced from 25.9°C to 13.2°C in 2010 and from 25.3°C to 14.4°C in 2011 (Table 1).

During 6:00am–6:00pm, PAR$_0$ peaked at midday, shading significantly reduced PAR$_0$ by 18%–25% for *CRLR* 80% and 35%–44% for *CRLR* 60% treatments as experimental design (Figure 1); air temperature peaked at 12:00am–2:00pm and was not significantly different between shading and normal light treatments, except one or two determination points and their numbers of each planting date with only small deviations of no more than 1.5°C (Figure 2); mean relative humidity decreased until 12:00am–2:00pm and thereafter increased (Figure 2), and shading treatments was statistically different around 10:00am–2:00pm, but the deviation of these treatments were less than 7% (Figure 2). Field microclimate data measured at 15, 30 and 45 DPA in Table S1 also showed that difference of air temperature or mean relative humidity among different shading treatments were no more than 1°C or 5%, respectively. These small differences in temperature and relative humidity would probably only be a minor effect on carbohydrate concentrations compared with the effect of 20%–40% PAR reduction, which reduced the carbohydrate levels in fiber under shading by lower photosynthetic rates because of the lower light levels as shown by Pettigrew [33]. Therefore, the decline of PAR$_0$ was the key reason for the adverse effect on cotton fiber development caused by shading.

### Sucrose, cellulose and callose contents in cotton fiber

During fiber development, sucrose contents in cotton fiber declined from 10 DPA under NPD$_{25-Apr}$, but there were a peak value which occurred at 17 and 24 DPA under LPD$_{25-May}$ and LPD$_{10-Jun}$, respectively (Figure 3). LP enhanced sucrose content in cotton fiber compared to the normal planting. Compared to *CRLR* 100%, under NPD$_{25-Apr}$, sucrose contents of *CRLR* 80% and *CRLR* 60% decreased. Sucrose in cotton fiber after 38 DPA under NPD$_{25-Apr}$ had already been depleted, but there was surplus sucrose in the developing fiber of 59 DPA in LPD$_{10-Jun}$ (Figure 3). Under LPD, fiber sucrose contents of *CRLR* 80% and *CRLR* 60% decreased before 31 DPA, but after 31 DPA the sucrose contents of *CRLR* 100% were lower than that of *CRLR* 80% and *CRLR* 60% (Figure 3). The maximum sucrose content in cotton fiber can reflect the amount of available sucrose in developing fiber, and the minimum sucrose content show the residual sucrose content in mature fiber [5]. Shading could decrease the maximum sucrose content by 3.5%–15.7%, while the maximum sucrose and minimum sucrose content under LPD increased 10.4%–48.5%

**Table 2.** Primers, T$_m$, extension time and cycles in RT-PCR program.

| Gene | Accession No. | Primer | T$_m$ (°C) | No. of cycles | Extension time (s) | Length of amplified DNA (bp) |
|---|---|---|---|---|---|---|
| *GhCesA1* | U58283 | Forward: 5′- TGGGTTGAATGTTAATGGT-3′ | 58 | 30 | 60 | 632 |
| | | Reverse: 5′- CAGGATACCACTTAGGGAACT-3′ | | | | |
| *GhCesA2* | U58284 | Forward: 5′- CTGGCTTTGGTTCACTTGC-3′ | 58 | 30 | 60 | 529 |
| | | Reverse: 5′- CCGCCATTATCGTTGCTTA-3′ | | | | |
| *Gh18srR* | U42827 | Forward: 5′-CTGAGAAACGGCTACCACAT-3′ | 53 | 25 | 30 | 500 |
| | | Reverse: 5′-CTATGAAATACGAATGCCCC-3′ | | | | |

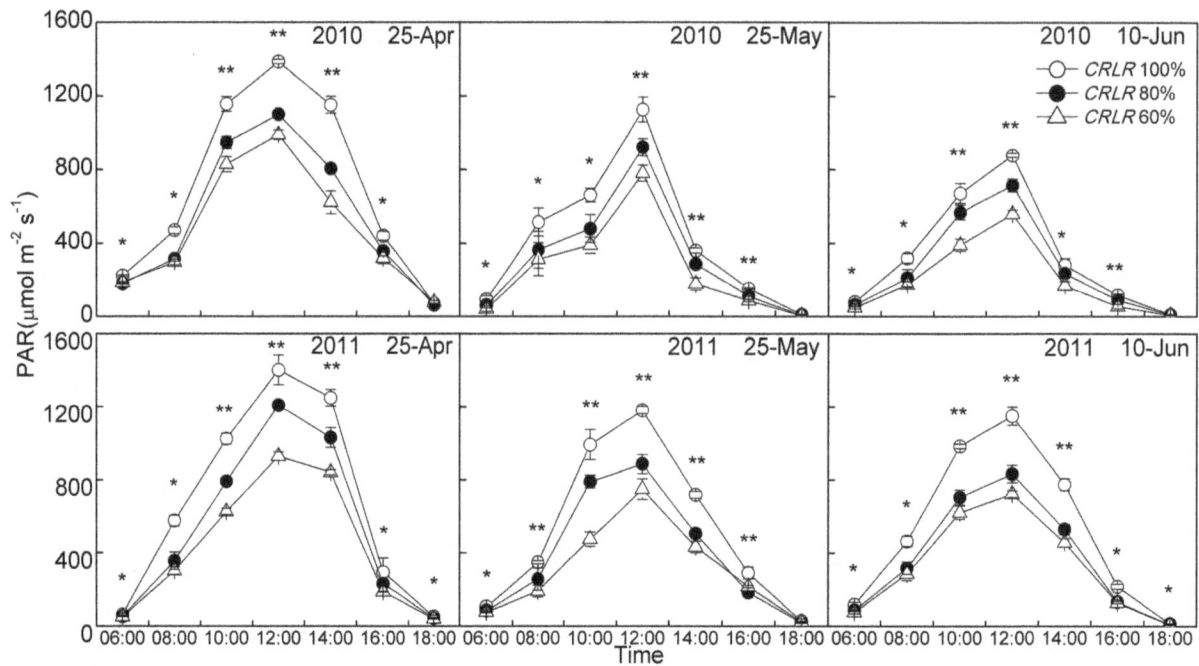

**Figure 1. Changes of photosynthetically active radiation measured at the position about 0.2 m above the canopy (PAR$_0$) at 30 DPA under the coupling of planting date and shading in 2010 and 2011.** * and ** mean significant difference among three shading treatments at 0.05 and 0.01 probability levels, respectively.

and 125.8%–1349.0%, respectively, compared to normal planted cotton (Table 3, Table 4). The maximum sucrose content under the coupling of LP and shading did not increase as much as that under the same planting date without shading. In contrast, the minimum sucrose content under LPD$_{10-Jun}$ plus *CRLR* 60% reached the highest (Table 3). The CVs of the maximum and minimum sucrose contents in two cultivars caused by LP were higher than that caused by shading, it was indicated that the effect of LP on fiber sucrose content was greater than that of shading, and the CVs caused by the coupling of LP and shading were similar to LP. Compared to the maximum sucrose content, the minimum sucrose content was more susceptible to LP or shading,

and the trend was consistent between two cultivars (Table 3). The CVs and its changing amplitude (Δ%) of the maximum sucrose content response to the coupling of LP and shading in Sumian 15 was higher than that of Kemian 1, but the CVs and its Δs of the minimum sucrose content between two cultivars was different in two years (Table 3, Table 4).

Carbon from sucrose can be converted at high rate to both cellulose and callose (β-1,3-glucan) [14]. In this study, β-1,3-glucan content in cotton fiber was low at 10 DPA and rose abruptly at approximately the time as the onset of secondary wall cellulose synthesis as also reported by Maltby et al. [45]. Callose content increased from 10 DPA to 17–24 DPA when the peak appeared

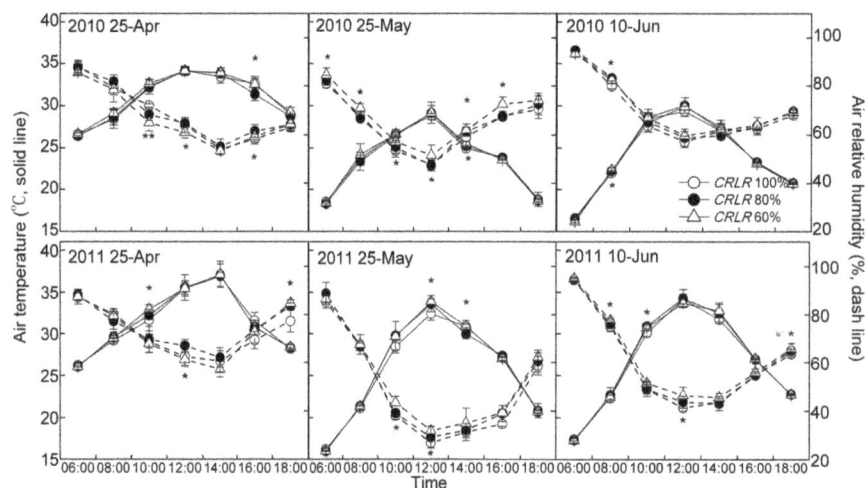

**Figure 2. Changes of mean air temperature and mean relative humidity at 30 DPA under the coupling of planting date and shading in 2010 and 2011.** * and ** mean significant difference among three shading treatments at 0.05 and 0.01 probability levels, respectively.

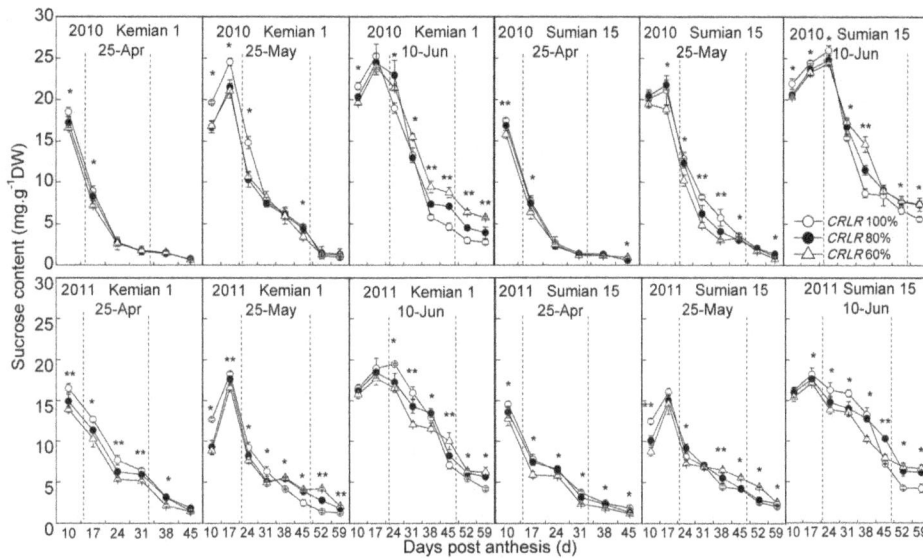

**Figure 3. Changes of sucrose contents in cotton fiber of two cultivars under the coupling of planting date and shading in 2010 and 2011.** The fiber growing period delineated by the two dotted lines stand for the period of fiber secondary wall rapid-thickening (FSWR). * and ** mean significant difference among three shading treatments at 0.05 and 0.01 probability levels, respectively.

(Figure 4), and then declined significantly. LP and shading could increased fiber callose content, which reached the maximum under the coupling of LP and shading (Figure 4, Table 3). The period of FSWR was 17–31 DPA under $NPD_{25-Apr}$, and were 24–45 DPA and 24–52 DPA under $LPD_{25-May}$ and $LPD_{10-Jun}$, respectively. Meanwhile, the fiber cellulose rapid-accumulation was almost the same (Figure 4). Cellulose content in cotton fiber increased from 10 DPA, and compared to $NPD_{25-Apr}$ which the initiation of fiber cellulose fast-accumulation began at 17 DPA, $LPD_{10-Jun}$ delayed to 24 DPA and restrained cellulose synthesis. Shading under all planting dates decreased fiber cellulose contents (Figure 4), but the decreasing extents were different in different fiber development stages and planting dates. Shading under $NPD_{25-Apr}$ had greater impacts on cellulose synthesis during 24–31 DPA, while shading under LPD were during 38–45 DPA. At the mature stage (59 DPA), $CRLR$ 80% and $CRLR$ 60% decreased cellulose content by 0.7%–2.8% and 1.1%–5.6% under $NPD_{25-Apr}$, respectively, while $LPD_{25-May}$ and $LPD_{10-Jun}$ decreased by 6.7%–10.2% and 15.0%–20.9%, respectively (Table 4). The final cellulose content under the coupling of $LPD_{10-Jun}$ and $CRLR$ 60% was the lowest which decreased 18.2%–25.5% (Table 3). The CVs of the maximum callose contents of two cultivars caused by shading were higher than that caused by LP and the coupling of LP and shading, it was indicated that the effects of shading on maximum callose contents were greater than LP and the coupling of LP and shading. In contrast, the effects of LP on final cellulose contents were greater than shading, but close to the coupling of LP and shading (Table 3). The CVs and Δs of the final cellulose content response to the coupling of LP and shading in Sumian 15 were higher than Kemian 1 (Table 3, Table 4), it was indicated that cellulose content of Sumian 15 was more sensitive to the coupling of LP and shading than Kemian 1.

## Activity changes of related cellulose synthesis enzymes during fiber secondary wall development

Fiber SuSy activity decreased during fiber development (Figure S1). During FSWR, SuSy activity was the lowest under $LPD_{25-May}$ and rose again under $LPD_{10-Jun}$ (Table 5). Shading decreased the SuSy activities under all planting dates. In the period of FSWR, the decline caused by shading under $NPD_{25-Apr}$ and $LPD_{10-Jun}$ were larger than $LPD_{25-May}$ (Figure S1) and the effect of shading under $LPD_{10-Jun}$ on the Δs of SuSy activities were opposite in two years (Table 6).

Acidic INV in cotton fiber was higher than alkaline invertase, from the point of view of CVs, acidic INV was more susceptible to LP or shading, and both of their activities decreased during fiber development (Figure S2–S3, Table 5). Under LPD, the fiber acidic and alkaline INV activities increased and remained at a high rate at the end of fiber development. During FSWR, shading decreased acidic and alkaline INV activities of Kemian 1, shading also decreased acidic and alkaline INV activities of Sumian 15 under planting date of 25 April and 25 May, but increased the activities under $LPD_{10-Jun}$ (Figure S2–S3).

Fiber SPS activity increased and peaked at 24 or 31 DPA, and then decreased during fiber development. LP delayed the peaks to 38, 45 or 52 DPA and decreased the peak values (Figure S4). During FSWR, SPS activity decreased under shading and the peak was delayed under $LPD_{25-May}$ in 2010. SPS activity under $NPD_{25-Apr}$ was similar to $LPD_{25-May}$ during FSWR, however decreased notably under $LPD_{10-Jun}$ (decreasing 19.6%–37.8%). Under the coupling of shading and $LPD_{10-Jun}$, SPS activities decreased by 24.3%–43.0%, which were more than shading or LP alone, and the decreasing amplitude of Sumian 15 (decreasing 37.8%–43.0%) was larger than that of Kemian 1 (decreasing 24.3%–37.4%) (Figure S4, Table 5, Table 6).

After analysing the CVs of mean sucrose metabolism enzyme activities (such as SuSy, SPS and acidic/alkaline INV), it was found that the effects of LP on sucrose metabolism enzymes were greater than shading during FSWR, and were similar to the effects of the coupling of LP and shading. Among the four kinds of sucrose metabolism enzymes, SPS was the most significant affected by shading, and acidic INV was the most significant affected by LP and by the coupling of LP and shading (Table 5).

**Table 3.** Maximum/minimum sucrose, maximum callose and final cellulose contents in cotton fiber and their analysis of variance under the coupling of planting date and shading in 2010 and 2011.

| Planting dates | CRLR | Maximum sucrose content ($mg\ g^{-1}DW$) | | Minimum sucrose content ($mg\ g^{-1}DW$) | | Maximum callose content ($mg\ g^{-1}DW$) | | Final cellulose content (%) | |
|---|---|---|---|---|---|---|---|---|---|
| (dd–mm) | (%) | 2010 | 2011 | 2010 | 2011 | 2010 | 2011 | 2010 | 2011 |
| **Kemian 1** | | | | | | | | | |
| 25-Apr | 100 | 18.5 a | 16.5 a | 0.7 a | 1.5 a | 7.7 b | 5.6 c | 92.2 a | 91.6 a |
| | 80 | 17.3 b | 14.9 b | 0.7 a | 1.9 a | 7.6 b | 9.6 b | 91.2 a | 89.0 a |
| | 60 | 16.6 c | 13.9 c | 0.6 a | 1.4 a | 11.4 a | 11.5 a | 87.0 b | 90.6 a |
| 25-May | 100 | 24.5 a | 18.3 a | 4.6 a | 3.9 a | 8.3 b | 7.0 c | 85.3 a | 82.3 a |
| | 80 | 21.5 b | 17.6 b | 4.4 a | 4.1 a | 8.1 b | 9.7 b | 85.0 a | 77.0 b |
| | 60 | 21.1 b | 16.6 c | 3.4 a | 2.5 b | 12.2 a | 12.6 a | 83.1 b | 76.8 b |
| 10-Jun | 100 | 25.2 a | 19.5 a | 4.6 c | 7.1 c | 9.4 b | 7.0 c | 78.4 a | 76.6 a |
| | 80 | 24.4 b | 19.0 a | 7.0 b | 8.2 b | 9.3 b | 11.6 b | 76.8 b | 73.3 b |
| | 60 | 23.9 c | 17.7 b | 8.7 a | 10.0 a | 12.8 a | 13.3 a | 75.4 c | 73.1 b |
| $CV_P$(%) | | 16.23 | 8.25 | 68.92 | 67.66 | 10.39 | 13.09 | 8.09 | 9.07 |
| $CV_S$(%) | | 5.60 | 8.64 | 6.97 | 14.50 | 24.11 | 34.11 | 3.90 | 1.45 |
| $CV_{P\times S}$(%) | | 15.53 | 10.66 | 73.90 | 70.52 | 20.86 | 27.90 | 7.09 | 9.19 |
| **>Sumian 15** | | | | | | | | | |
| 25-Apr | 100 | 17.5 a | 14.6 a | 0.6 a | 1.9 a | 8.4 b | 8.1 b | 86.2 a | 88.0 a |
| | 80 | 16.9 b | 13.6 b | 0.6 a | 1.3 a | 12.7 a | 8.2 b | 85.6 ab | 87.4 a |
| | 60 | 15.8 c | 12.8 c | 0.9 a | 1.2 a | 13.1 a | 13.0 a | 84.7 b | 86.0 a |
| 25-May | 100 | 21.2 a | 16.1 a | 3.6 a | 4.3 b | 9.3 c | 9.4 b | 80.4 a | 81.1 a |
| | 80 | 21.8 a | 15.0 b | 3.0 a | 4.2 b | 14.2 b | 9.2 b | 77.2 ab | 79.1 ab |
| | 60 | 19.5 b | 14.6 b | 3.3 a | 5.6 a | 15.1 a | 13.8 a | 76.5 b | 77.1 b |
| 10-Jun | 100 | 26.0 a | 18.2 a | 8.4 a | 7.2 b | 10.7 c | 10.3 c | 72.0 a | 69.6 a |
| | 80 | 24.8 b | 17.6 ab | 9.1 a | 10.3 a | 14.7 b | 11.1 b | 70.0 ab | 66.5 ab |
| | 60 | 24.4 b | 17.1 b | 9.0 a | 8.0 b | 16.2 a | 14.4 a | 68.0 b | 65.6 b |
| $CV_P$(%) | | 19.74 | 11.31 | 94.45 | 59.86 | 12.20 | 12.13 | 8.98 | 11.68 |
| $CV_S$(%) | | 5.05 | 6.57 | 30.95 | 24.40 | 22.86 | 28.57 | 0.88 | 1.18 |
| $CV_{P\times S}$(%) | | 17.73 | 12.07 | 84.28 | 65.09 | 21.34 | 21.96 | 8.83 | 11.29 |

DW, dry weight; Values followed by the different letters within a column are significantly different at 0.05 probability level; $CV_S$ was calculated from the data of three shading treatments in normal planting date of 25 April; $CV_P$ was calculated from the data of three planting dates and $CV_{P\times S}$ was calculated from the data of all treatments; * and ** mean significant difference at 0.05 and 0.01 probability levels, respectively; NS means non-significant differences.

**Table 4.** The changing amplitude (Δ%) of maximum/minimum sucrose, maximum callose and final cellulose contents in cotton fiber under the coupling of planting date and shading in 2010 and 2011.

| Treatments | Maximum sucrose content | | Minimum sucrose content | | Maximum callose content | | Final cellulose content | |
|---|---|---|---|---|---|---|---|---|
| | 2010 | 2011 | 2010 | 2011 | 2010 | 2011 | 2010 | 2011 |
| **Kemian 1** | | | | | | | | |
| CRLR 80% | −6.8 | −9.6 | 9.2 | 25.6 | −1.3 | 73.0 | −1.1 | −2.8 |
| CRLR 60% | −10.4 | −15.7 | −4.7 | −2.6 | 47.6 | 106.7 | −5.6 | −1.1 |
| 25-May | 32.4 | 10.7 | 584.6 | 161.0 | 7.2 | 26.6 | −7.5 | −10.2 |
| 25-May+CRLR 80% | 16.3 | 6.8 | 554.0 | 174.4 | 4.5 | 75.0 | −7.8 | −15.9 |
| 25-May+CRLR 60% | 13.8 | 0.5 | 405.7 | 68.2 | 58.4 | 127.8 | −9.9 | −16.2 |
| 10-Jun | 36.4 | 18.0 | 584.9 | 376.5 | 22.4 | 26.8 | −15.0 | −16.4 |
| 10-Jun+CRLR 80% | 32.0 | 14.8 | 944.7 | 455.2 | 20.4 | 108.4 | −16.7 | −20.0 |
| 10-Jun+CRLR 60% | 29.2 | 7.3 | 1191.3 | 573.4 | 66.8 | 138.9 | −18.2 | −20.2 |
| **Sumian 15** | | | | | | | | |
| CRLR 80% | −3.5 | −6.6 | −0.9 | −28.9 | 51.7 | 1.8 | −0.7 | −0.7 |
| CRLR 60% | −9.5 | −12.3 | 64.5 | −36.1 | 55.6 | 60.6 | −1.7 | −2.3 |
| 25-May | 21.1 | 10.4 | 513.6 | 125.8 | 10.8 | 16.6 | −6.7 | −7.8 |
| 25-May+CRLR 80% | 24.8 | 3.3 | 422.1 | 121.8 | 69.8 | 13.5 | −10.4 | −10.1 |
| 25-May+CRLR 60% | 11.6 | 0.2 | 466.5 | 195.3 | 80.5 | 70.2 | −11.3 | −12.4 |
| 10-Jun | 48.5 | 25.2 | 1349.0 | 281.9 | 27.3 | 27.7 | −16.5 | −20.9 |
| 10-Jun+CRLR 80% | 41.9 | 21.2 | 1468.0 | 446.5 | 75.5 | 37.0 | −18.8 | −24.4 |
| 10-Jun+CRLR 60% | 39.5 | 17.7 | 1445.5 | 322.1 | 92.9 | 77.9 | −21.1 | −25.5 |

Δ% = (Treatment-Control)/Control × 100%, control is the plot of 25 April + *CRLR* 100%.

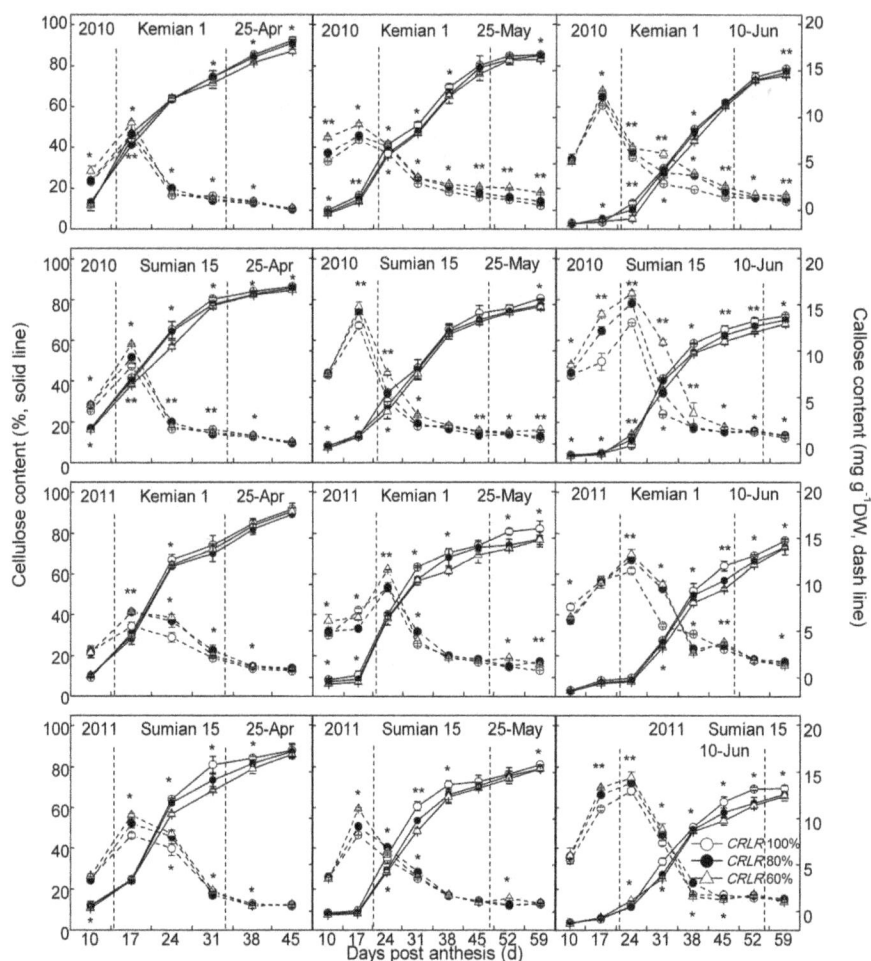

**Figure 4. Changes of cellulose and callose contents in cotton fiber of two cultivars under the coupling of planting date and shading in 2010 and 2011.** The fiber growing period delineated by the two dotted lines stand for the period of fiber secondary wall rapid-thickening (FSWR). * and ** mean significant difference among three shading treatments at 0.05 and 0.01 probability levels, respectively.

### Gene expression of cellulose synthase in cotton fiber

The peak of *GhCesA1* and *GhCesA2* expression under $NPD_{25-Apr}$ appeared around 17 DPA, and delayed to 24 DPA under LPD (Figure 5). Under $NPD_{25-Apr}$ and $LPD_{25-May}$, shading delayed the *GhCesA1* expression in Kemian 1, and then shading restrained the expression in $LPD_{10-Jun}$ (Figure 5–6). Compared to Kemian 1, shading delayed the *GhCesA1* expression of Sumian 15 under $NPD_{25-Apr}$ and shading restrained the expression under LPD.

Gene expression intensity and duration of *GhCesA1* were lower than that of *GhCesA2* in two cultivars, it was indicated that the coupling of LP and shading might have a more adverse effect on *GhCesA1* than *GhCesA2*. Under $NPD_{25-Apr}$, shading delayed the *GhCesA2* expression, and appeared to restrain the *GhCesA2* expression under LPD. The postponing or restraining extent under *CRLR* 60% was more serious than under *CRLR* 100% and *CRLR* 80% (Figure 5–6).

### Discussion

Cotton fiber development is affected by cool temperature [5,36,46] and low light [19,20,33,47], which often occur together [7,48]. In order to fit the coupling situation of declining temperature and low light which cotton generally suffers from in the cotton-rapeseed or cotton-wheat double cropping systems, the

couping of LP and shading experiments were designed and carried out in 2010 and 2011. The results showed that MDTmin was the primary environmental factor affected by LP during FSWD (Table 1), and shading under three planting dates significantly reduced $PAR_0$ by 18%–25% and 35%–44% under *CRLR* 80% and *CRLR* 60%, respectively (Figure 1). $LPD_{25-May}$ and $LPD_{10-Jun}$ ($MDTmin_{FSWD}$ of 18.5 and 13.2°C in 2010, 16.6 and 14.4°C in 2011) plus *CRLR* 80% and *CRLR* 60% resulted in an extending of cotton fiber development period and affected cellulose synthesis (Tables 1 and 3).

Toward the end of fiber elongation, secondary wall deposition begins via enhanced cellulose synthesis, while fiber callose synthesis reaches the peak in this period [2,14,45]. Sucrose is the substrate that is required for high-rate callose and secondary-wall cellulose synthesis, and sucrose metabolism is sensitive to cool temperature or shading [2,14,41,33]. In this study, the coupling of LP and shading had more adverse impacts on cellulose synthesis during 38–45 DPA than before 31 DPA (Figure 4). Under LPD, shading decreased sucrose content before 31 DPA, and then increased it after 31 DPA in fiber (Figure 3, Table 3). The initial decline of sucrose content under shading was probably caused by lower photosynthetic rates [33], and the next increasing of sucrose content under LPD was a self-regulating phenomenon and the reduced sucrose transformation rate [5]. In developing fiber

**Table 5.** Average activities of sucrose phosphate synthase (SPS), sucrose synthase (SuSy), acidic invertase (Acidic INV) and alkaline invertase (Alkaline INV) in cotton fiber during fiber secondary wall rapid-thickening (FSWR) and their analysis of variance under the coupling of planting date and shading in 2010 and 2011.

| Planting dates | CRLR | SPS | | SuSy | | Acidic INV | | Alkaline INV | |
|---|---|---|---|---|---|---|---|---|---|
| (dd-mm) | (%) | (mg sucrose g⁻¹FW h⁻¹) | | (mg fructose g⁻¹FW h⁻¹) | | (mg glucose g⁻¹FW h⁻¹) | | (mg glucose g⁻¹FW h⁻¹) | |
| | | 2010 | 2011 | 2010 | 2011 | 2010 | 2011 | 2010 | 2011 |
| **Kemian 1** | | | | | | | | | |
| 25-Apr | 100 | 13.9 a | 10.0 a | 71.2 a | 72.8 a | 63.7a | 78.1 a | 54.8 a | 70.8 a |
| | 80 | 14.2 a | 8.6 b | 68.4 b | 63.8 b | 62.2 b | 74.9 b | 52.6 ab | 65.1 b |
| | 60 | 12.7 b | 7.8 c | 64.3 c | 63.1 b | 57.8 c | 71.2 c | 50.8 b | 64.8 b |
| 25-May | 100 | 13.7 a | 9.8 a | 51.7 a | 50.4 a | 70.9 a | 73.7 a | 55.9 a | 69.5 a |
| | 80 | 13.6 a | 9.2 b | 51.8 a | 45.1 b | 66.1 b | 68.4 b | 52.7 b | 66.2 b |
| | 60 | 12.9 b | 8.4 c | 48.0 b | 43.6 c | 61.6 c | 64.1 c | 52.0 c | 63.3 c |
| 10-Jun | 100 | 11.2 a | 7.8 a | 94.6 a | 66.7 a | 118.3 a | 104.4 a | 73.8 a | 88.6 a |
| | 80 | 10.5 a | 6.8 b | 88.7 b | 59.8 b | 110.0 b | 101.3 b | 69.7 b | 85.5 b |
| | 60 | 9.8 b | 6.3 c | 88.8 b | 59.7 b | 103.2 c | 95.6 c | 68.7 b | 85.6 b |
| CV_P(%) | | 11.68 | 13.30 | 29.61 | 18.35 | 35.20 | 19.46 | 17.37 | 14.02 |
| CV_S(%) | | 6.04 | 12.50 | 5.11 | 8.14 | 4.95 | 4.65 | 3.77 | 5.06 |
| CV_{P×S}(%) | | 12.86 | 15.38 | 25.33 | 17.06 | 30.21 | 18.51 | 15.30 | 14.04 |
| **Sumian 15** | | | | | | | | | |
| 25-Apr | 100 | 14.4 a | 9.6 a | 66.3 a | 73.1 a | 73.2 a | 77.0 a | 50.3 a | 70.7 a |
| | 80 | 12.7 b | 9.3 a | 62.6 b | 69.5 b | 71.0 b | 71.3 b | 48.4 b | 68.3 b |
| | 60 | 11.0 c | 8.3 b | 59.5 c | 66.0 c | 67.6 c | 66.7 c | 47.3 b | 68.1 b |
| 25-May | 100 | 12.8 a | 9.6 a | 50.1 a | 49.6 a | 73.4 a | 81.4 a | 61.5 a | 65.3 a |
| | 80 | 12.5 ab | 9.1 b | 47.2 b | 47.9 b | 73.9 a | 77.0 b | 59.8 b | 62.4 b |
| | 60 | 11.9 b | 9.1 b | 47.0 b | 43.9 c | 67.6 b | 70.4 c | 57.0 c | 62.4 b |
| 10-Jun | 100 | 9.0 a | 6.6 a | 88.0 a | 68.6 a | 105.6 a | 130.2 a | 71.0 c | 77.5 c |
| | 80 | 8.8 ab | 5.9 b | 84.7 a | 63.8 b | 107.3 b | 124.2 b | 72.8 b | 79.8 b |
| | 60 | 8.2 b | 5.8 b | 74.8 b | 59.4 c | 108.5 a | 123.9 b | 75.3 a | 84.4 a |
| CV_P(%) | | 23.24 | 19.66 | 27.95 | 19.54 | 22.16 | 30.68 | 17.01 | 8.59 |
| CV_S(%) | | 13.53 | 7.15 | 5.38 | 5.08 | 4.00 | 7.23 | 3.12 | 2.09 |
| CV_{P×S}(%) | | 19.12 | 19.36 | 23.98 | 17.61 | 21.85 | 28.98 | 17.76 | 11.08 |

FW, fresh weight; Values in the same planting date followed by the different letters within a column are significantly different at 0.05 probability level; CV_S was calculated from the data of three shading treatments in normal planting date of 25 April, CV_P was calculated from the data of three planting dates and CV_{P×S} was calculated from the data of all treatments; * and ** mean significant difference at 0.05 and 0.01 probability levels, respectively; NS means non-significant differences.

**Table 6.** The changing amplitude (Δ%) of average activities of sucrose phosphate synthase (SPS), sucrose synthase (SuSy), acidic invertase (Acidic INV) and alkaline invertase (Alkaline INV) in cotton fiber during fiber secondary wall rapid-thickening (FSWR) under the coupling of planting date and shading in 2010 and 2011.

| Treatments | SPS | | SuSy | | Acidic INV | | Alkaline INV | |
|---|---|---|---|---|---|---|---|---|
| | 2010 | 2011 | 2010 | 2011 | 2010 | 2011 | 2010 | 2011 |
| **Kemian 1** | | | | | | | | |
| CRLR 80% | 1.8 | −13.5 | −3.9 | −12.4 | −2.4 | −4.1 | −3.9 | −8.1 |
| CRLR 60% | −9.2 | −21.8 | −9.7 | −13.4 | −9.2 | −8.9 | −7.2 | −8.5 |
| 25-May | −1.7 | −1.5 | −27.3 | −30.8 | 11.4 | −5.7 | 2.0 | −1.9 |
| 25-May+CRLR 80% | −2.3 | −8.4 | −27.3 | −38.1 | 3.8 | −12.4 | −3.8 | −6.6 |
| 25-May+CRLR 60% | −7.5 | −15.7 | −32.6 | −40.1 | −3.2 | −18.0 | −5.1 | −10.7 |
| 10-Jun | −19.6 | −21.9 | 32.9 | −8.4 | 85.8 | 33.7 | 34.7 | 25.1 |
| 10-Jun+CRLR 80% | −24.3 | −32.1 | 24.6 | −17.9 | 72.7 | 29.6 | 27.2 | 20.7 |
| 10-Jun+CRLR 60% | −29.9 | −37.4 | 24.7 | −18.1 | 62.0 | 22.4 | 25.5 | 20.9 |
| **Sumian 15** | | | | | | | | |
| CRLR 80% | −11.9 | −2.4 | −5.6 | −4.9 | −3.0 | −7.4 | −3.8 | −3.4 |
| CRLR 60% | −23.8 | −12.8 | −10.2 | −9.7 | −7.7 | −13.4 | −6.0 | −3.7 |
| 25-May | −10.8 | 0.3 | −24.5 | −32.1 | 0.3 | 5.7 | 22.2 | −7.7 |
| 25-May+CRLR 80% | −13.4 | −4.9 | −28.7 | −34.5 | 1.0 | 0.0 | 18.9 | −11.8 |
| 25-May+CRLR 60% | −17.5 | −5.1 | −29.2 | −39.9 | −7.7 | −8.6 | 13.3 | −11.7 |
| 10-Jun | −37.8 | −30.5 | 32.8 | −6.1 | 44.3 | 69.0 | 41.2 | 9.5 |
| 10-Jun+CRLR 80% | −38.8 | −37.8 | 27.8 | −12.7 | 46.6 | 61.3 | 44.6 | 12.9 |
| 10-Jun+CRLR 60% | −43.0 | −39.5 | 12.9 | −18.7 | 48.2 | 60.9 | 49.8 | 19.4 |

Δ% = (Treatment-Control)/Control × 100%, control is the plot of 25 April + CRLR 100%.

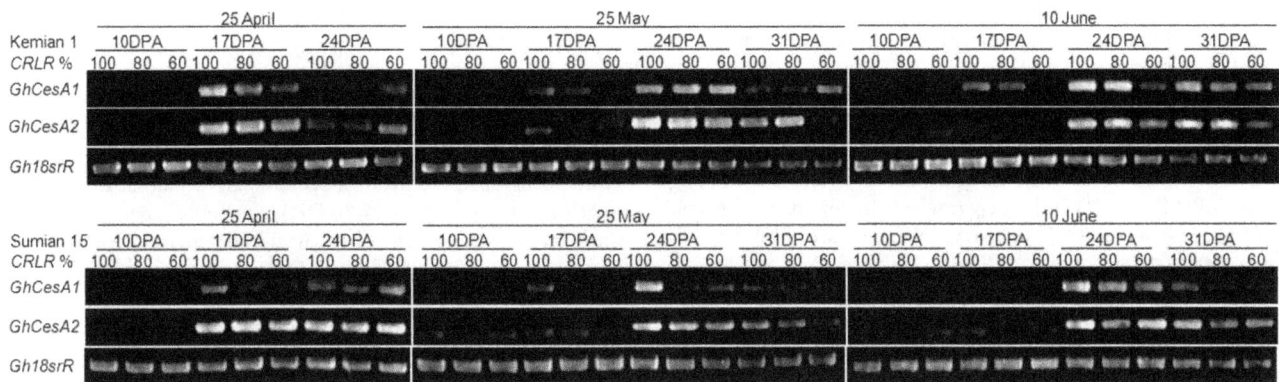

**Figure 5. Gene expressions of two cellulose synthase catalytic subunits (CesA1 and CesA2) in cotton fiber of two cultivars under the coupling of planting date and shading in 2011.**

during 38–45 DPA under the coupling of LP and shading, the greater increasing range of residual sucrose content and decreasing range of cellulose content indicated that compared to single stress, the sucrose transforming ability declined more under combined stress, and more obviously during 38–45 DPA. On the other hand, sucrose could not be effectively used for cellulose synthesis under cool temperature [5] and callose synthesis can replace cellulose synthesis after wounding [2]. Under NPD$_{25\text{-Apr}}$, the transformation from sucrose to cellulose and callose advanced simultaneously, and fiber cellulose started to accumulate rapidly before 17 DPA (cellulose contents were about 40%, Figure 4), but sucrose converted to callose more than to cellulose under shading. In contrast, under LPD, although fiber sucrose content was high during 10–17 DPA, the conversion rate of sucrose to callose was much greater than to cellulose, with lower cellulose contents about 10% before 24 DPA, shading under LPD exacerbated the situation (Figure 4). The result indicated that in the early stage of FSWD (before 24 DPA), compared to LP or shading, there was more abundant sucrose in cotton fiber under the coupling of LP and shading, however, carbon from sucrose was converted mainly to callose instead of cellulose. Whereas in the later stage of FSWD (after 38 DPA), carbon in fiber stagnated in the form of sucrose and the partitioning to cellulose synthesis decreased, and the coupling of LP and shading made it more serious.

The postponing or restraining trend of *GhCesA* expression under the coupling of LP and shading were corresponding with the downward trend of cellulose (Figure 4 and Table 4). The cellulose synthase complex was sensitive to adverse environmental effects and which directly affected cellulose synthesis. As to cool temperature-sensitive cultivar Sumian 15, shading appeared to restrain *GhCesA* expression under LPD$_{25\text{-May}}$, earlier than shading under LPD$_{10\text{-Jun}}$ restraining the expression in relative cool temperature-tolerant cultivar Kemian 1, it was indicated that the *GhCesA* in Sumian 15 was more sensitive to environmental change, and the postponing or restraining extent was more serious as increasing shading degree (Figure 5).

In cotton fiber, there were many sucrose metabolism enzymes contributing to cellulose synthesis besides CesA [2,12,49]. Sucrose is degraded to provide UDP-glucose for cellulose synthesis by SuSy, which is the critical partner in high-rate secondary-wall cellulose synthesis [2,12,49]. Only M-SuSy or CW-SuSy protein associated with CesA in the plasma membrane-associated rosette structure possess β-1,4-glucan (cellulose) synthesis activity [10,11]. Compared to NPD$_{25\text{-Apr}}$, the restrained activity of SuSy during FSWR under LPD$_{25\text{-May}}$ was caused by shading and declining temperature, and the increasing activity under the coupling of shading and LPD$_{10\text{-Jun}}$ (MDTmin was the lowest in this experiment) might be due to a large part of M-SuSy becoming

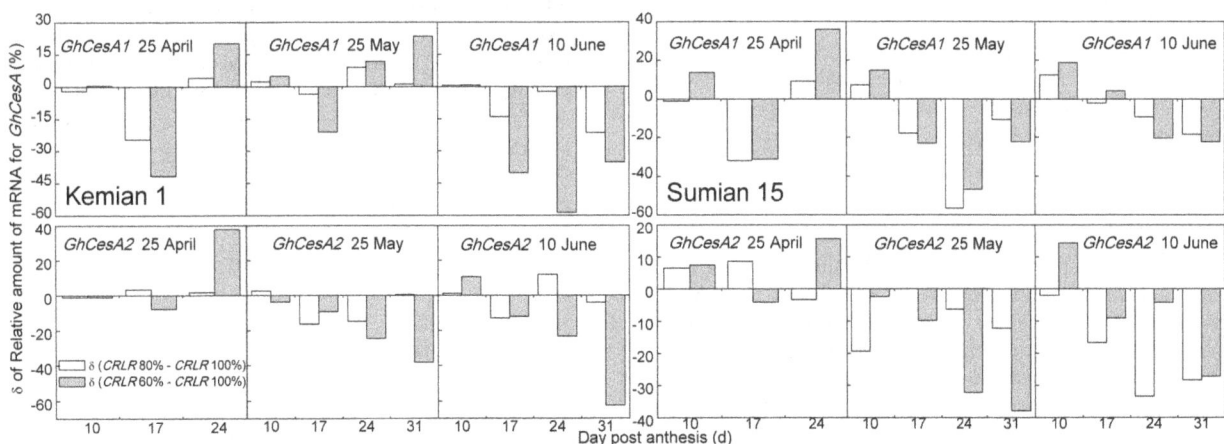

**Figure 6. δ of relative amount of mRNA for two cellulose synthase catalytic subunits (CesA1 and CesA2) in cotton fiber of two cultivars under the coupling of planting date and shading in 2011.** δ = (CRLR 80% - CRLR 100%) or (CRLR 60% - CRLR 100%).

S-SuSy. The enhanced S-SuSy degraded sucrose for maintenance and survival metabolism through glycolysis instead of contributing to cellulose synthesis [2,5,11].

Consistent with the results of Shu et al. [5], activities of another sucrose degrading enzymes acidic/alkaline INV in cotton fiber of two cultivars increased under LPD, and was reduced by shading under the planting date of 25 April and 25 May, but shading under LPD$_{10-Jun}$ (MDTmin$_{FSWD}$ reached the lowest of 16.6 and 14.4°C in 2010 and 2011) increased INV activities of Sumian 15. In contrast, shading under LPD$_{10-Jun}$ reduced INV activities of Kemian 1 (Table 5), the probable reason was that Sumian 15 was a cool temperature-sensitive cultivar, while acidic INV was the most affected enzyme by the coupling of LP and shading (Table 5), INV in fiber was more sensitive to the coupling of LP and shading than Kemian 1, the increasing extent of INV activity in Sumian 15 was greater as the coupling stress became heavier (Table 6). However, abundant fructose produced by the increased INV activity under the coupling of LP and shading would inhibit the ability of M-SuSy [2,5] and have an adverse effect on cellulose synthesis.

As the enzyme synthesizing sucrose in cotton fiber, SPS is very sensitive to cool temperature and the activity is hindered under adverse environmental conditions [2,5]. SPS was also the most sensitive to shading among four sucrose metabolism enzymes (SuSy, SPS, acidic/alkaline INV, Table 5). Flux from fructose to sucrose might be hindered due to the decline of SPS activity, leading to fructose increasing in fiber, which further suppressed M-SuSy activity, and resulting in an adverse effect on cellulose synthesis. Under the coupling of LP and shading, SPS activity in fiber decreased, and the activity peak was delayed to 38, 45 or 52 DPA, similar to the time when the coupling of LP and shading had greater effects on cellulose synthesis (Figure 4 and Figure S4), it was indicated that the decline of SPS activity and its peak delay may be the reason why cellulose synthesis was sensitive to the combined stress during the later stage of FSWD (38–45 DPA).

The basic mechanisms regulating cellulose synthesis in different cotton cultivars are believed to be similar [6], but cotton cultivars have different levels of sensitivity in response to adverse environmental stress [4,33]. It has been shown that Sumian 15 which was cool temperature-sensitive, and Kemian 1 was partially tolerant to cool temperature [6,35]. In our research, Sumian 15 had a higher negative response of fiber cellulose to the coupling of LP and shading compared to Kemian 1 (Table 4). The decreasing range of SPS activity in Sumian 15 was greater, and CesA was more sensitive to the coupling of LP and shading than Kemian 1 (Figure 5, Table 6). In contrast, Kemian 1 had a greater ability to form higher cellulose content under the coupling of LP and shading (Table 3) [5], it was indicated that the relative cool temperature-tolerant cultivar, such as Kemian 1, still had a greater adaptability to the coupling of LP and shading than the cool temperature-sensitive cultivar, such as Sumian 15.

## Conclusions

(1) The coupling of LP (mainly MDTmin decreased) and shading (CRLR 80% and CRLR 60%) affected the key enzymes activities (SuSy, SPS, acidic/alkaline INV and CesA) involved in fiber sucrose metabolism and cellulose synthesis and hindered the flux from sucrose to cellulose during FSWD. As for the four sucrose metabolism enzymes (SuSy, SPS, acidic/alkaline INV), effects of LP were greater than shading. The decline of SPS activity and its peak delay probably caused cellulose synthesis

being more sensitive to the coupling stress during the later stage of FSWD (38–45 DPA).

(2) LP and shading combined to produce a more severe impact on cellulose synthesis than either stress alone. In the earlier stage of cotton FSWD (before 24 DPA), sucrose contents in cotton fiber under the coupling of LP and shading were mainly used for synthesizing callose instead of synthesizing cellulose. In the later stage of cotton FSWD (after 38 DPA), carbon in fiber stagnated in the form of sucrose and the partitioning to cellulose synthesis decreased, and the coupling of LP and shading made it more serious.

(3) Due to a less sensitive INV, SPS and CesA, the relative cool temperature-tolerant cultivar Kemian 1 had a relatively higher tolerance to the coupling of LP and shading compared to the cool temperature-sensitive cultivar Sumian 15.

## Supporting Information

**Figure S1 Changes of sucrose synthase activities in cotton fiber of two cultivars under the coupling of planting date and shading in 2010 and 2011.** The fiber growing period delineated by the two dotted lines stand for the period of fiber secondary wall rapid-thickening (FSWR).

**Figure S2 Changes of acidic invertase activities in cotton fiber of two cultivars under the coupling of planting date and shading in 2010 and 2011.** The fiber growing period delineated by the two dotted lines stand for the period of fiber secondary wall rapid-thickening (FSWR).

**Figure S3 Changes of alkaline invertase activities in cotton fiber of two cultivars under the coupling of planting date and shading in 2010 and 2011.** The fiber growing period delineated by the two dotted lines stand for the period of fiber secondary wall rapid-thickening (FSWR).

**Figure S4 Changes of sucrose phosphate synthase activities in cotton fiber of two cultivars under the coupling of planting date and shading in 2010 and 2011.** The fiber growing period delineated by the two dotted lines stand for the period of fiber secondary wall rapid-thickening (FSWR).

**Table S1 Variance analysis of mean air temperature, mean relative humidity and photosynthetically active radiation (PAR) in the cotton field under the coupling of planting date and shading in 2010 and 2011.** Data in Table S1 are averaged by measurement data from 6:00am to 6:00pm and PAR was measured at the position about 0.2 m above the canopy. CRLR and DPA stand for crop relative light rates and days post anthesis, respectively. Values followed by a different small letter within the same column in the same planting date are significantly different at 0.05 probability level.

## Author Contributions

Conceived and designed the experiments: YW BC ZZ. Performed the experiments: JC FL JL Y. Ma. Analyzed the data: JC FL. Contributed reagents/materials/analysis tools: YW BC Y. Meng. Wrote the paper: JC DMO.

# References

1. Lee JJ, Woodward AW, Chen ZJ (2007) Gene expression changes and early events in cotton fibre development. Ann Bot 100: 1391–1401.
2. Haigler CH, Ivanova-Datcheva M, Hogan PS, Salnikov VV, Hwang S, et al. (2001) Carbon partitioning to cellulose synthesis. Plant Mol Biol 47: 29–51.
3. Kim HJ, Triplett BA (2001) Cotton fiber growth in planta and in vitro: models for plant cell elongation and cell wall biogenesis. Plant Physiol 127: 1361–1366.
4. Haigler CH (2007) Substrate supply for cellulose synthesis and its stress sensitivity in the cotton fiber. In: Brown RM Jr, Saxena IM, editors. Cellulose: Molecular and Structural Biology. New York: Springer. pp. 147–168.
5. Shu HM, Zhou ZG, Xu NY, Wang YH, Zheng M (2009) Sucrose metabolism in cotton (*Gossypium hirsutum* L.) fibre under low temperature during fibre development. Europ J Agron 31: 61–68.
6. Wang YH, Shu HM, Chen BL, McGiffen ME, Zhang WJ, et al. (2009) The rate of cellulose increase is highly related to cotton fibre strength and is significantly determined by its genetic background and boll period temperature. Plant Growth Regul 57: 203–209.
7. Zhao WQ, Wang YH, Zhou ZG, Meng YL, Chen BL, et al. (2012) Effect of nitrogen rates and flowering dates on fiber quality of cotton (*Gossypium hirsutum* L.). Am J Exp Agric 2: 133–159.
8. Meinert MC, Delemer DP (1977) Change in biochemical composition of the cell wall of the cotton fiber during development. Plant Physiol 59: 1088–1097.
9. Xie W, Trolinder NL, Haigler CH (1993) Cool temperature effects on cotton fibre initiation and elongation clarified using in vitro cultures. Crop Sci 33: 1258–1264.
10. Fujii S, Hayashi T, Mizuno K (2010) Sucrose synthase is an integral component of the cellulose synthesis machinery. Plant Cell Physiol 51: 294–301.
11. Ruan YL (2007) Rapid cell expansion and cellulose synthesis regulated by plasmodesmata and sugar: insights from the single-celled cotton fibre. Funct Plant Biol 34: 1–10.
12. Delmer DP, Amor Y (1995) Cellulose biosynthesis. Plant Cell 7: 987–1000.
13. Babb VM, Haigler CH (2001) Sucrose phosphate synthase activity rises in correlation with high-rate cellulose synthesis in three heterotrophic systems. Plant Physiol 127: 1234–1242.
14. Amor Y, Haigler CH, Johnson S, Wainscott M, Delmer DP (1995) A membrane-associated form of sucrose synthase and its potential role in synthesis of cellulose and callose in plants. Proc Natl Acad Sci 92: 9353–9357.
15. Salnikov VV, Grimson MJ, Seagull RW, Haigler CH (2003) Localization of sucrose synthase and callose in freeze-substituted secondary-wall-stage cotton fibers. Protoplasma 221: 175–184.
16. Tang GQ, Luscher M, Sturm A (1999) Antisense repression of vacuolar and cell wall invertase in transgenic carrot alters early plant development and sucrose partitioning. Plant Cell 11: 177–189.
17. Wäfler U, Meier H (1994) Enzyme activities in developing cotton fibers. Plant Physiol Biochem 32: 697–702.
18. Dong HZ, Li WJ, Tang W, Li Z, Zhang DM, et al. (2006) Yield, quality and leaf senescence of cotton grown at varying planting dates and plant densities in the Yellow River Valley of China. Field Crops Res 98: 106–115.
19. Wang QC, Wang ZL, Song XL, Li YJ, Guo Y, et al. (2005) Effects of shading at blossoming and boll-forming stages on cotton fiber quality. Chin J Appl Ecol 16: 1465-1468. (in Chinese with English abstract)
20. Wang QC, Sun XZ, Song XL, Guo Y, Li YJ, et al. (2006) Effect of shading at different developmental stages of cotton bolls on cotton fibre quality. Acta Agron Sin 32: 671–675. (in Chinese with English abstract)
21. Yeates SJ, Constable GA, McCumstie T (2010) Irrigated cotton in the tropical dry season. III: Impact of temperature, cultivar and planting date on fibre quality. Field Crops Res 116: 300–307.
22. Dai JL, Dong HZ (2014) Intensive cotton farming technologies in China: Achievements, challenges and countermeasures. Field Crops Res 155: 99–110.
23. Martin LK, Haigler CH (2004) Cool temperature hinders flux from glucose to sucrose during cellulose synthesis in secondary wall stage cotton fibers. Cellulose 11: 339–349.
24. Bradow JM, Davidonis GH (2010) Effects of environment on fiber quality. In: Stewart JMcD, Oosterhuis D, Heitholt JJ, Mauney JR, editors. Physiology of Cotton. New York: Springer. pp. 229–245.
25. Liaktas A, Roussopulos D, Whittington WJ (1998) Controlled-temperature effects on cotton yield and fiber properties. J Agric Sci 130: 463–471.
26. Eaton FM, Ergle DR (1953) Effects of shade and partial defoliation on carbohydrate levels and the growth, fruiting, and fiber properties of cotton plants. Plant Physiol 29: 39–49.
27. Pettigrew WT (1995) Cotton responses to shade at different growth stages: growth, lint yield and fiber quality. Agron J 87: 947–952.
28. Khayat E, Zieslin N (1987) Effect of night temperature on the activity of sucrose phosphate synthase, acid invertase, and sucrose synthase in source and sinktissues of Rosa hybrida cv Golden Times. Plant Physiol 84: 447–449.
29. Ren WJ, Yang WY, Xu JW, Fan GQ, Ma ZH (2003) Effect of low light on grains growth and quality in rice. Acta Agron Sin 29: 785–790. (in Chinese with English abstract)
30. Li T, Ohsugi R, Yamagishi T, Sasaki H (2006) Effects of low light on rice sucrose content and sucrose degradation enzyme activities at grain-filling stage. Acta Agron Sin 32: 943–945. (in Chinese with English abstract)
31. Zhang JW, Dong ST, Wang KJ, Hu CH, Liu P (2008) Effects of shading in field on key enzymes involved in starch synthesis of summer maize. Acta Agron Sin 34: 1470–1474. (in Chinese with English abstract)
32. Zhao D, Oosterhuis DM (2000) Cotton responses to shade at different growth stages: growth, lint yield and fibre quality. Expl Agric 36: 27–39.
33. Pettigrew WT (2001) Environmental effects on cotton fiber carbohydrate concentration and quality. Crop Sci 41: 1108–1113.
34. Bian HY, Wang YH, Chen BL, Shu HM, Zhou ZG (2008) Effects of the key enzymes activity on the fiber strength formation under low temperature condition. Sci Agric Sin 41: 1235–1241. (in Chinese with English abstract)
35. Wang YH, Shu HM, Chen BL, Xu NY, Zhao YC, et al. (2008) Temporal-spatial variation of cotton fiber strength and its relationship with temperature in different cultivars. Sci Agric Sin 41: 3865–3871. (in Chinese with English abstract)
36. Haigler CH, Rao NR, Roberts EM, Huang JY, Upchurch DR, et al. (1991) Cultured ovules as models for cotton fiber development under low temperatures. Plant Physiol 95: 88–96.
37. Updegraff DM (1969) Semimicro determination of cellulose inbiological materials. Anal Biochem 32: 420–424.
38. Hendrix DL (1993) Rapid extraction and analysis of nonstructural carbohydrates in plant tissues. Crop Sci 33: 1306–1311.
39. Köhle H, Jeblick W, Poten F, Blaschek W, Kauss H (1985) Chitosan-elicited callose synthesis in soybean cells as a $Ca^{2+}$-dependent process. Plant Physiol 77: 544–551.
40. King SP, Lunn JE, Furbank RT (1997) Carbohydrate content and enzyme metabolize in developing canola siliques. Plant Physiol 114: 153–160.
41. Winter H, Huber SC (2000) Regulation of sucrose metabolism in higher plants: localization and regulation of activity of key enzymes. Crit Rev Plant Sci 19: 31–67.
42. Jiang JX, Zhang TZ (2003) Extraction of total RNA in cotton tissues with CTAB-acidic phenolic method. Cott Sci 15: 166–167. (in Chinese with English abstract)
43. Wang L, Li XR, Lian H, Ni DA, He YK, et al. (2010) Evidence that high activity of vacuolar invertase is required for cotton fiber and Arabidopsis root elongation through osmotic dependent and independent pathway, respectively. Plant Physiol 154: 744–756.
44. Liu JR, Ma YN, Lv FJ, Chen J, Zhou ZG, et al. (2013) Changes of sucrose metabolism in leaf subtending to cotton boll under cool temperature due to late planting. Field Crops Res 144: 200–211.
45. Maltby D, Carpita NC, Montezinos D, Kulow C, Delmer DP (1979) β1,3-Glucan in Developing Cotton Fibers. Plant Physiol 63: 1158–1164.
46. Roberts EM, Rao NR, Huang JY, Trolinder NL, Haigler CH (1992) Effects of cycling temperatures on fiber metabolism in cultured cotton ovules. Plant Physiol 100: 979–986.
47. Zhao D, Oosterhuis DM (1998) Physiologic and yield responses of shaded cotton to the plant growth regulator PGR-IV. J Plant Growth Regul 17: 47–52.
48. Roussopoulos D, Liakatas A, Whittington WJ (1998) Cotton responses to different light-temperature regimes. J Agric Sci 131: 277–283.
49. Ruan YL, Llewellyn DJ, Furbank RT (2003) Suppression of sucrose synthase gene expression represses cotton fiber cell initiation, elongation and seed development. Plant Cell 15: 952–964.

# Gibberellin Overproduction Promotes Sucrose Synthase Expression and Secondary Cell Wall Deposition in Cotton Fibers

**Wen-Qin Bai[9], Yue-Hua Xiao[9], Juan Zhao, Shui-Qing Song, Lin Hu, Jian-Yan Zeng, Xian-Bi Li, Lei Hou, Ming Luo, De-Mou Li, Yan Pei***

Biotechnology Research Center, Southwest University, Beibei, Chongqing, China

## Abstract

Bioactive gibberellins (GAs) comprise an important class of natural plant growth regulators and play essential roles in cotton fiber development. To date, the molecular base of GAs' functions in fiber development is largely unclear. To address this question, the endogenous bioactive GA levels in cotton developing fibers were elevated by specifically up-regulating GA 20-oxidase and suppressing GA 2-oxidase via transgenic methods. Higher GA levels in transgenic cotton fibers significantly increased micronaire values, 1000-fiber weight, cell wall thickness and cellulose contents of mature fibers. Quantitative RT-PCR and biochemical analysis revealed that the transcription of sucrose synthase gene *GhSusA1* and sucrose synthase activities were significantly enhanced in GA overproducing transgenic fibers, compared to the wild-type cotton. In addition, exogenous application of bioactive GA could promote *GhSusA1* expression in cultured fibers, as well as in cotton hypocotyls. Our results suggested that bioactive GAs promoted secondary cell wall deposition in cotton fibers by enhancing sucrose synthase expression.

**Editor:** Baohong Zhang, East Carolina University, United States of America

**Funding:** This work was supported by the National Natural Science Foundation of China (31130039 to Y. P. and 31271769 to Y. H. X.) and Ph.D. Programs Foundation of Southwest University, China (Kb2009007 to W. Q. B.). The funders had no role in study design, data collection and analysis, decision to publish, or preparation of the manuscript.

**Competing Interests:** The authors have declared that no competing interests exist.

\* E-mail: peiyan3@swu.edu.cn

[9] These authors contributed equally to this work.

## Introduction

Cotton is the leading natural fiber for textile industry worldwide. Biologically, cotton fibers are extremely elongated single-celled trichomes originating from outermost layer of ovule epidermis [1–4]. The development of cotton fiber may be divided into 4 stages, i.e. initiation, elongation, secondary cell wall deposition and maturation. Secondary cell wall deposition starts at around 14–17 days post anthesis (dpa) and lasts for over 30d [1,3,5]. In this stage, cellulose is intensely deposited to form a thick secondary cell wall. At maturation, cotton fiber consists primarily of secondary cell wall and over 90% dry weight of fiber may exist as cellulose. Therefore, carbon partitioning to cellulose biosynthesis is a key determinant of fiber weight and qualities, such as fiber strength and fineness [1,3,5–8]. Many efforts have been taken to reveal the role of genes involved in the regulation of secondary cell wall deposition and to manipulate them by genetically modification for improvement of cotton yield and quality [1,6,7,9,10]. Recently, Jiang and coworkers showed that over-expressing a cotton sucrose synthase gene, *GhsusA1*, enhanced thickening of secondary cell wall and fiber qualities, suggesting an important role of sucrose synthase in controlling carbon partitioning to cellulose biosynthesis in cotton fibers [6].

Gibberellins (GA) are a class of important plant hormones involved in many physiological and developmental processes, including seed germination, cell elongation, photomorphogenesis, flowering and seed development [11]. In the last two decades, the molecular base of GA biosynthesis pathway and its regulation have been largely clarified in model plants [12–14]. Endogenous bioactive GA contents are regulated mainly through three 2-oxoglutarate-dependent dioxygenases, i.e. GA 20-oxidase (GA20ox), GA 3-oxidase (GA3ox) and GA 2-oxidase (GA2ox) [13,14]. GA20ox and GA3ox catalyze the last two steps to synthesize bioactive GAs, while GA2ox convert bioactive GAs and their precursors to inactive 2-hydroxylated forms. A wealth of evidence demonstrated that both up-regulating GA20ox and suppressing GA2ox could significantly increase endogenous bioactive GA levels and lead to GA overproduction phenotypes in plants [15–20].

Physiological and molecular studies have revealed that GAs played important roles in fiber development. Exogenous application of GAs *in vitro* and *in planta* promoted fiber initiation and elongation [21,22]. Recently, we showed that over-expression of *GhGA20ox1* in cotton significantly increased bioactive GA level and promote fiber initiation and elongation at early stage [17]. However, global up-regulation of GAs leaded to overgrowth of plant and somewhat negatively affected fiber development, especially at the late developmental stage. Instead, it is reasonable to elucidate GA roles in fiber development by tissue specific regulation of GA levels in developing fibers. To this end, we elevated the endogenous active GA levels in cotton fibers by tissue-specific up-regulation of GA20ox gene and down-regulation of GA2ox gene. We found that enhancement of GA production in fibers promoted sucrose synthase expression and secondary cell wall deposition. Our results implied that GAs might enhance carbon partitioning to cellulose and secondary cell wall synthesis via up-regulating sucrose synthase expression in cotton fibers.

## Materials and Methods

### Plant material and growth condition

Upland cotton (*Gossypium hirsutum* L. *cv.* Jimian No. 14) was used for cotton transformation and GA treatment. Cotton seedlings were grown in a greenhouse with a 16h/8h (light/dark) schedule and temperature kept at 26–30°C. Fibers and ovules were collected from field-grown cotton plants at growing season in Chongqing, China.

For $GA_3$ treatment of hypocotyls, 4-day-old seedlings were immersed in distilled water (pH6.0) or $GA_3$ solutions of various concentrations (0.05 mM, 0.1 mM and 0.5 mM, pH6.0) for 48 h. Then the hypocotyls were measured and collected for expression analyses and cellulose determination.

For GA treatment of *in vitro* fibers, cotton ovules were cultured as described by Beasley [23]. Cotton bolls were harvested at 0-dpa and surface-sterilized in 75% (v/v) ethanol for 1 min, rinsed in sterile water, then soaked in 0.1% w/v $HgCl_2$ solution for 12 min for sterilization, followed by rinsing with sterile water for six times. Ovules were separated, floated on BT media containing 5 μM IAA and $GA_3$ of various concentrations (0.5, 2.5, 10 and 25 μM), and then incubated in darkness at 32°C for 20d. Fibers were striped from ovules and used for RNA extraction.

### Vector construction and plant transformation

To construct specific expression vector of cotton GA20ox (*SCFP::GhGA20ox1*), the CaMV 35S promoter in an over-expression vector (*35S::GhGA20ox1*) [17] was replaced with SCFP promoter[24]. The BAN promoter was amplified from Arabidopsis with a forward primer (5'-TCTAGATAACAGAACCTTAC TGTAACACTATT-3') and a reverse primer (5'-ACTAGT-GATTGTACTTTTGAAATTACAGAG AT-3') and cloned into TA cloning vector pMD19-T (TaKaRa, Dalian, China). After sequencing, the BAN promoter was digested from the cloning vector by *Hind*III and *Bam*HI and inserted into a basic expression vector p5 vector [25] digested with the same enzymes to generate the vector p5-BAN. An intron-containing hairpin RNA construct of cotton GA20ox gene (*GhGA2ox2RNAi*) was amplified from cotton genomic DNA as previously described [26]. The 25-μl PCR mixture included 100 ng cotton genomic DNA, 10×Ex Taq buffer (TaKaRa), 200 μM each dNTPs, 2 mM $MgCl_2$, 400 nM flanking primer (5'-GTATTGGTCTGGTGGGACTG-3'), 40 nM bridge primer (5'-CAAGTATCTCACATGCC AAGACCC-GAATTCTCCTTG-3'), 1.5U Ex Taq DNA polymerase. The PCR thermo cycling parameters were as follows: 94°C for 5 min, followed by 35 cycles of 94°C for 30 s, 56°C for 30 s and 72°C for 30 s, and a final extension of 10 min at 72°C. The *GhGA2ox2RNAi* fragment was cloned and sequenced, then inserted into p5-BAN using *Bam*HI and *Kpn*I. Transgenic plants were generated using *Agrobacterium*-mediated transformation as described [25]. Based on expression analysis of target genes in transgenic cottons, two homologous transgenic lines were obtained by self-crossing, and their performances were documented at T3 and T4 generations in comparison with untransformed acceptor line (Jimian No. 14) grown in parallel in the field.

### RNA extraction and qRT-PCR analyses

Total RNA was extracted from roots, hypocotyls, leaves, petals, anthers, ovules and fibers using a rapid plant RNA extraction kit (Aidlab, Beijing, China). The single-stranded cDNAs were synthesized from total RNA using a cDNA synthesis kit (TaKaRa, Dalian, China). The gene-specific primers used for real-time PCR amplification were list in table S1. Cotton *histon3* gene (AF024716) was amplified as internal standard [27]. Real-time PCRs were performed on a CFX96 real-time PCR detection system with SYBR Green supermix (Bio-Rad, CA, USA). The thermocycling parameters were as follows: 95°C for 2 min, followed by 40 cycles of 95°C for 30 s, 56°C for 30 s and 72°C for 30 s, followed by a standard melting curve to monitor the specificity of PCR products. The reactions were duplicated for 3 times and data were analyzed using the software Bio-Rad CFX Manager 2.0 provided by the manufacturer.

### Determination of endogenous GA contents

Cotton fibers (200 mg FW) were ground to fine powder in liquid $N_2$, extracted overnight in 5 ml 80% methanol at −20°C and deuterium-labeled [17, 17-$^2H_2$] $GA_1$ and [17, 17-$^2H_2$] $GA_4$ (each 10 ng) from Prof. L. Mander (Australian National University) were added as internal standards. After centrifugation, supernatants were collected, dried in a rotavapor (BUCHI, Switzerland) at 40°C, and re-suspended in 3 ml 10% methanol. The extracts were applied on Oasis HLB extraction cartridges (60 mg, Waters) pretreated with 3 ml methanol and 3 ml water. After washing with 1 ml 10% methanol, GAs were eluted with 1 ml 90% methanol. The eluates were evaporated to dry, dissolved in 100 μL 10% methanol, and subjected to LC-MS assay. The procedures for LC-MS quantification of GAs were described previously [17].

### Measurement of fiber quality

Mature fibers were harvested from the field-grown cotton in the same period (Aug. 20 to Sep. 10). After ginning, fibers were mixed well and 6 repeats of 10 g fibers were randomly sampled for each material. Fiber sample were tested independently for fiber quality traits (fiber length, fiber strength, micronarie value) using at a HVI system (HFT 9000, Uster Technologies, Swiss) in Cotton Fiber Quality Inspection and Testing Center, Ministry of Agriculture of China (Anyang, Henan, China).

### Microscopic measurement of fiber cell wall thickness

Statistical analysis of cell wall thickness was performed according to Wang *et al.* [7]. After fixing in FAA (37% formaldehyde: acetic acid: ethanol: water, 10:5:50:35) at 25°C for 12 h, mature cotton fibers were dehydrated gradually in alcohol and tert-butyl alcohol series,and then infiltrated in tert-butyl alcohol/paraffin at 65°C and embedded in paraffin. The samples were sliced into 7-μm sections. The slices were mounted, stained with the Fast Green dye and photographed by a BX41TF light microscope (Olympus, Japan). Image-pro Plus program (Olympus) was employed to measure the thickness of cell wall and 1000 sections were measured for each sample.

### Determination of fiber weight

Fibers on seeds were combed straight and striped manually. Approximately 1.5 mg fibers were randomly bundled and weighed precisely (W1). The fiber number of each fiber bundle (N) was counted as described[28]. The weight of 1000 fibers (W2) was calculated from the following equation: W2 = 1000W1/N. For each material, the average 1000-fiber weight was calculated on the basis of 60 fiber bundles from different seeds.

## Sucrose synthase activity assays

The sucrose synthase was extracted according to Jiang et al. [6]. Fresh fibers (around 0.5 g) were ground to fine powder in liquid $N_2$. The grinding continued for 5 min in cold extraction buffer (25 mM Hepes–KOH (pH 7.3), 5 mM EDTA, 1 mM DTT, 0.1% soluble PVP, 20 mM β-mercaptoethanol, 1 mM PMSF and 0.01 mM leupeptin). The homogenate was separated by centrifugation (10000 g, 5 min, 4°C) and the supernatant was used as the crude extracts for assays. Protein concentrations were determined via Bradford method [29]and sucrose sythase activities were assayed as previously described [30,31].

## Analyses of soluble sugar contents

Fresh fibers (around 50 mg) were separated from developing bolls and ground to fine powder in liquid N2. The powder was extracted in 2 ml 80% (v/v) ethanol at 80°C?for 15 min. After centrifugation (3000 g, 10 min), supernatants were collected. The pellets were further extracted twice, and supernatants were combined and used for soluble sugar assays. The contents of glucose, fructose and sucrose were measured at 340 nm with a Synergy HT microplate reader (BioTek, Vermont, WS) as described [32].

## Determination of cellulose content

Cellulose contents were determined according to Wang *et al.* [7]. Around 0.1 g fiber samples were extracted in 10 ml boiling acetic/nitric reagent (80% acetic/nitric, 10:1) for 1 h, then rinsed three times with distilled water and once with ethanol. Residuals were dried at 105°C for 2 h. The weight ratio of residual to initial samples was regarded as cellulose content. Hypocotyls were excised from seeding (10 hypocotyls per sample) and ground to fine powder in liquid $N_2$. Samples were dried at 105°C for 2 h and extracted in 10 ml boiling acetic/nitric reagent (80% acetic/nitric, 10:1) for 1 h.Determination of cellulose content of per hypocotyls was carried out as described [33].

## Statistical analyses

Performances of transgenic materials were compared to wild-type control and statistical significance of divergence between averages was determined by t test. All statistical calculations were performed using Microsoft Excel.

## Results

### Enhancement of GA production in cotton fiber

We used two strategies to tissue-specially enhance GA production in fibers, i.e. to promote GA biosynthesis by up-regulation GA 20-oxidase (GA20ox) and suppressing GA deactivation by down-regulation of GA2-oxidase (GA2ox). To this end, we used a fiber-specific promoter (SCFP) [24] and a seed coat- and fiber-specific promoter BAN [28,34] to direct the expression of *GA20ox* (*GhGA20ox1*) [17], and *GA2ox*, respectively. Among *SCFP::GhGA20ox1* transgenic cottons (SG20), SG20-1 showed dramatically increase in *GhGA20ox1* expression level in developing fibers (Figure 1A and 1B).

We compared the expression pattern of six cotton GA 2-oxidase genes (*GhGA2ox1-6*, Figure S1~3). Among them, *GhGA2ox2* showed predominant expression in fibers. Thus we selected *GhGA2ox2* as RNAi target to suppress GA deactivation in fibers, and generated *GhGA2oxRNAi* transgenic cottons. Real-time RT-PCR revealed that the expression level of *GhGA2ox2* was reduced in *BAN::GhGA2oxRNAi* (BG2i) transgenic cottons (Figure 1C), in which transformant BG2i-2 showed most significant suppression of the target gene in fibers (Figure 1C and 1D).

To detect the effect of *GhGA20ox1* up-regulation and *GhGA2ox2* down-regulation on GA homeostasis in fibers, we determined the contents of endogenous bioactive GAs ($GA_1$ and $GA_4$) in 8- and- 20 dpa fibers of SG20-1 and BG2i-2 by LC-MS (Figure 1E and 1F). Compared to the wild-type control, $GA_4$ contents in the 8- and 20-dpa fibers of SG20-1 fibers increased 83.1% and 178.6%, respectively, while $GA_1$ content was moderately increased (24.0%) in the 20-dpa fibers(Figure 1E and 1F). In BG2i-2 fibers, $GA_1$ level was 21.6% and 65.9% higher than the control at 8- and 20-dpa respectively, whereas $GA_4$ contents remained almost unchanged compared to the control (Figure 1E and 1F).

### Effects of elevated GA levels on secondary cell wall thickening of cotton fiber

To clarify the effect of elevated GA levels on fiber development and fiber quality, we compared the agronomy performances of transgenic lines SG20-1 and BG2i-2 with the wild-type control in consecutive two-year field trails. No significant change in plant growth, yield traits and fiber length and strength (Figure 2A and S4; Table S2 and S3) was found between the transgenic lines and the wild type, except micronaire value. The micronaire values of SG20-1 and BG2i-2 fibers were significantly higher than that of the wild type (Figure 2C). Micronaire value is a composite measure of fiber maturity and fineness. To clarify whether the fiber fineness was increased in transgenic cotton, we measured the weight per 1000 fibers. The weights of SG20-1 and BG2i-2 mature fibers significantly enhanced in comparison with the control (2.0% and 5.7%, respectively; Figure 2E). Microscopic observation further confirmed that the cell walls of SG20-1 and BG2i-2 mature fibers were thicker (5.4% and 6.6%, respectively) than the wild-type control (Figure 2B and D). Considered that most of cell walls of mature cotton fibers consisted of secondary cell wall [2], it was reasonable that the fineness increase in SG20-1 and BG2i-2 fibers might be mainly attributed to promotion of secondary cell wall deposition. To prove this hypothesis, we determined the cellulose contents in fibers, and found that the contents of SG20-1 and BG2i-2 fibers were significantly higher than the control (Figure 2F). Taken together, these results suggested that elevating bioactive GA levels in cotton fibers promoted secondary cell wall deposition.

### Sucrose synthase expression in response to elevated GA levels in fibers and hypocotyls

To reveal the possible mechanism for GAs to control secondary cell wall deposition, we investigated transcript levels of six genes related to secondary cell wall biosynthesis, including *GhCesA1*, *GhCesA2*, *GhRac13*, *GhSusA1*, *GhADF1* and *GhCTL1* [6,7,35], in 20-dpa fibers. Only sucrose synthase gene (*GhSusA1*) showed significant increase in transgenic cottons (Figure 3A). The relative transcript levels of *GhSusA1* in SG20-1 and BG2i-2 fibers were 53% and 50% higher than the control, respectively. Biochemical analysis demonstrated that the sucrose synthase activities in 20-dpa fibers of SG20-1 and BG2i-2 increased 8.3% and 10.7%, respectively, compared to the control (Figure 3B). Meanwhile, the concentration of fructose, a direct product of sucrose synthase, was significantly higher in SG20-1 and BG2i-2 fibers (Figure 3C). Furthermore, we found *GhsusA1* transcript in cultured fibers was increased with $GA_3$ concentrations in ovule culture media (Figure 3D). The result of fiber culture, along with the observations on the mature fibers, implied that GA may promote cellulose biosynthesis and secondary cell wall deposition through up-regulation of the expression of sucrose synthase.

Like cotton fibers, hypocotyls that undergo rapid cell elongation require high-speed formation of cellulose. To investigate if same

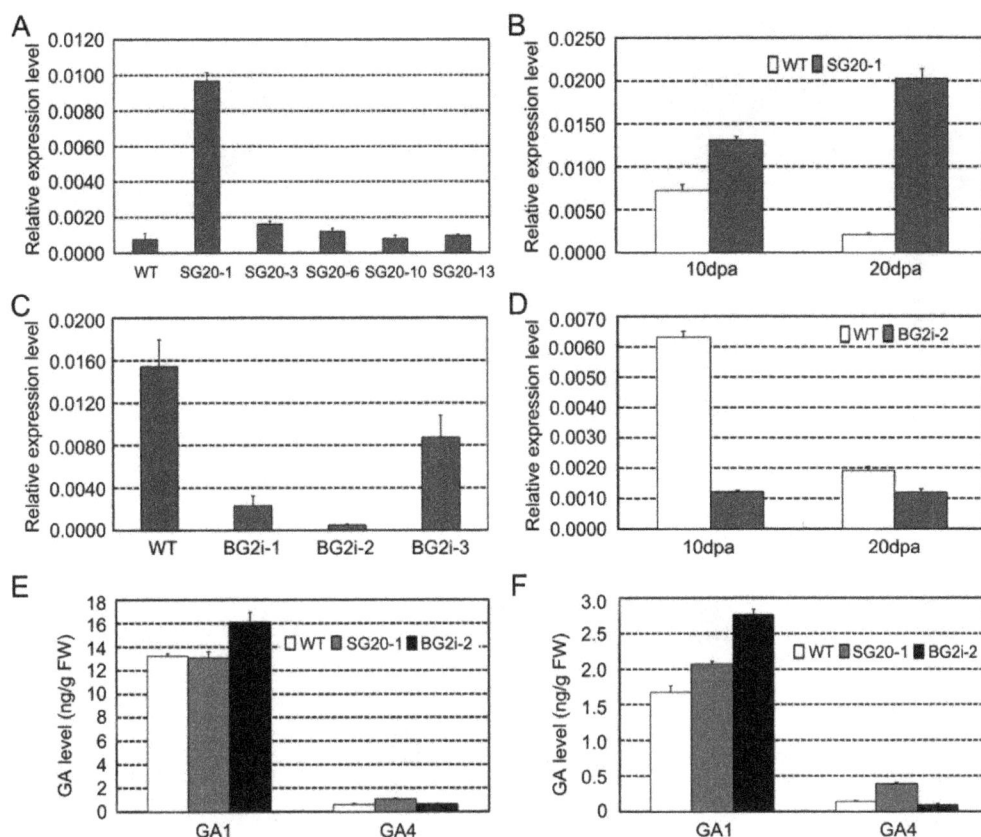

**Figure 1. Up-regulation of *GhGA20ox1* or suppression of *GhGA2ox* in cotton fibers.** (A) Quantitative RT-PCR analysis of *GhGA20ox1* in 5-dpa fibers of T₀SCPF::*GhGA20ox1* lines (SG20) and the wild-type cotton (WT). (B) Quantitative RT-PCR analysis of *GhGA20ox1* in 10-dpa and 20-dpa fibers of SG20-1 (T₃generation) and WT. (C) Quantitative RT-PCR analysis of *GhGA2ox2* in 5-dpa fibers of T₀BAN::*GhGA2oxRNAi* lines (BG2i) and WT. (D) Quantitative RT-PCR analysis of *GhGA2ox2* in 10-dpa and 20-dpa fibers of BG2i-2 (T₃generation) and WT. (E) Endogenous GA₁ and GA₄ content in 8-dpa fibers of SG20-1, BG2i-2 and WT. (F) Endogenous GA₁ and GA₄ contents in 20-dpa fibers of SG20-1, BG2i-2 and WT.

response takes place in hypocotyls, we detected the expression of sucrose synthase in the GA-treated hypocotyls. *GhSusA1* transcript levels in hypocotyls were significantly enhanced along with increase of GA₃ (Figure 4A). Meanwhile, elongation and cellulose deposition were accelerated (Figure 4B–D). These results further supported that GA promoted cellulose biosynthesis and secondary cell wall deposition through up-regulation of sucrose expression.

## Discussion

Genetically manipulating enzymes involved in GA biosynthesis and catabolism provided an effective strategy to regulate GA homeostasis in plants [15–20]. Our work demonstrated that in addition to up-regulation of *GA20ox*, down-regulation of *GA2ox* is another effective way to increase the active GA levels in plant tissues (Figure 1E and F). However, there may be a subtle difference between the two strategies. Up-regulation of GA20ox not only increased the GA level, but also changed the composition of some active GAs by promoting the non-13-hydroxylation pathway for producing 13-H GA₄ rather than 13-OH GA₁ (Figure 1E and F) [17,19,20]. On the contrary, in our study, down-regulation of GA2ox elevated the GA₁ instead of GA₄ (Figure 1E and F). The exact physiological effects of different bioactive GAs on cotton fiber development are still to be elucidated.

Cotton fibers that undergo rapid elongation and intense cellulose synthesis represent a strong sink that competes with the developing embryos and endosperms in a single ovule

[1,3,5,6,8,30,36,37]. It was revealed that sucrose synthase played important roles in carbon partition during fiber development. Suppression of sucrose synthase gene (*SS3*) inhibited fiber initiation and elongation [30], while over-expression of a potato sucrose synthase gene in cotton enhanced leaf expansion, early seed development and fiber elongation [5]. Recently, Jiang and coworkers cloned a novel cotton sucrose synthase gene (*GhSusA1*), which might be a key regulator of sink strength in cotton. Over-expression of *GhSusA1* significantly enhanced cell wall thickening during secondary wall formation stage, and improved fiber length and strength [6]. In this study, we revealed that enhancement of GA level in cotton fibers led to an increase of sucrose expression (*GhSusA1*) gene expression, and promoted cellulose biosynthesis and secondary cell wall deposition (Figure 2 and 3). Moreover, GA-induced *GhSusA1* up-regulation also found in cultured fibers and hypocotyls (Figure 4 and S5). GA has long been considered as an important regulator of sink strength in plants, but the molecular basis of how GA enhances the partitioning of carbon assimilates to sink tissues is still unknown [38,39].Our data offer an experimental evidence for the relationship between GAs and sucrose synthase.

Previous studies showed that exogenous application of GAs or constitutively increased endogenous GA levels promoted fiber initiation and elongation [17,21]. However, in this study we did not find significant improvement in fiber length of the transgenic GA-elevated fibers (TableS2). A possible explanation for this phenomenon is that the enhanced secondary cell wall deposition may fix the morphology of the fiber and, in turn, limit further

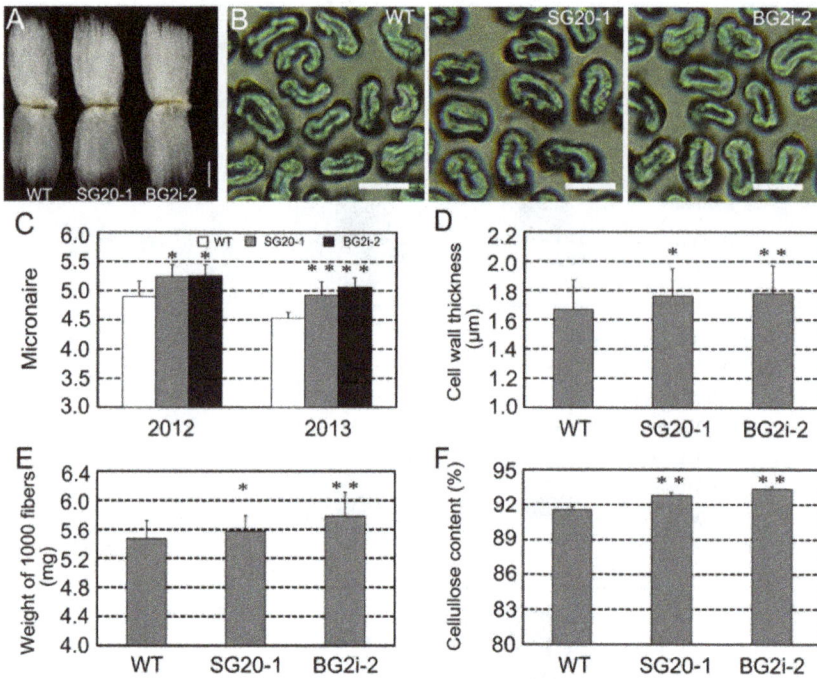

**Figure 2. Effects of GA overproduction on secondary cell wall deposition in cotton fibers.** (A) Cotton fibers on seeds; bar = 1 cm. (B) Cross-section of mature cotton fibers; Bar = 25 μm. (C) Micronaire values of cotton fibers in 2012 and 2013, n = 6. (D) Cell wall thickness of mature cotton fibers. (E) Weight of 1000 cotton fibers, n = 60. (F)Cellulose contents of mature cotton fibers, n = 6Asterisks (∗) and double asterisks (∗∗) represent significant differences(t test) at p = 0.05 and p = 0.01 compared with the wild type, respectively.

elongation of the GA-enhanced fibers at the late stage of fiber elongation. Nevertheless, the finding that GA related regulation of sucrose synthase gives useful information to reveal the mechanism of cotton fiber development and to improve fiber yield and quality for cotton breeders.

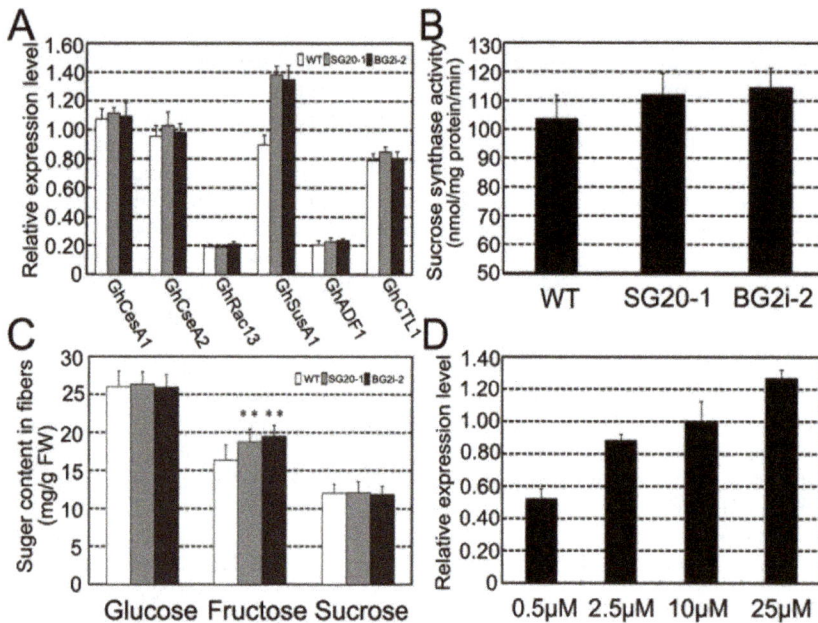

**Figure 3. Sucrose synthase expression in GA overproduction cotton fibers.** (A) Quantitative RT-PCR expression analyses of secondary cell wall related genes in 20-dpa fibers. *GhCesA1*, U58283; *GhCesA2*, U58284; *GhRac13*, S79308; *GhSusA1*, HQ702817; *GhADF1*, DQ088156 and*GhCTL1*, AY291285. (B) Sucrose synthase activity in 20-dpa cotton fibers. (C) Sugar contents in 20-dpa cotton fibers, n = 3. Double asterisks (**) represent significant differences (t test) at p = 0.01 compared with the wild type cotton. (D) *GhSusA1* expression levels of *in vitro* cultured fibers treated with GA₃ of various concentrations.

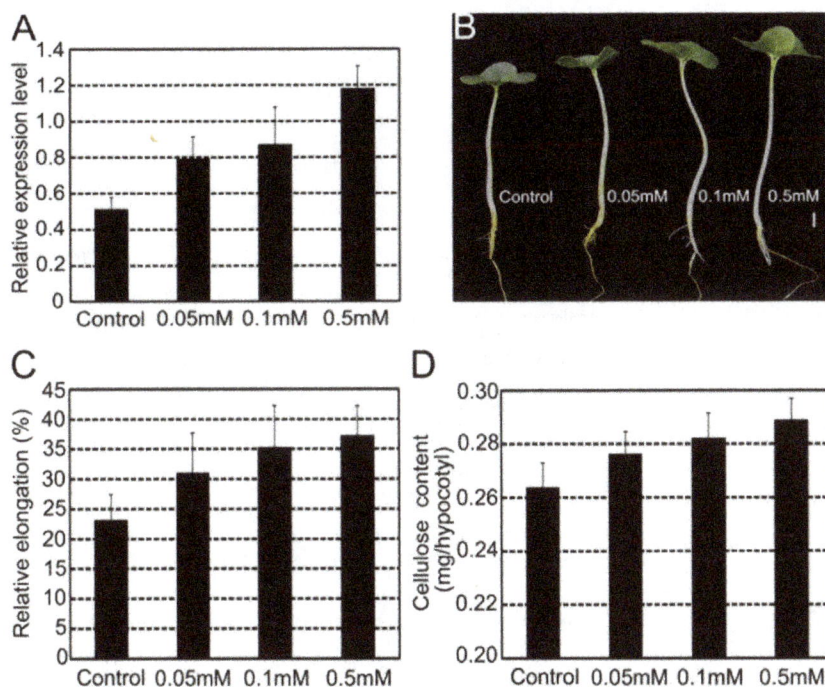

**Figure 4. Sucrose synthase expression and cellulose biosynthesis in GA-treated hypocotyls.** Seedlings wereimmersed in distilled water (pH6.0) or GA$_3$ solutions (0.05 mM, 0.1 mM and 0.5 mM, pH6.0).(A) Quantitative RT-PCR analysis of *GhSusA1* transcript in hypocotyls. (B) Effects of GA$_3$ treatment on cotton seedling growth. (C) Relative elongation of hypocotyls. (D) Cellulose amount in GA-treated hypocotyls.

## Supporting Information

**Figure S1 Alignment of cotton GA2ox proteins with homologous proteins.** The conserved amino acids are highlighted on black background, and similar amino acids are shown on gray background. The symbols (#) indicate Fe-binding sites, and the asterisks(*) indicateputative 2-oxoglutarate-interaction sites. GhGA2ox1-6, Cotton GA 2-oxidases (HQ891930-HQ891935, respectively); SoGA2ox1-3, Spinach GA 2-oxidases (AAN87571, AAN87572 and AAX14674, respectively); AtGA2ox1, 4 and 7, Arabidopsis GA 2-oxidases (CAB41007, AAG51528 and AAG50945, respectively).So, *Spinaciaolracea*; At, *Arabidopsis thaliana*.

**Figure S2 Phylogenetic relationship of GhGA2ox with other GA2ox, GA20ox and GA3ox.** GenBank accession nos. are as follows: GhGA2ox1-6, HQ891930-HQ891935, respectively; SoGA2ox1-3, AAN87571, AAN87572 and AAX14674, respectively; AtGA2ox1-4 and 6-8, CAB41007, CAB41008, CAB41009, AAG51528, AAG00891, AAG50945 and CAB79120, respectively; CmGA2ox, CAC83090; OsGA2ox3 and 5-9, AK101713, AK106859, AK107142, AK108802, AK101758 and AK059045, respectively; GhGA20ox1-3, AY603789, FJ623273 and FJ623274, respectively. So, *Spinaciaolracea*; At, *Arabidopsis thaliana*; Cm, *Cucurbita maxima*; Os, *Oryza sativa*.

**Figure S3 Expression pattern of *GhGA2ox* genes in cotton tissues.** The total RNA were prepared from different organs and tissues, including roots(Ro), hypocotyls (Hy), leaves (Le), petals (Pe), anthers (An), 0-dpa ovules (Ov), 6-dpa fibers (Fi).

**FigureS4 Field-grown plants of SG20-1, BG2i-2, *GhGA20ox1*-overexpressing (OG20) and wild-type (WT) cottons.** All the materials were transplanted to the field in parallel. The representative plants were photographed at 90d post germination. Bar = 30 cm.

**Figure S5 Quantitative RT-PCR analysis of fiber elongation related genes in 8-dpa fibers.** *GhACT1*, AF305723; *GhEXP1*, AF512539; *GhFLA1*, EF672627; *GhPEL*, DQ073046; *GhVLN1*, FL915120.

**Table S1 Primers used in Real-time PCR analyses.**

**Table S2 Fiber length and strength of mature fibers in two-year successive field trials.** Asterisk (*) and double asterisks (**) represent significant differences (*t* test, n = 6) at p = 0.05 and p = 0.01 compared with the wild type, respectively.

**Table S3 Plant height, Boll weight, seed index and lint index of transgenic and wild-type cotton in 2013.** Seed index, weight of 100 delinted seeds. Fiber index, weight of lints from 100 seeds. Data are shown as average±SD (n = 10).

## Author Contributions

Conceived and designed the experiments: YP YHX WQB. Performed the experiments: WQB YHX JZ SQS LH JYZ. Analyzed the data: YP WQB YHX XBL LH ML DML. Wrote the paper: YP YHX WQB.

## References

1. Haigler CH, Zhang D, Wilkerson CG (2005) Biotechnological improvement of cotton fibre maturity. Physiol Plantarum 124: 285–294.
2. Kim HJ, Triplett BA (2001) Cotton fiber growth in planta and in vitro. Models for plant cell elongation and eell wall biogenesis. Plant Physiol 127: 1361–1366.
3. Ruan Y (2007) Rapid cell expansion and cellulose synthesis regulated by plasmodesmata and sugar: insights from the single-celled cotton fibre. Funct Plant Biol 34: 1–10.
4. Qin YM, Zhu YX (2011) How cotton fibers elongate: a tale of linear cell-growth mode. Curr Opin Plant Biol 14: 106–111.
5. Xu SM, Brill E, Llewellyn DJ, Furbank RT, Ruan YL (2012) Overexpression of a potato sucrose synthase gene in cotton accelerates leaf expansion, reduces seed abortion, and enhances fiber production. Mol Plant 5: 430–441.
6. Jiang Y, Guo W, Zhu H, Ruan YL, Zhang T (2012) Overexpression of GhSusA1 increases plant biomass and improves cotton fiber yield and quality. Plant Biotechnol J 10: 301–312.
7. Wang H-Y, Wang J, Gao P, Jiao G-L, Zhao P-M, et al. (2009) Down-regulation of *GhADF1* gene expression affects cotton fibre properties. Plant Biotechnol J 7: 13–23.
8. Haigler CH, Ivanova-Datcheva M, Hogan PS, Salnikov VV, Hwang S, et al. (2001) Carbon partitioning to cellulose synthesis. Plant Mol Biol 47: 29–51.
9. Kurek I, Kawagoe Y, Jacob-Wilk D, Doblin M, Delmer D (2002) Dimerization of cotton fiber cellulose synthase catalytic subunits occurs via oxidation of the zinc-binding domains. P NATL ACAD SCI USA 99: 11109–11114.
10. Potikha TS, Collins CC, Johnson DI, Delmer DP, Levine A (1999) The involvement of hydrogen peroxide in the differentiation of secondary walls in cotton fibers. Plant Physiol 119: 849–858.
11. Olszewski N, Sun T-p, Gubler F (2002) Gibberellin signaling: Biosynthesis, catabolism, and response pathways. Plant Cell 14: S61–S80.
12. Magome H, Nomura T, Hanada A, Takeda-Kamiya N, Ohnishi T, et al. (2013) CYP714B1 and CYP714B2 encode gibberellin 13-oxidases that reduce gibberellin activity in rice. P NATL ACAD SCI USA 110: 1947–1952.
13. Hedden P, Thomas SG (2012) Gibberellin biosynthesis and its regulation. Biochem J 444: 11–25.
14. Yamaguchi S (2008) Gibberellin metabolism and its regulation. Annu Rev Plant Biol 59: 225–251.
15. Bhattacharya A, Ward DA, Hedden P, Phillips AL, Power JB, et al. (2012) Engineering gibberellin metabolism in Solanum nigrum L. by ectopic expression of gibberellin oxidase genes. Plant Cell Rep 31: 945–953.
16. Gou J, Ma C, Kadmiel M, Gai Y, Strauss S, et al. (2011) Tissue-specific expression of Populus C19 GA 2-oxidases differentially regulate above- and below-ground biomass growth through control of bioactive GA concentrations. New Phytol 192: 626–639.
17. Xiao Y-H, Li D-M, Yin M-H, Li X-B, Zhang M, et al. (2010) Gibberellin 20-oxidase promotes initiation and elongation of cotton fibers by regulating gibberellin synthesis. J Plant Physiol 167: 829–837.
18. Dayan J, Schwarzkopf M, Avni A, Aloni R (2010) Enhancing plant growth and fiber production by silencing GA 2-oxidase. Plant Biotechnol J 8: 425–435.
19. Vidal AM, Gisbert C, Talón M, Primo-Millo E, López-Díaz I, et al. (2001) The ectopic overexpression of a citrus gibberellin 20-oxidase enhances the non-13-hydroxylation pathway of gibberellin biosynthesis and induces an extremely elongated phenotype in tobacco. Physiol Plantarum 112:251–260. 112: 251–260.
20. Eriksson ME, Israelsson M, Olsson O, Moritz T (2000) Increased gibberellin biosynthesis in transgenic trees promotes growth, biomass production and xylem fiber length. Nat Biotechnol 18: 784–788.
21. Seagull RW, Giavalis S (2004) Pre- and post-anthesis application of exogenous hormones alters fiber production in Gossypium hirsutum L. cultivar Maxxa GTO. Journal of Cotton Science 8: 105–111.
22. Basra A, Saha S (1999) Growth regulation of cotton fibers. In: ASB, editor. Cotton fibers: developmental diology, quality improvement, and textile processing, New York: Food Products Press (an imprinting of Haworth Press). pp. 47–66.
23. Beasley CA (1973) Hormonal regulation of growth in unfertilized cotton ovules. Science 179: 1003–1005.
24. Hou L, Liu H, Li J, Yang X, Xiao Y, et al. (2008) SCFP, a novel fiber-specific promoter in cotton. Chinese Sci Bull 53: 2639–2645.
25. Luo M, Xiao Y, Li X, Lu X, Deng W, et al. (2007) GhDET2, a steroid 5α-reductase, plays an important role in cotton fiber cell initiation and elongation. Plant J 51: 419–430.
26. Xiao YH YM, Hou L, Pei Y (2006) Direct amplification of intron-containing hairpin RNA construct from genomic DNA. Biotechniques 41: 548–552.
27. Zhu Y-Q, Xu K-X, Luo B, Wang J-W, Chen X-Y (2003) An ATP-binding cassette transporter GhWBC1 from elongating cotton fibers. Plant Physiol 133: 580–588.
28. Zhang M, Zheng X, Song S, Zeng Q, Hou L, et al. (2011) Spatiotemporal manipulation of auxin biosynthesis in cotton ovule epidermal cells enhances fiber yield and quality. Nat Biotechnol 29: 453–458.
29. Bradford MM (1976) A rapid and sensitive method for the quantitation of microgram quantities of protein utilizing the principle of protein-dye binding. Anal Biochem 72: 248–254.
30. Ruan Y-L, Llewellyn DJ, Furbank RT (2003) Suppression of sucrose synthase gene expression represses cotton fiber cell initiation, elongation, and seed development. Plant Cell 15: 952–964.
31. Chourey P (1981) Genetic control of sucrose synthetase in maize endosperm. Mol Gen Genet 184: 372–376.
32. Zhao D, MacKown CT, Starks PJ, Kindiger BK (2010) Rapid analysis of nonstructural carbohydrate components in grass forage using microplate enzymatic assays. Crop Sci 50: 1537–1545.
33. Updegraff DM (1969) Semimicro determination of cellulose in biological materials. Anal Biochem 32: 420–424.
34. Debeaujon I, Nesi N, Perez P, Devic M, Grandjean O, et al. (2003) Proanthocyanidin-accumulating cells in Arabidopsis testa: Regulation of differentiation and role in seed development. Plant Cell 15: 2514–2531.
35. Singh B, Cheek HD, Haigler CH (2009) A synthetic auxin (NAA) suppresses secondary wall cellulose synthesis and enhances elongation in cultured cotton fiber. Plant Cell Rep 28: 1023–1032.
36. Ruan Y-L, Chourey PS (1998) A fiberless seed mutation in cotton is associated with lack of fiber cell initiation in ovule epidermis and alterations in sucrose synthase expression and carbon partitioning in developing seeds. Plant Physiol 118: 399–406.
37. Nolte KD, Hendrix DL, Radin JW, Koch KE (1995) Sucrose synthase localization during initiation of seed development and trichome differentiation in cotton ovules. Plant Physiol 109: 1285–1293.
38. Nadeau CD, Ozga JA, Kurepin LV, Jin A, Pharis RP, et al. (2011) Tissue-specific regulation of gibberellin biosynthesis in developing pea seeds. Plant Physiol 156: 897–912.
39. Iqbal N, Nazar R, Khan MIR, Masood A, Khan NA (2011) Role of gibberellins in regulation of source–sink relations under optimal and limiting environmental conditions. Current Science 100: 998–1007.

# Market Forces and Technological Substitutes Cause Fluctuations in the Value of Bat Pest-Control Services for Cotton

Laura López-Hoffman[1,2]*, Ruscena Wiederholt[1,2], Chris Sansone[3], Kenneth J. Bagstad[4], Paul Cryan[5], Jay E. Diffendorfer[4], Joshua Goldstein[6], Kelsie LaSharr[1], John Loomis[7], Gary McCracken[8], Rodrigo A. Medellín[9], Amy Russell[10], Darius Semmens[4]

1 School of Natural Resources & the Environment, The University of Arizona, Tucson, Arizona, United States of America, 2 Udall Center for Studies in Public Policy, The University of Arizona, Tucson, Arizona, United States of America, 3 Bayer CropScience, Research Triangle Park, North Carolina, United States of America, 4 United States Geological Survey, Geosciences and Environmental Change Science Center, Denver, Colorado, United States of America, 5 United States Geological Survey, Fort Collins Science Center, Fort Collins, Colorado, United States of America, 6 Department of Human Dimensions of Natural Resources, Colorado State University, Fort Collins, Colorado, United States of America, 7 Department of Agricultural and Resource Economics, Colorado State University, Fort Collins, Colorado, United States of America, 8 Department of Ecology and Evolutionary Biology, University of Tennessee, Knoxville, Tennessee, United States of America, 9 Instituto de Ecología, Universidad Nacional Autónoma de México, Distrito Federal, México, 10 Department of Biology, Grand Valley State University, Allendale, Michigan, United States of America

## Abstract

Critics of the market-based, ecosystem services approach to biodiversity conservation worry that volatile market conditions and technological substitutes will diminish the value of ecosystem services and obviate the "economic benefits" arguments for conservation. To explore the effects of market forces and substitutes on service values, we assessed how the value of the pest-control services provided by Mexican free-tailed bats (*Tadarida brasiliensis mexicana*) to cotton production in the southwestern U.S. has changed over time. We calculated service values each year from 1990 through 2008 by estimating the value of avoided crop damage and the reduced social and private costs of insecticide use in the presence of bats. Over this period, the ecosystem service value declined by 79% ($19.09 million U.S. dollars) due to the introduction and widespread adoption of Bt (*Bacillus thuringiensis*) cotton transgenically modified to express its own pesticide, falling global cotton prices and the reduction in the number of hectares in the U.S. planted with cotton. Our results demonstrate that fluctuations in market conditions can cause temporal variation in ecosystem service values even when ecosystem function – in this case bat population numbers – is held constant. Evidence is accumulating, however, of the evolution of pest resistance to Bt cotton, suggesting that the value of bat pest-control services may increase again. This gives rise to an economic option value argument for conserving Mexican free-tailed bat populations. We anticipate that these results will spur discussion about the role of ecosystem services in biodiversity conservation in general, and bat conservation in particular.

**Editor:** Robert B. Srygley, USDA-Agricultural Research Service, United States of America

**Funding:** This work was funded by a National Science Foundation award (DEB-1118975) to LLH. This work was supported by the U.S. Geological Survey's John Wesley Powell Center for Analysis and Synthesis working group, "Animal Migration and Spatial Subsidies: Establishing a Framework for Conservation Markets." The funders had no role in study design, data collection and analysis, decision to publish, or preparation of the manuscript.

* E-mail: lauralh@email.arizona.edu

## Introduction

The underlying goal of market-based, Payments for Ecosystem Services (PES) approaches to conservation is the creation of monetary incentives for the protection of critical ecological processes such as watershed functioning, pollination and natural pest control [1,2]. Within the conservation community, criticisms about market-based programs range from the ideological – e.g. unease that the approach diminishes nature's intrinsic value [3] – to apprehensions about the nature of the market [3,4]. The latter criticism stems from the worry there will be no reason to protect ecosystems when their services are no longer perceived to be valuable [3,4]. Two issues in particular – volatile market conditions and technological substitutes – are the main source of concern about the compatibility of market-based approaches and biodiversity conservation [4–6].

The first of these concerns is based on the economic principle that as the supply and demand curves for a market good change, the price of that good also changes. It follows that as the price of a market good fluctuates, the value of its ecosystem service inputs also will vary since ecosystem service values are derived from the demands of users of the services, in this case, cotton producers [7]. McCauley [6] illustrated this concern with the anecdote of a Costa Rican coffee plantation that was converted to pineapple production following world-wide declines in coffee prices. The monetary value to coffee production of the pollinators in the surrounding forest fragments previously had been estimated to be $60,000 USD per year [8]. Because pineapples are propagated and not pollinated, as the need for pollination services disappeared, McCauley [6] worried that the rationale for protecting the forest fragments might have disappeared as well.

The second concern about market-based approaches to conservation arises when manufactured capital is substituted for natural capital [9]. The story of the Zapp potato-chip factory in Louisiana (U.S.A.) illustrates this point. The factory once used a nearby wetland to filter its waste. But, as the potato chip business boomed, and as the volume of waste increased, the cost of using the wetland also increased and the company switched to technological forms of wastewater treatment [9].

In our study – using as an example the pest-control services provided by Mexican free-tailed bats (*Tadarida brasiliensis mexicana*) to cotton production in the U.S. – we demonstrate how bat pest-control values have changed in response to both changing market conditions for cotton and the adoption of a technological substitute for the service. To our knowledge, we present one of the first empirical, time-series analyses of the effects of changing market conditions and of the adoption of technological substitutes on the value of an ecosystem service.

As two-thirds of the more than 1,200 extant bat species are insectivorous [10], bats can provide significant complementary pest-control services, particularly by preying on pests early in the growing season before insecticide use has begun and preventing pest outbreaks [11]. Two studies have estimated the monetary value of the pest-control services of Mexican free-tailed bats in reducing crop damage and lowering the costs of insecticide use in cotton [12,13]. Using cotton price and acreage data from the mid-2000s, Cleveland et al. [12] estimated an annual pest-control value in an eight-county region of south-central Texas of $121,000 to $1,725,000 USD. Researchers have subsequently applied the estimates from the above studies to different regions in North America [14,15]. However, no previous study has considered how the value of bat ecosystem services has changed over time in response to shifts in commodity markets or advances in agricultural technology.

Over the last two decades, cotton prices and hectares planted in cotton in the U.S. have declined. The declines are generally attributed to global market forces – including trade barriers falling in the 1990s and increased production of cotton in developing countries [16]. In addition, in 1996, U.S. cotton growers started using Bt (*Bacillus thuringiensis*) cotton, which is transgenically modified to produce proteins that are toxic to susceptible insects [17]. As of 2012, 77% of all cotton grown in the U.S. was Bt-modified (www.ers.usda.gov/Data/BiotechCrops/). Bats have a lower pest-control impact on Bt than on conventional cotton [13]; however, over the past two years, mounting evidence from around the world suggests that insect pests are evolving resistance to Bt-modified crops [18]. While not yet widespread, Bt-resistant pests have been found in the field in India, China, and the U.S., and in laboratory studies [19–23].

Here we investigate the impacts of major global market factors – changes in cotton commodity price, the consequent change in the number of hectares planted with cotton, and the adoption of Bt cotton – on the value of the pest-control services provided by Mexican free-tailed bats in the U.S. We calculate service values each year from 1990 through 2008 by estimating both the value of avoided crop damage and the reduced social and private costs of insecticide use in the presence of bats. We assess pest-control services across all U.S. cotton-producing areas containing major Mexican free-tailed bat roosts. Due to the lack of time-series data on the size of bat populations, we could not consider the impact of fluctuations in population size on the value of bat pest control services; as an alternative we analyzed the sensitivity of pest-

control values to changes in bat population size. Currently, there is no market-based approach linking bat pest control services to cotton production, just estimates of value that might allow for the development of such mechanisms. As such, we conclude by addressing implications of our valuation research for the development of incentive-based approaches to the conservation of Mexican free-tailed bats in the context of the anticipated increase in Bt-resistant pests and the future value of bat ecosystem services.

## Results

From 1990 through 2008, the ecosystem service value of cotton pest-control services provided by Mexican free-tailed bats across the southwestern U.S. declined by 79%, from a high of $23.96 million in 1990 to a low of $4.88 million in 2008 (Fig. 1; mean values; values in 2011 USD). The value notably spiked in 1995 (Fig. 1) due to high prices for Pima and Upland cotton and the high number of hectares planted with cotton in that year (only 1991 had more cotton hectares). The mean annual value of pest control by bats over this time period was $12.24 million (s.d. = $6.04 million). The value of pest-control services expressed as a proportion of the total cotton crop value also varied over time, from 28% in 1991 to 6.5% in 2007. The presence of bats in cotton fields precluded, on average, the use of 32,046 kg of insecticide per year and damage to 131,385 kg of cotton per year.

Since Bt cotton produces its own insecticide, bats have less of an impact in controlling pests in Bt cotton than they do in conventional cotton [13]. To illustrate the decreased pest-control value of the bats resulting from the adoption of Bt cotton, we used the mean pest-control values (Fig. 1) to calculate what the pest-control values would have been had Bt not been adopted in 1996 (Fig. 2). To do this, we assumed that conventional cotton (i.e. non-Bt cotton) was planted from 1996 to 2008 (Fig. 2). In 2008, for example, the value of bat pest-control services was $2.66 million dollars (approximately 33%) less than what it might have been had all fields been planted with conventional cotton (Fig. 2). Our calculations also indicate that on a per-hectare basis the number of cotton bolls saved by the use of Bt technology was equivalent on average to the foraging efforts of about 27.5 bats.

To better understand temporal variability in ecosystem service values, we assessed the sensitivity of the annualized pest-control value to three factors: total area planted with cotton, the price of cotton, and the Mexican free-tailed bat population size. Notably, the mean pest-control values were equally sensitive to ecological and economic factors: both a ±10% change in the bat population size and a ±10% change in cotton prices caused a ±9.1% change in the mean pest-control values. However, altering the total area planted with cotton by ±10% only caused a ±0.9% change in the mean pest-control value over time.

## Discussion

The results of this study document that volatile market conditions and technological substitutes can affect the value of an ecosystem service [6]. The value of the pest-control services provided by Mexican free-tailed bats in the U.S. declined by 79% from 1990 through 2008 in response to declining global cotton prices, the consequent reduction in the number of hectares planted with cotton, and the introduction and widespread adoption of Bt cotton, a technological substitute to the natural pest-control services of the bats. Our analysis further indicates that these types of changes in market conditions may have as much impact on the

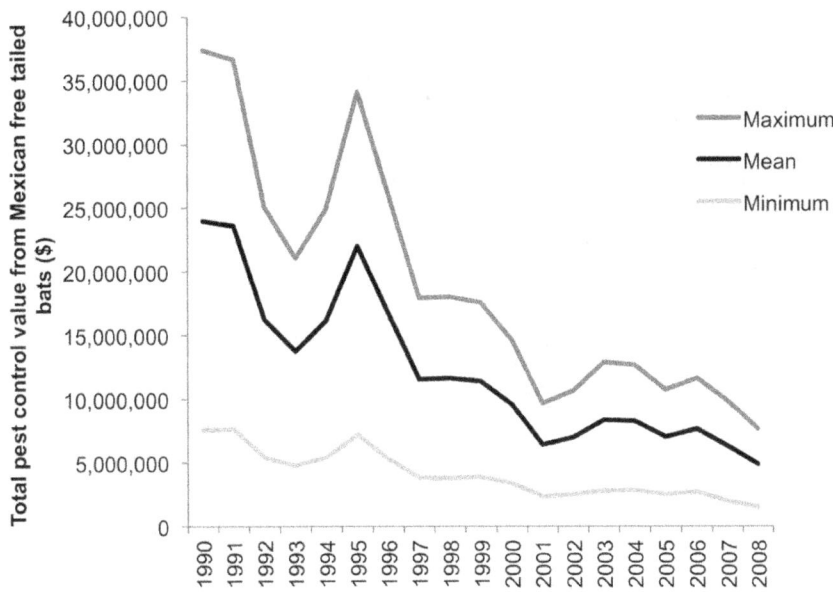

**Figure 1. Cotton pest-control value provided by Mexican free-tailed bats over time.** Maximum and minimum ecosystem service values for pest control represent calculations using the highest and lowest values, respectively, for several model parameters. From 1990 through 2008, the value of cotton pest-control services across the southwestern U.S. declined by 79%, from a high of $23.96 million in 1990 to a low of $4.88 million in 2008 (mean values). Values are indexed to 2011 U.S. dollars.

value of ecosystem services as changes in ecosystem function such as changes in bat population numbers.

Critics of the ecosystem service approach to biodiversity conservation contend that the risk of diminished service values – like the trends we document here – are a fundamental weakness to "economic benefits" arguments for conservation [3,4]. The fact that falling cotton prices and the adoption of Bt cotton caused the

value of bat cotton pest-control services to fall appears to confirm this concern. However, as the pest-control service of bats depreciates with investments in technology, so might the depreciation of manufactured capital. A recent meta-analysis of studies from five continents indicates that five of thirteen major insect crop pests have evolved Bt resistance in the last eight years, including cotton bollworms [18]. Resistance to Bt can arise in as

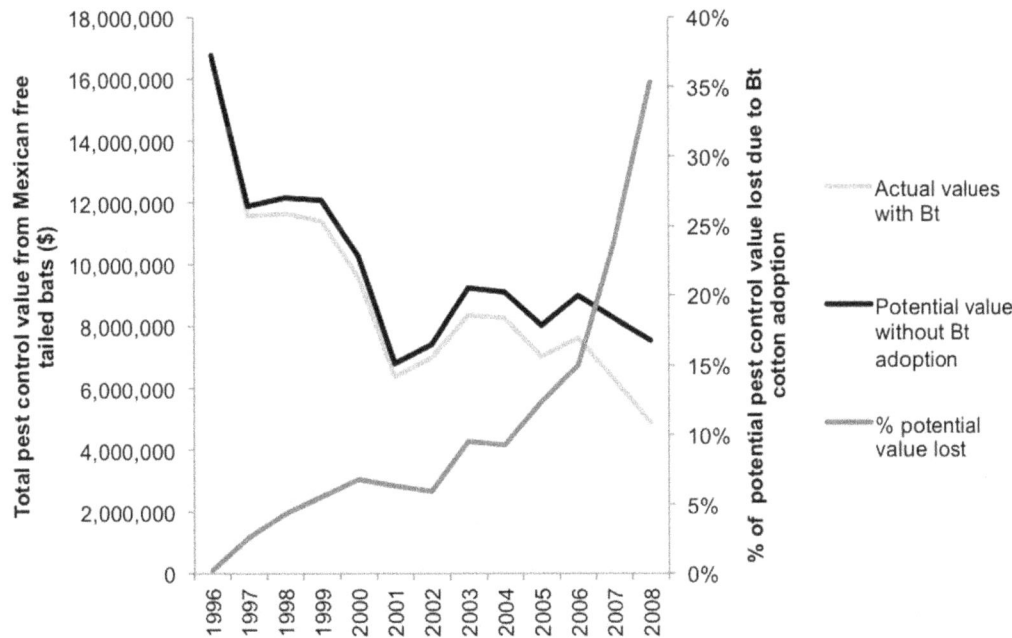

**Figure 2. Decreased pest-control value of the bats resulting from adoption of Bt cotton.** The light gray line shows actual pest control values from 1996 through 2008. The black line shows the potential value of bat pest control services if Bt had not been adopted. The dark gray line shows the percentage of the potential pest control value lost due to the adoption of Bt cotton. In 2008 the value of bat pest control services was $2.66 million dollars (approximately 33%) less than what it might have been.

few as two years depending on local conditions and how carefully growers hew to guidelines for maintaining refuges of non-Bt host plants [18,24]. Further, the efficacy of the second-generation Bt cotton (called pyramids because they produce more than one type of toxin), can be compromised if local pests are resistant to just one of the toxins [18]. This evidence of resistance evolution suggests that Bt may not be a permanent or even long-term solution to pest-related losses in the production of cotton and other crops, and that bats may again play a critical role in pest control. In fact, by preying on the individual insects that survive the Bt toxin – and preventing them from multiplying – bats may provide the additional service of slowing the evolution of resistance to Bt and other insecticides [13]. Indeed, bats and other natural enemies can play an important role in integrated pest management [25]. For example, Hagerty and colleagues [25] showed that Bollgard II cotton, a two-gene Bt product, experienced increased crop damage when natural predators were disrupted. Many agronomic researchers recommend that Bt crops be used in conjunction with other tactics, including natural predators, to avoid pest outbreaks and to delay the evolution of pest resistance [18,25].

## The option value of bats: an argument for conservation?

Option values in environmental economics can take one of two forms. The first is a user's willingness to pay for protecting a resource that is not currently in use so that it might be available for future use [26]. The exact value is influenced by the likelihood of needing the resource and the cost of replacing it should it be lost [27,28]. Alternatively, an option value may be the premium a decision-maker or society is willing to incur to avoid an irreversible loss of a resource by preserving it for future use [29]. In this case, it is possible that the pest-control services of Mexican free-tailed bats will become more valuable again, given that technological replacements for the service run the risk of being temporary. While more work is needed to assess the option value of protecting bats, the notion might provide an intriguing rationale to add to the existing list of reasons for enhancing bat conservation. Conservationists could argue that an investment in protecting bat populations now is an investment in protecting a future stream of potential pest-control services. They might also argue that bats provide pest control free of charge whereas pesticides and Bt cotton are costly to purchase, and their use carries a number of private and social costs. A final point might be that because bats are generalist predators, they provide a broad-spectrum pest control.

This line of reasoning suggests that it might be economically rational for decision-makers to promote bat conservation. However, there is often a disconnection between what is economically profitable and ecologically rational and what is implemented as public policy. Considerable investments have been made in developing Bt and similar pest-control methods and it is likely that such investments will continue [30]. However, we believe that bat conservation is not an alternative, but a rational strategy to use in concert with Bt technology given that neither is completely effective and generalist predators, such as bats, provide broad spectrum pest control and may help slow the evolution of resistance in pest species.

## Have we missed the full value of the ecosystem services provided by Mexican-free-tailed bats?

Critics have long pointed out that it is impossible to identify and measure, let alone value, all the ways functioning ecosystems provide benefits to society [31–33]. This leads us to an important caveat to this study – here we have assessed the impacts of one of 44 insectivorous bat species in the U.S. [34] on a single, albeit

major pest (*H. zea*) of only one important crop (cotton). Mexican free-tailed bats are generalist predators and can switch diet preferences very quickly [35,36]. When cotton prices fell and farmers planted other crops instead, the bats likely provided pest-control services to those other crops – indicating that we have not fully accounted for the total value of bats' pest-control services. Finally, a full accounting would consider the other types of services, such as ecotourism, provided by Mexican free-tailed bats [37].

## Conclusion

Valuation of ecosystem services has improved greatly in the last decade [38–41]. There is now a rich and growing literature showing how temporal and spatial variation in ecological functions can cause variation in the economic benefits provided by nature [42–46]. Our contribution is to demonstrate empirically how fluctuations in market conditions also cause temporal variation in ecosystem service values even when ecosystem function is held constant. Just as values from a particular region should not be blindly transferred to other regions [9,31,47,48], our study illustrates that ecosystem service valuations must also consider how changing market conditions and technological substitutes can alter service values over time. At a minimum, results like ours can be used to develop transfer functions for valuing agricultural pest-control services by accounting for the role of changing agricultural prices and practices in the value of these services [49,50].

We also hope that these and similar results will spur discussion about the role of ecosystem services in biodiversity conservation in general, and bat conservation in particular. While the value of bat pest-control services to cotton production in the U.S. did indeed decline in response to global market forces and advances in biotechnology, there is a possibility that as pest resistance to Bt cotton rises, bat pest-control service values will rise again. Although currently there are no market-based approaches linking bat pest control services to cotton production, we wonder if the hope of protecting bats now to preserve the "option value" of future services is a sufficient argument to develop incentive-based approaches to bat conservation? Or, is pinning conservation hopes on the notion of option values risky, since we have already witnessed one cycle of technological capital supplanting natural capital? These are the questions we are left to ponder.

## Materials and Methods

### Pest-control value

We employ the avoided-cost approach used by Cleveland et al. [12] to estimate the value of bat services in reducing crop damage and pesticide use on conventional cotton in an eight-county region of South Texas located west of San Antonio. We expand on the approach of Cleveland et al. [12] by considering the effect of the adoption of transgenic Bt cotton in 1996 on bat service values. Our analysis covers the two decades from 1990 through 2008, allowing us to understand how ecosystem service values vary over time as a function of changes in land-use practices and socioeconomic factors. In addition, we estimated bat service values across the southwestern U.S., rather than just a region of Texas. This increased geographic scope includes all cotton-producing areas near major Mexican free-tailed bat roosts in the U.S.

Estimating the avoided costs of crop damage involves the following steps: (a) estimating the number of insects consumed nightly by individual bats; (b) determining the hectares of cotton fields within proximity to bat roosts, which allows us to estimate the number of insects consumed nightly by bats in the area of the fields; and (c) the value of the crops that would have been damaged in the absence of bats. To determine the value of reducing insecticide use, we calculated both the reduced private costs to farmers of applying insecticides, and the reduced cost to society of releasing fewer insecticides into the environment. We modified the method of Cleveland et al. [12] to consider the social costs of only those insecticides that specifically target cotton bollworms (*Helicoverpa zea*) [51].

## Bat population estimates and roost locations

Our study area includes all U.S. counties that produce Pima or Upland cotton and that are located within 50 km (conservatively, the bats' nightly foraging distance [52]) of a major Mexican free-tailed bat roost. We obtained data about roosts (location and bat population censuses) from the U.S. Geological Survey's Bat Population Database [53] and our own literature search (Table S1 in File S1). We only considered large summer roosts (>7,000 individuals) because many smaller roosts lack good geospatial information, and because the combined populations of the largest summer colonies are thought to account for the majority (>99%) of the migratory Mexican free-tailed bat population [54]. These major roosts thus provide a reasonable estimate of the number of bats engaged in pest-control services in the U.S. We used only estimates obtained after 1970 to account for concerns that bat populations may have declined in the 1950s and 1960s due to DDT exposure [55,56]. We assumed that 90% of the adult bats in each colony were female and 10% were male, which is consistent with field data [13]. We did not model changes in the bat population size over time, as the data do not permit time-series analysis [13], but we did analyze the sensitivity of ecosystem service values to a 10% change in bat population numbers (see "Sensitivity analysis of pest-control values").

## Avoided crop damage calculation

**Number of pests consumed.** We first estimated the value of the crops that would have been lost in the absence of bats providing pest-control services. Conventional and molecular analyses show that moths comprise 30–60% of the bats' diet and indicate that each reproductively active female bat consumes 5–10 female adult bollworms (*Helicoverpa zea*) per night during periods of peak bollworm infestation [12,35,36]. Since bollworms also infest other crops in the area or migrate out of the region, we estimated that only 10–20% of the female moths consumed each night (approximately 1.5 individuals per bat) would have dispersed into cotton and laid eggs [12]. Due to high mortality rates during insect development (95–98%), the nightly consumption of 1.5 adult female moths would prevent 5 larvae from developing and damaging cotton crops [12,57]. Bollworm consumption by non-reproductive females and male bats was calculated as 32% lower than reproductive females due to the high metabolic costs of lactation [12,13].

For those bollworm that survive development to the larval stage, a single larva can damage 2–3 bolls of cotton over its lifetime [12]. However, because the value of the cotton bolls declines over the season – bolls produced during the first third of season generate about 50% of the harvest while bolls from the last third generate only 7% [58] – we estimated values separately for each third of the season. Further, because bats prevent damage to fewer bolls in Bt versus conventional cotton, we assumed that bats prevented approximately half (52.6%) the number of larvae from developing in Bt versus conventional cotton [13].

**Cotton locations.** We used data from the U.S. Department of Agriculture's National Agricultural Statistics Service (www.usda.nass.gov) and the National Cotton Council (www.cotton.org) on

number of cotton hectares planted per county (Table S2 in File S1). Data on numbers of hectares planted with cotton are at the county level, so we approximated locations of cotton fields using crop potential soil maps for each county. This approach assumes that cotton hectares are uniformly distributed over soils with high cotton potential. We used the U.S. General Soil Maps (STATSGO data) from the USDA Natural Resource Conservation Service (NRCS) for locations of soil types suitable for cotton production. For each year from 1990 through 2008, we assumed that the proportion of the cotton hectares planted per county within foraging distance of the bats was equal to the proportion of suitable cotton-growing soils for each county within their foraging range of 50 km from each roost. We also expected that bats disperse randomly from their roost, such that the percentage of the roost's bat population foraging in each cotton-growing area was equal to the percentage of the area each cotton-growing region composed of a roost's total foraging range. Because bats likely disperse non-randomly from their roosts and concentrate on high quality foraging grounds, our calculation is conservative.

## Cotton prices

We used data on cotton prices from 1990 through 2008 from the National Cotton Council. The prices were adjusted for inflation and reported in 2011 USD (SI Appendix, Table S2).

## Avoided insecticide costs calculation

Private costs savings for insecticides reflect the reduced cost to farmers of purchasing and applying chemicals. Data on costs of cotton insecticide applications from 1990 through 2008 were obtained from the Mississippi State University Department of Entomology and Plant Pathology's databases on cotton losses due to insects (http://www.entomology.msstate.edu/resources/cottoncrop.asp). Social cost savings arise from lowered public health impacts to the farm workers who apply the pesticides, and reduced environmental damage due to loss of beneficial pollinators and groundwater contamination [59]. We ascertained the insecticides in the U.S. that are used predominantly on cotton bollworms [60], and used data from Kovach et al. [61] and from Cornell University's Integrated Pest Management Program [62] to estimate the environmental and toxicological impacts of particular cotton insecticides. We then used a pesticide environmental accounting tool [51] to assign a social-cost value in dollars for each insecticide according to the degree of impact estimates. The pesticide accounting tool calculates detrimental impacts in six categories: human health, ground water contamination, aquatic systems (fish), birds, bees, and other beneficial insects. We used a weighted mean cost of insecticide applications per hectare over time.

## Numbers of insecticide applications avoided

Insecticides are generally applied to cotton fields when bollworm infestations reach a threshold of 20,000–25,000 larvae per hectare. The date at which the threshold is reached, which triggers the first insecticide application, varies by region. Regional estimates of dates of first insecticide application were provided by the following cotton pest experts: C Sansone (Texas, Oklahoma, and Kansas), D Munier (California), J Pierce (New Mexico), and P Ellsworth (Arizona). For fields planted with Bt cotton, the threshold is reached later because the bollworm population growth rate is <10% of that in conventional cotton [63], resulting in a lower number of avoided insecticide applications in Bt cotton. We estimated the number of insecticide applications that were avoided in the bats' presence by calculating the number of times the threshold would have been reached without bat predation from the first date of cotton flowering (and susceptibility to bollworms) to the first date of insecticide application. We used a uniform insecticide application rate of 0.29 kg/ha [60]. Finally, we estimated the value of these avoided applications by summing the private and social costs [12,61]. Data on the cotton season (e.g., mean planting and harvest dates) for different regions were obtained from the USDA's National Agricultural Statistics Service.

## Sources of uncertainty in pest-control estimates

We arrived at high and low estimates of total pest-control services provided by bats using ranges of several parameters for which we did not have accurate estimates. We used the following at their maximum and minimum value: the insecticide application threshold (20,000–25,000 larvae/ha), the number of bolls consumed by a larva over its lifetime (2–3 bolls/larva), and the number of adult female moths dispersing into cotton (0.5–2 individual per bat per night).

## Sensitivity analysis of pest-control values

To better understand factors influencing the ecosystem service values over time, we analyzed the sensitivity of the annualized mean pest-control value over our study period (1990–2008). We altered the following parameters by ±10%: total area planted with cotton, Mexican free-tailed bat population size, and price of cotton. We measured the effect of the parameter alterations on the annualized mean pest control value.

## Impact of Bt on value of pest-control services

Bt cotton was introduced for the Upland variety of cotton in 1996, but is not available for the Pima variety, which accounts for less than 5% of cotton production in the U.S. [64]. Information on the timing of adoption of transgenic Bt Upland cotton was

obtained from the Mississippi State University Department of Entomology and Plant Pathology's database on cotton crop losses [64]. To better understand the influence of the adoption of Bt cotton on the bat's pest-control service value, we used the mean pest-control values to calculate potential pest-control values had Bt not been adopted in 1996. To do this, we assumed that conventional cotton (i.e. non-Bt cotton) was planted from 1996 through 2008 and recalculated the pest-control values.

## Acknowlegments

We thank D. Munier, J. Pierce, and P. Ellsworth for providing information on insecticide use in cotton. LLH thanks R. Merideth, J.A. Quijada, E. Valdez, C. Kremen, N. Brozovic and an anonymous reviewer for their comments on the paper. Any use of trade product or firm names is for descriptive purposes only and does not imply endorsement by the US Government.

## Author Contributions

Conceived and designed the experiments: RW LLH JL JD CS DS JG. Performed the experiments: RW LLH KL CS KB. Analyzed the data: RW LLH CS JD KB. Contributed reagents/materials/analysis tools: GFM KL RM AR PC. Wrote the paper: LLH RW GFM. Developed the code to analyze the data: RW.

## References

1. Daily G (1997) Nature's services: societal dependence on natural ecosystems. Washington D.C.: Island Press.
2. Pagiola S, Von Ritter K, Bishop J (2004) Assessing the economic value of ecosystem conservation. Washington D.C.: World Bank, Environment Department.
3. Redford KH, Adams WM (2009) Payment for ecosystem services and the challenge of saving nature. Conserv Biol 23: 785–787.
4. Berck P, Ligon E (1996) The swamp and shopping center: an interest rate parable. Ecol Model 92: 275.
5. Rodriguez JP, Beard TD, Bennett EM, Cumming GS, Cork SJ, et al. (2006) Trade-offs across space, time, and ecosystem services. Ecol Soc 11.
6. McCauley DJ (2006) Selling out on nature. Nature 443: 27–28.
7. Brown T, Bergstrom J, Loomis J (2007) Defining, Valuing and Providing Ecosystem Goods and Services. Nat Resour J 47: 331–376.
8. Ricketts TH, Daily GC, Ehrlich PR, Michener CD (2004) Economic value of tropical forest to coffee production. Proc Natl Acad Sci U S A 101: 12579–12582.
9. Plummer ML (2009) Assessing benefit transfer for the valuation of ecosystem services. Front Ecol Environ 7: 38–45.
10. Kunz TH, de Torrez EB, Bauer D, Lobova T, Fleming TH (2011) Ecosystem services provided by bats. Ann N Y Acad Sci1223: 1–38.
11. Wiedenmann RN (2011) Contributions of Robert (Bob) J. O'Niel to the field of predaceous Heteroptera. Biol Control 59.
12. Cleveland CJ, Betke M, Federico P, Frank JD, Hallam TG, et al. (2006) Economic value of the pest control service provided by Brazilian free-tailed bats in south-central Texas. Front Ecol Environ 4: 238–243.
13. Federico P, Hallam TG, McCracken GF, Purucker ST, Grant WE, et al. (2008) Brazilian free-tailed bats as insect pest regulators in transgenic and conventional cotton crops. Ecol Appl 18: 826–837.
14. Boyles JG, Cryan PM, McCracken GF, Kunz TH (2011) Economic importance of bats in agriculture. Science 332: 41–42.
15. Gándara Fierro G, Correa Sandoval A, Hernández Cienfuegos C (2006) Valoración económica de los servicios ecológicas que prestan los murciélagos Tadarida brasiliensis como controladores de plagas en el norte de México.
16. Cátedra de Integración Económica y Desarrollo Social. USDA (2013) Cotton: World Markets and Trade.: U.S. Department of Agriculture.
17. Marvier M, McCreedy C, Regetz J, Kareiva P (2007) A meta-analysis of effects of Bt cotton and maize on nontarget invertebrates. Science 316: 1475–1477.

18. Tabashnik BE, Brevault T, Carriere Y (2013) Insect resistance to Bt crops: lessons from the first billion acres. Nat Biotechnol 31: 510–521.
19. Tabashnik BE, Gassmann AJ, Crowder DW, Carriere Y (2008) Insect resistance to Bt crops: evidence versus theory. Nat Biotechnol 26: 199–202.
20. Bagla P (2010) INDIA Hardy Cotton-Munching Pests Are Latest Blow to GM Crops. Science 327: 1439–1439.
21. Tabashnik BE, Wu KM, Wu YD (2012) Early detection of field-evolved resistance to Bt cotton in China: Cotton bollworm and pink bollworm. Journal of Invertebrate Pathology 110: 301–306.
22. Gassmann A (2012) Field-evolved resistance to Bt maize by western corn rootworm: Predictions from the laboratory and effects in the field. J Invertebr Pathol 110: 287–293.
23. Kaur P, Dilawari VK (2011) Inheritance of resistance to Bacillus thuringiensis Cry1Ac toxin in Helicoverpa armigera (Hubner)(Lepidoptera: Noctuidae) from India. Pest manag sci 67: 1294–1302.
24. Kruger M, Van Rensburg JBJ, Van den Berg J (2012) Transgenic Bt maize: farmers' perceptions, refuge compliance and reports of stem borer resistance in South Africa. J Appl Entomol 136: 38–50.
25. Hagerty AM, Kilpatrick AL, Turnipseed SG, Sullivan MJ, Bridges WC (2005) Predaceous arthropods and lepidopteran pests on conventional, Bollgard, and Bollgard II cotton under untreated and disrupted conditions. Environ Entomol 34: 105–114.
26. Weisbrod BA (1964) Collective-Consumption Services of Individual-Consumption Goods. Q J Econ 78: 471–477.
27. Walsh RG, Loomis JB, Gillman RA (1984) Valuing option, existence and bequest demands for nature. Land Econ 60: 14–29.
28. Brookshire DS, Eubanks LS, Randall A (1983) Estimating option prices and existence values for wildlife resources. Land Econ 59: 1–15.
29. Kruitlla J, Fisher A (1976) Economics of Natural Environments. Baltimore, MD.: Johns Hopkins University Press.
30. McDougall P (2011) The cost and time involved in the discovery, development and authorisation of a new plant biotechnology derived trait. Consultancy Study for Crop Life International.
31. Spash CL (2008) How much is that ecosystem in the window? The one with the bio-diverse trail. Environ Value 17: 259–284.
32. Peterson MJ, Hall DM, Feldpausch-Parker AM, Peterson TR (2010) Obscuring Ecosystem Function with Application of the Ecosystem Services Concept. Conserv Biol 24: 113–119.

33. Muradian R, Rival L (2012) Between markets and hierarchies: The challenge of governing ecosystem services. Ecosystem Services 1: 93–100.
34. Smithsonian Institute (1980) Bat facts. In: Smithsonian Institute, editors. Science and Technology.
35. McCracken GF, Westbrook JK, Brown VA, Eldridge M, Federico P, et al. (2012) Bats Track and Exploit Changes in Insect Pest Populations. PLoS One 7: e43839.
36. Lee Y-F, McCracken GF (2005) Dietary variation of Brazilian free-tailed bats linked to migratory populations of pest insects. J Mammal 86: 67–76.
37. Bagstad K, Wiederholt R (2013) Tourism values for Mexican free-tailed bat (Tadarida brasiliensis mexicana) viewing. Hum Dimens Wildl 18: 307–311.
38. Barbier EB (2011) Progress and Challenges in Valuing Coastal and Marine Ecosystem Services. Rev Environ Econ Policy 6: 1–19.
39. Turner RK, Paavola J, Cooper P, Farber S, Jessamy V, et al. (2003) Valuing nature: lessons learned and future research directions. Ecol Econ 46: 493–510.
40. Lu Y, Fu B, Feng X, Zeng Y, Liu Y, et al. (2012) A Policy-Driven Large Scale Ecological Restoration: Quantifying Ecosystem Services Changes in the Loess Plateau of China. PLoS One 7: e31782.
41. Kovacs K, Polasky S, Nelson E, Keeler BL, Pennington D, et al. (2013) Evaluating the Return in Ecosystem Services from Investment in Public Land Acquisitions. PLoS One 8: e62202.
42. Kozak J, Lant C, Shaikh S, Wang G (2011) The geography of ecosystem service value: The case of the Des Plaines and Cache River wetlands, Illinois. Appl Geogr 31: 303–311.
43. Barbier EB, Hacker SD, Kennedy C, Koch EW, Stier AC, et al. (2011) The value of estuarine and coastal ecosystem services. Ecol Monogr 81: 169–193.
44. Aburto-Oropeza O, Ezcurra E, Danemann G, Valdez V, Murray J, et al. (2008) Mangroves in the Gulf of California increase fishery yields. Proc Natl Acad Sci U S A 105: 10456–10459.
45. Koch EW, Barbier EB, Silliman BR, Reed DJ, Perillo GME, et al. (2009) Non-linearity in ecosystem services: temporal and spatial variability in coastal protection. Front Ecol Environ 7: 29–37.
46. Bagstad KJ, Johnson GW, Voigt B, Villa F (2013) Spatial dynamics of ecosystem service flows: A comprehensive approach to quantifying actual services. Ecosyst Serv 4: 117–125.
47. Fisher B, Naidoo R (2011) Concerns about extrapolating right off the bat. Science 335: 287.
48. Eigenbrod F, Armsworth PR, Anderson BJ, Heinemeyer A, Gillings S, et al. (2010) Error propagation associated with benefits transfer-based mapping of ecosystem services. Biol Conserv 143: 2487–2493.
49. Bergstrom JC, Taylor LO (2006) Using meta-analysis for benefits transfer: Theory and practice. Ecol Econ 60: 351–360.
50. Loomis JB (1992) The evolution of a more rigorous approach to benefit transfer: benefit function transfer. Water Resour Res 28: 701–705.
51. Leach A, Mumford J (2008) Pesticide Environmental Accounting: A method for assessing the external costs of individual pesticide applications. Environ pollut 151: 139–147.
52. Williams TC, Ireland LC, Williams JM (1973) High altitude flights of the free-tailed bat, Tadarida brasiliensis, observed with radar. J Mammal 54: 807–821.
53. Ellison L, O'Shea T, Bogan M, Everette A, Schneider D (2003) Existing data on colonies of bats in the United States: summary and analysis of the US Geological Survey's bat population database. USGS/BRD/ITR–2003-0003. USGS Information and Technology Report Fort Collins, CO, USA: USGS Fort Collins Science Center. pp. 127–171.
54. McCracken G (2003) Estimates of population sizes in summer colonies in Brazilian free-tailed bats (Tadarida brasiliensis). USGS/BRD/ITR–2003-0003. In: O. S. T. B MA, editor editors. Monitoring trends in bat populations of the United States and territories: problems and prospects. Fort Collins, CO, USA: U.S. Geological Survey, Fort Collins, CO. pp. 21–29.
55. Lewis SE (1995) Roost fidelity of bats - a review Journal of Mammalogy 76: 481–496.
56. Betke M, Hirsh DE, Makris NC, McCracken GF, Procopio M, et al. (2008) Thermal Imaging Reveals Significantly Smaller Brazilian Free-Tailed Bat Colonies Than Previously Estimated. J Mammal 89: 18–24.
57. Sansone C, Smith J (2001) Natural mortality of Helicoverpa zea (Lepidoptera: Noctuidae) in short-season cotton. Biol Control 30: 113–122.
58. Sansone CG, Isakeit T, Lemon R, Warrick B (2002) Texas cotton production: Emphasizing integrated pest management. In: T. C. E. Publication, editor editors. College Station, TX.
59. Pimentel D, Acquay H, Biltonen P, Rice S, Silva M, et al. (1991) Environmental and economic effects of reducing pesticide use. BioScience 42: 750–760.
60. Gianessi L, Reigner N (2006) Pesticide use in U.S. crop production. Washington D.C.: CropLife Foundation.
61. Kovach J, Petzoldt C, Degni J, Tette J (1992) A method to measure the environmental impact of pesticides. New York Food Life 129: 1–8.
62. Cornell University's Integrated Pest Management Program (2012) New York State Integrated Pest Management Program.
63. Jackson RE, Bradley Jr JR, Van Duyn JW (2003) Field performance of transgenic cottons expressing one or two Bacillus thuringiensis endotoxins against bollworm, Helicoverpa zea (Boddie). J Cotton Sci 7: 57–64.
64. Mississippi State University Department of Entomology and Plant Pathology United States of America, http://www.entomology.msstate.edu/resources/cottoncrop.asp, Accessed 11 January 2014.

# The *Li₂* Mutation Results in Reduced Subgenome Expression Bias in Elongating Fibers of Allotetraploid Cotton (*Gossypium hirsutum* L.)

**Marina Naoumkina**[1]*, **Gregory Thyssen**[1], **David D. Fang**[1], **Doug J. Hinchliffe**[2], **Christopher Florane**[1], **Kathleen M. Yeater**[3], **Justin T. Page**[4], **Joshua A. Udall**[4]

**1** Cotton Fiber Bioscience Research Unit, USDA-ARS, Southern Regional Research Center, New Orleans, Louisiana, United States of America, **2** Cotton Chemistry & Utilization Research Unit, USDA-ARS, Southern Regional Research Center, New Orleans, Louisiana, United States of America, **3** USDA-ARS-Southern Plains Area, College Station, Texas, United States of America, **4** Plant and Wildlife Science Department, Brigham Young University, Provo, Utah, United States of America

## Abstract

Next generation sequencing (RNA-seq) technology was used to evaluate the effects of the Ligon lintless-2 (*Li₂*) short fiber mutation on transcriptomes of both subgenomes of allotetraploid cotton (*Gossypium hirsutum* L.) as compared to its near-isogenic wild type. Sequencing was performed on 4 libraries from developing fibers of *Li₂* mutant and wild type near-isogenic lines at the peak of elongation followed by mapping and PolyCat categorization of RNA-seq data to the reference $D_5$ genome (*G. raimondii*) for homeologous gene expression analysis. The majority of homeologous genes, 83.6% according to the reference genome, were expressed during fiber elongation. Our results revealed: 1) approximately two times more genes were induced in the $A_T$ subgenome comparing to the $D_T$ subgenome in wild type and mutant fiber; 2) the subgenome expression bias was significantly reduced in the *Li₂* fiber transcriptome; 3) *Li₂* had a significantly greater effect on the $D_T$ than on the $A_T$ subgenome. Transcriptional regulators and cell wall homeologous genes significantly affected by the *Li₂* mutation were reviewed in detail. This is the first report to explore the effects of a single mutation on homeologous gene expression in allotetraploid cotton. These results provide deeper insights into the evolution of allotetraploid cotton gene expression and cotton fiber development.

**Editor:** Tianzhen Zhang, Nanjing Agricultural University, China

**Funding:** This research was funded by United States Department of Agriculture-Agricultural Research Service project number 6435-21000-017-00D and Cotton Incorporated project number 58-6435-2-663. The funders had no role in study design, data collection and analysis, decision to publish, or preparation of the manuscript.

* E-mail: marina.naoumkina@ars.usda.gov

## Introduction

Cotton is the major source of natural fibers used in the textile industry. There are four cultivated species: AA genome diploids, *Gossypium arboretum* L. and *G. herbaceum* L.; and AADD genome allotetraploids, *G. hirsutum* L. and *G. barbadense* L. Upland cotton (*G. hirsutum*) represents about 95% of world cotton production [1]. Allotetraploid species originated around 1–2 million years ago from inter-specific hybridization between an AA-genome diploid native to Africa and Mexican DD-genome diploid [1,2].

Cotton fibers are single-celled trichomes that emerge from the ovule epidermal cells. About 25–30% of the seed epidermal cells differentiate into spinnable fibers [3]. Fiber length ranges from short (fuzz <6 mm) to long (lint). Lint fibers of economically important *G. hirsutum* generally grow up to about 30–40 mm in length. Cotton fiber development undergoes four distinctive but overlapping stages: initiation, elongation, secondary cell wall biosynthesis, and maturation [4]. The rate and duration of each developmental stage is important to the quality attributes of the mature fiber. Cell elongation is crucial for fiber length, whereas secondary cell wall thickening is important for fiber fineness and strength.

Cotton fiber mutants are useful tools to elucidate biological processes of cotton fiber development. A cotton plant with abnormally short lint fibers was discovered in a breeding nursery of the Texas Agricultural Experiment Station in 1984. This mutant had short lint fibers (<6 mm) visually similar to those produced by Ligon lintless-1 (*Li₁*); however, unlike the stunted and deformed vegetative morphology caused by the *Li₁* mutation, this fiber mutant had normal vegetative growth. The trait was controlled by one dominant gene named Ligon lintless-2 (*Li₂*) [5]. This gene was mapped to chromosome 18 ($D_T$ subgenome of *G. hirsutum*) using several approaches [6–8]. In a fiber developmental study, Kohel *et al.* [9] observed that elongation is restricted in *Li₂* fibers, however secondary wall development proceeds normally in proportion to fiber length. Two near-isogenic lines (NILs) of *Li₂* with the Upland cotton variety DP5690 were developed in a backcross program at Stoneville, MS [6]. Morphological evaluation of developing fibers did not reveal apparent differences between WT and *Li₂* NILs during initiation or early elongation up to 5 days post-anthesis (DPA). Transcript and metabolite evaluations revealed significant changes in biological processes associated with cell expansion in the *Li₂* mutant line at peak of fiber elongation, including reactive oxygen species, hormone homeostasis, nitrogen metabolism, carbohydrate biosynthesis, cell wall biogenesis, and cytoskeleton [6,10]. Therefore, the $Li_2$ mutation can be considered as a factor affecting cotton fiber elongation process, making it an excellent model system to study cotton fiber elongation.

In previous reports, we used microarray techniques to investigate global gene expression in $Li_2$ NILs [6,10]. However, by using the genome sequence of G. raimondii [11], RNA-seq can provide a more comprehensive and accurate transcriptome analysis based on the reference DNA sequences [12]. RNA-seq offers a larger dynamic range of quantification, reduced technical variability, and higher accuracy for distinguishing and quantifying expression levels of homeologous copies than DNA microarrays [12]. Because of the limited sequence divergence between the $A_T$ and $D_T$ subgenomes in cotton [13], a pipeline was developed to map and categorize RNA-seq reads as originating from the $A_T$ or $D_T$ subgenomes [14].

In the present study we compared quantitative gene expression levels of RNA-seq data between developing fibers of $Li_2$ and its WT NILs. We investigated the $Li_2$ mutation's effect on global transcriptional changes in subgenomes and on the functional distribution of homeologous genes during fiber elongation. These results provide deeper insights into the evolution of allotetraploid cotton gene expression.

## Results

### RNA-seq of Wild Type and $Li_2$ Developing Fibers at Peak of Elongation

Considering the cost of deep sequencing, only one time point, at the peak of elongation, was selected for RNA-seq analysis, including two biological replicates for wild type and mutant NILs. The time points 8 and 12 days post-anthesis (DPA) represent peak rates of fiber elongation. The time point 8 DPA was selected because: 1) our earlier research revealed significant transcript and metabolite changes between the $Li_2$ and wild type NILs during this time of fiber development [6,10]; 2) the transcript level of the elongation stage-related gene GhExp1 significantly decreased in $Li_2$ mutant fiber at 8 DPA [6].

A total of 639 million reads (each 101 bp in length) from 4 libraries were obtained by paired-end Illumina sequencing. Approximately 2.3% more reads were obtained from $Li_2$ than wild type fiber transcriptomes. From 84.4% to 90.2% of the reads were mapped to the $D_5$-genome reference sequence of G. raimondii (Table 1). Not all the reads mapped to the reference genome sequence, probably since some of the genes were not included in the 13 large pseudo-molecules and transcripts mapped to genomic regions were outside of annotated genes. Of the mapped reads, between 29.3%–31.4% were mapped to the $A_T$ subgenome and between 23.4%–25.1% were mapped to the $D_T$ subgenomes of G. hirsutum. If the mapped reads overlapped a homeologous SNP position (SNPs between the $A_T$ and $D_T$ subgenomes), they were categorized as belonging to one of the two subgenomes or as a chimeric read (A-reads, D-reads, and X-reads, respectively; [14]). If a read did not overlap a homeologous SNP position, the read was unable to be categorized as originating from either the $A_T$ or $D_T$ subgenome (N reads; Table 1). Notably, more reads from each library were aligned to the $A_T$ subgenome than to the $D_T$ subgenome. Among the 37,223 genes on the 13 chromosomes of the G. raimondii genome, 34,692 genes (93%) had at least one mapped read from developing fibers at peak of elongation (Table 2).

## Differential Gene Expression in Developing Fibers

Counts of mapped reads were evaluated in wild type and mutant fiber transcriptomes. Genes were considered to be expressed if they had $\geq 10$ reads mapped in one sample. Genes that were not considered to be expressed were not included in further analyses. Approximately 3% more expressed genes were detected in $Li_2$ than in wild type. Of the 37,223 genes on the 13 chromosomes of the G. raimondii $D_5$ reference genome 29,603 (79.5%) genes were expressed in wild type and 30,842 (82.9%) genes were expressed in $Li_2$ fiber (Table 2).

Many genes had altered expression levels as a result of the $Li_2$ mutation (Table S1 in File S1). Some genes were expressed in one treatment (such as $Li_2$) but not the other treatment. For example, expressions of genes annotated as SAUR-like auxin-responsive protein (Gorai.005G257000), bHLH (Gorai.003G034700) and NAC domain transcription factor (Gorai.009G170700) were only detected in wild-type fiber. Cytokinin response factor 6 (Gorai.007G105600), UGT73C14 (Gorai.002G107900), cystein proteinase (Gorai.007G329600), MYB-like 102 (Gorai.012G132200) and WRKY transcription factor (Gorai.009G157300) were only detected in $Li_2$ fiber. The majority of these genes have not yet been functionally characterized in cotton, except of glycosyltransferase UGT73C14, which has been shown recently to be involved in ABA homeostasis [15].

The quantitative levels of mapped reads were evaluated for differential expression between elongating fibers of $Li_2$ and wild type NILs. A gene-by-gene ANOVA determined that 7,163 of 31,114 expressed genes were differentially expressed (FDR corrected $p$-value $<0.05$) and had $\geq 2$-fold difference in at least one of the following comparisons: fiber type ($Li_2$ versus wild type), $A_T$/$D_T$ subgenomes, and combinations of these factors (statistical data for significantly regulated genes are provided in Data S1). The highest numbers of significantly differentially expressed genes were identified between the $A_T$ and $D_T$ subgenomes in wild type and $Li_2$; whereas approximately 3 times fewer differentially expressed genes were detected between fiber type comparisons (Figure 1A). Of the 29,603 expressed homeologous pairs in wild type, 4,578 (wtA/wtD, 15.5%) showed significantly different expression level between subgenomes; whereas in mutant fiber of the 30,842 expressed genes, 3,967 (LiA/LiD, 12.9%) were differentially expressed between subgenomes. Therefore, the homeolog expression bias was significantly (p-value $<0.0001$; Chi square) reduced in $Li_2$ fiber transcriptome.

In general, the $A_T$ subgenome contributed more differentially expressed genes to fiber transcriptome than did the $D_T$ subgenome. Approximately two times more genes were differentially expressed in the $A_T$ subgenome compared to the $D_T$ subgenome in wild type ($A_T$ - 2958 vs. $D_T$ - 1620; Figure 1B) and mutant fiber ($A_T$ - 2574 vs. $D_T$ - 1393). Comparison between fiber types showed more genes were upregulated in $Li_2$ versus wild type in both subgenomes (Figure 1B). It should be noted that only about 38% (583 genes out of 1,536 in $A_T$ and 1,511 in $D_T$ subgenome) of significantly regulated genes between mutant and wild type overlapped between subgenomes (Figure 1A).

### Mutation Effects on Transcriptome of $A_T$ and $D_T$ Subgenomes of Allotetraploid G. hirsutum

The effect of $Li_2$ mutation on the transcriptome of each subgenome was evaluated. The genes significantly (FDR corrected $p$-value $<0.05$) up-regulated ($\geq 2$-fold) in one subgenome of wild type were considered to have biased expression. Of the 2,958 $A_T$ biased genes, 26.5% (784) had significantly changed the expression levels in both subgenomes of $Li_2$ as a result of mutation. However, of the 1,620 $D_T$ biased genes, 35.9% (582) had significantly

**Table 1.** Results of mapping reads.

| Library | $Li_2$ BR1 | | $Li_2$ BR2 | | WT BR1 | | WT BR2 | |
|---|---|---|---|---|---|---|---|---|
| | count | % | count | % | count | % | count | % |
| Number reads | 155,057,542 | 100.0 | 168,114,870 | 100.0 | 149,831,738 | 100.0 | 165,935,842 | 100.0 |
| A reads | 45,382,923 | 29.3 | 51,937,609 | 30.9 | 47,115,502 | 31.4 | 52,072,768 | 31.4 |
| D reads | 36,234,551 | 23.4 | 41,617,562 | 24.8 | 37,675,249 | 25.1 | 41,512,224 | 25.0 |
| X reads | 7,972,844 | 5.1 | 9,130,944 | 5.4 | 8,068,891 | 5.4 | 8,896,820 | 5.4 |
| N reads | 41,252,612 | 26.6 | 46,157,152 | 27.5 | 42,291,972 | 28.2 | 46,818,360 | 28.2 |
| Mapped total | 130,842,930 | 84.4 | 148,843,267 | 88.5 | 135,151,614 | 90.2 | 149,300,172 | 90.0 |

changed the expression levels in both subgenomes of the $Li_2$ mutant (Figure 2). Therefore $Li_2$ has a significantly greater effect ($p$-value <0.0001; Chi square) on $D_T$ biased genes than $A_T$ biased genes. Importantly, the majority of biased genes had significantly reduced expression levels (8.6% and 12.4%), whereas only a small portion of genes increased expression levels (1.6% and 2.7%) in the mutant. However, more genes, which were down-regulated in homeologous subgenome in wild type, had increased expression levels in the mutant fiber: 11.9% and 14% were up-regulated, whereas 4.4% and 6.8% were down-regulated (Figure 2).

Furthermore, a few homeolog pairs had reciprocal expression biases between two subgenomes as a result of mutation. Expression levels for three of these genes were tested by RT-qPCR across eight developmental time points from DOA to 20 DPA, representing initiation, elongation, and beginning of secondary cell wall biosynthesis stages (Figure 3). Interestingly, the direction of expression bias changed between developmental stages in these three genes. For example, expression of homeolog pair was biased in favor of the $D_T$ subgenome for Gorai.002G223800 at initiation (1 DPA), but switched to favor the $A_T$ subgenome at elongation (5–16 DPA) in wild type developing fibers, whereas expression was biased in favor of the $D_T$ subgenome across all evaluated time points in mutant fibers. These results demonstrate that the $Li_2$ mutation had a greater effect on the $D_T$ subgenome and also influenced direction of expression bias for some genes across developmental stages.

## Mutation Effects on Functional Distribution of Homeologous Genes during Fiber Elongation

The greater effect of $Li_2$ on $D_T$ biased genes was observed in overall transcript data. In general, the subgenomes contributed unequally to different biological processes [16]; therefore diverse mutation effects could be expected on different functional categories of genes. To determine which biological processes were affected by the mutation, MapMan ontology was used (Data S2).

The distribution of genes from the $A_T$ and $D_T$ subgenomes with significantly changed expression levels in the mutant were categorized into MapMan functional categories (Figure 4; Table S2 in File S1). Relative gene frequencies in functional categories were represented in percents of biased genes in each subgenome (2,958 $A_T$ biased genes and 1,620 $D_T$ biased genes). Most functional categories were biased in favor of the $D_T$ subgenome with the exception of photosynthesis and redox, which only contained $A_T$ homeologs. Two functional categories were significantly ($p$-value <0.05; Fisher's exact test) biased in enrichment among $D_T$ biased genes: secondary metabolism and stress (Table S2 in File S1). These results demonstrate that different biological processes were unequally affected by $Li_2$ mutation.

**Transcriptional regulators.** Transcriptional regulators (TRs) were identified in the *G. raimondii* genome based on similarity to Arabidopsis TRs and categorized into 76 families. Among them, 229 homeolog pairs were $A_T$ biased and 111 were $D_T$ biased in elongating cotton fibers. Of the 229 $A_T$ biased TRs, 21 (9.2%) of them changed transcription level, whereas of the 111 $D_T$ biased TRs 14 (12.6%) of them changed transcription level (Table 3), but this difference was statistically insignificant. Expression levels for the majority of subgenome biased homeologs decreased as the result of $Li_2$ mutation. However, six TRs (including both homeologs) had increased expression levels in mutant fibers. Three classes of TRs were the most abundant, including Aux/IAA (6 members), bHLH (5 members) and MYB (3 members). Interestingly, two of the three members of MYB TRs had increased expression due to mutation.

**Cell wall.** In the cell wall functional category, 60 homeologs were $A_T$ biased and 40 were $D_T$ biased in fiber transcriptome. Ten (16.7%) of the $A_T$ biased homeologs changed expression levels; whereas 12 (30%) of the $D_T$ biased homeologs changed expression levels (Table 4), indication a higher, but statistically insignificant effect on $D_T$ biased homeologs. Interestingly, more $D_T$ homeologs (11) than $A_T$ homeologs (4) increased transcript levels as a result of the $Li_2$ mutation. Genes encoding enzymes involved in polysac-

**Table 2.** Count of expressed genes.

| Sample | $Li_2$ BR 1 | $Li_2$ BR 2 | $Li_2$ Total | WT BR1 | WT BR2 | WT Total | Total |
|---|---|---|---|---|---|---|---|
| Expressed (≥1) | 32,413 | 31,845 | 33,553 | 32,170 | 32,481 | 33,460 | 34,692 |
| % of annotated genes of *G. raimondii* | 87.1 | 85.6 | 90.1 | 86.4 | 87.3 | 89.9 | 93.2 |
| Expressed (≥10) | 29,919 | 29,820 | 30,842 | 28,796 | 29,041 | 29,603 | 31,114 |
| % of annotated genes of *G. raimondii* | 80.4 | 80.1 | 82.9 | 77.4 | 78.0 | 79.5 | 83.6 |

**A**

**B**

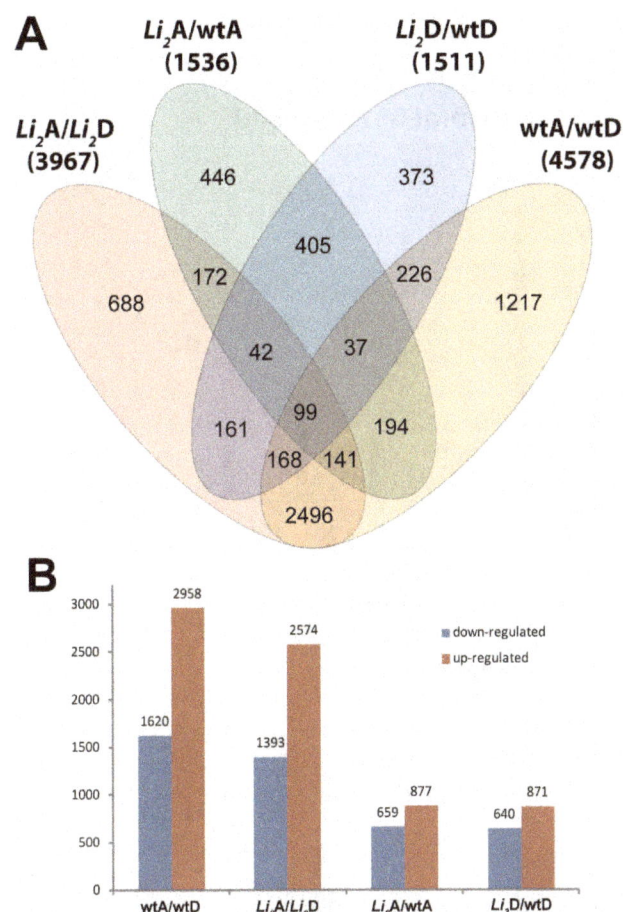

**Figure 1. Overview of significantly regulated genes in developing fiber of $Li_2$ NILs across all comparisons.** (A)Venn diagram of regulated genes in $Li_2$ versus wild type in $A_T/D_T$ subgenomes. Total number of significantly regulated genes in each comparison is indicated in parentheses. (B) The chart represents up- and down- regulated genes between subgenomes and fiber type comparisons.

charide degradation (14 genes) and cell wall proteins (9 genes) were the most abundant classes.

## Validation of Illumina RNA-seq Expression and Subgenome Specific Categorization of Reads by RT-qPCR Analysis

To test the reliability of Illumina sequencing and SNP-based categorization of reads to the $A_T$ or $D_T$ subgenome of allotetraploid *G. hirsutum*, RT-qPCR analysis was performed for a subset of 8 genes (selected from Table S1 in File S1) expressed only in WT or $Li_2$ NILs, and for a subset of 11 genes (selected from Tables 3 and 4) that showed subgenome biased expression. Overall, the results of RT-qPCR analysis were consistent with results of RNA-seq analysis for 19 selected genes (Figures S1 and S2). RT-qPCR analysis confirmed silencing or activation of the expression by the $Li_2$ mutation for the subset of 8 genes (Figure S1). Correlation analysis of the expression patterns revealed strong correlations between RNA-seq and RT-qPCR data. In the subset of 11subgenome biased genes, 7 genes showed 100% correlation ($p$-value <0.05) and 4 genes showed 99% correlation ($p$-value > 0.05; Figure S2).

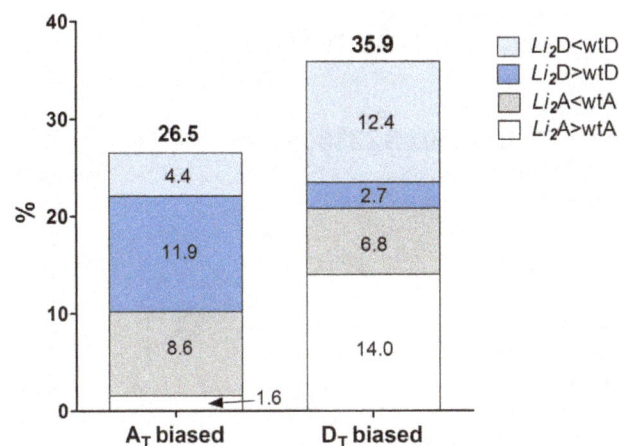

**Figure 2. Mutation effects on $A_T$ and $D_T$ subgenomes of allotetraploid *G. hirsutum*.** The bar chart represents percent of $A_T$ and $D_T$ biased genes regulated in subgenomes of $Li_2$ mutant fiber.

## Discussion

Our results demonstrate that the $A_T$ subgenome in general contributed approximately two times more significantly induced genes to the fiber transcriptome than the $D_T$ subgenome; however, the $Li_2$ mutation had greater effects on the $D_T$ subgenome than the $A_T$ subgenome.

### Global Transcript Changes in Subgenomes of *G. hirsutum* Following $Li_2$ Mutation

The role of the $A_T$ and $D_T$ subgenomes in determination of fiber quality in allotetraploid cotton has been extensively discussed in the literature. Allopolyploidization resulted in significant improvements in the desirable agronomic fiber traits in the allotetraploid species in comparison with the diploid progenitors [17,18]. The first evidences showing that QTLs for fiber quality (including length, strength and fineness) were associated with DNA markers mapped to the $D_T$ subgenome rather than the $A_T$ subgenome was published by Paterson's group [18]. Review of numerous QTLs published from 1998 to 2007 confirmed the observation that the $D_T$ subgenome plays a larger role in genetic control of fiber traits [19]. A microarray study published by Wendel's group found that the homeolog expression in *G. hirsutum* was biased in favor of the $D_T$ subgenome in fiber cells [16]. Similar results were reported by Lacape and coauthors utilizing deep sequencing approach to analyze the fiber transcriptome of two allotetraploid species *G. hirsutum* and *G. barbadense* [20]. From an evolutionary point of view, these observations are surprising since the genes responsible for improved fiber properties evolved in the diploid AA genome before polyploidization [21]. None of the DD genome species produce spinnable fibers [22].

There are discrepancies in the literature regarding homeolog bias in contribution to fiber traits. Using a core set of 111 RFLP markers, Ulloa and coauthors revealed that the $A_T$ subgenome exhibited 68% of QTLs from the five chromosomes, whereas the $D_T$ subgenome exhibited only 32% of QTLs from the three chromosomes [23]. Another study utilizing combinations of markers found more fiber trait QTLs in the $A_T$ subgenome than in the $D_T$ subgenome [24]. The expression analysis of ESTs derived from immature ovules of *G. hirsutum* TM-1 revealed significant enrichment in all functional categories for $A_T$ subgenome ESTs [25].

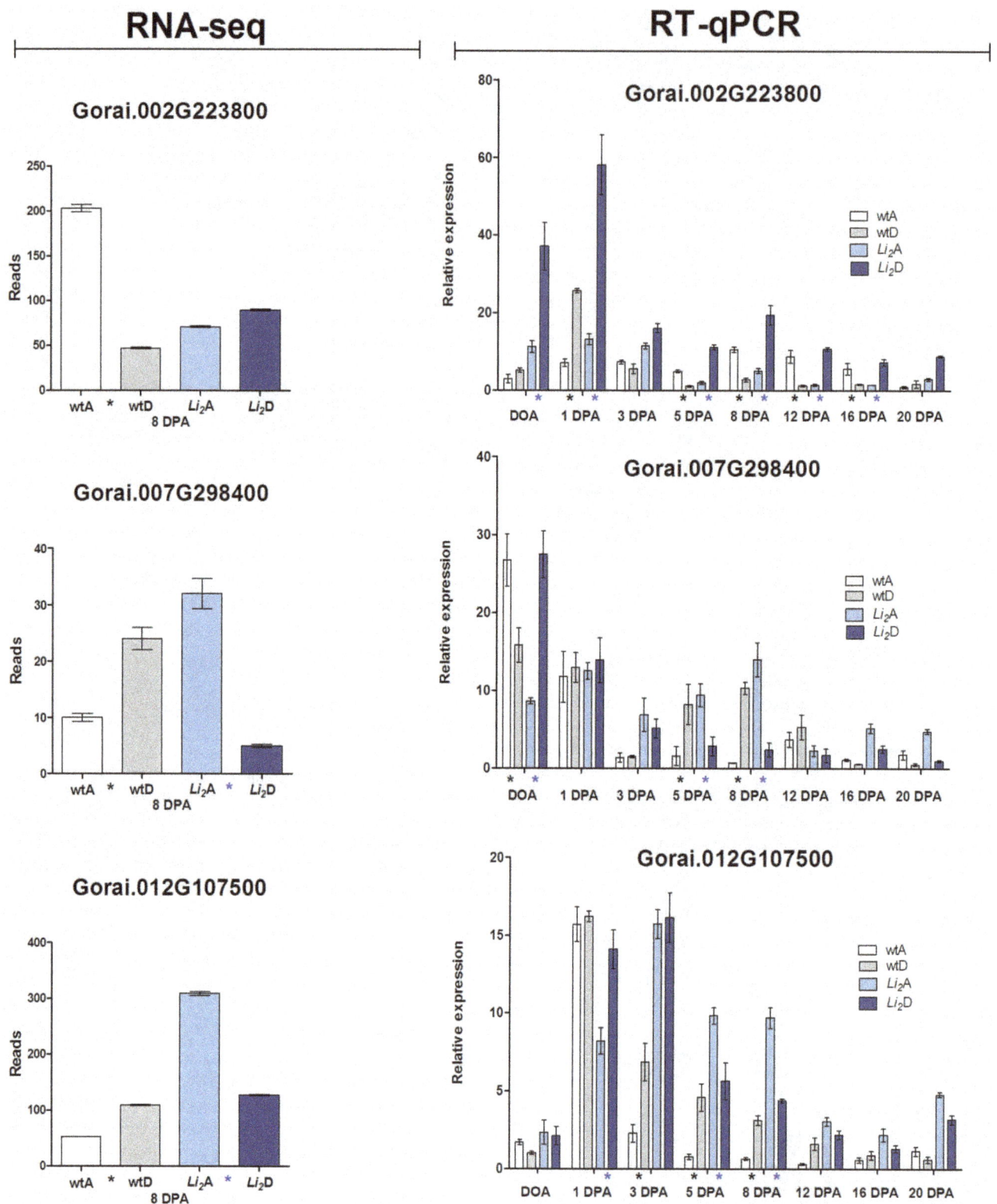

**Figure 3. RNA-seq and rt-qPCR analysis detected reciprocal expression biases as a result of mutation.** Original RNA-seq data are shown on left. Asterisks indicate significant ($p$-value <0.05) difference in gene expression level between $A_T$ vs. $D_T$ subgenomes in wild type (black) and mutant (blue) developing fibers. Error bars represent standard deviation from two biological replicates for RNA-seq data and three biological replicates for RT-qPCR.

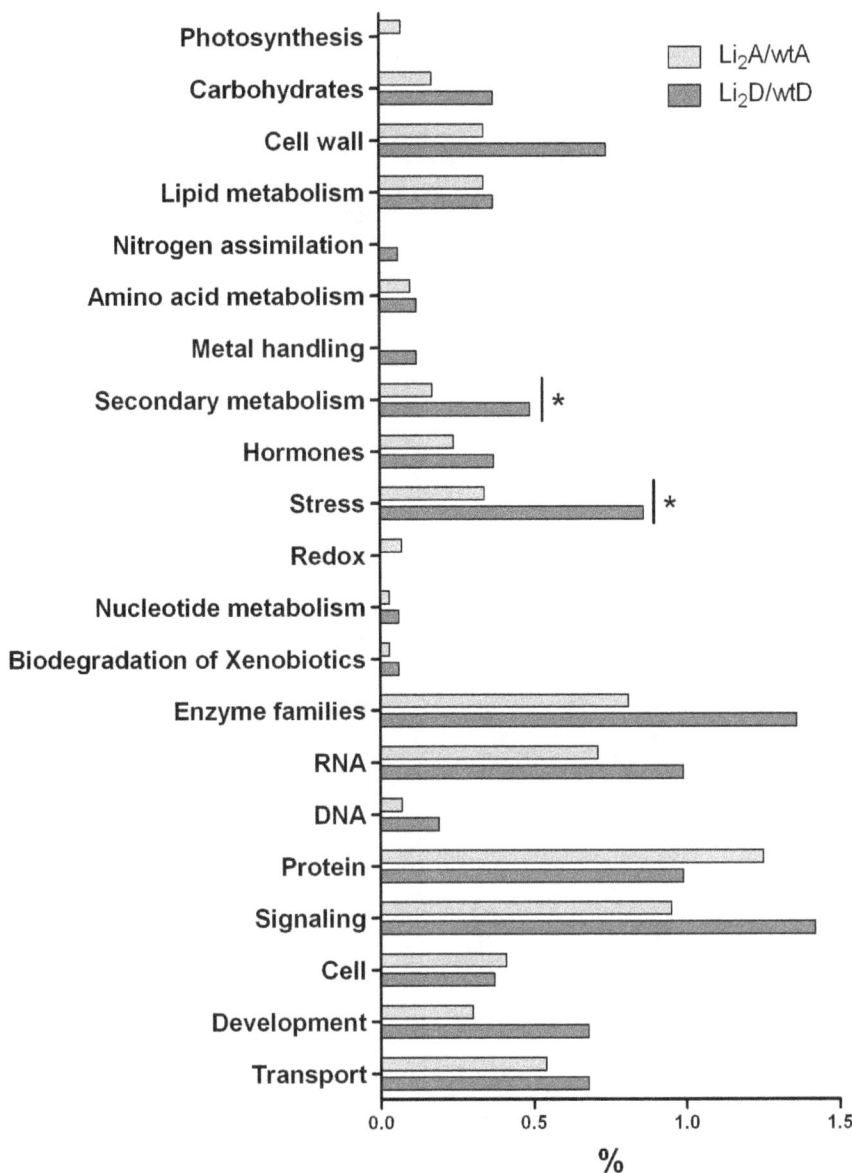

**Figure 4. Mutation effects on functional distribution of homeolog genes during fiber elongation.** Relative gene frequencies in functional categories were represented in percents from amount of biased genes in each subgenome. Asterisks indicate significant (*p*-value <0.05; Fisher's exact test) enrichment of functional category between subgenomes with genes that changed expression in result of mutation. Table S2 in File S1 provides Fisher's exact test results. MapMan BIN structure was used for functional categorization of genes regulated by $Li_2$ mutation. Only functional categories with more than 0.06% gene representation are shown here. Carbohydrates combine 6 BIN classes, including major and minor carbohydrates, glycolysis, fermentation, gluconeogenesis and oxidative pentose phosphate pathway.

These inconsistencies could be explained by technical limitations. The QTL studies reported in the current literature are detecting only a small subset of the genes related to fiber traits that may not cover the whole genome and could be insufficient to conclude which subgenome more significantly contributes to fiber properties [19]. The microarray studies evaluated a limited number of homeologous gene pairs, resulting in limited statistical power [16,26]. Lacape and coauthors used next generation DNA sequencing technology for fiber transcriptome analysis; however, they evaluated only 617,000 good quality reads from four libraries without biological replication [20]. Unlike previous studies we obtained ~160 million reads per sample for each of two biological replicates (Table 1), providing ~5.6 times coverage of the *G. hirsutum* genome (~2.83 Gb per haploid [27]), which is more than

enough to deliver statistically powerful transcriptional analysis. Our observation of higher expression of $A_T$ than $D_T$ genes in the fiber transcriptome is consistent with the results of cotton ovules ESTs analysis [25] and reflects the evolutionary role of the AA diploid progenitor in fiber quality traits of allotetraploid cotton.

It is interesting to note that the $Li_2$ mutation coincides with an increase in the number of expressed genes, but the homeolog expression bias was significantly decreased in $Li_2$ fiber. How expression of homeologous genes is regulated in polyploids is still unclear, although it could involve altered regulatory interactions and rapid genetic and epigenetic changes in subgenomes [28]. The evolution homeolog-specific expression after polyploidization has been extensively studied in allotetraploid cotton. Higher rates of homeolog expression bias in natural allotetraploids than in hybrid

**Table 3.** Subgenome biased transcriptional regulators affected by mutation.

| Class | Identifier | TAIR10 best hit | TAIR 10 symbol | $A_T$ biased (wtA/wtD) | $Li_2$A/wtA | $D_T$ biased (wtD/wtA) | $Li_2$D/wtD |
|---|---|---|---|---|---|---|---|
| AP2/EREBP | Gorai.006G222600 | AT4G13620 | | **3.78** | **0.35** | **0.26** | 1.47 |
| AP2/EREBP | Gorai.007G010200 | AT5G52020 | | **2.22** | **0.28** | **0.45** | 0.68 |
| bHLH | Gorai.007G005700 | AT1G73830 | BEE3 | **0.14** | 1.39 | **7.36** | **0.37** |
| bHLH | Gorai.006G115500 | AT2G27230 | LHW | **2.71** | **2.45** | **0.37** | **2.87** |
| bHLH | Gorai.007G157700 | AT3G07340 | | **5.10** | **0.50** | **0.20** | 1.77 |
| bHLH | Gorai.001G275600 | AT3G26744 | ICE1,SCRM | **5.82** | **0.22** | **0.17** | 0.64 |
| bHLH | Gorai.009G219100 | AT4G33880 | RSL2 | **0.37** | 1.53 | **2.68** | **0.31** |
| C2C2(Zn) CO-like | Gorai.009G065600 | AT5G24930 | COL4 | **2.38** | **2.17** | **0.42** | **2.23** |
| C2H2(Zn) | Gorai.009G092800 | AT2G29660 | | **3.56** | **0.40** | **0.28** | 0.77 |
| C3H(Zn) | Gorai.011G126400 | AT5G51980 | | **31.56** | **0.44** | **0.03** | 1.88 |
| CPP(Zn) | Gorai.013G121300 | AT3G22780 | TSO1 | **0.33** | 1.27 | **2.99** | **0.38** |
| G2-like | Gorai.008G224400 | AT1G32240 | KAN2 | **0.31** | 0.93 | **3.20** | **0.41** |
| Homeobox | Gorai.003G094500 | AT3G01470 | HAT5,HB-1 | **0.27** | 0.76 | **3.76** | **0.47** |
| MYB | Gorai.006G195700 | AT4G22680 | MYB85 | **12.21** | **0.31** | **0.08** | 0.56 |
| MYB | Gorai.001G015200 | AT5G35550 | MYB123,TT2 | **3.05** | **2.13** | **0.33** | **2.08** |
| MYB | Gorai.007G037100 | AT5G62470 | MYB96 | **4.72** | **2.30** | **0.21** | **5.78** |
| WRKY | Gorai.009G066900 | AT4G31550 | WRKY11 | **6.87** | **0.44** | **0.15** | 0.77 |
| bZIP | Gorai.013G258300 | AT3G44460 | DPBF2 | **0.27** | 0.78 | **3.68** | **0.37** |
| AS2, LOB | Gorai.013G016900 | AT3G11090 | LBD21 | **0.29** | 1.67 | **3.48** | **0.18** |
| Aux/IAA | Gorai.009G132300 | AT3G15540 | IAA19,MSG2 | **0.33** | 0.97 | **3.07** | **0.41** |
| Aux/IAA | Gorai.010G227800 | AT3G15540 | IAA19,MSG2 | **0.10** | 0.20 | **9.65** | **0.21** |
| Aux/IAA | Gorai.001G242900 | AT3G23050 | AXR2,IAA7 | **0.24** | **0.15** | **4.23** | **0.06** |
| Aux/IAA | Gorai.006G246000 | AT4G14550 | IAA14,SLR | **4.08** | **0.28** | **0.24** | **0.43** |
| Aux/IAA | Gorai.001G043900 | AT4G32280 | IAA29 | **2.39** | **2.20** | **0.42** | **2.60** |
| Aux/IAA | Gorai.007G277000 | AT5G43700 | IAA4 | **0.19** | **0.32** | **5.28** | **0.25** |
| General | Gorai.011G240900 | AT3G52270 | | **3.94** | **0.40** | **0.25** | 0.40 |
| GATA type (Zn) | Gorai.005G230900 | AT1G10200 | WLIM1 | **3.86** | **0.30** | **0.26** | 0.62 |
| TAZ domain | Gorai.007G056400 | AT5G67480 | BT4 | **0.09** | 0.99 | **11.08** | **0.19** |
| GRP | Gorai.008G029100 | AT4G39260 | GRP8 | **3.23** | **0.47** | **0.31** | 0.74 |
| unclassified | Gorai.013G029900 | AT2G01818 | | **3.23** | **0.26** | **0.31** | **0.39** |
| unclassified | Gorai.013G258400 | AT2G27580 | | **2.85** | **0.38** | **0.35** | 0.82 |
| unclassified | Gorai.012G146800 | AT3G18870 | | **0.47** | 1.31 | **2.14** | **0.45** |
| unclassified | Gorai.008G221500 | AT4G19050 | | **2.08** | **2.23** | **0.48** | **2.01** |
| unclassified | Gorai.010G120300 | AT5G10770 | | **0.06** | 1.17 | **17.63** | **0.36** |
| unclassified | Gorai.004G118800 | AT5G23750 | | **2.79** | **0.48** | **0.36** | **0.47** |

The genes significantly (FDR corrected $p$-value <0.05) down-regulated or up-regulated more than 2-fold are shown in boldface and underlined.

and synthetic polyploid cottons suggested that the extent of homeolog expression bias increases over time from hybridization through evolution [26,29,30]. The $Li_2$ mutation is negative for desirable fiber quality traits, resulting in extremely short lint fiber. Significant reduction of homeolog expression bias in short fiber suggests that the extent of homeolog expression bias is also important for fiber quality characteristics.

We observed a reciprocal switch for some genes in expression bias between homeologs during fiber developmental stages in the mutant. A high degree of expression differences between homeologous genes that are developmentally and stress regulated was reported in cotton [17,31,32]. A high-resolution genome-specific study of expression profiling for 63 gene pairs in 24 tissues in

allopolyploid and their diploid progenitor cotton species demonstrated that the majority of expression differences between homeologs are caused by cis-regulatory divergence between the diploid progenitors; however, some degree of transcriptional neofunctionalization was detected as well [32].

The $Li_2$ mutation was mapped to the $D_T$ subgenome [6–8]; however, the mutated gene and the nature of mutation are currently unknown. The greater mutation effect on the $D_T$ than on the $A_T$ subgenome observed here suggests two possible mechanisms. The network of regulatory interactions may have been interrupted by a mutation in the $D_T$ subgenome resulting in transactivation or repression of individual gene expression levels and expression cascades. Alternatively an epigenetic modulation

**Table 4.** Subgenome biased genes encoding enzyme involved in cell wall biosynthesis changed expression level as a result of $Li_2$ mutation.

| Description | Identifier | TAIR10 best hit | $A_T$ biased (wtA/wtD) | $Li_2$A/wtA | $D_T$ biased wtD/wtA | $Li_2$D/wtD |
|---|---|---|---|---|---|---|
| **Hemicellulose synthesis** | | | | | | |
| Exostosin family protein | Gorai.011G272700 | AT3G45400 | **4.82** | 1.16 | **0.21** | **2.03** |
| Xyloglucan β-galactosyltransferase; MUR3 | Gorai.007G150100 | AT2G20370 | **3.36** | **0.41** | **0.30** | 0.97 |
| Xylosyltransferase; IRX9 | Gorai.006G168500 | AT2G37090 | **0.28** | 0.64 | **3.63** | **0.42** |
| **Cell wall proteins** | | | | | | |
| Fasciclin-like arabinogalactan 7 | Gorai.001G219000 | AT2G04780 | **9.65** | **0.41** | **0.10** | 1.72 |
| Arabinogalactan protein 16 | Gorai.007G025300 | AT2G46330 | **3.07** | **0.42** | **0.33** | **0.44** |
| Fasciclin-like arabinogalactan | Gorai.003G132100 | AT3G46550 | **0.05** | 0.25 | **20.68** | **0.43** |
| Fasciclin-like arabinogalactan 10 | Gorai.007G092300 | AT3G60900 | **0.09** | 2.30 | **11.31** | **2.30** |
| Arabinogalactan protein 18 | Gorai.013G048900 | AT4G37450 | **3.01** | 0.91 | **0.33** | **2.51** |
| Fasciclin-like arabinogalactan 17 | Gorai.006G147500 | AT5G06390 | **2.08** | **0.42** | **0.48** | 0.59 |
| Fasciclin-like arabinogalactan 1 | Gorai.013G152900 | AT5G55730 | **3.71** | **2.62** | **0.27** | 1.65 |
| Proline-rich protein 2 | Gorai.002G193500 | AT2G21140 | **0.23** | 0.89 | **4.35** | **0.37** |
| Reversibly glycosylated polypeptide 1 | Gorai.001G090200 | AT3G02230 | **0.33** | **2.04** | **3.07** | 1.99 |
| **Degradation** | | | | | | |
| Peptidoglycan-binding LysM domain-containing protein | Gorai.008G281400 | AT5G62150 | **19.29** | **0.35** | **0.05** | 1.68 |
| Glycosyl hydrolase 9B8 | Gorai.007G170600 | AT2G32990 | **0.10** | 1.57 | **10.48** | **0.37** |
| Glycosyl hydrolase | Gorai.006G118100 | AT5G20950 | **0.17** | 1.33 | **6.02** | **0.49** |
| α-L-fucosidase 1 | Gorai.008G282700 | AT2G28100 | **4.69** | 0.86 | **0.21** | **2.36** |
| β-Xylosidase 1 | Gorai.011G198200 | AT5G49360 | **2.01** | 1.88 | **0.50** | **2.25** |
| β-Xylosidase 2 | Gorai.008G140600 | AT1G02640 | **0.31** | 0.71 | **3.18** | **0.22** |
| Rhamnogalacturonate lyase | Gorai.002G138200 | AT1G09890 | **3.63** | 1.60 | **0.28** | **3.51** |
| Rhamnogalacturonate lyase | Gorai.007G231600 | AT2G22620 | **0.40** | 2.06 | **2.53** | **2.33** |
| Rhamnogalacturonate lyase | Gorai.008G089300 | AT2G22620 | **11.16** | **0.20** | **0.09** | **6.32** |
| Polygalacturonase 2 | Gorai.002G161400 | AT1G70370 | **0.35** | **0.46** | **2.85** | **0.38** |
| Polygalacturonase 2 | Gorai.005G136300 | AT1G70370 | **5.03** | **0.39** | **0.20** | 1.55 |
| Pectin lyase-like | Gorai.004G138900 | AT4G23500 | **0.36** | 1.48 | **2.81** | **2.50** |
| Pectin methylesterase 1 | Gorai.009G147800 | AT1G53840 | **11.39** | **2.27** | **0.09** | 3.18 |
| Pectinacetylesterase | Gorai.004G176200 | AT4G19410 | **0.38** | **2.11** | **2.66** | 1.89 |
| **Modification** | | | | | | |
| Expansin A4 | Gorai.003G131000 | AT2G39700 | **2.97** | 0.51 | **0.34** | **0.48** |
| Expansin A4 | Gorai.007G376300 | AT2G39700 | **2.07** | 1.66 | **0.48** | **2.27** |
| Expansin A4 | Gorai.012G104400 | AT2G39700 | **2.10** | 1.56 | **0.48** | **2.50** |
| Expansin A8 | Gorai.012G014400 | AT2G40610 | **0.24** | **0.32** | **4.11** | **0.23** |
| Xyloglucan endotransglucosylase/hydrolase | Gorai.004G030500 | AT4G25810 | **4.32** | **0.40** | **0.23** | 0.43 |
| Xyloglucan endotransglucosylase/hydrolase, GhXTH1 | Gorai.007G057400 | AT4G37800 | **0.16** | 0.97 | **6.06** | **0.41** |

The genes significantly (FDR corrected $p$-value <0.05) down-regulated or up- regulated more than 2-fold are shown in boldface and underlined.

may preferentially target the $D_T$ subgenome. It has been shown that small RNAs can control gene expression and epigenetic regulation in response to hybridization [33–36]. For example, miRNAs in allopolyploid Arabidopsis triggered unequal degradation of parental target genes [33]. Similarly, in rice hybrids small RNA populations inherited from parents were responsible for biased expression [36]. Additional investigations of epigenetic and chromatin level modifications will provide insights into causes of gene expression variation between subgenomes.

## TRs and Cell Wall Functional Categories of Genes Regulated by $Li_2$ Mutation

Previous transcriptomics and metabolomics studies have shown that the $Li_2$ mutation terminated the cotton fiber elongation process [6,10]; therefore, genes with changed expression level in the mutant could be involved in elongation. In the present work, we described in detail TRs and cell wall functional categories, which are critical for fiber developmental processes. Many genes in this list (Table 3 and Table 4) were not functionally characterized in cotton; although, based on sequence similarity to genes characterized in Arabidopsis, they could be involved with fiber elongation and represent candidates for further functional analysis in cotton.

**Transcriptional regulators.** Many TRs regulated by $Li_2$ mutation are involved in hormonal signaling and development. Particularly, two AP2/EREBPs and six Aux/IAAs were in the pool of TRs affected by $Li_2$ mutation (Table 3). Plant hormones are important for fiber development. It is well documented that exogenous applications of auxins and gibberellic acid stimulate the differentiation of fibers and promote elongation, while abscisic acid and cytokinins inhibit fiber growth in an *in vitro* cotton ovule culture system [37,38]. Among the auxin responsive genes, Gorai.009G132300 and Gorai.010G227800, whose transcript abundances were significantly reduced in the $D_T$ genome of $Li_2$, showed sequence similarity to grapevine VvIAA19 regulator [39]. Transgenic Arabidopsis plants over expressing VvIAA19 exhibited faster growth, including root elongation and floral transition, than the control, suggesting that grape Aux/IAA19 protein is likely to play a crucial role as a plant growth regulator. In the group of bHLH family of TRs, Gorai.007G005700 transcript abundance was significantly reduced in the $D_T$ genome of $Li_2$ and showed sequence similarity to Arabidopsis *BEE3*, one of several redundant positive regulators of brassinosteroids signaling required for normal growth and development [40].

The actin cytoskeleton plays an important role in cell morphogenesis; down-regulation of *GhACT1* disrupted the actin cytoskeleton network in fibers that resulted in inhibition of fiber elongation [41]. A GATA type TR Gorai.005G230900, a homolog of Arabidopsis *WLIM1*, was down-regulated in the $D_T$ subgenome of $Li_2$; a recent study revealed that plant LIM-domain containing proteins (LIMs) define a highly specialized actin binding protein family, which contributes to the regulation of actin bundling in virtually all plant cells [42].

**Cell wall.** The plant cell wall has a dual role during elongation: to sustain the large mechanical forces caused by cell turgor and to permit controlled polymer extension generating more space for protoplast enlargement [43]. The active biosynthesis of matrix polysaccharides along with increased activity of cell wall loosening enzymes has been considered to be associated with cell wall extension [44–48]. Expression levels of genes encoding enzymes involved in xyloglucan and glucuronoxylan biosynthesis were decreased as a result of $Li_2$ mutation. Particularly, xyloglucan β-galactosyltransferase (Arabidopsis homolog, *MUR3* [49]) and xylosyltransferase (*IRX9* [50]) were down-regulated in the $A_T$ or $D_T$ subgenomes of mutant fibers (Table 4).

Among cell wall proteins arabinogalactans were the most abundant members. Arabinogalactan-proteins have been implicated in many processes involved in plant growth and development, including cell expansion [51,52].

Primary cell wall expansibility and strength is in part mediated by a group of enzymes that comprise a large family of cell wall modifying proteins, the xyloglucan endotransglycosylase/hydrolases (XTHs). XTHs are apoplast-localized enzymes that cleave and reattach xyloglucan polymers [53,54]. The role of XTHs in cotton fiber elongation has been demonstrated: transgenic over-expression of *GhXTH1* in cotton increased fiber length up to 20% [55]. $D_T$ biased Gorai.007G057400 corresponding to *GhXTH1* was down-regulated in mutant fiber.

## Conclusion

Repeated polyploidization over evolutionary time has played a significant role in adding genetic variation to the genomes of plant species. The evolution of the homeolog expression after polyploidization has been extensively studied in cotton comparing expression profiling between parental diploids and natural and synthetic allopolyploid species. This is the first report that explored the effects of a single mutation on the homeolog expression of allotetraploid cotton. Our results showed that significant reduction of the homeolog expression bias in mutant fiber correlates with negative fiber traits, indicating that the extent of homeolog expression bias is important for fiber quality characteristics. In addition, we observed significantly greater mutation effects on the $D_T$ than on the $A_T$ subgenome that might be explained by localization of the mutated gene. Additional studies using numerous naturally occurring cotton fiber mutations are needed to confirm these observations. This work will lead to an understanding of how gene regulation between $A_T$ and $D_T$ homeologs contributes to enhanced fiber morphology in cultivated cotton allopolyploids.

## Materials and Methods

### Plant Material and RNA Isolation

The cotton short fiber mutant $Li_2$ was developed as a near-isogenic line (NIL) with the WT upland cotton line DP5690 as described before [6]. Growth conditions and fiber sampling were previously described [6]. Cotton bolls were harvested at the following time-points during development: day of anthesis (DOA), 1, 3, 5, 8, 12, 16, and 20 days post-anthesis (DPA). Cotton fibers were isolated from developing ovules using a glass bead shearing technique to separate fibers from the ovules [56]. Total RNA was isolated from detached fibers using the Sigma Spectrum Plant Total RNA Kit (Sigma-Aldrich, St. Louis, MO) with the optional on column DNase1 digestion according to the manufacturer's protocol. The concentration of each RNA sample was determined using a NanoDrop 2000 spectrophotometer (NanoDrop Technologies Inc., Wilmington, DE). The RNA quality for each sample was determined by RNA integrity number (RIN) using an Agilent Bioanalyzer 2100 and the RNA 6000 Nano Kit Chip (Agilent Technologies Inc., Santa Clara, CA) with 250 ng of total RNA per sample.

### RT-qPCR Analysis

The experimental procedures and data analysis related to RT-qPCR were performed according to the Minimum Information for Publication of Quantitative Real-Time PCR Experiments (MIQE) guidelines [57]. Eight fiber developmental time-points mentioned above were used for RT-qPCR analyses of homeolog pairs which showed reciprocal expression biases. Only one time point, 8 DPA, was used for RT-qPCR confirmation of RNA-seq data of selected genes. The detailed description of reverse transcription, qPCR and calculation were previously reported [6]. Single nucleotide polymorphisms that distinguish the $A_T$ and $D_T$ subgenome copies of the selected genes were identified by aligning reads from the RNA-seq data to the *G. raimondii* reference mRNA sequences [11]. These homeologous SNPs were used to design subgenome specific primers by the SNAPER approach, whereby an additional mismatch is included near the end of the SNP-specific primers to increase stringency [58]. Primer sequences are provided in Table S3 and Table S4 in File S1. Correlations of biased expression patterns between RNA-seq and RT-qPCR data were calculated using GraphPad Prism 5 software (Pearson test).

## Library Preparation and Sequencing

RNA samples from $Li_2$ and wild type cotton fiber at 8 DPA (in two biological replicates) were subjected to paired-end Illumina mRNA sequencing (RNA-seq). Library preparation and sequencing were conducted by Data2Bio LLC (2079 Roy J. Carver Co-Laboratory, Ames, Iowa). Indexed libraries were prepared using the Illumina protocol outlined in the TruSeq RNA Sample Prep Guide (Part# 15008136 Rev. A, November 2010). The library size and concentration were determined using an Agilent Bioanalyzer. The indexed libraries were combined and seeded onto one lane of the flowcell. The libraries were sequenced using 101cycles of chemistry and imaging, resulting in paired end (PE) sequencing reads with length of $2 \times 101$ bp. The raw reads were submitted to the Sequence Read Archive (accession number SRP026301).

## Processing of Illumina RNA-Seq Reads and Mapping to $A_T$ and $D_T$ Subgenomes of *Gossypium hirsutum*

The reads were trimmed with SICKLE (https://github.com/najoshi/sickle) using a quality score cutoff of 20. Mapping the reads (in pairs where both reads of a pair passed trimming) to the 13 chromosomes of the *G. raimondii* genome $D_5$ v2 reference sequence was performed using GSNAP [59]. Default parameters were used, but with the flags "-n 1 -Q" which means that only a single mapping was reported for each read, and reads with multiple equally good hits were thrown away rather than randomly mapped. We used a cotton SNP index generated between DD genome *G. raimondii* and the AA genome *G. arboreum* to categorize reads of the allotetraploid *G. hirsutum* as belonging to the $A_T$ or $D_T$ subgenomes according to the method reported previously [14].

## Digital Gene Expression Analysis

The comparison of the number of reads mapped to the genes of *G. raimondii* reference genome was used as an indicator of the relative digital gene expression (DGE). The JMP/Genomics 6.0 (SAS, Cary, NC, USA) was used for data normalization and statistical analysis. The data was normalized using TMM (Trimmed Mean of M component) method [60]. Genes with less than 10 reads in one sample were removed before normalization; from 37,223 genes assigned to chromosomes, 31,114 genes passed filtering conditions and were processed for normalization. The ANOVA process was fit to the normalized data, with the data following a Poisson distribution. This was accomplished with a generalized linear mixed model for each gene: $Y_{ij} = T_i + G_j + TG_{ij} + E_{ijk}$, where $T$ is the treatment effect for the $i$th biological treatment ($Li_2$ or wild-type fiber), $G$ is the specific subgenome type effect for the $j$th subgenome type ($A_T$, $D_T$, X and N categorized reads), their interaction ($TG$), and the error term ($E$). The linear model was used to test the null hypothesis that expression of a given gene was not different. Specifically, multiple comparisons were made between fiber type ($Li_2$ versus wild type) and $A_T/D_T$ subgenomes as well as combinations of these factors, such as fiber type in $A_T$ and $D_T$ subgenomes. We identified genes for which the difference in expression levels within these *a priori* questions were significantly different (false discovery rate≤0.05) [61].

## Functional Categorization of Genes

Functional categorization of genes was performed using MapMan ontology [62]; the MapMan mapping for *G. raimondii* is available at http://mapman.gabipd.org/. Fisher's exact test was used to estimate enrichment or depletion relative to background of functional categories with differentially regulated genes.

## Supporting Information

**Figure S1   RT-qPCR confirmation of silencing or activation of genes as a result of mutation.** Bar charts represent RNA-seq and RT-qPCR data (side by side) at 8 DPA of fiber development for 8 randomly selected genes from Table S1 in File S1. Error bars indicate standard deviation from two biological replicates for RNA-seq data and three biological replicates for RT-qPCR.

**Figure S2   RT-qPCR confirmation of biased expression of homeolog pairs.** Bar charts represent RNA-seq and RT-qPCR data (side by side) at 8 DPA of fiber development for 11 randomly selected genes from Table 3 and Table 4. Pearson correlation (GraphPad Prism 5 software) of expression patterns for selected genes between RNA-seq and RT-qPCR data is provided in the table; correlation coefficients with p-value less than 0.05 are shown in boldface and underlined. Error bars indicate standard deviation from two biological replicates for RNA-seq data and three biological replicates for RT-qPCR.

**File S1   Supporting tables. Table S1.** Silencing or activation of genes as a result of mutation. **Table S2.** Mutation effects on functional distribution of homeolog genes. Fisher's exact test results. **Table S3.** Primer's sequences for detection expression of homeolog pairs. **Table S4.** Primer's sequences.

**Data S1   Statistical data for significantly regulated genes.**

**Data S2   $A_T/D_T$ biased genes annotated by MapMan ontology.**

## Acknowledgments

We thank Data2Bio for sequencing service. Mention of trade names or commercial products in this publication is solely for the purpose of providing specific information and does not imply recommendation or endorsement by the U.S. Department of Agriculture that is an equal opportunity provider and employer.

## Author Contributions

Conceived and designed the experiments: MN DDF. Performed the experiments: MN DJH CF. Analyzed the data: MN GT KMY JTP JAU. Contributed reagents/materials/analysis tools: DJH JTP JAU. Wrote the paper: MN GT DDF DJH JTP JAU. Read and approved the final manuscript: MN GT DDF DJH CF KMY JTP JAU.

## References

1. Wendel J, Cronn RC (2002) Polyploidy and the evolutionary history of cotton. Advances in Agronomy 78: 139–186.
2. Wendel JF (1989) New World tetraploid cottons contain Old World cytoplasm. Proceedings of the National Academy of Sciences of the United States of America 86: 4132–4136.
3. Basra AS, Malik AC (1984) Development of the cotton fiber. International Review of Cytology 89: 65–113.
4. Kim HJ, Triplett BA (2001) Cotton fiber growth in planta and in vitro. Models for plant cell elongation and cell wall biogenesis. Plant physiology 127: 1361–1366.

5. Narbuth EV, Kohel RJ (1990) Inheritance and Linkage Analysis of a New Fiber Mutant in Cotton. Journal of Heredity 81: 131–133.

6. Hinchliffe DJ, Turley RB, Naoumkina M, Kim HJ, Tang Y, et al. (2011) A combined functional and structural genomics approach identified an EST-SSR marker with complete linkage to the Ligon lintless-2 genetic locus in cotton (*Gossypium hirsutum* L.). BMC genomics 12: 445.

7. Kohel RJ, Stelly DM, Yu J (2002) Tests of six cotton (*Gossypium hirsutum* L.) mutants for association with aneuploids. The Journal of Heredity 93: 130–132.

8. Rong J, Pierce GJ, Waghmare VN, Rogers CJ, Desai A, et al. (2005) Genetic mapping and comparative analysis of seven mutants related to seed fiber development in cotton. Theoretical and applied genetics 111: 1137–1146.

9. Kohel RJ, Narbuth EV, Benedict CR (1992) Fiber development of Ligon lintless-2 mutant of cotton. Crop Science 32: 733–735.

10. Naoumkina M, Hinchliffe D, Turley R, Bland J, Fang D (2013) Integrated metabolomics and genomics analysis provides new insights into the fiber elongation process in ligon lintless-2 mutant cotton (*Gossypium hirsutum* L.). BMC genomics 14: 155.

11. Paterson AH, Wendel JF, Gundlach H, Guo H, Jenkins J, et al. (2012) Repeated polyploidization of Gossypium genomes and the evolution of spinnable cotton fibres. Nature 492: 423–427.

12. Wang Z, Gerstein M, Snyder M (2009) RNA-Seq: a revolutionary tool for transcriptomics. Nature reviews Genetics 10: 57–63.

13. Doyle JJ, Flagel LE, Paterson AH, Rapp RA, Soltis DE, et al. (2008) Evolutionary genetics of genome merger and doubling in plants. Annual review of genetics 42: 443–461.

14. Page JT, Gingle AR, Udall JA (2013) PolyCat: a resource for genome categorization of sequencing reads from allopolyploid organisms. G3: Genes|-Genomes|Genetics 3: 517–525.

15. Gilbert MK, Bland JM, Shockey JM, Cao H, Hinchliffe DJ, et al. (2013) A transcript profiling approach reveals an abscisic acid specific glycosyltransferase (UGT73C14) induced in developing fiber of Ligon lintless-2 mutant of cotton (*Gossypium hirsutum* L.). PloS one 8: e75268.

16. Hovav R, Udall JA, Chaudhary B, Rapp R, Flagel L, et al. (2008) Partitioned expression of duplicated genes during development and evolution of a single cell in a polyploid plant. Proceedings of the National Academy of Sciences of the United States of America 105: 6191–6195.

17. Adams KL, Cronn R, Percifield R, Wendel JF (2003) Genes duplicated by polyploidy show unequal contributions to the transcriptome and organ-specific reciprocal silencing. Proceedings of the National Academy of Sciences of the United States of America 100: 4649–4654.

18. Jiang C, Wright RJ, El-Zik KM, Paterson AH (1998) Polyploid formation created unique avenues for response to selection in Gossypium (cotton). Proceedings of the National Academy of Sciences of the United States of America 95: 4419–4424.

19. Chee PW, Campbell BT (2009) Bridging classical and molecular genetics of cotton fiber quality and development. In: Paterson AH, editor. Genetics and genomics of cotton. New York: Springer. 283–311.

20. Lacape JM, Claverie M, Vidal RO, Carazzolle MF, Guimaraes Pereira GA, et al. (2012) Deep sequencing reveals differences in the transcriptional landscapes of fibers from two cultivated species of cotton. PloS one 7: e48855.

21. Brubaker CL, Paterson AH, Wendel JF (1999) Comparative genetic mapping of allotetraploid cotton and its diploid progenitors. Genome 42: 184–203.

22. Kohel R, Yu J, Park Y-H, Lazo G (2001) Molecular mapping and characterization of traits controlling fiber quality in cotton. Euphytica 121: 163–172.

23. Ulloa M, Saha S, Jenkins JN, Meredith WR, McCarty JC, et al. (2005) Chromosomal Assignment of RFLP Linkage Groups Harboring Important QTLs on an Intraspecific Cotton (*Gossypium hirsutum* L.) Joinmap. Journal of Heredity 96: 132–144.

24. Mei M, Syed NH, Gao W, Thaxton PM, Smith CW, et al. (2004) Genetic mapping and QTL analysis of fiber-related traits in cotton (Gossypium). Theoretical and applied genetics 108: 280–291.

25. Yang SS, Cheung F, Lee JJ, Ha M, Wei NE, et al. (2006) Accumulation of genome-specific transcripts, transcription factors and phytohormonal regulators during early stages of fiber cell development in allotetraploid cotton. The Plant journal 47: 761–775.

26. Flagel L, Udall J, Nettleton D, Wendel J (2008) Duplicate gene expression in allopolyploid Gossypium reveals two temporally distinct phases of expression evolution. BMC biology 6: 16.

27. Grover CE, Kim H, Wing RA, Paterson AH, Wendel JF (2004) Incongruent patterns of local and global genome size evolution in cotton. Genome research 14: 1474–1482.

28. Osborn TC, Pires JC, Birchler JA, Auger DL, Chen ZJ, et al. (2003) Understanding mechanisms of novel gene expression in polyploids. Trends in genetics : TIG 19: 141–147.

29. Flagel LE, Wendel JF (2010) Evolutionary rate variation, genomic dominance and duplicate gene expression evolution during allotetraploid cotton speciation. The New phytologist 186: 184–193.

30. Yoo MJ, Szadkowski E, Wendel JF (2013) Homoeolog expression bias and expression level dominance in allopolyploid cotton. Heredity 110: 171–180.

31. Dong S, Adams KL (2011) Differential contributions to the transcriptome of duplicated genes in response to abiotic stresses in natural and synthetic polyploids. The New phytologist 190: 1045–1057.

32. Chaudhary B, Flagel L, Stupar RM, Udall JA, Verma N, et al. (2009) Reciprocal silencing, transcriptional bias and functional divergence of homeologs in polyploid cotton (Gossypium). Genetics 182: 503–517.

33. Ha M, Lu J, Tian L, Ramachandran V, Kasschau KD, et al. (2009) Small RNAs serve as a genetic buffer against genomic shock in Arabidopsis interspecific hybrids and allopolyploids. Proceedings of the National Academy of Sciences of the United States of America 106: 17835–17840.

34. Madlung A, Tyagi AP, Watson B, Jiang H, Kagochi T, et al. (2005) Genomic changes in synthetic Arabidopsis polyploids. The Plant journal 41: 221–230.

35. Salmon A, Ainouche ML, Wendel JF (2005) Genetic and epigenetic consequences of recent hybridization and polyploidy in *Spartina* (Poaceae). Molecular Ecology 14: 1163–1175.

36. He G, Zhu X, Elling AA, Chen L, Wang X, et al. (2010) Global epigenetic and transcriptional trends among two rice subspecies and their reciprocal hybrids. The Plant cell 22: 17–33.

37. Beasley CA, Ting IP (1973) The effects of plant growth substances on in-vitro fiber development from fertilized cotton ovules. American Journal of Botany 60: 130–139.

38. Beasley CA, Ting IP (1974) Effects of plant growth substances on in-vitro fiber development from unfertilized cotton ovules. American Journal of Botany 61: 188–194.

39. Kohno M, Takato H, Horiuchi H, Fujita K, Suzuki S (2012) Auxin-nonresponsive grape Aux/IAA19 is a positive regulator of plant growth. Molecular biology reports 39: 911–917.

40. Friedrichsen DM, Nemhauser J, Muramitsu T, Maloof JN, Alonso J, et al. (2002) Three redundant brassinosteroid early response genes encode putative bHLH transcription factors required for normal growth. Genetics 162: 1445–1456.

41. Li XB, Fan XP, Wang XL, Cai L, Yang WC (2005) The cotton ACTIN1 gene is functionally expressed in fibers and participates in fiber elongation. The Plant cell 17: 859–875.

42. Papuga J, Hoffmann C, Dieterle M, Moes D, Moreau F, et al. (2010) Arabidopsis LIM proteins: a family of actin bundlers with distinct expression patterns and modes of regulation. The Plant cell 22: 3034–3052.

43. Cosgrove DJ (2001) Wall structure and wall loosening. A look backwards and forwards. Plant physiology 125: 131–134.

44. Takeda T, Furuta Y, Awano T, Mizuno K, Mitsuishi Y, et al. (2002) Suppression and acceleration of cell elongation by integration of xyloglucans in pea stem segments. Proceedings of the National Academy of Sciences of the United States of America 99: 9055–9060.

45. Park YW, Baba K, Furuta Y, Iida I, Sameshima K, et al. (2004) Enhancement of growth and cellulose accumulation by overexpression of xyloglucanase in poplar. FEBS letters 564: 183–187.

46. Hayashi T, Delmer DP (1988) Xyloglucan in the cell walls of cotton fiber. Carbohydrate research 181: 273–277.

47. Huwyler HR, Franz G, Meier H (1979) Changes in the composition of cotton fiber cell walls during development. Planta 146: 635–642.

48. Shimizu Y, Aotsuka S, Hasegawa O, Kawada T, Sakuno T, et al. (1997) Changes in levels of mRNAs for cell wall-related enzymes in growing cotton fiber cells. Plant & cell physiology 38: 375–378.

49. Madson M, Dunand C, Li X, Verma R, Vanzin GF, et al. (2003) The MUR3 gene of Arabidopsis encodes a xyloglucan galactosyltransferase that is evolutionarily related to animal exostosins. The Plant cell 15: 1662–1670.

50. Lee C, Zhong R, Ye ZH (2012) Arabidopsis family GT43 members are xylan xylosyltransferases required for the elongation of the xylan backbone. Plant & cell physiology 53: 135–143.

51. Lee KJ, Sakata Y, Mau SL, Pettolino F, Bacic A, et al. (2005) Arabinogalactan proteins are required for apical cell extension in the moss Physcomitrella patens. The Plant cell 17: 3051–3065.

52. Yang J, Showalter AM (2007) Expression and localization of AtAGP18, a lysine-rich arabinogalactan-protein in Arabidopsis. Planta 226: 169–179.

53. Nishitani K, Tominaga R (1992) Endo-xyloglucan transferase, a novel class of glycosyltransferase that catalyzes transfer of a segment of xyloglucan molecule to another xyloglucan molecule. The Journal of biological chemistry 267: 21058–21064.

54. Fry SC, Smith RC, Renwick KF, Martin DJ, Hodge SK, et al. (1992) Xyloglucan endotransglycosylase, a new wall-loosening enzyme activity from plants. The Biochemical journal 282 (Pt 3): 821–828.

55. Lee J, Burns TH, Light G, Sun Y, Fokar M, et al. (2010) Xyloglucan endotransglycosylase/hydrolase genes in cotton and their role in fiber elongation. Planta 232: 1191–1205.

56. Taliercio EW, Boykin D (2007) Analysis of gene expression in cotton fiber initials. BMC plant biology 7: 22.

57. Bustin SA, Benes V, Garson JA, Hellemans J, Huggett J, et al. (2009) The MIQE guidelines: minimum information for publication of quantitative real-time PCR experiments. Clinical chemistry 55: 611–622.

58. Drenkard E, Richter BG, Rozen S, Stutius LM, Angell NA, et al. (2000) A simple procedure for the analysis of single nucleotide polymorphisms facilitates map-based cloning in Arabidopsis. Plant physiology 124: 1483–1492.

59. Wu TD, Nacu S (2010) Fast and SNP-tolerant detection of complex variants and splicing in short reads. Bioinformatics 26: 873–881.

60. Robinson MD, Oshlack A (2010) A scaling normalization method for differential expression analysis of RNA-seq data. Genome Biol 11: R25.

61. Benjamini Y, Hochberg Y (1995) Controlling the false discovery rate: a practical and powerful approach to multiple testing. Journal of the Royal Statistical Society Series B (Methodological) 57: 289–300.

62. Thimm O, Blasing O, Gibon Y, Nagel A, Meyer S, et al. (2004) MAPMAN: a user-driven tool to display genomics data sets onto diagrams of metabolic pathways and other biological processes. The Plant journal 37: 914–939.

# 6

# Ecoinformatics Reveals Effects of Crop Rotational Histories on Cotton Yield

**Matthew H. Meisner[1,2]\*, Jay A. Rosenheim[3]**

1 Department of Evolution and Ecology, University of California Davis, Davis, California, United States of America, 2 Department of Statistics, University of California Davis, Davis, California, United States of America, 3 Department of Entomology and Nematology, University of California Davis, Davis, California, United States of America

## Abstract

Crop rotation has been practiced for centuries in an effort to improve agricultural yield. However, the directions, magnitudes, and mechanisms of the yield effects of various crop rotations remain poorly understood in many systems. In order to better understand how crop rotation influences cotton yield, we used hierarchical Bayesian models to analyze a large ecoinformatics database consisting of records of commercial cotton crops grown in California's San Joaquin Valley. We identified several crops that, when grown in a field the year before a cotton crop, were associated with increased or decreased cotton yield. Furthermore, there was a negative association between the effect of the prior year's crop on June densities of the pest *Lygus hesperus* and the effect of the prior year's crop on cotton yield. This suggested that some crops may enhance *L. hesperus* densities in the surrounding agricultural landscape, because residual *L. hesperus* populations from the previous year cannot continuously inhabit a focal field and attack a subsequent cotton crop. In addition, we found that cotton yield declined approximately 2.4% for each additional year in which cotton was grown consecutively in a field prior to the focal cotton crop. Because *L. hesperus* is quite mobile, the effects of crop rotation on *L. hesperus* would likely not be revealed by small plot experimentation. These results provide an example of how ecoinformatics datasets, which capture the true spatial scale of commercial agriculture, can be used to enhance agricultural productivity.

**Editor:** Raul Narciso Carvalho Guedes, Federal University of Viçosa, Brazil

**Funding:** Funding Sources: 1. California State Support Committee of Cotton Incorporated. URL: http://www.ccgga.org/cotton_research/cssc.htm. 2. University of California Statewide IPM Program. URL: http://www.ipm.ucdavis.edu. 3. USDA-NRICGP (Grant 2006-01761). URL: http://www.csrees.usda.gov/funding/rfas/nri_rfa. html. 4. California Department of Pesticide Regulation. URL: http://www.cdpr.ca.gov. 5. National Science Foundation GRFP (Grant DGE-1148897). URL: http://www. nsfgrfp.org. The funders had no role in study design, data collection and analysis, decision to publish, or preparation of the manuscript.

**Competing Interests:** The authors have declared that no competing interests exist.

\* E-mail: mhmeisner@ucdavis.edu

## Introduction

Maximizing agricultural crop yield is an important goal for several reasons. First, a growing worldwide population will generate increased demand for agricultural resources [1]. Since expanding the land area devoted to agriculture is often unfeasible, or would involve the destruction of sensitive landscapes such as forests and wetlands, the only way to meet this demand will be to increase the crop yield generated from existing farmland. Second, there are substantial economic incentives for profit-seeking farmers to maximize the yield of their crops, especially given the low profit margins typical of commercial agriculture [2].

Farmers make a wide range of decisions regarding the management of their crops, involving pest management, planting/harvest dates, fertilization, irrigation, and, as we focus on in this study, crop rotation. These decisions are, along with external factors that fall outside farmers' control, such as weather, likely to affect crop performance and yield substantially. A rigorous quantitative understanding of the factors, including farmer management decisions, that affect crop yield is an essential prerequisite for developing management strategies that maximize yield.

A critical factor known to affect crop yield in a given field is the crop rotational history of that field [3]. There are several possible mechanisms by which the crops previously grown in a field can

affect crop yield. First, different crops have different effects on the nutrient composition of the soil, so the identities of crops previously grown in a field can affect nutrient availability and crop yield [3]. For example, nitrogen-limited crops can benefit from rotation with nitrogen-fixing legumes [4], and phosphorus nutrition in California cotton is shaped by whether or not the previous crop received phosphorus fertilizer [5]. Second, certain crops may increase the local abundance of particular insect pests and pathogens [6–8]. Since different crops are often susceptible and resistant to different pathogens and pests, the identities of the crops recently grown in a field can affect yield. For example, if one crop increases local abundances of an insect pest that also attacks a second crop, planting the second crop immediately following the first may lead to decreased yield resulting from attack from the built up local pest population. In contrast, such a yield depression could potentially be averted if the second crop were planted following a crop that does not lead to local accumulation of the pest. In monocultures of wheat, substantial yield declines have been noted and attributed to the buildup of the soil-borne fungal pathogen *Gaeumannomyces graminis* [9]. Third, many studies have shown that a field's crop rotational history can strongly affect weed densities [10]. Numerous other mechanistic explanations for the yield effects of crop rotation have also been suggested [3].

Crop rotation has been practiced for thousands of years; evidence for its inception dates back to ancient Roman and Greek societies [11,12]. Experimental studies on the effects of crop rotation first appeared in the early 20th century, revealing that growing crops in rotation led to increased crop yields of up to 100% compared to continuous planting of a single crop [13,14]. Interest in the yield effects of crop rotation waned during the middle of the 20th century, due to the increasing availability of cheap fertilizers, insecticides, and herbicides [3,14]. However, crop rotation continues to be a relevant and important practice; low-input farming remains desirable due to the costs of fertilizers and pesticides, and fertilizer and pesticide applications can often not fully compensate for the benefits afforded by crop rotation [3]. In addition, the significant environmental and public health concerns surrounding fertilizer and pesticide use [1,15] highlight the desirability of methods of increasing crop yield through alternative methods such as crop rotation.

The effects of rotational histories on yield are well understood for some crops, such as corn, where rotation is recognized to be crticial in avoiding the buildup of corn rootworms [16]. However, for many crops, the direction, magnitude, and mechanism of the effect of crop rotational histories on crop yield remain poorly understood [3]. Cotton is one such crop. Experimental field studies of the effect of crop rotation on cotton yield have demonstrated increased cotton yield, compared to continuous cultivation of cotton, when cotton is grown in rotation with sorghum [17,18], corn [19], and wheat [20,21]. Despite these useful results, only a small subset of possible rotations has been studied, experiments have been restricted to plots significantly smaller than typical commercial cotton fields, and mechanisms for these effects remain poorly understood. To help address these limitations, we seek to expand upon this work by exploring the effects of crop rotational histories on yield in commercial cotton fields in California, using an "ecoinformatics" approach [22] capitalizing on existing observational data gathered by growers and professional agricultural pest consultants.

In recent years, there has been a surge in research and interest involving the rapidly emerging field of "big data." The big data movement has been fueled by several developments, including a dramatic increase in the magnitude of data generation, an improved ability to cheaply store, manipulate, and explore massive datasets, and the development of new analytic methods [23]. Most importantly, the movement has been driven by a growing realization that existing data, and data generated as a byproduct of our everyday lives, can be leveraged to explore key questions about nature and human behavior, even if the data were not collected for this purpose [24]. Ecoinformatics is a nascent field focused on harnessing the power of big data to address questions in environmental biology. Ecoinformatics approaches typically involve the analysis of large datasets, the synthesis of diverse data sources, and the analysis of pre-existing, observational datasets [22]. In some commercial agricultural settings, farmers, along with hired consultants, collect a great deal of regular data about their fields that are used to guide real-time crop management decisions, such as the timing of pesticide applications. By capitalizing on data that are already generated as a byproduct of commercial agriculture, ecoinformatics provides a low-cost means of obtaining a large dataset that can be used to explore key questions in agricultural biology, some of which might be too difficult or too costly to explore experimentally. Furthermore, the large size of datasets created for ecoinformatics can afford greater statistical power than could possibly be generated through experimental work.

Experimentally studying the yield effects of crop rotational histories is challenging for several reasons. There are a plethora of possible rotational histories, which means that a large number of treatments would be required to explore the space of possible rotational histories thoroughly. Furthermore, experimentally studying effects of crop rotations requires experiments spanning several growing seasons, which may be logistically challenging. Finally, in order to maintain realism and applicability to commercial fields, which are typically quite large, sizeable experimental plots would be required, especially in light of research suggesting that landscape composition as far as 20 km from a focal field can affect the densities of agricultural pests in that field [25]. While yield effects of non-mobile factors such as soil characteristics may be readily detected through small plot experimentation, the effects of highly mobile arthropods may only be detected at much larger spatial scales.

An ecoinformatics approach offers attractive solutions to these challenges. Since we analyze a large preexisting dataset that includes over a thousand records, a diversity of the possible crop rotational histories already exists in the dataset. In addition, our dataset spans 11 years of data, so the data span the temporal scale necessary to ask questions regarding effects of multi-year rotational histories. And, since the data come from the exact setting where we wish to apply our results, the data are realistic and capture the appropriate spatial scale of commercial agriculture.

First, we sought to identify which crop rotational histories are associated with increased and decreased cotton yield, and to quantify these yield effects. We then explored possible explanations for the yield effects identified in the previous step by examining the associations between crop rotational histories and pest abundance.

## Materials and Methods

### Dataset

The dataset was constructed by collecting existing crop records from commercial cotton fields in California's San Joaquin Valley. The data were shared by both growers and pest control advisors (PCAs), professional consultants hired to monitor field conditions and provide crop management recommendations. The dataset contains records of 1498 unique field-year instances from 566 unique fields, ranging from 1997 to 2008. Growers and PCAs collect and maintain detailed records of the conditions in their fields; numerous variables were recorded for each field-year record, and the following were used in our analyses:

1. Cotton yield. Measured once for each field-year instance, cotton lint yield was measured in bales/acre (converted to kg/ha for our analyses) and recorded for 1240 of the 1498 total records.

2. Crop rotational histories. The identity of the crop grown in the same field in previous growing seasons was recorded. For some fields, records extended back for 10 years. However, the vast majority of fields did not have records extending this far into the past. There were 15 unique crops that appeared in rotational histories: alfalfa, barley, carrots, corn, cotton, garbanzo beans, garlic, lettuce, melons, onions, potatoes, safflower, sugarbeets, tomatoes, and wheat.

3. Surrounding crops. For 1026 of the 1498 crops, we had data on the identity of the crop grown in each of the 8 fields immediately adjoining the focal field (to the North, Northeast, East, Southeast, South, Southwest, West, and Northwest).

4. Cotton variety. The database consisted of records of two different cotton species: *Gossypium barbadense* L. ("Pima cotton") and *Gossypium hirsutum* L. ("upland cotton").

5. *Lygus hesperus* densities. The plant bug *L. hesperus* is one of the most damaging pests of cotton, and a frequent target of

insecticide applications [26,27]. PCAs measured *L. hesperus* densities approximately weekly, primarily during June and July. The PCAs' sampling procedure consisted of 50 swings of a sweep net across the top of the plant canopy. Since not all PCAs sampled on the same days or at exactly the same intervals for all fields, we transformed successive samples into mean *L. hesperus* density estimates by calculating the area under the linear curve of *L. hesperus* density versus time and dividing by the number of days in the sampling interval.

## Modeling approach

We employed a hierarchical Bayesian modeling approach, fitting linear mixed models to explore our questions about the effects of crop rotational histories on cotton yield. Mixed models combine the use of random effects and fixed effects, making them ideally suited for analysis of data that are structured, or clustered, in some known way, such that separate observations from within clusters are expected to be similar to one another [28]. When we model a source of clustering using a random effect, we assume that each cluster-specific parameter was drawn from a common distribution, and we estimate the parameters of this distribution from the data. We use this common distribution as the prior when calculating the posterior distribution of each cluster-specific parameter. The parameters (often called hyperparameters) of the distribution of cluster-specific parameters have posteriors that are estimated from the data, typically after assuming uninformative priors for the hyperparameters [28]. Using a common, empirical prior for all cluster-specific parameters allows pooling of information across clusters, so that data from all clusters can help inform estimates of every other per-cluster parameter. Assuming all clusters are the same introduces high bias and tends to underfit the data, whereas estimating fixed effects for each cluster introduces high variance and tends to overfit the data; however, using a random effect provides an optimal compromise between introducing bias and introducing variance [28]. In this dataset, there are several plausible sources of clustering.

1. First, we expect the data to be clustered by field, since there likely exist field-specific factors that affect yield, such as soil characteristics, local climate, and grower agronomic and pest management practices. We controlled for variable yield potential between fields by including field identity as a random effect in our models. Random effects allow pooling of information across clusters, so they are particularly useful when there are few observations from some clusters - a situation in which it is difficult to accurately estimate each per-cluster parameter with only the data from that one cluster [28]. Since there are three or fewer records for 78% of the fields in our database, we feel that including field as a random effect was preferable to trying to estimate field-specific fixed effects with very few observations per field.

Additionally, including field as a random effect provides a straightforward way to make predictions for fields not represented in our database. Since modeling field as a random effect involves estimating a distribution of per-field parameters, we can simply sample a field-specific parameter from this distribution if we wish to make predictions about a previously unobserved field. Uncertainty in this field-specific parameter can be propagated by simulating many samples from this distribution, while simultaneously accounting for uncertainty in the parameters of this distribution. However, if we were to model field as a fixed effect, we would not estimate a distribution of field-specific parameters. We would only estimate parameters for the specific

fields in our database, leaving us with no obvious way to make inferences about new fields.

2. Second, we expect that our data are clustered by year, since there is substantial between-year variability in climate, particularly in the winter and early spring. Climatic variables can affect crop performance, planting date, and insect pest populations, all of which can in turn affect cotton yield. To control for and quantify variation in yield due to year-specific factors, we included year as a random effect in our models. Our reasons for including year as a random effect are the same as those for field: there are few observations from some years, and we may wish to make predictions for future years not covered by the existing database.

All models were fit using a No-U-Turn Sampler variant of Hamiltonian Markov Chain Monte Carlo [29] implemented in Stan version 1.3.0, accessed through the rstan packing in R [30,31]. We ran three chains from random initializations, each with 10,000 samples, and discarded the first 5,000 samples from each as burn-in. Inferences were based upon the remaining 15,000 samples. We checked convergence by making sure that $\hat{R}$, an estimate of the potential scale reduction of the posterior if sampling were to be infinitely continued, was near 1 [32].

## Models

**Model 1.** To explore the yield effects of the crop grown in the same field the previous year, we fit a linear mixed model with yield as the response variable. The predictor variable of primary interest was the identity of the crop grown in that field the previous year, which was included as a fixed effect.

Given that we are working with an observational dataset, a critical step in order to make meaningful inferences about the variable of primary interest - the crop grown the year before - was to control, to the extent possible, for potentially confounding variables that could generate spurious correlations and taint the validity of our inferences about crop rotation. To control for variable yield potential between fields and years, field and year were included in the model as random effects. The field terms control for the possibility that some fields may have higher yield potential due to their location, soil characteristics, or growing practices; the year terms control for the substantial year-to-year variation in cotton yield, which likely results from yearly weather differences. A term indicating cotton species (Pima or upland) was included in the model to account for yield differences between cotton species. Cotton species was modeled as a fixed effect, since there are only two possible categories - not enough to meaningfully estimate a random effects distribution [28]. We also included 15 real-valued fixed effect predictor variables that indicate the number of fields, out of the 8 surrounding fields, planted with each of the 15 crops we analyzed. The goal was to control for effects of the surrounding landscape, and thereby avoid spurious correlations between rotational history (which may be correlated with the crops surrounding the focal crop) and yield.

Our Bayesian modeling approach required the specification of priors for all parameters whose posteriors were estimated using MCMC. Noninformative priors (normal distributions with mean 0 and standard deviation of 100) were used for all fixed effects. The random effects for both field and year were assumed to follow a normal distribution with mean 0 (allowing means of these distributions to be estimated from the data would lead to nonidentifiability with the fixed effects for prior crop identity) and variance hyperparameters estimated from the data. Since the support of variance parameters is constrained to positive real numbers, noninformative inverse gamma distributions with shape and scale parameters set to 0.001 were used as the prior for the variance parameter of the top-level stochastic node, and as the priors for the variance hyperparameters of the field and year random effects distributions.

**Model 2.** To help us understand whether any effects of the crop grown in the field the previous year on cotton yield could be due to effects on *L. hesperus*, we fit the same model as Model 1, but with average June *L. hesperus* abundance as the response variable.

**Model 3.** Next, to formally assess whether there was an association between the effects of crop rotation on yield and the effects of crop rotation on *L. hesperus* abundance, we performed a linear regression of the estimated effects on yield (measured as the posterior means from Model 1) against the estimated effects on *L. hesperus* abundance (measured as posterior means from Model 2). Noninformative $\mathcal{N}(0,100^2)$ priors were used for the mean and intercept, and a noninformative inverse gamma distribution with shape and scale parameters set to 0.001 was used as the prior for the variance.

**Model 4.** A great deal of experimental evidence has demonstrated that crop rotation leads to increased yield compared to successive plantings of a single crop [3]; therefore, we explored whether or not a yield loss was incurred by cotton crops grown in fields where cotton was grown in previous years. For the 782 fields that had complete crop rotational records for the previous 4 years, we calculated the number of consecutive cotton plantings (from 1 to 4) in the 4 years preceding the focal cotton crop. We then fit a model, with yield as the response variable, using the number of consecutive prior cotton plantings as a predictor (again with the same noninformative prior of $\mathcal{N}(0,100^2)$). Field, year, and cotton type were included as they were in Models 1 and 2. Since the number of prior consecutive cotton plantings could be correlated with the number of cotton fields in the surrounding landscape during the focal year, we avoided a possible spurious correlation between consecutive cotton plantings and yield by also including a fixed effect for the number of cotton fields in the 8 fields adjacent to the focal field. We chose not to explore rotational histories of specific crops (and instead just grouped all crops into "cotton" or "not cotton") for longer than one previous year, since the number of possible rotational histories becomes very large and the number of records for each possible history becomes too small to allow for robust statistical analysis.

**Model 5.** To see if the number of consecutive years of cotton cultivation preceding the focal year was associated with June *L. hesperus* densities, we fit the same model as Model 4, but with June *L. hesperus* as the response variable.

## Results

### Model 1

Using our samples from the joint posterior of Model 1, we calculated, for each crop other than cotton, the posterior distribution of the difference in mean cotton yield in fields where that crop was grown the year before compared to mean yield in fields where cotton was grown the year before. The posterior means of these comparisons, as well as 95% highest posterior density intervals (HPDIs), are displayed in Figure 1A. Highest posterior density intervals are a Bayesian analogue of frequentist confidence intervals; they denote the narrowest region of parameter space containing 95% of the posterior probability [28]. Three crops had 95% HPDIs that did not overlap 0. Garlic (lower limit = 42.0 kg/ha, upper limit = 213.7 kg/ha), tomatoes (57.9 kg/ha 178.1 kg/ha), and melons (92.9 kg/ha, 793.7 kg/ha) had entirely positive 95% HPDIs, suggesting that previous

cultivation of these crops was associated with increased cotton yield. While no crops had entirely negative 95% HPDIs, the posterior probability of safflower and sugarbeets having negative effects on yield was 96% and 95%, respectively, suggesting that cultivation of these crops the previous year was associated with decreased cotton yield. Yield was 153.0 kg/ha higher, with a 95% HPDI of (115.0 kg/ha, 192.8 kg/ha), for upland cotton than for Pima cotton.

### Model 2

Using the joint posterior of Model 2, we calculated the posterior distribution of the difference in mean June *L. hesperus* densities between fields where cotton was the year grown before and where other specific crops were grown the year before. The posterior means of these comparisons, as well as 95% HPDIs, are displayed in Figure 1B. Corn (0.10 insects/sweep, 1.41 insects/sweep), onions (0.09 insects/sweep, 0.76 insects/sweep), and garlic (0.06 insects/sweep, 0.50 insects/sweep) all had 95% HPDIs that were entirely positive, suggesting that previous cultivation of these crops was associated with increased *L. hesperus* abundance. June *L. hesperus* density was 0.35 insects/sweep lower, with a 95% HPDI for this decrease of (0.26 insects/sweep, 0.44 insects/sweep), for upland cotton than for Pima cotton.

### Model 3

While there were exceptions, we noticed that there was a trend for crops associated with increased pest abundances to also be associated with decreased yield. To more rigorously quantify this trend, for the 14 crops other than cotton, we regressed the posterior mean of the yield difference from cotton against the posterior mean of the *L. hesperus* difference from cotton. There was a negative slope with posterior mean −0.49 and 95% HPDI of (−1.16, 0.15) that marginally overlapped 0; the posterior probability of there being a negative slope was 93.4%. This provided evidence that crops associated with increased June *L. hesperus* densities were also associated with negative effects on yield (Figure 2).

### Model 4

Model 4 suggested that every additional consecutive year of prior cotton cultivation in a field led to reduced cotton yield. Figure 3A displays the posterior distribution of the change in yield for each additional year that cotton was consecutively grown in the field prior to the focal year; the posterior mean for this change in yield was −40.9 kg/ha, with 95% HPDI (−57.5,−23.4 kg/ha). This translates to a mean of the percentage change in yield of −2.4% per year and 95% HPDI of (−1.4%,−3.4%) per year. We refit Model 4 without the term for consecutive cotton plantings; boxplots of the residuals are plotted against consecutive cotton plantings in Figure 3B, where a decreasing trend can be observed. Yield was 169.2 kg/ha higher, with a 95% HPDI of (123.4 kg/ha, 211.5 kg/ha), for upland cotton than for Pima cotton.

### Model 5

Model 5 revealed a positive association between the number of preceding consecutive cotton plantings and June *L. hesperus* densities; the posterior mean of the slope regressing June *L. hesperus* on consecutive cotton plantings was 0.037 insects/sweep with a 95% HPDI that slightly overlapped 0 of (−0.007,0.079). The posterior probability of there being a positive relationship between consecutive cotton plantings and *L. hesperus* densities was 95.3%. June *L. hesperus* density was 0.32 insects/sweep lower, with a 95% HPDI of (0.20 insects/sweep, 0.43 insects/sweep), for upland cotton than for Pima cotton.

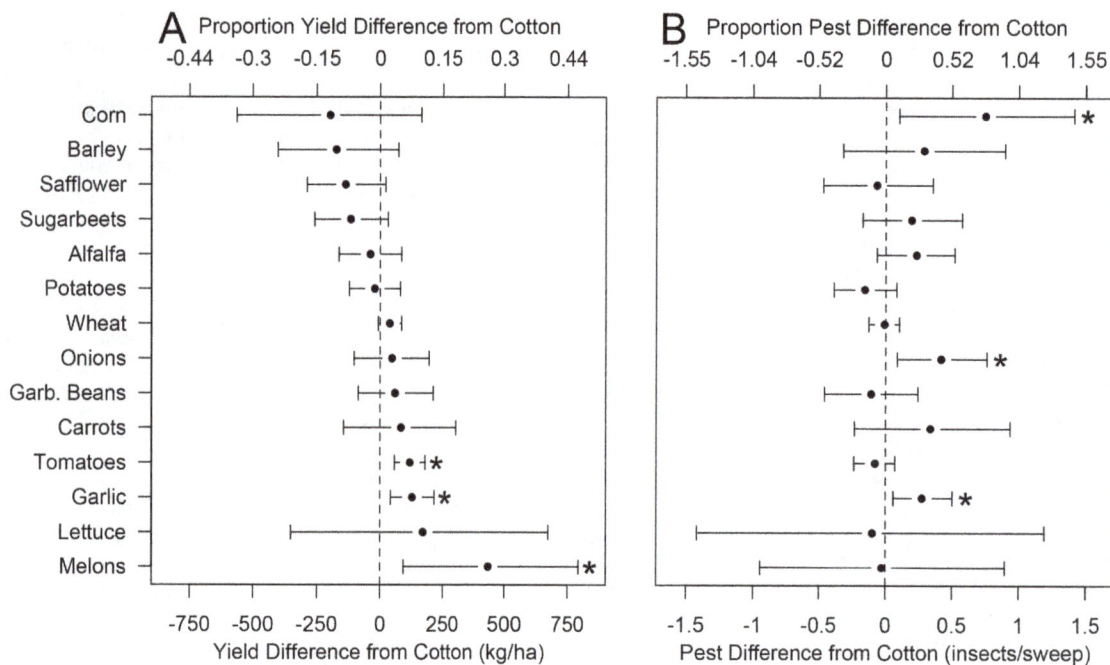

**Figure 1. Means and 95% HPDIs of the differences in mean yield (A) and mean June *L. hesperus* density (B) between fields where a certain crop was grown the previous year and where cotton was grown the previous year.** 95% HPDIs that do not overlap 0 are marked with a (∗).

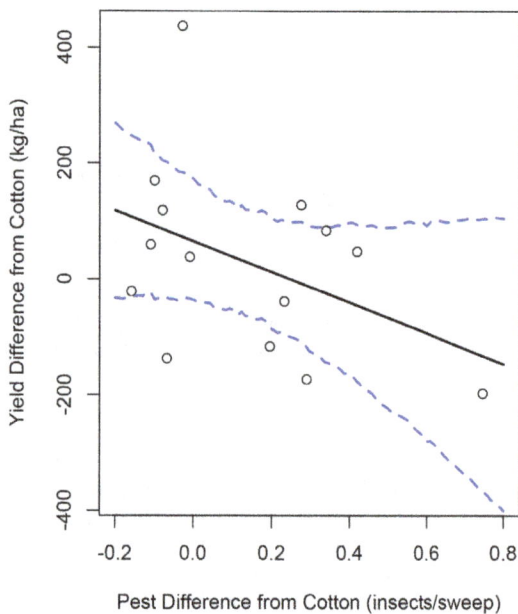

**Figure 2. For each of the 14 crops other than cotton, we calculated the posterior mean of the mean difference in yield when that crop was grown in the field the year before compared to when cotton was grown in the field the year before (y-axis; these estimates are also displayed in Figure 1A).** We also calculated the posterior mean of the difference in mean June *L. hesperus* densities between fields where a specific crop was grown the year before and where cotton was grown the year before (x-axis; these estimates are also displayed in Figure 1B). These estimates are plotted above (open circles). Then, we fit a linear model by regressing the mean yield differences on the mean *L. hesperus* differences. The posterior mean of the model fit (solid black) and 95% HPDI (dashed blue) are overlaid.

## Discussion

Capitalizing on a large existing set of crop records from commercial cotton fields in California, we employed an ecoinformatics approach to explore the effects of crop rotational histories on cotton yield. Our hierarchical Bayesian analyses revealed evidence that several crops, when grown in the same field the year before the focal cotton planting, were associated with either decreased or increased cotton yield (Figure 1A), and either increased or decreased early season densities of the pest *L. hesperus* (Figure 1B). Furthermore, crops associated with decreased yield were generally also associated with increased *L. hesperus* densities, while those associated with increased yield were also associated with decreased *L. hesperus* densities (Figure 2).

These results suggest a possible mechanism for the observed yield effects of these rotational histories. Since *L. hesperus* preferentially attacks certain crops [33], a field cultivated with a crop that is heavily attacked by *L. hesperus* may, if *L. hesperus* disperse from the focal field, increase the abundance of *L. hesperus* in nearby fields. These populations may subsequently attack the crop planted in the focal field the following year, explaining the increase in early-season *L. hesperus* densities that we detected following certain crops. In turn, these increased *L. hesperus* populations may exert strong herbivorous pressure on focal cotton crops, possibly explaining the corresponding decrease in yield.

We believe that the effect of rotational history on early-season *L. hesperus* likely operates at a landscape scale that is larger than the within-field scale. If cotton was grown in a field the previous year, then farmers in the San Joaquin Valley are required to maintain a 90-day plant-free period prior to 10 March of the following year [27]. This prevents *L. hesperus*, which overwinter as adults on live host plants, from overwintering in a focal field where cotton was grown the year before. If a crop other than cotton was grown the previous year, then it could be possible for *L. hesperus* to overwinter in the focal field on residual plant or weed populations; however,

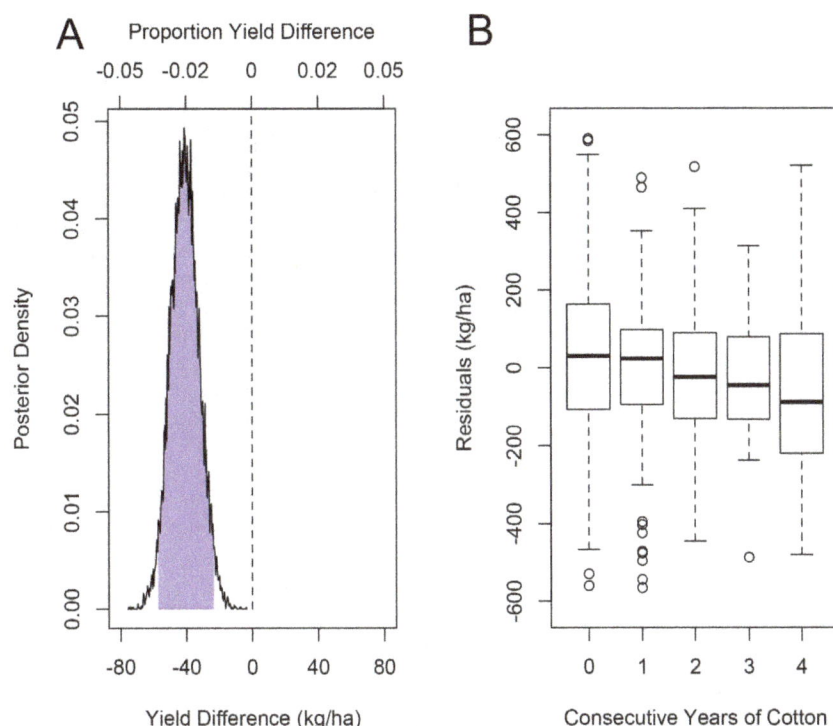

**Figure 3. The posterior distribution of the change in yield for every additional year that cotton was grown consecutively in a field prior to the focal year (A).** We refit the model without consecutive cotton plantings in the model and display boxplots of the residuals vs. consecutive cotton plantings, where a decreasing trend can be observed (B).

since fields are completely plowed prior to planting cotton in the spring, *L. hesperus* adults would still need to temporarily leave the focal field. Therefore, we believe that the preferred host crops for *L. hesperus* increase *L. hesperus* populations at a landscape scale. Then, when cotton, another target of *L. hesperus*, is planted in the same field the following year, the cotton field is attacked by this regional population. If regional populations are already large due to lingering effects from crops grown the previous year, *L. hesperus* populations may move into cotton early in the growing season; this could be particularly damaging given research suggesting that cotton yield is particularly sensitive to *L. hesperus* densities early in the growing season [34]. Using our data, we were not able to determine at exactly what scale the effects of rotation on *L. hesperus* likely operate. We do not believe a within-field scale is plausible, but determining a more precise spatial scale for these effects could be an interesting topic for future research.

Our findings match expectations of crop yield effects based on previous research on *L. hesperus* host crop preferences, lending support to our hypothesis that yield effects of crop rotational histories are, at least partially, mediated by effects on *L. hesperus*. Alfalfa and sugarbeets, both crops for which we found negative effects on yield and positive effects on *L. hesperus* when grown in a field the previous growing season, are all considered preferred hosts for *L. hesperus* [27], and have been shown to also increase *L. hesperus* populations in nearby cotton fields during an individual growing season [33,35]. Presumably, this effect is due to these crops supporting large *L. hesperus* populations. Large *L. hesperus* populations are known to build up in alfalfa [36], and their dispersal following alfalfa harvesting can threaten nearby cotton crops [27,37]. *L. hesperus* is also known to emigrate to nearby cotton fields when safflower begins to dry in mid-summer [38]. While the potential for nearby alfalfa [27,33] and safflower [33,35]

fields to increase *L. hesperus* populations in cotton fields in a given year has been recognized, our results are the first indication that these landscape effects may extend temporally, affecting *L. hesperus* populations, and yield, in the next growing season. Tomatoes, associated with increased yield and decreased pest abundance in our data, have likewise been shown to decrease *L. hesperus* abundances in nearby cotton fields within a given year [33].

While previous experimental work has examined the effects of crop rotations on cotton yield [14,17,18,20,21], our work expands on these studies in several ways. First, we explore a much wider diversity of possible crop rotational histories, providing quantitative estimates of the cotton yield effects of cultivating 14 different crops the previous year. Second, since we analyze records from commercial cotton fields, our data have the potential to capture yield effects (such as those due to highly mobile arthropods) that could only be detected at this realistic spatial scale. Third, since we have collected data on pest abundances, not only yield, we have also been able to use our data to generate and build evidence for a hypothesized mechanistic explanation of the yield effects we identify.

We also found that farmers incurred a decline in cotton yield of about 2.4% for every additional year cotton was grown consecutively in a field preceding the focal season (Figure 3). This is consistent with previous research suggesting that continuous cultivation of cotton in the same location can reduce yield compared to interspersing cotton with other crops [14,17,18, 20,21]. We also found some evidence that the number of years cotton was grown consecutively in a field was associated with higher June *L. hesperus* densities: the posterior probability of there being a positive association was about 95%. Identifying the actual mechanism underlying this yield effect is beyond the scope of this study, but would be an interesting avenue for future research. It is possible that the yield decline is not caused by changes in *L. hesperus* densities, and instead results from the buildup of soil pathogens, especially in light of previous research showing that continuous cotton cultivation increases the densities of fungal pathogens in the soil [18].

When interpreting our results, it is important to remain cognizant of the challenges of drawing causal inferences from observational data. The key assumption required to make causal inferences from regression coefficients is that all variables that affect both the treatment assignment (crop rotation, in our analyses) and the response variable (yield and *L. hersperus* density, in our analyses) are included in the model; this ensures that the probability of receiving each treatment becomes, conditional on the predictor variables included in the model, conditionally independent of the response variable [28]. In experimental studies, the treatment assignment is typically controlled by the experimenter, so one can be confident that the only difference between treatment and control groups is in fact the treatment. However, in observational studies, it is impossible to prove definitively that there was no other factor that affected both the treatment assignment and the response variable (thus spuriously suggesting a treatment effect).

As such, we want to be very clear that our hypothesis that the effects of rotation on yield are mediated by effects on *L. hesperus* densities is exactly that - a hypothesis. While our data do support a negative *association* between effects on *L. hesperus* and effects on yield, we cannot prove with observational data that the varying effects on yield are *caused* by the varying effects on *L. hesperus*. This could be a fruitful topic for future experimental work.

Although causality is impossible to prove using observational data, ecoinformatics paves the way for implementing data-driven agricultural strategies and allows us to mine large datasets to explore important questions that are difficult to address experimentally. While by no means a replacement for experimentation, ecoinformatics can be a cost-effective and realistic complementary approach. In particular, our result identifying the effects of crop rotation on *L. hesperus* density would have been extremely difficult to reach experimentally. Since *L. hesperus* readily disperse across spatial scales of more than 1000 meters [37], an experimental study would have required massive plots comparable to the size of

commercial fields in order to adequately capture their spatial dynamics.

Our results have numerous practical applications for commercial cotton growers. Growers with knowledge of the crop rotations associated with depressed cotton yield could make more informed decisions, selecting the sequence of crop cultivations that lead to maximized yield. When feasible, cotton plantings could be avoided following crops that decrease cotton yield, and instead limited to fields where crops that increase cotton yield were previously planted. In some cases, market conditions may lead a grower to plant cotton following a yield-depressing crop, even given the knowledge of likely yield loss. In those situations, our results may still be helpful, as an early warning sign of a potential pest problem in a particular field could allow the grower and PCA to focus pest detection efforts on that field and provide time to eliminate the problem before severe yield loss was incurred.

Our results suggest that the yield effects of crop rotational histories in cotton are relatively modest in magnitude: the posterior means for effects of any specific crop were mostly under 15%. However, given the tight profit margins of commercial agriculture, a 15% change in yield could translate into a far greater percentage change in profit, and could therefore be of substantial economic significance to a grower. As we seek to feed a growing worldwide population while doing minimal harm to the environment, crop management practices that increase yield while reducing the need for costly and damaging pesticides and fertilizers are of great value. Crop rotation is one such method, and we are optimistic that ecoinformatics approaches may be helpful in elucidating the details of how to optimally implement crop rotation.

## Acknowledgments

We would like to sincerely thank the growers and PCAs who generously donated their data and time to help make this research possible, and R.F. Denison for valuable feedback on this manuscript.

## Author Contributions

Conceived and designed the experiments: MM JR. Performed the experiments: MM JR. Analyzed the data: MM JR. Contributed reagents/materials/analysis tools: JR. Wrote the paper: MM JR.

## References

1. Godfray HCJ, Beddington JR, Crute IR, Haddad L, Lawrence D, et al. (2010) Food security: the challenge of feeding 9 billion people. Science 327: 812–818.
2. Crookston RK (1984) The rotation effect: What causes it to boost yields? Crops Soils 36: 12–14.
3. Karlen DL, Varvel GE, Bullock DG, Cruse RM (1994) Crop rotations for the 21st century. Adv Agron 53: 1–45.
4. Russelle MP, Hesterman OB, Shaeffer CC, Heichel GH (1987) Estimating nitrogen and rotation effects in legume-corn rotations. In: Power JF, editor, The role of legumes in conservation tillage systems, Soil Conserv Soc Am. pp. 41–42.
5. Forbes AA, Rosenheim JA (2011) Plant responses to insect herbivore damage are modulated by phosphorus nutrition. Entomol Exp Appl 139: 242–249.
6. Benson GO (1985) Why the reduced yields when corn follows corn and possible management responses. Proc 40th Corn Sorghum Res Conf: 161–174.
7. Cook RJ (1984) Root health: Importance and relationship to farming practices. In: Bezdicek DF, editor, Organic Farming: Current Technology and its Role in a Sustainable Agriculture, Am Soc Agron. pp. 111–127.
8. Edwards JH, Thurlow DL, Eason JT (1988) Inuence of tillage and crop rotation on yields of corn, soybean, and wheat. Agron J 80: 76–80.
9. Rothamsted Research Center (2006) Guide to the classical and other long-term experiments, datasets and sample archive. Technical report, Rothamsted Research Center.
10. Liebman M, Dyck E (1993) Crop rotation and intercropping strategies for weed management. Ecol Appl 3: 92–122.
11. White KD (1970) Fallowing, crop rotation, and crop yields in Roman times. Agric Hist 44: 281–290.
12. White KD (1970) Roman farming. Ithaca: Cornell University Press.
13. Johnson TC (1927) Crop rotation in relation to soil productivity. J Am Soc Agron 19: 518–527.
14. Mitchell CC, Westerman RL, Brown JR, Peck TR (1991) Overview of long-term agronomic re-search. Agron J 83: 24–29.
15. Rosner D, Markowitz G (2013) Persistent pollutants: a brief history of the discovery of the widespread toxicity of chlorinated hydrocarbons. Environ Res 120: 126–133.
16. Peairs FB, Pilcher SD (2013) Western Corn Rootworm. Colorado State University Extension. URL http://www.ext.colostate.edu/pubs/insect/05570.html. Accessed 6 November 2013.
17. Bordovsky JP, Mustian JT, Cranmer AM, Emerson CL (2011) Cotton-grain sorghum rotation under extreme deficit irrigation conditions. Appl Eng Agric 27: 359–371.
18. Wheeler TA, Bordovsky JP, Keeling JW, Mullinix BG, Woodward JE (2012) Effects of crop ro-tation, cultivar, and irrigation and nitrogen rate on verticiullium wilt in cotton. Plant Dis 96: 985–989.
19. Mitchell CC, Delaney DP, Balkcom KS (2008) A historical summary of Alabama's Old Rotation (circa 1896): the world's oldest, continuous cotton experiment. Agron J 100: 1493–1498.
20. Constable GA, Rochester IJ, Daniells IG (1992) Cotton yield and nitrogen requirement is modi_ed by crop rotation and tillage method. Soil Tillage Res 23: 41–59.
21. Bordovsky JP, Lyle WM, Keeling JW (1994) Crop rotation and tillage effects on soil water and cotton yield. Agron J 86: 1–6.
22. Rosenheim JA, Parsa S, Forbes AA, Krimmel WA, Law YH, et al. (2011) Ecoinformatics for integrated pest management: Expanding the applied insect ecologist's tool-kit. J Econ Entomol 102: 331–342.
23. Streibich K (2013) Using big data to drive commercial advantage: excellent processes, excellent results. Database Netw J 43: 12–13.

24. Gobble MM (2013) Big data: The next big thing in innovation. Res-Technol Manage 56: 64–66.

25. O'Rourke ME, Rienzo-Stack K, Power AG (2011) A multi-scale, landscape approach to predicting insect populations in agroecosystems. Ecol Appl 21: 1782–1791.

26. Rosenheim JA, Steinmann K, Langellotto GA, Zink AG (2006) Estimating the impact of Lygus hesperus on cotton: the insect, plant, and human observer as sources of variability. Environ Entomol 35: 1141–1153.

27. Godfrey LD, Goodell PB, Natwick ET, Haviland DR, Barlow VM (2013) UC IPM pest management guidelines: cotton. University of California Division of Agriculture and Natural Resources. URL http://www.ipm.ucdavis.edu/PMG/r114301611.html. Accessed 17 September 2013.

28. Gelman A, Hill J (2009) Data analysis using regression and multilevel/hierarchical models. Cam- bridge: Cambridge University Press.

29. Hoffman MD, Gelman A (2013) The No-U-Turn sampler: Adaptively setting path lengths in Hamiltonian Monte Carlo. J Mach Learn Res: In press.

30. Stan Development Team (2013) Stan: A C++ library for probability and sampling, Version 1.3. URL http://mc-stan.org/. Accessed 17 September 2013.

31. Stan Development Team (2013) Stan modeling language user's guide and reference manual, Version 1.3. URL http://mc-stan.org/. Accessed 17 September 2013.

32. Gelman A, Rubin DB (1992) Inference from iterative simulation using multiple sequences. Stat Sci 7: 457–511.

33. Carriere Y, Goodell PB, Ellers-Kirk C, Larocque G, Dutilleul P, et al. (2012) Effects of local and landscape factors on population dynamics of a cotton pest. PLOS ONE 7: e39862.

34. Rosenheim JA, Meisner MH (2013) Ecoinformatics can reveal yield gaps associated with crop-pest interactions: a proof-of-concept. PLOS ONE 8: e80518.

35. Carriere Y, Ellsworth P, Dutilleul P, Ellers-Kirk C, Barkley C, et al. (2006) A GIS-based approach for areawide pest management: the scales of Lygus hesperus movements to cotton from alfalfa, weeds, and cotton. Entomol Exp Appl 118: 203–210.

36. Sevacherian B, Stern VM (1975) Movement of Lygus bugs between alfalfa and cotton. Environ Entomol 4: 163–165.

37. Sivakoff FS, Rosenheim JA, Hagler JR (2012) Relative dispersal ability of a key agricultural pest and its predators in an annual agroecosystem. Biol Control 63: 296–303.

38. Mueller AJ, Stern VM (1974) Timing of pesticide treatments on safflower to prevent Lygus from dispersing to cotton. J Econ Entomol 67: 77–80.

# 7

# Thirdhand Cigarette Smoke: Factors Affecting Exposure and Remediation

**Vasundhra Bahl[1,2], Peyton Jacob III[3], Christopher Havel[3], Suzaynn F. Schick[4], Prue Talbot[1]***

**1** Department of Cell Biology and Neuroscience, University of California Riverside, Riverside, California, United States of America, **2** Environmental Toxicology Graduate Program, University of California Riverside, Riverside, California, United States of America, **3** Department of Clinical Pharmacology, University of California San Francisco, San Francisco, California, United States of America, **4** Department of Medicine, Division of Occupational and Environmental Medicine, University of California San Francisco, San Francisco, California, United States of America

## Abstract

Thirdhand smoke (THS) refers to components of secondhand smoke that stick to indoor surfaces and persist in the environment. Little is known about exposure levels and possible remediation measures to reduce potential exposure in contaminated areas. This study deals with the effect of aging on THS components and evaluates possible exposure levels and remediation measures. We investigated the concentration of nicotine, five nicotine related alkaloids, and three tobacco specific nitrosamines (TSNAs) in smoke exposed fabrics. Two different extraction methods were used. Cotton terry cloth and polyester fleece were exposed to smoke in controlled laboratory conditions and aged before extraction. Liquid chromatography-tandem mass spectrometry was used for chemical analysis. Fabrics aged for 19 months after smoke exposure retained significant amounts of THS chemicals. During aqueous extraction, cotton cloth released about 41 times as much nicotine and about 78 times the amount of tobacco specific nitrosamines (TSNAs) as polyester after one hour of aqueous extraction. Concentrations of nicotine and TSNAs in extracts of terry cloth exposed to smoke were used to estimate infant/toddler oral exposure and adult dermal exposure to THS. Nicotine exposure from THS residue can be 6.8 times higher in toddlers and 24 times higher in adults and TSNA exposure can be 16 times higher in toddlers and 56 times higher in adults than what would be inhaled by a passive smoker. In addition to providing exposure estimates, our data could be useful in developing remediation strategies and in framing public health policies for indoor environments with THS.

**Editor:** Ruby John Anto, Rajiv Gandhi Centre for Biotechnology, India

**Funding:** This work was supported by the California Consortium on Thirdhand Smoke, California Tobacco-Related Disease Research Program (trdrp.org) grant 20PT-0184 and California Tobacco-Related Disease Research Program grant 21 ST-011, the National Institute on Drug Abuse P30 DA012393, the National Center for Research Resources S10 RR026437, the National Center for Advancing Translational Sciences, National Institutes of Health (nih.gov), through UCSF-CTSI Grant Number UL1 TR000004. Its contents are solely the responsibility of the authors and do not necessarily represent the official views of the funding agencies. VB was supported on a Deans Pre-doctoral Fellowship and a California Tobacco-Related Disease Research Program pre-doctoral fellowship. The funders had no role in study design, data collection and analysis, decision to publish, or preparation of the manuscript.

**Competing Interests:** The authors have declared that no competing interests exist.

* Email: talbot@ucr.edu

## Introduction

Thirdhand smoke (THS) consists of residual tobacco smoke that sorbs to indoor surfaces and remains after the majority of the airborne components of the smoke have cleared. THS raises the concentration of nicotine and other smoke constituents in indoor environments occupied by smokers [1], [2]. During aging, the chemicals in THS can desorb back into the air or react to form new chemicals. For example, nicotine reacts with ambient nitrous acid (HONO) to form tobacco specific nitrosamines (TSNAs) [3], [4]. Exposure to THS and remediation of buildings and vehicles contaminated with THS have received little attention in the past and are important, especially in light of recent health-related studies that indicate the potentially hazardous nature of THS [5]–[8]. Because THS affects individuals with unknown or unwanted exposure, it is an issue with public health implications [2].

The negative health effects of active smoking and secondhand smoke exposure have been analyzed *in vitro*, in animals, and studies of human volunteers and populations [9]–[13]. Active smoking and secondhand smoke exposure adversely affect health across all age groups [9], [14], [15]. In contrast, little is known about the level of human exposure to THS and the resulting health effects. THS exposure can occur through the skin, by ingestion, and by inhalation. Infants and small children could be at greater risk than adults because their skin is thinner, their surface to volume ratio is higher, and because they spend more time in contact with THS-contaminated surfaces and where they can mouth THS-contaminated objects. If ingested, the fraction of THS that is soluble in saliva and digestive fluids will be available for intake (passage into the body but not across absorptive barriers) [16]. The extent of intake will depend on the concentration of THS chemicals, the fraction of THS that is in the air and on surfaces, and their solubility in saliva or sweat. The concentration of THS chemicals will vary with the number of cigarettes smoked in the room, the air exchange rate, and the time elapsed since smoking. Therefore, when evaluating exposure, it is important to consider that THS is dynamic and that aging can change the composition of THS over time.

Remediation, which is the removal of THS residue from surfaces in indoor environments or the safe containment of THS, is another important aspect of THS contamination that needs study [2], [17]. Methods of remediation will depend upon the level of contamination as well as the type of material. The materials commonly found indoors, such as natural and synthetic fibers, carpets, paper and wall board, each differ in their capacities to adsorb, absorb, bind, and release THS chemicals (unpublished data).

As a first step to understanding the persistence of THS in indoor environments, potential human exposures, and options for remediation, we repeatedly exposed cotton and polyester fabrics to cigarette smoke in an experimental chamber, stopped exposure and aged the fabrics for up to 19 months, then measured the concentrations of nicotine, nicotine-related alkaloids and tobacco-specific nitrosamines in extracts of fabrics. We tested chemical concentrations in both organic and aqueous solvent extracts, and then used the resulting data to model exposures that toddlers and adults could receive in environments containing THS.

## Methods

### Exposure of fabric to cigarette smoke

100% cotton terry cloth, and 100% polyester fleece were purchased at retail and washed three times in a domestic washing machine using an unscented, enzyme-free laundry detergent (Country Save powdered laundry detergent, Arlington, WA) in hot water with two rinses/cycle, and washing again with no detergent. These fabrics were chosen as they are commonly used in household products and in clothing. After line drying, fabrics were hung in a 6 m$^3$ stainless steel chamber at UCSF and exposed to cigarette smoke as described previously in detail [18]. Briefly, smoke generated by an automatic smoking machine (Model TE-10z, Teague Enterprises, Woodland, California, USA) was diluted into conditioned, filtered air, and conducted through a 6 m$^3$ stainless steel smoke aging chamber. The aging chamber contained three vertically staggered baffles and two internal fans to promote mixing. The cloth samples were hung on the baffles. Marlboro Red cigarettes were smoked according to ISO protocol 3308: 2012. Marlboro Red was chosen as it is the best-selling cigarette in the United States and is popular worldwide. Particle concentration at the outlet of the smoke aging chamber was measured using a laser photometer calibrated gravimetrically (Dusttrak II, model 8530, TSI Inc., Shoreview MI). Aerosol flow rates through the chamber were measured using an air velocity transmitter (model 641-b Dwyer Instruments, Michigan City IN). After smoking, air flow was turned off, and the chamber was closed while it still contained detectable levels of smoke. Smoke was generated 0–8 times/month according to the needs of ongoing clinical research between experiments. The time that the cloth sample was in the chamber, the number of hours of smoke, the average particle concentration for each experiment and the air velocity through the chamber were logged. The total particle mass that the fabric was exposed to was calculated as $sm = a*b*c*60\frac{m}{h}*\frac{1\ liter}{1000\ cubic\ meters}$ where a = air velocity in liters/minute, b = hours of smoke, and c = average smoke particle concentration.

A sheet of cotton terry cloth and a sheet of polyester fleece were exposed to smoke. The terry cloth was exposed to smoke containing 1329 mgs of particles for 114 hours over 1 year. The polyester fleece was exposed to 1846 mgs of smoke particles for 257 hours over 10 months. Both fabrics were folded and stored in separate polyethylene bags at room temperature in the dark. The terry cloth was stored 8 months and the polyester 1 month prior to

shipment. Samples were wrapped in aluminum foil, placed in polyethylene bags and shipped at ambient temperature, overnight, to the Talbot Laboratory at UCR. Upon receipt, samples were transferred to amber glass bottles and stored at room temperature (RT) in the dark.

### Organic solvent extractions

Samples of fabric were incubated at RT overnight in 50% MeOH/1% HCl then vortexed for 3–5 minutes at RT. Solvent was removed by squeezing the fabric in the vial with a spatula, fibers and dust were removed by centrifugation, and the extract was analyzed as described below.

### Aqueous extractions

Aqueous extracts of THS were prepared in Dulbecco's Modified Eagle Medium (DMEM). Terry cloth and polyester fleece were weighed and cut into very small pieces using scissors. Either 0.05 or 0.125 g of fabric/ml were extracted in DMEM in 15 ml conical tubes on a rotating shaker. The medium was recovered by placing the fabric in a syringe and centrifuging at 4,000 g for 5 minutes The recovered medium was passed through a 0.22 μm filter, aliquoted into 1.5 ml vials, and stored at −80°C.

To examine the effect of repeated aqueous extraction on chemical yield, three samples of terry cloth and two samples of polyester were extracted five times, serially at RT with media being replaced every hour for 5 hours. To determine the effect of time and temperature, the terry cloth was extracted under four conditions: RT for 1 hour; RT for 2 hours; 4°C for 1 hour; and 4°C for 2 hours. For extraction at 4°C, tubes were placed in a beaker of ice on a rocker shaker. To examine the effect of aging, extraction was done after storing the terry cloth for 11, 16 and 19 months and the polyester for 11 and 19 months in amber glass jars at RT.

### Chemical Analysis of THS extracts

1 mL extracts of THS were shipped to UCSF on dry ice where they were analyzed using liquid chromatography-tandem mass spectrometry (LC-MS/MS) [19], [20]. The method was modified to include NNA in the analysis, by treating the extract with pentafluorophenylhydrazine (PFPH) to convert NNA to the pentafluorophenylhydrazone derivative which enhances sensitivity of detection [21].

### LC-MS/MS

The samples were analyzed on a Thermo Scientific Vantage LC-MS/MS with an Accela UPLC system using a 3×150 mm 2.6 micron Phenomenex Kinetex PFP column as detailed in [20].

### Limits of quantification

The limits of quantitation for each of the chemicals analyzed are as follows: nicotine: 1.02 ng/ml; myosmine: 0.305 ng/ml; 2,3′-bipyridine: 0.914 ng/ml; cotinine: 0.914 ng/ml; N-formylnornicotine: 0.305 ng/ml; nicotelline: 0.030 ng/ml; NNN: 0.030 ng/ml; NNK; 0.0130 ng/ml; NNA 0.010 ng/ml.

### Statistical analyses

The concentrations of chemicals in aqueous extracts were converted to grams/gram of fabric. Averages of four samples in each group were then calculated using Microsoft Excel. ANOVA (one way analysis of variance) was performed using GraphPad Prism to determine if the chemical concentrations in extracts made under different conditions varied significantly. ANOVA was also used to analyze extracts made from terry cloth after 11, 16 and 19 months of aging. Groups differing significantly (p<0.05) from the 11 month samples were identified using Dunnett's posthoc test. Data were checked to determine if they satisfied the assumptions of ANOVA (normal distribution and homogeneity of variances). T-tests were used to determine if the chemical concentrations in aqueous extracts were different from those in methanol/HCL extracts.

## Results

### Fabrics used for extraction

THS was extracted from 100% cotton terry cloth and 100% polyester fleece. Terry cloth is a loosely knit natural fabric with many thin fibers that provide a large surface area for absorption of chemicals. One surface of polyester has numerous short highly packed fibers while the other is comprised of a large tightly woven mesh of fibers (Fig. 1).

### Aqueous and methanol:HCl solvents extracted THS chemicals from cotton fabric

The concentrations of nicotine and related chemicals in the aqueous extracts of THS from cotton terry cloth after 31 months of aging were similar to those in methanol:HCl extracts (Figs. 2A, B). Negligible amounts of nicotine and related chemicals were recovered when aqueous extraction was followed by methanol:HCl extraction. Nicotine (50–60 μg/gram of fabric) was the most abundant of the chemicals analyzed. Myosmine, bipyridine, formylnornicotine and cotinine were present in 1–2 μg/gm of fabric quantities, while the TSNAs and nicotelline were the least abundant (nanogram/gram of fabric) of the chemicals analyzed in THS extracts from terry cloth.

### Extraction of polyester fabric yielded lower concentrations of THS chemicals

The concentrations of all chemicals tested were lower in extracts of polyester fleece than in extracts of cotton terry cloth (Fig. 2C, D). As an example, in aqueous extracts approximately 40 times less nicotine was extracted from polyester than from terry cloth. For polyester fleece, methanol:HCl and aqueous extracts had similar concentrations of nicotine and other chemicals. However, when aqueous extraction was followed by methanol:HCl extraction, higher concentrations of myosmine and 2,3′-bipyridine were obtained than with aqueous extractions alone. All other chemicals were retrieved at lower concentrations in the methanol:HCl extract that followed the aqueous extraction. This suggests two possibilities: that polyester binds less nicotine, nicotine- related alkaloids and TSNAs than cotton or that these compounds are harder to extract from polyester than from cotton.

### Serial aqueous extractions from terry cloth and polyester

To determine if all nine chemicals were removed from terry cloth and polyester during 1 hour of aqueous extraction, the same fabric samples were extracted five times. Each extraction lasted one hour (Fig. 3). All of the chemicals extractable by water were successfully removed from cotton terry cloth during the first hour of extraction. Concentrations of some chemicals (e.g., nicotine, myosmine and nicotelline) were very similar from batch to batch, while others, such as cotinine, NNA, and NNN, varied somewhat in concentration among batches. For polyester, cotinine was found only in the first hour extracts. Nicotine and N-formylnornicotine were found in the first and the second hour extracts.

### One hour of aqueous extraction at RT removes THS chemicals from cotton terry cloth

The effects of temperature and time on the concentration of chemicals recovered by aqueous extraction was tested (Fig. 4). Extracts were made at RT and at 4°C for 1 or 2 hours. Chemical concentrations appeared to be similar for each extract. When tested by ANOVA, no significant differences in chemical concentrations were found between extraction conditions. Data for each chemical were therefore combined in Table 1, which also includes the combined data for polyester. These data confirm that 1 hour at RT is sufficient time to achieve the maximum yield of each chemical from cotton terry cloth using aqueous medium and that changing the time or temperature does not improve extraction efficiency. All chemicals were more abundant in extracts of terry cloth than in polyester, and NNN and NNA were not detected in the extracts of polyester.

### Effect of aging on the concentrations of THS chemicals in extracts from terry cloth and polyester

In extracts of terry cloth, nicotine concentrations (105.8, 112.9, and 69.6 μg/gram of fabric) at 11, 16, and 19 months of aging (Fig. 5) were not significantly different when evaluated by ANOVA (p = 0.0595). Extracts of polyester made after 11 and 19 months of aging had very low amounts of nicotine (557 ng/g fabric and 168.8 ng/g fabric) in contrast to terry cloth that aged for similar times (Fig. 5A).

Myosmine, N-formylnornicotine, 2,3′-bipyridine, and cotinine were present in extracts of terry cloth at μg/gram of fabric concentrations (Fig. 5B). The concentrations of extractable myosmine (p<0.0001), 2,3′-bipyridine (p<0.0001) and cotinine (p = 0.0001) decreased significantly after 19 months of aging (January 2013). The concentration of N-formylnornicotine decreased significantly (p<0.0001) after 16 months of aging, but did not decrease further by 19 months. For this group of chemicals, the extract of polyester which aged 11 months

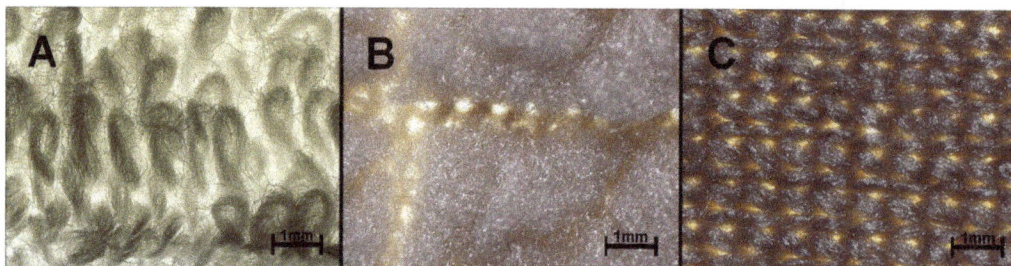

**Figure 1. Micrographs of fabrics used for THS extraction.** (A) Terry cloth is a loosely knit fabric made of loops of cotton which increase its surface area tremendously and contribute to the absorption of THS. (B) Polyester is a more tightly knit fabric with one fuzzy surface and (C) one compact tightly woven surface.

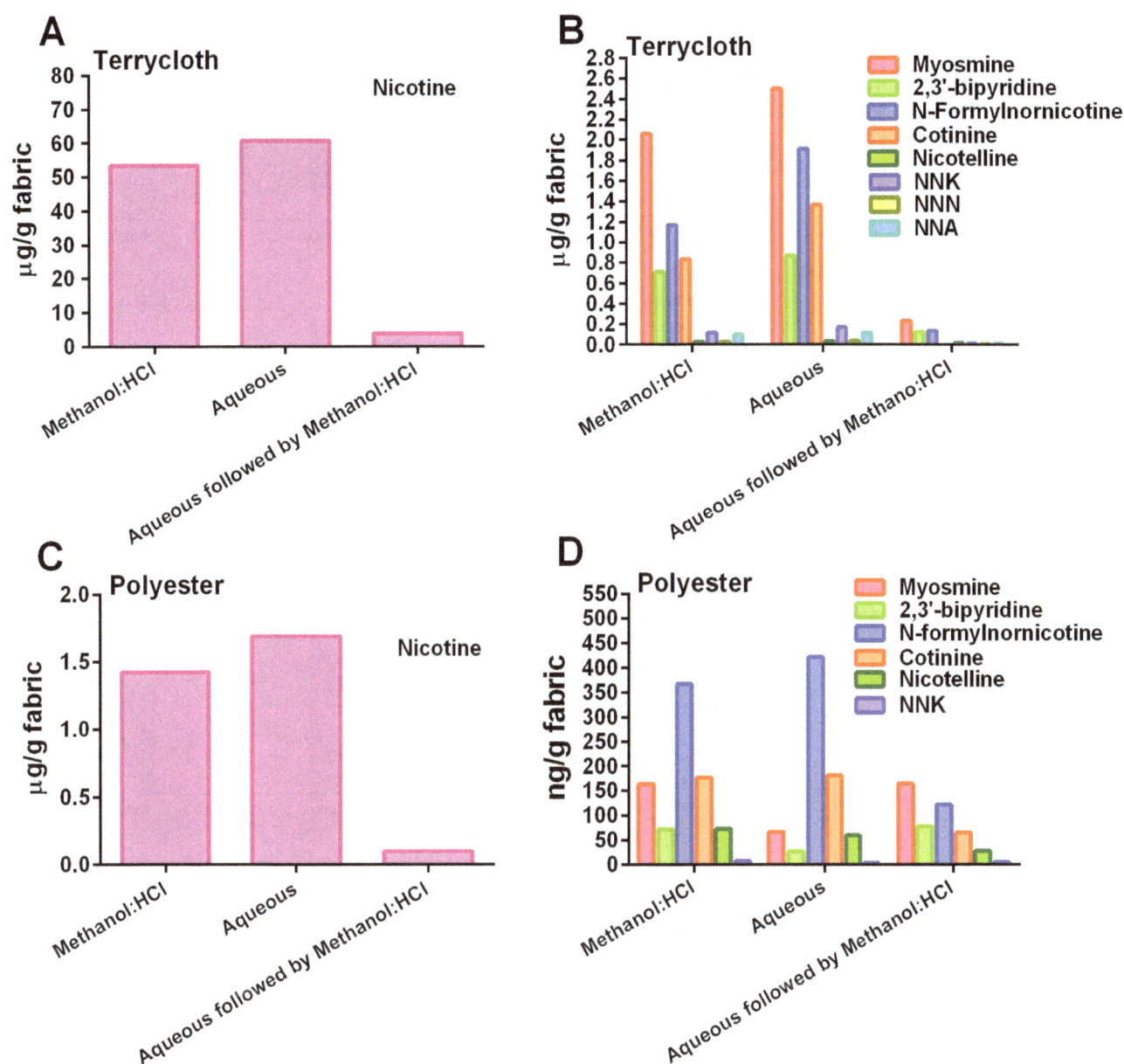

**Figure 2. Comparison of chemical concentrations in aqueous and methanol:HCl extracts of terry cloth and polyester exposed to THS.** Terry cloth was extracted after 31 months of aging and polyester was extracted after 19 months of aging. Results for aqueous extracts are an average of three experiments, and all other groups are averages of two experiments.

contained only N-formylnornicotine and cotinine, and these were present in low concentrations compared to the corresponding terry cloth sample (11 months) (Fig. 3B). Extracts of polyester made after 19 months of aging had very low levels of myosmine, 2,3'-bypiridne, N-formylnornicotine and cotinine. All the chemicals in extracts of polyester were present at concentrations less than 1 µg/gram of fabric (Fig. 5B).

In extracts of terry cloth, concentrations of nicotelline, NNA, NNK and NNN were in the ng/gram of fabric range (Fig. 5C). Of these chemicals, only nicotelline did not decrease in concentration with aging, supporting its use as a tracer for tobacco smoke particulate matter [19]. The concentration of NNA decreased significantly by 19 months of aging (p = 0.0108). For both NNN and NNK, there was a slight but significant increase in concentration at 16 months of aging (p = 0.0020 and 0.0004

respectively), followed by a significant decrease in NNK (p = 0.0421) at 19 months.

In extracts of polyester made after 11 months of aging, nicotelline was detected in very small amounts, but the three TSNAs were absent. After 19 months of aging, very small amounts of NNK were also detected (Fig. 5C). Statistical analysis was not performed for extracts of polyester since the extract prepared after 19 months of aging had only two experiments.

## Discussion

While the concentrations of some extractable THS chemicals in cotton terry cloth and polyester fleece changed during aging, in general THS chemicals remained on these fabrics for over 1.5 years after the last exposure to smoke. Nicotine and its derivatives,

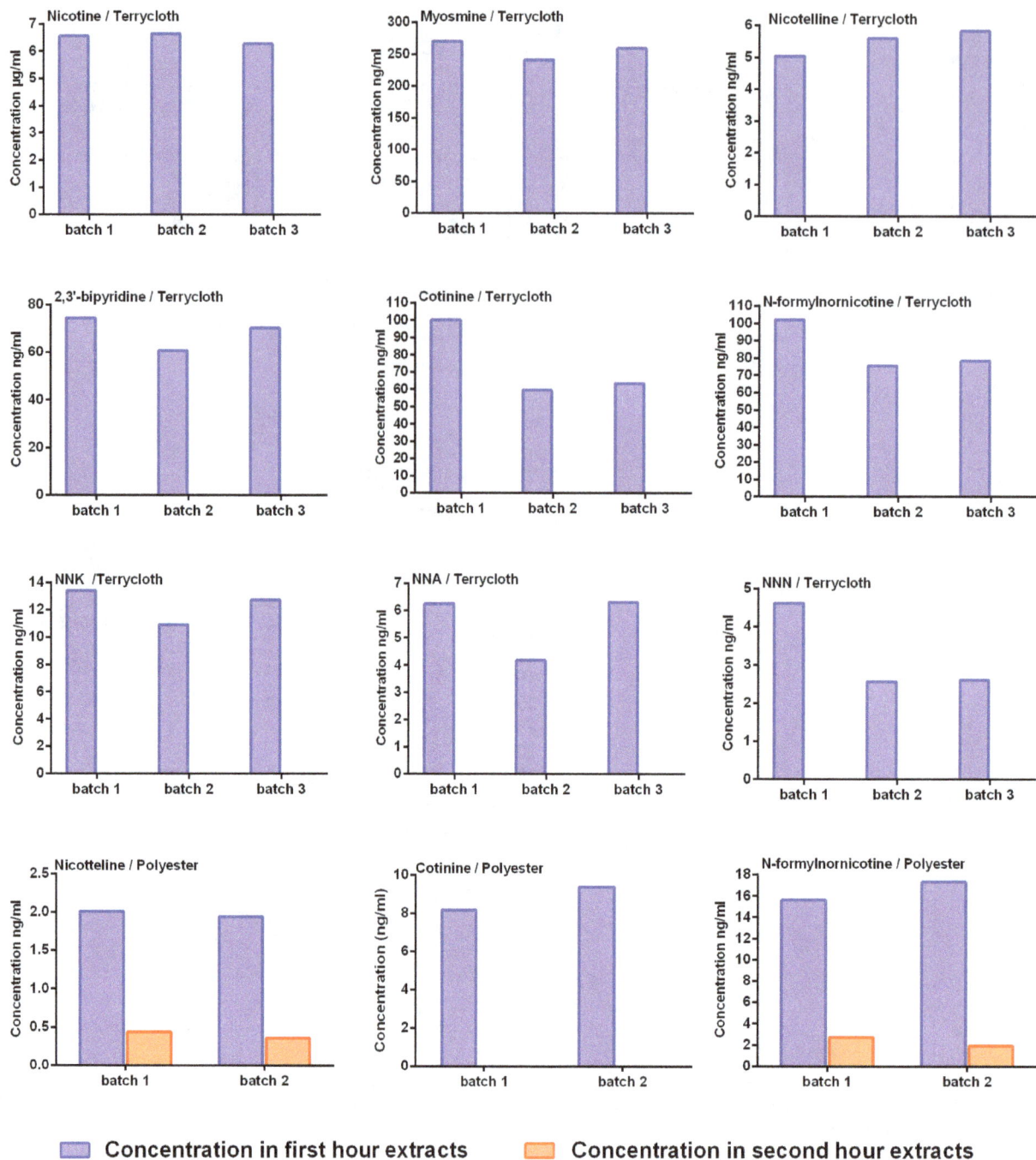

**Figure 3. Iterative aqueous extractions from terry cloth and polyester.** Extractions were done for a total 5 hours with extraction medium being replaced every hour. After every hour, extracts were analyzed for nicotine and its derivatives. Graphs represent chemical concentrations in three different batches of extracts. No chemicals were found in extracts after 1 hour for terry cloth and after 2 hours for polyester.

including NNK, a known carcinogen, were rapidly extracted from cotton fabric in an aqueous medium that is similar in composition to saliva and sweat and has a physiological pH. This implies that an infant who mouths cloth that has been exposed to cigarette smoke will be exposed to significant amounts of cigarette smoke toxicants. There was a large difference in the quantity of chemicals extracted from cotton cloth and polyester cloth, showing that natural and synthetic fibers have different abilities to bind and release THS chemicals. These observations are important in understanding human exposure to THS, devising strategies for

remediation of contaminated environments, and in developing regulatory policies for indoor use of tobacco products.

Changes in the concentration of an individual THS chemicals of on a surface depend on multiple processes including sorbtion, desorbtion and chemical reactions. Whether a chemical remains on a surface or rapidly desorbs and is removed by ventilation depends on its volatility and chemical properties. Whether a chemical reacts or remains intact depends on its chemical properties and the availability of other chemicals in the environment. With the exception of nicotelline, the chemicals we

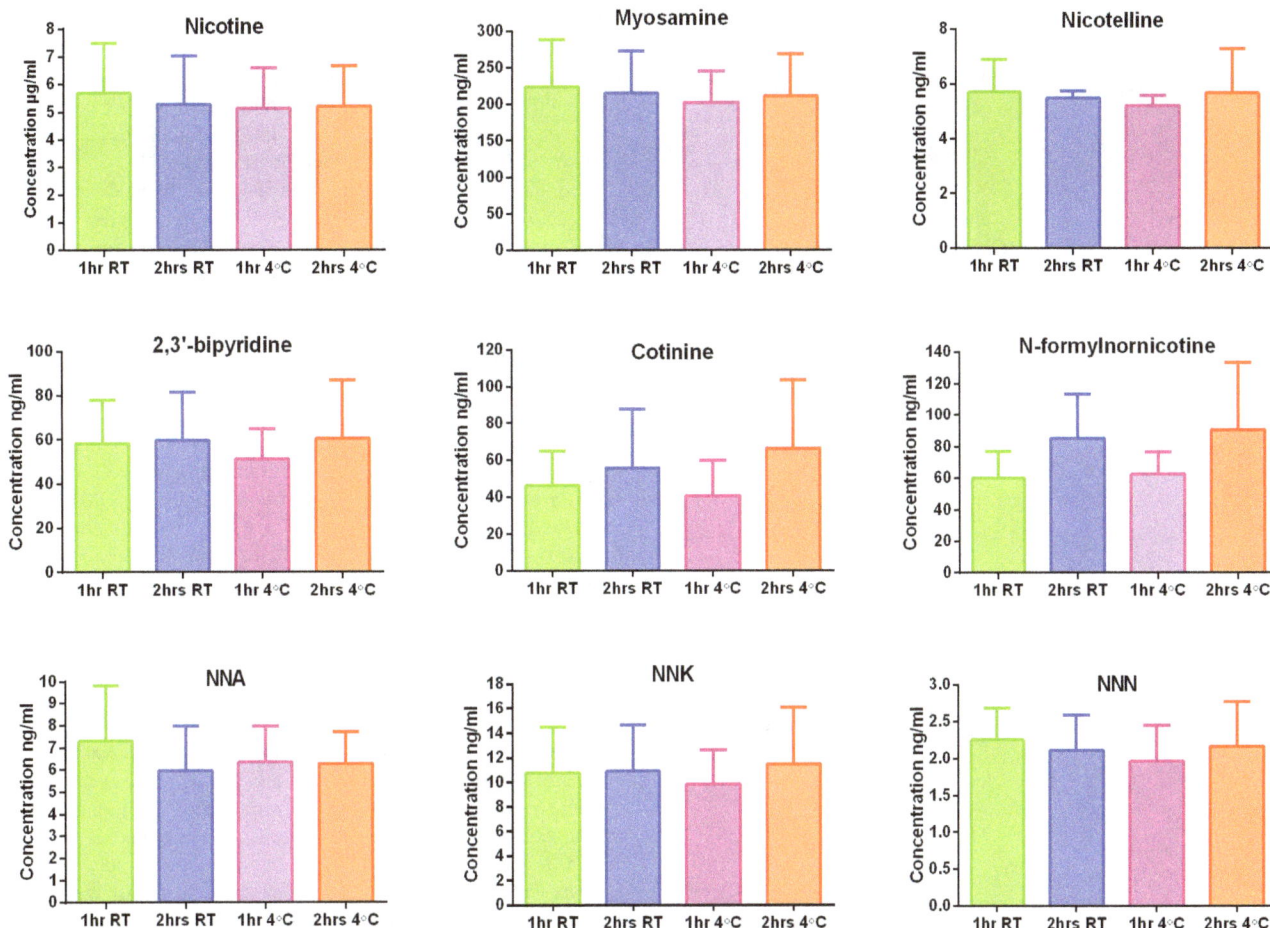

**Figure 4. Concentration of chemicals in aqueous extracts of THS from terry cloth when temperature and time of extraction were varied.** Extracts were made at RT and at 4°C for 1 and 2 hours. Each bar is the mean ± standard deviation of three experiments. Chemical concentrations did not vary significantly with temperature or time of extraction when tested by ANOVA.

analyzed are semivolatile organic compounds, which means they will be present in both the gas phase and solid phase at normal indoor temperatures. For terry cloth, myosmine, 2′,3′;-bipyridine, N-formylnornicotine and cotinine decreased significantly during aging, possibly due to breakdown into other chemicals, volatilization, or conversion reactions with the ambient environment.

The increased concentrations of both NNN and NNK at 16 months of aging followed by a decrease at 19 months could be due to formation of fresh TSNAs from settled nicotine before reaching

**Table 1.** Chemicals identified in aqueous THS extracts.

| Chemical | Terrycloth THS aqueous extract | | | Polyester THS aqueous extract | |
|---|---|---|---|---|---|
| | May 2012 | Oct 2012 | Jan 2013 | Jan 2013 | Sept 2013 |
| Nicotine µg/g | 105.8±25.5 | 112.92±8.59 | 69.6±30.4 | 0.557±0.82 | 1.689 |
| Cotinine µg/g | 0.899±0.13 | 1.04±0.14 | 0.446±0.04 | 0.269±0.06 | 0.18 |
| N-formylnor-nicotine µg/g | 3.9±0.72 | 1.138±0.13 | 1.047±0.27 | 0.427±0.09 | 0.420 |
| Myosmine µg/g | 4.844±0.31 | 4.518±0.25 | 3.010±0.12 | X | 0.066 |
| 2,3′-bipyridine µg/g | 1.242±0.08 | 1.196±0.03 | 0.681±0.06 | X | 0.026 |
| Nicotelline ng/g | 105.8±25.5 | 113.22±20.61 | 104.95±10.9 | 36.35±10.6 | 59.61 |
| NNA ng/g | 229.3±95.6 | 218.8±16.4 | 88.3±8.8 | X | X |
| NNK ng/g | 169.5±27.2 | 218.8±16.38 | 132.36±9.76 | X | 3.2 |
| NNN ng/g | 37.10±5.18 | 45.84±3.08 | 31.25±3.46 | X | X |

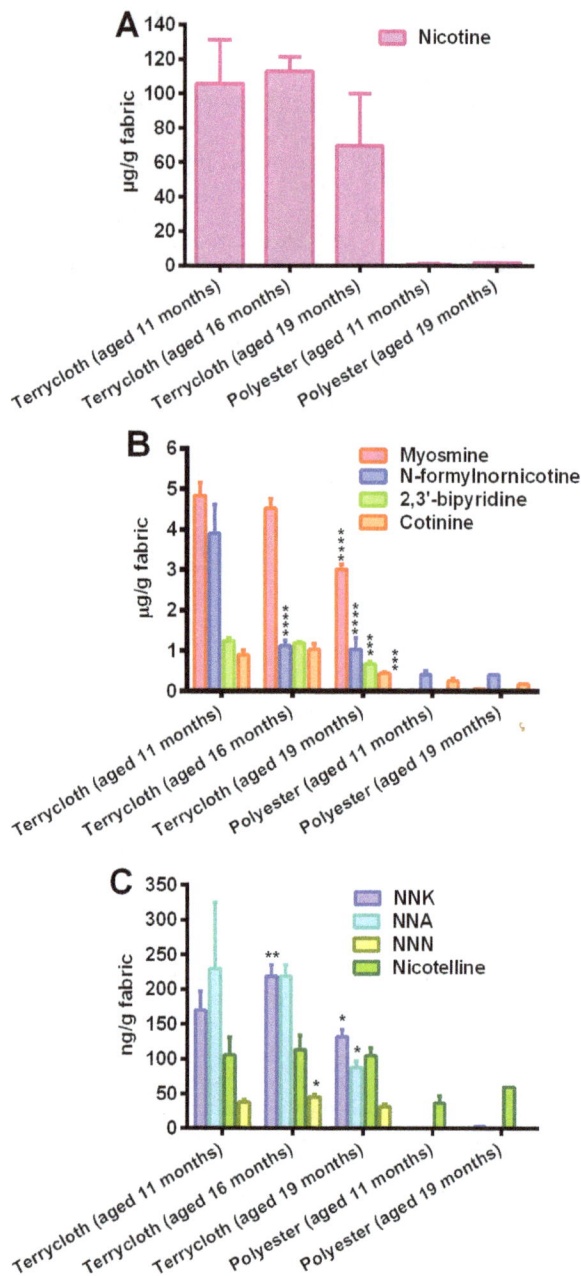

**Figure 5. Effect of aging on the concentration of nicotine and its derivatives in aqueous extracts of terry cloth and polyester exposed to THS.** Aqueous extracts of THS were made from terry cloth after 11, 16 and 19 months of aging and from polyester after 11 and 19 months of aging. (A) Nicotine and (B) Myosmine, N-formylnornicotine, bipyridine and cotinine were present in higher concentration than (C) nicotelline, NNK, NNA and NNN. Each bar is the average ± standard deviation of three extracts, except polyester (aged 19 months) which had only two experiments. ANOVA was used for testing significance followed by Dunnett's posthoc test in which comparisons were made to the samples aged for 11 months. **** $p < 0.0001$; *** $0.0001 < p < 0.001$; ** $0.001 < p < 0.01$; * $p < 0.05$.

other chemicals during aging. For polyester, the concentrations of all chemicals in aqueous extracts were very low. The increase in the number of chemicals that were present in the polyester sample that aged 19 months vs. 11 months may be an artifact caused by analyzing chemicals close to their lower limit of quantification (0.01 ng/ml to 1 ng/ml for different chemicals).

The difference in the concentrations of chemicals extracted from cotton terry cloth and polyester fleece may be due to their surface chemistry. Our data are in agreement with prior studies showing that polar substances like nicotine and dyes do not bind well to polyester [22], [23]. Cotton, which is made of cellulose, has three free hydroxyl groups/glucose monomer that can form hydrogen bonds with the polar groups on nicotine and its derivatives (Fig. 6) [22], [24], [25]. In contrast, polyester which is a polymer of terepthalic acid and ethylene glycol, is highly oleophilic [26], and its hydrophobicity tends to repel polar compounds. However, it is possible that these THS compounds adsorb more strongly to polyester than to cotton and the DMEM or methanol:HCl extractions we used are not rigorous enough to fully extract them from polyester. THS contains thousands more chemicals than we analyzed, including many non-polar, non-water soluble chemicals. The interactions of other classes of chemicals with indoor surface materials will need to be characterized in future studies.

Our data clearly show that fabrics found in indoor environments act as reservoirs for THS smoke chemicals. Although the samples used in this study received relatively light exposure to cigarette smoke, significant amounts of nicotine and related chemicals were extractable from cotton cloth 19 months after smoke exposure had stopped. In studies where smoke is released into a large chamber and allowed to mix with the air, age and interact with surfaces before particle concentrations are measured, the mass of particles emitted by a single cigarette ranges from 7–22 mg, with averages between 8 and 14 mg [27]–[30]. Using an average emission factor of 10 mg per cigarette, the cotton cloth

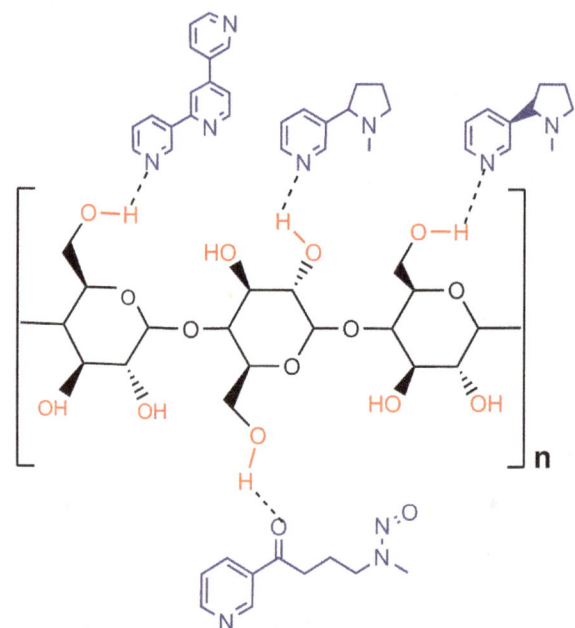

**Figure 6. Chemical interactions of nicotine and its derivatives with terry cloth through hydrogen bonds.** Terry cloth absorbs nicotine and related chemicals as these are polar in nature and can form hydrogen bonds with the free hydroxyl groups in terry cloth.

a threshold and starting to decrease due to further conversions. Although NNA concentrations did not change in our extracts, the formation of NNA may have occurred prior to our first extraction, after 11 months of aging. Also, NNA, being an aldehyde, is more reactive than NNN and NNK and could have combined with

**Table 2.** Estimated nicotine and TSNAs exposure to a toddler.

| Chemical | Inhalation Exposure Active Smoker | Inhalation Exposure Passive Smoker | Estimated Oral Exposure to THS |
|---|---|---|---|
| | Adult[a] | Toddler[b] | Toddler[c] |
| Nicotine | 22,000 µg/day [36] | 77.76 µg/day [37] | **529 µg/day** |
| TSNA | 7.2 µg/day [38] | 0.137 µg/day [39] | **2.2 µg/day** |

[a] Based on smoking 1 pack/day.
[b] Based on respiration rate of 30 breaths/minute and tidal volume of 60 ml in a room with 30 µg/m$^3$ nicotine and 53 ng/m$^3$ of TSNA.
[c] Based on 1 hour of mouthing 5 grams of terry cloth exposed to 133 cigarettes and aged 19 months.

was exposed to the equivalent of 133 cigarettes and the polyester was exposed to the equivalent of 185 cigarettes. These exposures translate to 7–9 days of exposure in a room where 20 cigarettes are smoked per day or 27–37 days of exposure in a room where 5 cigarettes are smoked per day. Exposure in our study occurred in a steel chamber and therefore THS chemicals did not have an opportunity to be removed by ventilation, open doors or open windows. Therefore while our experiment was done under controlled laboratory conditions, it does not exactly duplicate a real world situation. However, the fact we used a low number of cigarettes (approximately 133 for terry cloth) supports the idea that in a real world situation the concentrations of the chemicals studied could be much higher than reported here. For example in a home where one individual smokes a pack a day for one year, the total number of cigarettes consumed would be 7,280 in contrast to approximately 133 used in our study. While the health effects of these chemicals in THS residue are not yet known, it may become desirable or even necessary in the future to remediate property with THS residue before it is rented or sold [2]. Our data demonstrate that nicotine and related compounds, including two carcinogens, can easily be removed from cotton fabrics by standard washing methods.

Since indoor surfaces act as reservoirs of THS, toddlers and infants could be exposed to THS chemicals by sucking on household fabrics, and all age groups could be exposed dermally by touching contaminated surfaces. To evaluate the exposure that could be received from cotton fabric containing THS residue, we examined a hypothetical scenario for dermal exposure to an adult. An adult wearing a 500 g cotton outfit containing THS residue from 20 cigarettes will be exposed to about 7,894 µg of nicotine/day and 32.7 µg of TSNAs/day, with a small fraction of this contributing to intake, assuming that the outfit would be washed frequently and could reasonably contain THS from 20 cigarettes before being washed.

A more accurate scenario can be developed for ingestion exposure to a toddler, where the intake will be roughly equal to the exposure. The main source of THS exposure to a toddler would be through mouthing fabrics used in toys, drapes and upholstery that are not frequently washed and have long-term accumulation of THS. For terry cloth containing THS from about 133 cigarettes (as used in this study), a 12 kg toddler mouthing and sucking 5 grams of cloth for 1 hour would be exposed to 529 µg of nicotine/day and 2.2 µg of TSNAs/day. Since the exposure and intake are equal, the toddler would receive 44 µg/kg body weight of nicotine and 0.183 µg/kg body weight of TSNAs per day. These intake values for the toddler would be less than those received by an active smoker but higher than respiratory exposure in passive smokers (6.8× higher for nicotine and 16× higher for TSNAs) (Table 2). While information on the effects of pre and postnatal nicotine exposure comes largely from animal models and women on nicotine replacement therapy, data consistently show links between nicotine exposure early in life and subsequent cognitive impairment, attention deficit disorders as well as obesity, hypertension, type-2 diabetes, respiratory dysfunction and impaired fertility [31]–[33]. Although the intake value for TSNAs is much less than doses known to cause tumors in rodent models [34], the above scenarios may underestimate exposure if significant levels of chemicals were lost during the first 11 months of aging or if THS accumulates from more than 133 cigarettes. TSNAs contribute to pancreatic cancer [35]. It will be interesting to determine in future studies if there is a correlation between TSNA exposure during infancy and the recent increase in pancreatic or other types of cancer in adults. Exposure of toddlers to nicotine and TSNAs in THS is therefore a matter of concern and may need regulation.

## Conclusions

Our data show that under controlled laboratory conditions fabrics exposed to cigarette smoke retain significant concentrations of THS chemicals long after smoking has ceased. Estimated exposure to and uptake of nicotine and TSNAs from residual THS are above what toddlers would receive by inhaling environmental tobacco smoke. These observations coupled with recent reports linking THS exposure to adverse health effects support the idea that THS residues on indoor surfaces are a public health concern. Since THS chemicals do not spontaneously disappear from indoor surfaces, it may be important to actively remove them to reduce risk from THS exposure. Our data show that nicotine, nicotine-related alkaloids and TSNAs could be readily removed from cotton fabrics by washing, which could become a simple remediation procedure.

This study focused on THS that had aged in fabrics that are often used in homes and clothing. Studies are in progress to determine the levels of chemicals in freshly exposed household fabrics, such as carpets, drapes and upholstery, as well as the actual intake and uptake levels of THS chemicals in humans and if these concentrations are high enough to produce harm.

## Acknowledgments

We thank Pura Tech for her help handling the samples.

## Author Contributions

Conceived and designed the experiments: VB PJ SS PT. Performed the experiments: VB CH SS. Analyzed the data: VB PJ CH SS PT. Contributed reagents/materials/analysis tools: VB PJ SS PT. Wrote the paper: VB PJ SS PT.

## References

1. Fortmann AL, Romina AR, Sklar M, Pham V, Zakarian J, et al. (2010) Residual Tobacco Smoke in Used Cars: Futile Efforts and Persistent Pollutants. Nicotine Tob Res 12(10):1029–1036; doi: 10.1093/ntr/ntq144

2. Matt GE, Quintana PJ, Destaillats H, Gundel LA, Sleiman M, et al. (2011) Thirdhand tobacco smoke: emerging evidence and arguments for a multidisciplinary research agenda. Environ Health Perspect 119(9): 1218–1226; doi: 10.1289/ehp.1103500

3. Petrick LM, Sleiman M, Dubowski Y, Gundel LA, Destaillats H (2011) Tobacco smoke aging in the presence of ozone: A room-sized chamber study. Atmos Environ 45(28): 4959–4965; doi:10.1016/j.atmosenv.2011.05.076

4. Sleiman M, Gundel LA, Pankow JF, Jacob III P, Singer BC, et al. (2010) Formation of carcinogens indoors by surface-mediated reactions of nicotine with nitrous acid, leading to potential thirdhand smoke hazards. Proc Natl Acad Sci U S A 107: 6576–6581; doi: 10.1073/pnas.0912820107

5. Hammer TR, Fischer K, Mueller M, Hoefer D (2011) Effects of cigarette smoke residues from textiles on fibroblasts, neurocytes and zebrafish embryos and nicotine permeation through human skin. Int J Hyg Environ Health 214: 384–391; doi: 10.1016/j.ijheh.2011.04.007

6. Hang B, Sarker AH, Havel C, Saha S, Hazra TK, et al. (2013) Thirdhand smoke causes DNA damage in human cells. Mutagenesis 28(4): 381–391; doi: 10.1093/mutage/get013

7. Martins-Green M, Adhami N, Frankos M, Valdez M, Goodwin B, et al. (2014) Cigarette smoke toxicants deposited on surfaces: Implications for human health. PLoS ONE 9(1): e86391; doi:10.1371/journal.pone.0086391

8. Rehan VK, Sakurai R, Torday JS (2011) Thirdhand Smoke: A New Dimension to the Effects of Cigarette Smoke on the Developing Lung. Am J Physiol Lung Cell Mol Physiol 301(1): L1–8; doi: 10.1152/ajplung.00393.2010

9. CDC (Centers for Disease Control and Prevention) (2014) Surgeon General's Report: The Health Consequences of Smoking—50 Years of Progress. Available: http://www.cdc.gov/tobacco/data_statistics/sgr/50th-anniversary/index.htm. Accessed 28 January 2014.

10. DiCarlantonio G, Talbot P (1999) Inhalation of mainstream and sidestream cigarette smoke retards embryo transport and slows muscle contraction in oviducts of hamsters (Mesocricetus auratus). Biol of Reprod 61(3): 651–656; doi: 10.1095/biolreprod61.3.651

11. Gieseke C, Talbot P (2005) Cigarette smoke inhibits hamster oocyte pickup by increasing adhesion between the oocyte cumulus complex and oviductal cilia. Biol of Reprod 73(3): 443–451; doi 10.1095/biolreprod.105.041152

12. Riveles K, Tran V, Roza R, Kwan D, Talbot P (2007) Smoke from traditional commercial, harm reduction and research brand cigarettes impairs oviductal functioning in hamsters (Mesocricetus auratus) in vitro. Hum Reprod 22 (2): 346–355; doi: 10.1093/humrep/del380

13. Talbot P, Lin S (2011) Cigarette smoke's effect on fertilization and pre-implantation development: assessment using animal models, clinical data, and stem cells. Biol Res 44: 89–194.

14. Hofhuis W, de Jongste JC, Merkus PJFM (2003) Adverse health effects of prenatal and postnatal tobacco smoke exposure on children. Arch Dis Child 88:1086–1090; doi: 10.1136/adc.88.12.1086

15. Öberg M, Jaakkola MS, Woodward A, Peruga A, Prüss-Ustün A (2011) Worldwide burden of disease from exposure to second-hand smoke: a retrospective analysis of data from 192 countries. Lancet 377: 139–46; doi:10.1016/S0140-6736(10)61388-8

16. Lehman-McKeeman LD (2007) Absorption, distribution and excretion of toxicants. In: Klaassen CD editors. Casarett & Doull's Toxicology the Basic Science of Poisons. McGraw Hill Professional. pp. 131–160.

17. Matt GE, Quintana PJE, Fortmann AL, Zakarian JM, Galaviz VE, et al. (2013) Thirdhand smoke and exposure in California hotels: Non-smoking rooms fail to protect non-smoking hotel guests from tobacco smoke exposure. Tob Control 0: 1–9; doi:10.1136/tobaccocontrol-2012-05082

18. Schick SF, Farraro KF, Fang J, Nasir S, Kim J, et al. (2012) An Apparatus for Generating Aged Cigarette Smoke for Controlled Human Exposure Studies. Aerosol Sci Technol 46(11): 1246–1255; doi:10.1080/02786826.2012.708947

19. Jacob P 3rd, Goniewicz ML, Havel C, Schick CF, Benowitz NL (2013) Nicotelline: A Proposed Biomarker and Environmental Tracer for Particulate Matter Derived from Tobacco Smoke. Chem Res Toxicol 26(11): 1615–31; doi: 10.1021/tx400094y

20. Sarker AH, Chatterjee A, Williams M, Lin S, Havel C, et al. NEIL2 protects against oxidative DNA damage induced by sidestream smoke in human cells. PlosOne. In Press.

21. Pang X, Lewis AC (2011) Carbonyl compounds in gas and particle phases of mainstream cigarette smoke. Sci Total Environ 409(23): 5000–50009; doi: 10.1016/j.scitotenv.2011.07.065

22. Petrick L, Destaillats H, Zouev I, Sabach S, Dubowski Y (2010) Sorption, desorption, and surface oxidative fate of nicotine. Phys Chem Chem Phys 12: 10356–10364; doi:10.1039/c002643c

23. Koh J (2011) Dyeing with Disperse Dyes. In: Textile Dyeing (Hauser PJ, eds). Intech, 195–222; doi: 10.5772/800

24. Nishiyama Y, Langan P, Chanzy H (2002) Crystal Structure and Hydrogen-Bonding System in Cellulose Iβ from Synchrotron X-ray and Neutron Fiber Diffraction. J Am Chem Soc 124: 9074–9082.

25. Senthilkumar L, Ghanty TK, Kolandaivel P, Ghosh SK (2012) Hydrogen-bonded complexes of nicotine with simple alcohols. Int J Quantum Chem 112: 2787–2793; doi: 10.1002/qua.23304

26. Bendak A, El-Marsafi SM (1991) Effects of chemical modifications on polyester fibers. Journal of Islamic Academy of Sciences 4: 4, 275–284.

27. Daisey JM, Mahanama KR, Hodgson AT (1998) Toxic volatile organic compounds in simulated environmental tobacco smoke: emission factors for exposure assessment. J Expo Anal Environ Epidemiol 8(3): 313–34.

28. Klepeis NE, Apte MG, Gundel LA, Sextro RG, Nazaroff W (2003) Determining size-specific emission factors for environmental tobacco smoke. Aerosol Sci Tech 37(10): 780–90.

29. Leaderer PP, Hammond SK (1991) Evaluation of vapor phase nicotine and respirable suspended particle mass as markers for environmental tobacco smoke. Environ Sci Technol 25: 770–777.

30. Repace JL (2007) Exposure to secondhand smoke. In: Exposure Analysis (Ott WR, Steinemann AC, Wallace LA, ed). Boca Raton FL:CRC Press, 201–35.

31. Yolton K, Dietrich K, Auinger P, Lanphear BP, Hornung R (2005) Exposure to Environmental Tobacco Smoke and Cognitive Abilities among U.S. Children and Adolescents. Environ Health Perspect 113(1): 98–103.

32. Dwyer JB, McQuown SC, Leslie FM (2009) The Dynamic Effects of Nicotine on the Developing Brain. Pharmacol Ther 122(2): 125–139; doi:10.1016/j.pharmthera.2009.02.003

33. Bruin J, Gerstein HC, Holloway AC (2010) Long-Term Consequences of Fetal and Neonatal Nicotine Exposure: A Critical Review. Toxicol Sci 116 (2): 364–374; doi: 10.1093/toxsci/kfq103

34. Brown BG, Borschke AJ, Dolittle DJ (2003) An analysis of the role of tobacco-specific nitrosamines in the carcinogenicity of tobacco smoke. Nonlinearity Biol Toxicol Med 1(2): 179–198.

35. Edderkaoui M, Thrower E (2013) Smoking and pancreatic disease. J Cancer Ther 4(10A): 34–40; doi:10.4236/jct.2013.410A005

36. Foulds J, Delnevo C, Ziedonis DM, Steinberg MB (2008) Health Effects of Tobacco, Nicotine, and Exposure to Tobacco Smoke Pollution. In: Brick J editor. Handbook of the Medical Consequences of Alcohol and Drug Abuse. New York: Haworth Press. pp. 423–459.

37. Okoli CTC, Kelly T, Hahn EJ (2007) Secondhand smoke and nicotine exposure: A brief review. Addict Behav 32: 1977–1988; doi:10.1016/j.addbeh.2006.12.024

38. Ashley DL, O'Connor RJ, Bernert JT, Watson CH, Polzin GM, et al. (2010) Effect of Differing Levels of Tobacco-Specific Nitrosamines in Cigarette Smoke on the Levels of Biomarkers in Smokers. Cancer Epidemiol Biomarkers Prev 19: 1389–1398; doi: 10.1158/1055-9965.EPI-10-0084

39. Hecht SS (2003) Carcinogen derived biomarkers: applications in studies of human exposure to secondhand tobacco smoke. Tob Control 13: i48–i56; doi: 10.1136/tc.2002.002816

# The Effects of Fruiting Positions on Cellulose Synthesis and Sucrose Metabolism during Cotton (*Gossypium hirsutum* L.) Fiber Development

**Yina Ma, Youhua Wang, Jingran Liu, Fengjuan Lv, Ji Chen, Zhiguo Zhou***

Key Laboratory of Crop Growth Regulation, Ministry of Agriculture, Nanjing Agricultural University, Nanjing, Jiangsu Province, PR China

## Abstract

Cotton (*Gossypium hirsutum* L.) boll positions on a fruiting branch vary in their contribution to yield and fiber quality. Fiber properties are dependent on deposition of cellulose in the fiber cell wall, but information about the enzymatic differences in sucrose metabolism between these fruiting positions is lacking. Therefore, two cotton cultivars with different sensitivities to low temperature were tested in 2010 and 2011 to quantify the effect of fruit positions (FPs) on fiber quality in relation to sucrose content, enzymatic activities and sucrose metabolism. The indices including sucrose content, sucrose transformation rate, cellulose content, and the activities of the key enzymes, sucrose phosphate synthase (SPS), acid invertase (AI) and sucrose synthase (SuSy) which inhibit cellulose synthesis and eventually affect fiber quality traits in cotton fiber, were determined. Results showed that as compared with those of FP1, cellulose content, sucrose content, and sucrose transformation rate of FP3 were all decreased, and the variations of cellulose content and sucrose transformation rate caused by FPs in Sumian 15 were larger than those in Kemian 1. Under FP effect, activities of SPS and AI in sucrose regulation were decreased, while SuSy activity in sucrose degradation was increased. The changes in activities of SuSy and SPS in response to FP effect displayed different and large change ranges between the two cultivars. These results indicate that restrained cellulose synthesis and sucrose metabolism in distal FPs are mainly attributed to the changes in the activities of these enzymes. The difference in fiber quality, cellulose synthesis and sucrose metabolism in response to FPs in fiber cells for the two cotton cultivars was mainly determined by the activities of both SuSy and SPS.

**Editor:** Jinfa Zhang, New Mexico State University, United States of America

**Funding:** This work was supported by the National Natural Science Foundation of China (Grant number 30971735) (http://isisn.nsfc.gov.cn). The funders had no role in study design, data collection and analysis, decision to publish, or preparation of the manuscript.

**Competing Interests:** The authors have declared that no competing interests exist.

* E-mail: giscott@njau.edu.cn

## Introduction

Cotton (*Gossypium hirsutum* L.) fiber is an important raw material for the textile industry. Thus, its yield and quality are the key criteria for cotton fiber value evaluation. The cotton plant has a prominent main stem and an indeterminate growth habit [1]. It has been shown that bolls at different fruiting positions (FPs) can produce different yield and fiber quality [2] [3].

Cotton yield is mainly determined by boll number and boll size [4]. Previous reports indicated that the position of the first node on a sympodial branch (FP1) contributed more to higher yield than the other FPs did on the same sympodial branch, and the relative contribution of FPs 1, 2, and 3 accounted for about 60%, 30% and 10% of the total yield of seed cotton, respectively [1] [2] [5] [6]. However, in some fields where cotton yield was high (i.e. 7657Kg·ha-1) in the Yangtze River Valley of China, the boll retention rate on the distal sites (e.g. FP3 and greater) reached as high as 58.8% [7]. Thus, the increased boll retention rate of distal FPs might be the key for improving fiber yield in the fields where cotton yield potential is high.

Fiber quality depends on complex interactions among the genetic and physiological factors. The effect of the growth environment on the genetic potential of a genotype modulates fiber properties to varying degrees [8] [9] [10]. For example, application of water or fertilizer and the inevitable seasonal shifts such as temperature, day length, and insolation all could realize the changes of genetic potential [11] [12] [13] [14]. Previous documents on fiber qualities under environmental factors were mainly concentrated on the adaxial fruiting positions (e.g. FP1 and FP2) [3] [15] [16] [17], but nothing has been conducted on fiber qualities of the distal's (e.g. FP3 and greater).

Up to 90% of mature cotton fiber is consisted of cellulose. Thus, the process of cotton fiber formation primarily is a process of cellulose synthesis [18]. Sucrose is the initial carbon source for cellulose synthesis and supplies the UDP-glucose (UDP-G) as the immediate substrate for cellulose polymerization [19] [20]. A number of enzymes are involved in the sucrose metabolism and can be classified into two types, sucrose-synthesis enzymes and sucrose-decomposition enzymes [21]. Sucrose phosphate synthase (SPS), which is consider as a major regulator in organs and tissues adapted to cold and drought stresses, regulates sucrose synthesis [22] [23]. Invertases, especially the acid invertase (AI) catalyse the irreversible hydrolysis of sucrose to glucose and fructose [24] [25]. In addition to AI, sucrose synthase (SuSy) can both degrade and synthesize sucrose, but its function in cotton fiber is primarily on reversible sucrose cleavage [18] [26].

**Table 1.** Mean daily temperature, mean daily maximum temperature, mean daily minimum temperature, mean diurnal temperature difference, total solar radiation and cumulative photo-thermal index during cotton fiber development period from flowering date to boll opening date on FP1 and FP3 of two cultivars during 2010–2011.

| Years | Fruiting positions (FPs) | Planting date (dd-mm) | Flowering dates (dd-mm) | Boll opening dates (dd-mm) | Fiber development period (d) | MDT[a] (°C) | MDT max[b] (°C) | MDT min[c] (°C) | MDT dif[d] (°C) | MDSR[e] (MJ·m$^{-2}$) | PTI[f] (MJ·m$^{-2}$) |
|---|---|---|---|---|---|---|---|---|---|---|---|
| 2010 | FP1 | 25-Apr | 29-Jul | 12-Sep | 46 | 29.0[g] | 33.1 | 25.8 | 7.3 | 17.1 | 683.4 |
| | FP3 | 25-Apr | 11-Aug | 27-Sep | 48 | 26.6 | 30.6 | 23.7 | 6.9 | 14.3 | 549.9 |
| | | | | | CV[h], % | 4.27 | 3.87 | 4.3 | 2.49 | 9.17 | 10.83 |
| 2011 | FP1 | 25-Apr | 27-Jul | 14-Sep | 50 | 26.7 | 30.7 | 24.0 | 6.7 | 14.2 | 630.5 |
| | FP3 | 25-Apr | 11-Aug | 1-Oct | 52 | 24.5 | 28.3 | 21.6 | 6.7 | 12.7 | 479.8 |
| | | | | | CV, % | 4.38 | 3.97 | 5.18 | 0.11 | 5.84 | 13.58 |

[a]MDT, mean daily temperature.
[b]MDTmax, mean daily maximum temperature.
[c]MDTmin, mean daily minimum temperature.
[d]MDTdif, mean diurnal temperature difference.
[e]MDSR, mean daily solar radiation.
[f]PTI, cumulative photo-thermal index during cotton fiber development period.
[g]Weather data were provided by Nanjing Weather Station, which was located nearby the experimental site.
[h]CV, coefficient of variation.

Base on the background described above, we designed an experiment to study the response of related enzymes involved in sucrose metabolism at various stages of development in cotton fibers on different FPs, aiming at 1) finding the sensitive enzymes to FP effect in sucrose metabolism for the two cultivars; 2) clarifying the relationship between sucrose metabolism, cellulose synthesis and fiber qualities. Finally, elucidate the differences and their physiological mechanisms among fiber cells of the bolls at different FPs. This study could provide further supplement and extension for the research of fruiting positions and would be valuable for cotton cultivators to improve cotton yield and fiber quality by distal FPs.

## Materials and Methods

### Plant material and experimental design

Field experiments were conducted at the Pailou experimental station of the Nanjing Agricultural University, Nanjing (32°02'N and 118°50'E), Jiangsu (the Yangtze River Valley), China, in the Yangtze River Valley from 2010 to 2011. The soil at the experimental site was clayed, mixed, thermic, Typic alfisols (udalfs; FAO luvisol) in 20 cm depth of the soil profile, and the nutrient contents of soil before planting cotton contained 18.5 and 16.3 g·kg$^{-1}$ organic matter, 1.1 and 1.0 g·kg$^{-1}$ total N, 64.4 and 50.2 mg·kg$^{-1}$ available N, 17.9 and 16.8 mg·kg$^{-1}$ available P, and 102.3 and 96.4 mg·kg$^{-1}$ available K. Cotton cultivars had different sensitivities to low temperature, 14 diverse cultivars which were widely grown in the Yangtze River Valley in China were studied by Wang et al. [27] with different mean daily temperature during fiber development period. Based on the variance of fiber strength, these cultivars were clustered into three groups as a temperature-sensitive group (typical for Sumian 15), a moderately sensitive group (typical for NuCOTN 33B) and a temperature-insensitive group (typical for Kemian 1). Bolls from different FPs in the same sympodial branches with different flowering dates, therefore, Kemian 1 (temperature-insensitive) and Sumian 15 (temperature-sensitive) were selected for this study. Cotton seeds were sown in a nursery bed on 25 April, and seedlings with three true leaves were transplanted to field. Each plot size was 5.6 m wide by 6 m long, with row spacing of 80 cm and interplant spacing of 25 cm. Three replications for each cultivar were assigned randomly in the field. The nitrogen fertilizer applied 40% before transplanting and 30%, 30% applied at first flowering and peak flowering for each cultivar, respectively. Furrow-irrigation was applied as needed to minimize the moisture stress during each season.

### Sampling and processing

Cotton flowers were labeled at anthesis with a tag listing the date at FPs 1 and 3 of the 7[th] sympodial branches, respectively (hereafter FP1 and FP3). The 7[th] sympodial branch is located in the middle of cotton canopy, which has better fiber quality than the bottom and upper branches do. It is usually used as the typical branch to research in document [28] [29] [30]. White flowers were tagged for each cultivar on the same day, no more than 3 days after the start of tagging, to ensure that the tagged flowers were of equivalently metabolic and developmental ages for each cultivar. The 17, 31, and 45 days post anthesis (DPA) are usually regarded as the key representative stages to study the physiological characteristics during cotton fiber development [18] [21] [31] [32]. Therefore, subsequently, the labeled boll samples (about 6–8 bolls in each cultivar) were collected at the 17[th], 31[st] and 45[th] DPA, respectively. Cotton boll samples were harvested at 9:00–11:00 am, and fibers were excised from the bolls with a scalpel and were immediately put into liquid nitrogen for subsequent measurement. Tagged bolls (about 10–15 bolls) in each replications were hand harvested after bolls opened and ginned in individual groups according to each fruiting position.

### Fiber quality, cellulose content and sucrose content quantification

Ginned fiber from each group was sent to the Cotton Quality Supervision, Inspection, and Testing Center of China Ministry of Agriculture for quality analysis. Fiber quality including fiber upper-half mean length (UHML), uniformity index (UI), strength

**Table 2.** Fiber properties of field-grown cotton on FP1 and FP3 of two cultivars during 2010–2011.

| Years | Cultivars | Fruiting positions (FPs) | UHML[a] (mm) | UI[b] (%) | MIC[c] | EL[d] (%) | ST[e] (cN tex$^{-1}$) |
|---|---|---|---|---|---|---|---|
| 2010 | Kemian 1 | FP1 | 30.5a[f] | 84.3a | 4.9a | 6.4a | 30.3a |
| | | FP3 | 29.6b | 84.2a | 4.7b | 6.4a | 29.3b |
| | | CV[g],% | 1.58 | 0.11 | 2.77 | 0.39 | 1.78 |
| | Sumian 15 | FP1 | 29.8a | 84.0a | 4.7a | 6.2a | 29.6a |
| | | FP3 | 28.7b | 83.5a | 4.4b | 6.2a | 28.4b |
| | | CV,% | 1.77 | 0.34 | 4.12 | 0.28 | 2.07 |
| 2011 | Kemian 1 | FP1 | 30.7a | 85.1a | 4.9a | 6.3a | 31.4a |
| | | FP3 | 29.8b | 84.9a | 4.6b | 6.2a | 29.9b |
| | | CV,% | 1.38 | 0.10 | 3.31 | 0.34 | 2.42 |
| | Sumian 15 | FP1 | 30.0a | 84.6a | 4.7a | 6.1a | 30.5a |
| | | FP3 | 28.8b | 84.3a | 4.3b | 6.1a | 28.7b |
| | | CV,% | 2.04 | 0.18 | 4.36 | 0.25 | 3.04 |
| Source of variation | | | | | | | |
| Years | | | ns [h] | ns | ns | ns | ns |
| Cultivars | | | ** [h] | ns | ** | * [h] | * |
| Fruiting positions | | | ** | ns | ** | ns | ** |
| Years × Cultivars | | | ns | ns | ns | ns | ns |
| Years × Fruiting positions | | | ns | ns | ns | ns | ns |
| Cultivars × Fruiting positions | | | ns | ns | ns | ns | ns |
| Years × Cultivars × Fruiting positions | | | ns | ns | ns | ns | ns |

[a]UHML, fiber upper-half mean length.
[b]UI, uniformity index.
[c]MIC, micronaire value.
[d]EL, elongation percentage.
[e]ST, fiber strength.
[f]Values followed by a different letter between fruiting positions are significantly different at $P=0.05$ probability level. Each value represents the mean of three replications.
[g]CV, coefficient of variation.
[h]* and ** indicate significant differences at $P \le 0.05$ and 0.01 probability levels, respectively. ns, not significant ($P \ge 0.05$).

(ST), elongation (EL) and micronaire (MIC) of each lint sample was read with a high volume instrument (HVI). Sucrose and glucose were extracted by a modified method of Pettigrew [33]. The sucrose and glucose assay was conducted according to the method described by Hendrix [34]. Sucrose transformation rate was calculated according to Shu et al. [21]. Fibers were digested in an acetic-nitric reagent, and the cellulose content was measured with anthrone according to the method described by Updegraff [35].

### Enzymatic analyses

Enzyme extracts were prepared essentially as described by King et al. [36]. Soluble acid invertase (AI) was measured by incubation of 100 µl of extract with 1 M sucrose in 200 mM acetic acid-NaOH (pH 5.0) in a total volume of 2.5 ml [36]. Reactions were started by incubating at 30°C for 30 min. The reactions were stopped by adding 1 ml of 3,5-dinitrosalicylic acid (DNS), and then boiled for 5 min. Glucose content was determined by relating the spectrophotometer metrically at 540 nm. Sucrose synthase (SuSy) activity was assayed by measuring the level of fructose formed from the cleavage of sucrose [36]. Each reaction contained 20 mM Pipes–KOH (pH 6.5), 100 mM sucrose, 2 mM UDP, and 200 µl of extract in a total volume of 650 µl. Reactions were

started by incubating at 30°C for 30 min. The reactions were stopped by adding 250 µl of 0.5 M Tricine–KOH (pH 8.3), and then boiled for 10 min. The amount of fructose in SuSy reactions was determined and calculated from a standard curve of fructose at 540 nm. Sucrose phosphate synthase (SPS) activity was assayed by measuring the synthesis of sucrose-6-P [23]. Each reaction contained 14 mM UDP-glucose, 50 mM fructose-6-P, 50 mM extraction buffer, 50 mM MgCl$_2$ and 200 µl of extract in a total volume of 650 µl. The reaction was started by incubating the enzyme extract at 30°C for 30 min. The reaction was stopped by adding 100 µl of 2 N NaOH and by boiling for 10 min at 100°C to destroy unreacted hexoses and hexose phosphates. The 1 ml of 0.1% (w/v) resorcin in 95% (v/v) ethanol was added and then incubated at 80°C for 30 min. Sucrose-6-P content was calculated based on a standard curve measured at 480 nm. UDPG and F-6-P were purchased from Sigma-Aldrich.

### Weather data and data analysis

Weather data across the two-year study period were collected from an established local weather station (Nanjing Weather Station) located near the experimental site (about 10 m away from the field)(Table 1). Analysis of variance was conducted with SPSS statistic package Version 17.0 and the difference between mean

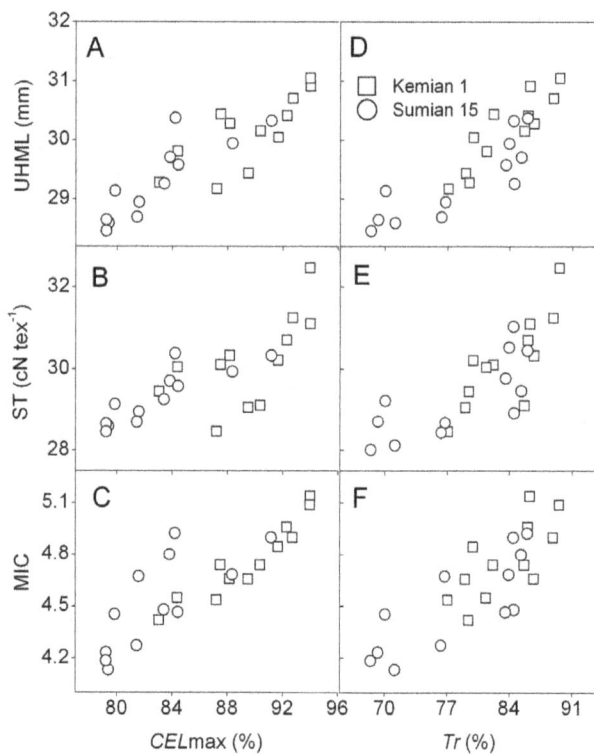

**Figure 1. Correlation coefficients among parameters in fibers on FP1 and FP3 of two cultivars during 2010–2011.** The coefficient between UHM and *CEL*max (A), ST and *CEL*max (B), MIC and *CEL*max (C), UHM and *Tr* (D), ST and *Tr* (E), and MIC and *Tr* (F) in fibers on FP1 and FP3 of two cultivars. UHML-fiber upper-half mean length, MIC-micronaire value, ST-fiber strength, *CEL*max-maximum cellulose content, *Tr*-sucrose transformation rate. * and **, significant differences at $P=0.01$ and $P=0.05$ probability levels, respectively. $n=12$, $R_{0.05}=0.576$, $R_{0.01}=0.707$.

values greater than the LSD ($P=0.05$) was determined as significant. The coefficient of variation (CV, %) was calculated as the ratio of the standard deviation to the mean. According to

Zhao et al. [37], cumulative photo-thermal index (*PTI*), which represents the effect of temperature and radiation during cotton fiber development period, was calculated by equation (Eq)(1).

$$PTI_i = RTE_i \times PAR_i \qquad (1)$$

where RTE refers to daily relative thermal effectiveness, according to the non-linear response curves of boll development to temperature [38], $RTE_i$ is calculated by Eq. (2), and the relative temperature effect (RTE (*T*)) is calculated by Eq. (3). PAR is daily photosynthetically active radiation.

$$RTE_i = \\ 0.5 \times RTE(T_{avg}) + 0.25 \times RTE(T_{max}) + 0.25 \times RTE(T_{min}) \qquad (2)$$

$$RTE(T) = \begin{cases} 0, & T > T_c, or, T < T_b \\ \left(\dfrac{T-T_b}{T_o-T_b}\right)^{1+\left(\frac{T_o-T}{T_o-T_b}\right)} \times \left(\dfrac{T_c-T}{T_c-T_o}\right)^{\frac{T_c-T_o}{T_o-T_b}}, & T_b \le T \le T_o \\ \left(\dfrac{T-T_b}{T_o-T_b}\right) \times \left(\dfrac{T_c-T}{T_c-T_o}\right)^{\frac{T_c-T_o}{T_o-T_b}}, & T_o < T \le T_c \end{cases} \qquad (3)$$

where $T_{avg}$, $T_{max}$, and $T_{min}$ refer to the daily average, maximum, and minimum air temperature, respectively. $T_b$, $T_o$, and $T_c$ are the cardinal temperature values (base, optimum, and ceiling temperatures) for development. They were retained as 15, 30, and 35°C, respectively.

## Results and Discussion

### Environmental conditions

The flowering date and cotton fiber development period (CFDP), i.e. from flowering date to boll opening date, were affected by the boll positions (Table 1). Generally, at the same planting date, the flowering date of bolls on FP1 was about two weeks earlier than that on FP3 in the same sympodial branches

**Table 3.** Cellulose content, sucrose content and sucrose transformation rate in cotton fibers on FP1 and FP3 of two cultivars during 2010–2011.

| Years | Cultivars | Fruiting positions (FPs) | Cellulose content (%) | | | Sucrose content (mg g⁻¹ DW) | | | $Tr^a$ |
|---|---|---|---|---|---|---|---|---|---|
| | | | 17DPA[b] | 31DPA | 45DPA | 17DPA | 31DPA | 45DPA | (%) |
| 2010 | Kemian 1 | FP1 | 46.88a[c] | 74.55a | 92.22a | 11.91a | 6.68a | 1.67a | 85.93a |
| | | FP3 | 38.79b | 68.80b | 89.46b | 9.80b | 6.17b | 2.09a | 78.69b |
| | Sumian15 | FP1 | 43.18a | 70.25a | 83.82a | 10.70a | 6.21a | 1.58a | 85.25a |
| | | FP3 | 34.91b | 63.41b | 80.81b | 8.19b | 5.47b | 2.08a | 74.71b |
| 2011 | Kemian 1 | FP1 | 41.93a | 74.16a | 91.62a | 12.70a | 6.40a | 1.49a | 88.31a |
| | | FP3 | 30.85b | 66.20b | 84.98b | 9.95b | 5.52b | 1.89b | 81.04b |
| | Sumian15 | FP1 | 40.82a | 71.04a | 87.99a | 11.79a | 5.75a | 1.89a | 84.01a |
| | | FP3 | 27.38b | 62.20b | 79.43b | 8.19b | 4.59b | 2.52b | 69.23b |

[a]*Tr*, sucrose transformation rate.
[b]DPA, days post anthesis.
[c]Values followed by a different letter between fruiting positions are significantly different at $P=0.05$ probability level. Each value represents the mean of three replications.

**Table 4.** Correlation coefficient of environmental factors during cotton fiber development period with maximum/minimum sucrose content and sucrose transformation rate in cotton fibers on FP1 and FP3 of two cultivars during 2010–2011.

| Correlation with | MDT | MDTmax | MDTmin | MDTdif | MDSR | PTI |
|---|---|---|---|---|---|---|
| SUCmax[a] | 0.512 | 0.514 | 0.558 | 0.168 | 0.446 | 0.740* [d] |
| SUCmin[b] | −0.666 | −0.667 | −0.690 | −0.423 | −0.625 | −0.778* |
| Tr[c] | 0.625 | 0.626 | 0.658 | 0.336 | 0.572 | 0.781* |

[a]SUCmax, maximum sucrose content.
[b]SUCmin, minimum sucrose content.
[c]Tr, sucrose transformation rate.
[d]n = 8, $R_{0.05} = 0.707$, $R_{0.01} = 0.834$. *, significant differences at $P = 0.05$ probability level.

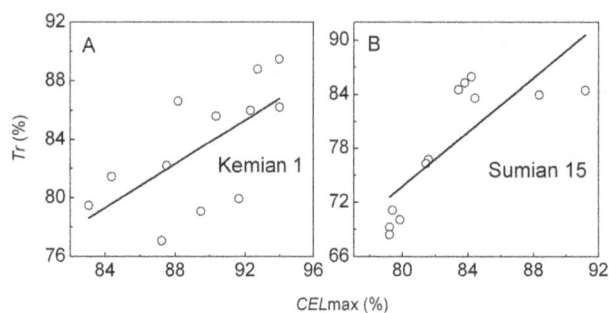

**Figure 2. Correlation coefficients between cellulose content and sucrose transformation rate in fibers during 2010–2011.** The coefficient between CELmax and Tr in fibers on FP1 and FP3 of Kemian 1 (A), CELmax and Tr in fibers on FP1 and FP3 of Sumian 15 (B). CELmax-maximum cellulose content, Tr-sucrose transformation rate. * and **, significant differences at $P = 0.01$ and $P = 0.05$ probability levels, respectively. n = 12, $R_{0.05} = 0.576$, $R_{0.01} = 0.707$.

[5]. But the CFDPs differed by only one or two-days between FP1 and FP3. Thus, the differences between FP1 and FP3 were primarily due to the changes in environmental conditions at the various flowering dates. The coefficients of variance of cumulative photo-thermal index (PTI) during CFDP were higher than those of mean daily temperature (MDT), mean daily maximum (MDT max), mean daily minimum (MDT min), mean diurnal temperature difference (MDTdif), and mean daily solar radiation (MDSR) in both 2010 and 2011 (Table 1). This indicated that the difference in environmental conditions during the fiber development period for different FPs was primarily on PTI. The PTI of FP3 was 19.5–23.9% lower than that of FP1.

### Fiber quality in cotton fiber

A number of the fiber quality traits including UHML, ST and MIC were significantly affected by FPs in both 2010 and 2011 (Table 2). Compared to those of FP1, UHML, ST and MIC of FP3 were all decreased in both Kemian 1 and Sumian 15. However, EL and UI were not affected by FPs in any year. When both years were combined and fiber qualities were analyzed, there was no significant effect of year and interaction of years × cultivars, years × fruiting positions and cultivars × fruiting positions for all fiber qualities. However, there were highly significant differences ($P<0.01$) in UHML and MIC, and significant differences ($P<0.05$) in EL and ST between cultivars but there was no significant difference ($P>0.05$) of cultivars for UI. The two cultivars, Kemian 1 and Sumian 15, had different ranges of UHML, ST and MIC between FP1 and FP3. The CVs of UHML, ST and MIC for Sumian 15 was greater than that of Kemian 1. These results indicated that cotton fiber quality was more easily affected by FPs in Sumian 15 than in Kemian 1.

Previous studies have documented that fiber length is largely dependent on genetic factors, while fiber maturity properties, which are dependent on deposition of photosynthates in the fiber cell wall, are more sensitive to changes in the growth environment [9] [39]. However, several studies have indicated that fiber length could be significantly affected by cool temperature or planting dates [10] [16] [40] [41]. Thus, fiber qualities such as length, strength, and micronaire etc. are probably affected by the changing growth environment. Previous reports about FP effects on fiber quality mainly concentrated on longitudinal direction of the main stem [42] [43], but few studies paid attentions to the horizontal direction. In term of the horizontal direction, Heitholt

[44] found that FP had no effect on fiber strength, maturity and micronaire values, whereas Pettigrew [15] documented the opposite results. In addition, Davidonis et al. [3] indicated that boll position of the horizontal direction could affect the fiber quality indices such as fiber length. Hence, FP effects on fiber quality traits are not consistent, and the possible reason for this disparity might be due to the differences in cultivars and the various environment conditions during the CFDP. Our results showed that fiber length, strength and micronaire values from FP1 were significantly higher ($P<0.05$) than those from FP3 (Table 2), which were consistent with those reported by Pettigrew [15] on fiber strength and by Davidonis [3] on fiber length and micronaire values. Since the difference in the effects caused by environmental conditions during the CFDP for different FPs was primarily on PTI (Table 1), thus, our results have indicated that the shorter and weaker fibers from FP3 are correlated with its lower PTI.

### Cellulose content and sucrose content in cotton fiber

Cellulose content in cotton fibers increased from 17 DPA (Table 3). In both cultivars Kemian 1 and Sumian 15, the cellulose content at FP1 was significantly higher ($P<0.05$) than that at FP3. Finally, at the mature stage, the cellulose content of the fibers from FP1 was 3–10% higher than that from FP3. Cotton fiber quality is predominantly determined by the process of cellulose synthesis [31] [45]. And environmental factors such as cool temperature can alter fiber strength by reducing cellulose content within secondary cell walls [46] [47]. Comparing with that of FP1, FP3 had a lower PTI restrained cellulose synthesis. The correlation coefficient of cellulose content with the key fiber properties indicated that fiber length was positively correlated with ultimate cellulose content, and so did fiber strength and micronaire values (Fig. 1A, Fig. 1B and Fig. 1C). The coefficients for the correlations between fiber length and maximum cellulose content, fiber strength and maximum cellulose content, and fiber micronaire values and maximum cellulose content were 0.760**, 0.634* and 0.938** (**$P<$ 0.01, *$P<0.05$), respectively, in Kemian 1, and were 0.857**, 0.894** and 0.758** (**$P<0.01$), respectively, in Sumian 15. These results indicate that fiber quality traits are correlated with the maximum cellulose content and that these correlations are influenced by the effects of FPs.

During the cotton fiber development period, sucrose content in cotton fibers declined from 17 DPA to 45 DPA at both FP1 and FP3 (Table 3). Compared to that of FP3, sucrose content in bolls on FP1 was significantly higher ($P<0.05$) in both 17 DPA and 31

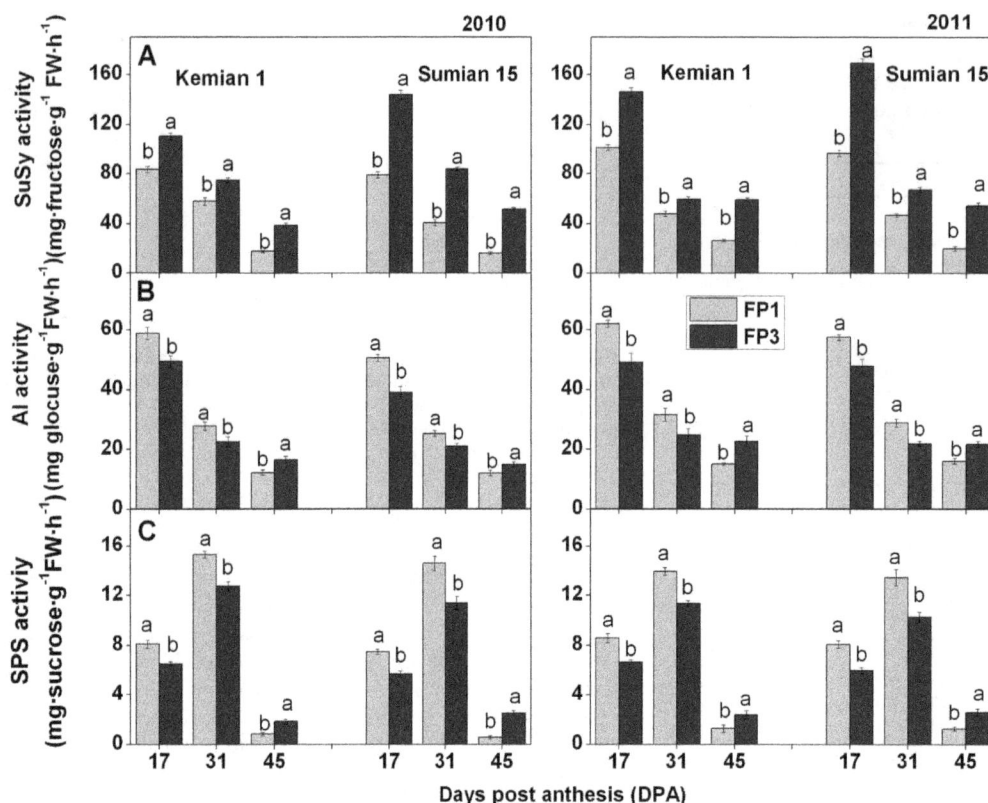

**Figure 3. Changes of enzymes activities in fibers on FP1 and FP3 of two cultivars during 2010–2011.** The activities of SuSy (A), AI (B) and SPS (C) in fibers on FP1 and FP3 of two cultivars. SuSy-sucrose sythase, AI-acid invatse and SPS-sucrose phosphate sythase. Values followed by a different letter between fruiting positions are significantly different at $P = 0.05$ probability level. Each value represents the mean of three replications.

DPA, but was lower at 45 DPA. This indicated that the decreased rate of sucrose content in bolls on FP3 was slower than that on FP1. According to Shu et al. [21], we used the maximum sucrose content and minimum sucrose content in cotton fibers to calculate the sucrose transformation rate, which reflected the sucrose transformation capacity during cotton fiber development in our experiment. The result showed that the FP effect significantly decreased the sucrose transformation rate ($P<0.05$) in both cultivars (Table 3). During cotton fiber development, the correlations of environment factors with maximum/minimum sucrose content and sucrose transformation rate were analyzed (Table 4). Sucrose transformation rate in cotton fiber were positively correlated to PTI ($P<0.05$), but MDT, MDTmin, MDTmax, MDTdif and MDSR were not significantly related to it ($P>0.05$). This indicates that sucrose metabolism of fiber during CFDP of FPs are determined primarily by PTI.

The basic mechanisms regulating cellulose synthesis in different plant species are believed to be similar [18] [19] [20]. Sucrose metabolism is the pivotal process for cellulose synthesis and is sensitive to environment conditions. In many plants, the sucrose metabolism was restrained when they were subjected to low temperature [48] [49]. In our results, there was a positive correlation between maximum cellulose content and sucrose transformation rate in fibers when subjected to FP effect ($P<0.05$) (Fig. 2A and Fig. 2B). And the coefficient of this correlations was 0.653[*] in Kemian 1 ([*]$P<0.05$) and 0.801[**] in Sumian 15 ([**]$P<0.01$), respectively. This analysis indicates that fiber development and cellulose synthesis are determined primarily by sucrose metabolism as it is influenced by the FP effect.

Moreover, the coefficients for the correlation between sucrose transformation rate with the key fiber properties indicated that fiber length was positively correlated with sucrose transformation rate, and so did the fiber strength and micronaire values (Fig. 1D, Fig. 1E and Fig. 1F). The coefficients for the correlations between fiber length and sucrose transformation rate, fiber strength and sucrose transformation rate, and fiber micronaire values and sucrose transformation rate were 0.893[**], 0.808[**] and 0.719[**] ([**]$P<0.01$) in Kemian 1, were 0.850[**], 0.750[**] and 0.798[**] in Sumian 15 ([**]$P<0.01$), respectively. Based upon these results we indicate that lower PTI, which is caused by FP effect, affect the sucrose transformation process from sucrose to cellulose to weaken the fiber qualities.

The coefficients of variance of the sucrose transformation rate between FP1 and FP3 were different in the two cultivars. In Sumian 15, the CVs of the sucrose transformation rate were 6.59% in 2010 and 9.63% in 2011, respectively, and in Kemian 1, they were 4.40% in 2010 and 4.29% in 2011, respectively. These results indicated that FPs had more effects on cotton fiber sucrose metabolism in cultivar Sumian 15 than it did in cultivar Kemian 1.

## Changes in activities of sucrose metabolism enzymes

It has been documented that SuSy, AI, and SPS are the critical enzymes involved in sucrose metabolism [50] [51]. In our study, we observed that these enzymes were affected by the FP effect, which might be the reason why sucrose transformation and cellulose synthesis in cotton fiber were restrained on distal FPs (e.g. FP3). The total SuSy activity including the activities of both soluble SuSy (S-SuSy) and the membrane-associated SuSy(M-SuSy) in

**Table 5.** Comparisons of effect indices and coefficients of variations on the activities of sucrose metabolism enzymes in cotton fibers on FP1 and FP3 of two cultivars during 2010–2011.

| Years | DPA[a] | SuSy activity | | SPS activity | | AI activity | |
|---|---|---|---|---|---|---|---|
| | | Kemian 1 | Sumian15 | Kemian 1 | Sumian15 | Kemian 1 | Sumian15 |
| EI[b], (%) | | | | | | | |
| 2010 | 17 | −31.75 | −82.30 | 19.68 | 23.69 | 16.11 | 23.10 |
| | 31 | −28.92 | −107.51 | 16.58 | 21.93 | 19.08 | 16.76 |
| | 45 | −116.47 | −215.25 | −123.89 | −346.97 | −34.84 | −24.55 |
| 2011 | 17 | −44.62 | −75.11 | 22.53 | 26.08 | 20.71 | 16.85 |
| | 31 | −23.68 | −43.73 | 18.50 | 23.72 | 21.73 | 24.72 |
| | 45 | −123.88 | −173.71 | −86.08 | −103.44 | −50.6 | −34.49 |
| CV[c], (%) | | | | | | | |
| 2010 | 17 | 13.70 | 29.15 | 10.92 | 13.43 | 8.76 | 13.06 |
| | 31 | 12.63 | 34.96 | 9.04 | 12.32 | 10.54 | 9.15 |
| | 45 | 36.80 | 51.84 | 38.25 | 63.43 | 14.83 | 10.93 |
| 2011 | 17 | 18.24 | 27.30 | 12.69 | 15.00 | 11.55 | 9.20 |
| | 31 | 10.59 | 17.94 | 10.19 | 13.46 | 12.19 | 14.10 |
| | 45 | 38.25 | 46.48 | 30.09 | 34.09 | 20.19 | 14.71 |

[a]DPA, day post anthesis.
[b]EI, the effect indices values, which were calculated as $EI_X = (X_{FP1}-X_{FP3})/X_{FP1}*100\%$. If $EI>0$, it is a decrease variation in FP3 compared with FP1. If $EI<0$, it is an increase variation in FP3 compared with FP1.
[c]CV, coefficient of variation (%).

cotton fibers declined from 17 DPA to 45 DPA (Fig. 3A). FP effect enhanced SuSy activity in cotton fiber, and changes in response to FP effect were affected by cotton fiber developmental age. In both Kemian 1 and Sumian 15, SuSy activity at FP3 was significantly higher ($P<0.05$) than that on FP1, and the largest variation was seen at 45 DPA (Table 5). SuSy, especially M-SuSy, is the critical partner in high-rate secondary-wall cellulose synthesis [52]. The suppression of SuSy activity by 70% or more in the ovule epidermis led to a fiberless phenotype and it is supposed that the increase or decrease in SuSy activity is associated with the increment or decrement of cellulose synthesis [53]. However, in this study, we observed that this high level of SuSy activity at FP3 did not enhance cellulose synthesis. This may be due to the reason that while SuSy contained both M-SuSy and S-SuSy in cotton fiber, a part of M-SuSy became S-SuSy under stress [18] [50]. In this study on distal FPs (like FP3), the increased SuSy activity was likely due to the enhanced S-SuSy activity, which supplied UDP-glucose for general metabolic needs rather than having a major regulatory role in partitioning of carbon to cellulose [18] [52] [54] [55].

The activity of AI decreased with the distal FPs in the immature fiber stage (17 DPA and 31 DPA), but increased at 45 DPA (Fig. 3B). Usually, isoforms of invertase in the cell wall and vacuole is AI, which may act as the major player in response to biotic and abiotic stresses [56] [57] [58]. For example, AI preferred to direct hydrolysis of sucrose to provide energy for maintaining fiber and winter oat development under cool temperature and cold hardening [59] [60]. In addition, AI affects the size of intracellular sucrose pools, but it appears to be mostly involved in regulating plant processes such as phloem unloading [61]. Therefore, our results indicate that this decreased AI activity by FPs effect affects the source-to-sink unloading of sucrose and ultimately weakens sink strength.

SPS is a key enzyme for sucrose synthesis and regulates sucrose accumulation [27]. The changing trends of SPS were similar in all cotton fibers, the peak values occurred at 31 DPA (Fig. 3C). Compared to those of FP1, the peak activity values were lower and the activity declined slowly at FP3. The largest differences in decreased extent of the activity between FP1 and FP3 were observed at 17 DPA (Table 5). Haigler et al. [62] indicated that under controlled environmental conditions, in cotton plants over-expressing SPS, sucrose synthesis and fiber quality was enhanced. Therefore, we presumed that the decreased SPS activity in cotton fiber on FP3 would hinder the flux of sucrose from glucose, which reduced the sucrose synthesis and cellulose synthesis.

In addition, according to Table 5, we found that the CVs of the activities of SuSy, SPS and AI at 45 DPA were all higher than those at 17 and 31 DPA. Moreover, the CVs of the activities of SuSy and SPS were much higher than that of AI activity. These results showed that the activities of SuSy and SPS were easily affected by FPs involved in sucrose metabolism. Besides, the activities of SuSy and SPS in the two cultivars had different sensitivities to FP effect, with higher CV in Sumian 15 than in Kemian 1, indicating that they are more sensitive to FP effect in Sumian 15 (Table 5). These indicated that the differences in SuSy and SPS activities in two cultivars in response to FP effect might be the reason for them to possess different sensitivities to boll positions for cellulose synthesis and sucrose metabolism.

This study revealed that both cellulose synthesis and sucrose metabolism were restrained by fruiting position effects. Hence, alternative strategies for alleviating the unfavorable effects of FPs need to be considered. Previous researches on the removal from specific fruiting positions in cotton indicated that fiber qualities on the second sympodial fruiting positions could be increased if the nutrition was sufficient [15] [44]. Nitrogen, phosphorus and potassium fertilization, planting density and abscisic acid may allow the crop to compensate partially for the potential yield losses and increase plant tolerance against abiotic stress such as cool temperature and low irradiance [12] [16] [37] [41] [63]. Therefore, elucidating the different mechanisms among different FPs in cotton fibers and improving the quality from distal positions are urgently needed. In this study, we found that FP effect restrained cellulose synthesis and sucrose utilization, and reduced the activities of the key enzymes involved in sucrose metabolism. Meanwhile, the examination of the effect of FPs on fiber development for two cultivars revealed that Sumian 15 was more sensitive to the FP effect than Kemian 1 was.

## Conclusions

In this study, we observed that FP effect affected cotton fiber development was primarily on cumulative photo-thermal index. Fiber length, fiber strength, and micronaire values, which are the critical fiber properties in textile processing, were all sensitive to FP effect. Highly correlated coefficients between these critical fiber quality traits and cellulose content revealed the close relationship between them under FP effect. Since sucrose metabolism is the pivotal process for cellulose synthesis, correlations between cellulose content and sucrose transformation rate of FP effect were analyzed and these coefficients were found to be highly significant. These results have indicated that sucrose metabolism determines cellulose synthesis under FP effect and both SuSy and SPS are the critical enzymes involved in sucrose metabolism and they also vary in cotton fibers on FP3 as compared to that on FP1. These variations indicate a potentially important reason for the decrease in sucrose and cellulose synthesis.

## Author Contributions

Conceived and designed the experiments: YM YW ZZ. Performed the experiments: YM JL FL JC. Analyzed the data: YM JL FL JC. Contributed reagents/materials/analysis tools: YM JL FL JC. Wrote the paper: YM.

## References

1. Jenkins JN, McCarty JC, Parrott WL (1990) Fruiting efficiency in cotton: Boll size and boll set percentage. Crop Sci 30: 857–860.
2. Anjum R, Soomro A, Chang M, Memon A (2001) Effect of fruiting positions on yield in American cotton, Pak J Biol Sci 4: 960–962.
3. Davidonis GH, Johnson AS, Landivar JA, Fernandez CJ (2004) Cotton fiber quality is related to boll location and planting date. Agron J 96: 42–47.
4. Boquet DJ, Moser EB (2003) Boll retention and boll size among intrasympodial fruiting sites in cotton. Crop Sci 43: 195–201.
5. Oosterhuis DM (1991) Growth and development of a cotton plant. In: Miley WN, Oosterhuis DM, editors. Nitrogen Nutrition of Cotton: Practical Issues. 1–24.
6. Heitholt JJ (1993) Cotton boll retention and its relationship to lint yield. Crop Sci 33: 485–490.
7. Gu LL, Wang XS, Zhou ZG, Chen DH, Xu LH, et al. (2010) Researches of high yield cotton cultivations in Jiangsu province. China Cotton (in Chinese) 4: 14–16.
8. May OL (1996) Genetic variation in fiber quality. In: Basra AS, editor. Cotton fibers. 183–229.
9. Bradow JM, Davidonis GH (2000) Quantitation of fiber quality and the cotton production-processing interface: a physiologist's perspective. J Cotton Sci 4: 34–64.
10. Yeates S, Constable G, McCumstie T (2010) Irrigated cotton in the tropical dry season. III: Impact of temperature, cultivar and sowing date on fibre quality. Field Crop Res 116: 300–307.
11. Bradow JM, Bauer PJ (1997) Fiber-quality variations related to cotton planting date and temperature. P. 1491–1495. In Proc. Beltwide Cotton Conf., New Orleans, LA. 6–10 Jan. 1997. Natl. Cotton Counc. Of Am., Memphis, TN.
12. Read JJ, Reddy KR, Jenkins JN (2006) Yield and fiber quality of upland cotton as influenced by nitrogen and potassium nutrition. Eur J Agron 24: 282–290.
13. Dağdelen N, Başal H, Yılmaz E, Gürbüz T, Akcay S (2009) Different drip irrigation regimes affect cotton yield, water use efficiency and fiber quality in western Turkey. Agr Water Manage 96: 111–120.
14. Thomasson J, Manickavasagam S, Mengüç M (2009) Cotton fiber quality characterization with light scattering and Fourier transform infrared techniques. Applied Spectrosc 63: 321–330.
15. Pettigrew WT (1995) Source-to-sink manipulation effects on cotton fiber quality. Agron J 87: 947–952.
16. Dong H, Li W, Tang W, Li Z, Zhang D, et al. (2006) Yield, quality and leaf senescence of cotton grown at varying planting dates and plant densities in the Yellow River Valley of China. Field Crop Res 98: 106–115.
17. Baraiya B, Barde S, Singhal H (2009) Genetic evaluation of Gossypium hirsutum genotypes for yield, drought parameters and fiber quality. Ann Plant Physiol 23: 144–148.
18. Haigler CH, Ivanova-Datcheva M, Hogan PS, Salnikov VV, Hwang S, et al. (2001) Carbon partitioning to cellulose synthesis. Plant Mol Biol 47: 29–51.
19. Delmer DP, Haigler CH (2002) The regulation of metabolic flux to cellulose, a major sink for carbon in plants. Metab Eng 4: 22–28.
20. Williamson RE, Burn JE, Hocart CH (2002) Towards the mechanism of cellulose synthesis. Trends Plant Sci 7: 461–467.
21. Shu H, Zhou Z, Xu N, Wang Y, Zheng M (2009) Sucrose metabolism in cotton (Gossypium hirsutum L.) fibre under low temperature during fibre development, Eur J Agron 31: 61–68.
22. Huber SC, Huber JL (1996) Role and regulation of sucrose-phosphate synthase in higher plants. Annu Rev Plant Bio 47: 431–444.
23. Winter H, Huber SC (2000) Regulation of sucrose metabolism in higher plants: localization and regulation of activity of key enzymes. Crit Rev Plant Sci 19: 31–67.
24. Thaker V, Saroop S, Vaishnav P, Singh Y (1992) Physiological and biochemical changes associated with cotton fiber development V. Acid invertase and sugars. Acta Physiol. Plant 14: 11–18.
25. Barratt D, Derbyshire P, Findlay K, Pike M, Wellner N, et al. (2009) Normal growth of Arabidopsis requires cytosolic invertase but not sucrose synthase. P Natl Acad Sci 106: 13124–13129.
26. Huber SC, Akazawa T (1986) A novel sucrose synthase pathway for sucrose degradation in cultured sycamore cells. Plant Physiol 81: 1008–1013.
27. Wang YH, Shu HM, Chen BL, Xu NY, Zhao YC, et al. (2008) Temporal-spatial variation of cotton fiber strength of different cultivars and its relationship with temperature, Sci Agric Sin (in Chinese) 41: 3865–3871.
28. Xiangbin G, Youhua W, Zhiguo Z, Oosterhuis DM (2012) Response of cotton fiber quality to the carbohydrates in the leaf subtending the cotton boll. J Plant Nutr Soil Sci 175: 152–160.
29. WenQing Z, YouHua W, ZhiGuo Z, YaLi M, BingLin C, et al. (2012) Effect of nitrogen rates and flowering dates on fiber quality of cotton (Gossypium hirsutum L.)[J]. Am J Exp Agr 2: 133–159.
30. Zheng M, Wang Y, Liu K, Shu H, Zhou Z (2012) Protein expression changes during cotton fiber elongation in response to low temperature stress. J Plant Physiol 169: 399–409.
31. Haigler CH (2007) Substrate supply for cellulose synthesis and its stress sensitivity in the cotton fiber. In: Brown RM, Saxena IM, editors. Cellulose: Molecular and Structural Biology. 147–168.
32. Liu J, Ma Y, Lv F, Chen J, Zhou Z, et al. (2013) Changes of sucrose metabolism in leaf subtending to cotton boll under cool temperature due to late planting, Field Crop Res 144: 200–211.
33. Pettigrew WT (2001) Environmental effects on cotton fiber carbohydrate concentration and quality. Crop Sci 41: 1108–1113.
34. Hendeix DL (1993). Rapid extraction and analysis of nonstructural carbohydrates in plant tissues. Crop Sci 33: 1306–1311.
35. Updegraff DM (1969) Semimicro determination of cellulose in biological materials. Anal Biochem 32: 420–424.
36. King SP, Lunn JE, Furbank RT (1997) Carbohydrate content and enzyme metabolize in developing canola siliques. Plant Physiol 114: 153–160.
37. Zhao W, Meng Y, Li W, Chen B, Xu N, et al. (2012) A model for cotton (Gossypium hirsutum L.) fiber length and strength formation considering temperature-radiation and N nutrient effects. Ecol Model 243: 112–122.
38. Li W, Zhou Z, Meng Y, Xu N, Fok M (2009) Modeling boll maturation period, seed growth, protein, and oil content of cotton (Gossypium hirsutum L.) in China. Field Crop Res 112: 131–140.
39. Pettigrew WT (2008) The effect of higher temperatures on cotton lint yield production and fiber quality. Crop Sci 48: 278–285.
40. Liakatas A, Roussopoulos D, Whittington W (1998) Controlled-temperature effects on cotton yield and fibre properties. J Agr Sci 130: 463–471.
41. Wrather J, Phipps B, Stevens W, Phillips A, Vories E (2008) Cotton Planting Date and Plant Population Effects on Yield and Fiber Quality in the Mississippi Delta. J Cotton Sci 12:1–7.
42. ZhiGuo Z, YaLi M, YouHua W, BingLin C, XinHua Z, et al. (2011) Effect of planting date and boll position on fiber strength of cotton (Gossypium hirsutum L.). Am J Exp Agr 1: 331–342.
43. Zhao W, Wang Y, Shu H, Li J, Zhou Z (2012) Sowing date and boll position affected boll weight, fiber quality and fiber physiological parameters in two cotton (Gossypium Hirsutum L.) cultivars. Afr J Agr Res 7: 6073–6081.
44. Heitholt J (1997) Floral bud removal from specific fruiting positions in cotton: Yield and fiber quality. Crop Sci 37: 826–832.
45. Kim HJ, Triplett BA (2001). Cotton fiber growth in Planta and in vitro models for plant cell elongation and cell wall biogenesis. Plant Physiol 127: 1361–1366.

The Effects of Fruiting Positions on Cellulose Synthesis and Sucrose Metabolism during Cotton...

77

46. Jiang G, Zhou Z, Chen B, Meng Y (2006) Effect of cotton physiological age on the fiber thickening development and fiber strength formation, Sci Agric Sin (in Chinese) 39: 265–273.

47. Wang Y, Shu H, Chen B, McGiffen ME, Zhang W, et al. (2009) The rate of cellulose increase is highly related to cotton fibre strength and is significantly determined by its genetic background and boll period temperature, Plant Growth Regul 57: 203–209.

48. Lingle SE (2004) Effect of transient temperature change on sucrose metabolism in sugarcane internodes. J Am Soc Sugar Cane Tech 24: 132–140.

49. Marangoni AG, Duplessis PM, Lencki RW, Yada RY (1996) Low-temperature stress induces transient oscillations in sucrose metabolism in *Solanum tuberosum*. Biophy Chem 61: 177–184.

50. Salnikov VV, Grimson MJ, Seagull RW, Haigler CH (2003) Localization of sucrose synthase and callose in freeze substituted secondary wall stage cotton fibers. Protoplasma 221: 175–184.

51. Wang YH, Feng Y,Xu NY, Chen BL, Ma RH, et al. (2009) Response of the enzymes to nitrogen applications in cotton fiber (*Gossypium hirsutum* L.) and their relationships with fiber strength. Sci China Ser C, 52: 1065–1072.

52. Amor Y, Haigler CH, Johnson S, Wainscott M, Delmer DP (1995) A membrane-associated form of sucrose synthase and its potential role in synthesis of cellulose and callose in plants. P Natl Acad Sci 92: 9353–9357.

53. Ruan YL, Llewellyn DJ, Furbank RT (2003) Suppression of sucrose synthase gene expression represses cotton fiber cell initiation, elongation, and seed development, Plant Cell, 15: 952–964.

54. Delmer DP (1999) Cellulose biosynthesis: exciting times for a difficult field of study, Annual Rev Plant Biol 50: 245–276.

55. Kutschera U, Heiderich A (2002) Sucrose metabolism and cellulose biosynthesis in sunflower hypocotyls, Physiol Plantarum 114: 372–379.

56. Roitsch T, Balibrea M, Hofmann M, Proels R, Sinha A (2003) Extracellular invertase: key metabolic enzyme and PR protein, J Exp Bot 54: 513–524.

57. McLaughlin JE, Boyer JS (2004) Sugar-responsive gene expression, invertase activity, and senescence in aborting maize ovaries at low water potentials, Ann Bot-London 94: 675–689.

58. Essmann J, Schmitz-Thom I, Schön H, Sonnewald S, Weis E, et al. (2008) RNA interference-mediated repression of cell wall invertase impairs defense in source leaves of tobacco, Plant Physiol 147: 1288–1299.

59. Livingston III DP, Henson CA (1998) Apoplastic sugars, fructans, fructan exohydrolase, and invertase in winter oat: responses to second-phase cold hardening. Plant Physiol 116: 403–408.

60. Martin LK, Haigler CH (2004) Cool temperature hinders flux from glucose to sucrose during cellulose synthesis in secondary wall stage cotton fibers. Cellulose 11: 339–349.

61. Sturm A, Tang GQ (1999) The sucrose-cleaving enzymes of plants are crucial for development, growth and carbon partitioning, Trends Plant Sci 4: 401–407.

62. Haigler CH, Singh B, Zhang D, Hwang S, Wu C, et al. (2007) Transgenic cotton over-producing spinach SPS showed enhanced leaf sucrose synthesis and improved fiber quality under controlled environmental conditions. Plant Mol Biol 63: 815–832.

63. Battal P, Erez ME, Turker M, Berber I (2008) Molecular and physiological changes in maize (*Zea mays*) induced by exogenous NAA, ABA and MeJa during cold stress. Ann. Bot. Fenn. 45: 173–185.

# Feeding and Dispersal Behavior of the Cotton Leafworm, *Alabama argillacea* (Hübner) (Lepidoptera: Noctuidae), on Bt and Non-Bt Cotton: Implications for Evolution and Resistance Management

**Francisco S. Ramalho[1]\*, Jéssica K. S. Pachú[1], Aline C. S. Lira[1], José B. Malaquias[1], José C. Zanuncio[2], Francisco S. Fernandes[1]**

1 Unidade de Controle Biológico, Embrapa Algodão, Campina Grande, Paraíba, Brazil, 2 Departamento de Entomologia, Universidade Federal de Viçosa, Minas Gerais, Brazil

## Abstract

The host acceptance of neonate *Alabama argillacea* (Hübner) (Lepidoptera: Noctuidae) larvae to Bt cotton plants exerts a strong influence on the potential risk that this pest will develop resistance to Bt cotton. This will also determine the efficiency of management strategies to prevent its resistance such as the "refuge-in-the-bag" strategy. In this study, we assessed the acceptance of neonate *A. argillacea* larvae to Bt and non-Bt cotton plants at different temperatures during the first 24 h after hatching. Two cotton cultivars were used in the study, one a Bt DP 404 BG (Bollgard) cultivar, and the other, an untransformed isoline, DP 4049 cultivar. There was a greater acceptance by live neonate *A. argillacea* larvae for the non-Bt cotton plants compared with the Bt cotton plants, especially in the time interval between 18 and 24 h. The percentages of neonate *A. argillacea* larvae found on Bt or non-Bt plants were lower when exposed to temperatures of 31 and 34°C. The low acceptance of *A. argillacea* larvae for Bt cotton plants at high temperatures stimulated the dispersion of *A. argillacea* larvae. Our results support the hypothesis that the dispersion and/or feeding behavior of neonate *A. argillacea* larvae is different between Bt and non-Bt cotton. The presence of the Cry1Ac toxin in Bt cotton plants, and its probable detection by the *A. argillacea* larvae tasting or eating it, increases the probability of dispersion from the plant where the larvae began. These findings may help to understand how the *A. argillacea* larvae detect the Cry1Ac toxin in Bt cotton and how the toxin affects the dispersion behavior of the larvae over time. Therefore, our results are extremely important for the management of resistance in populations of *A. argillacea* on Bt cotton.

**Editor:** Youjun Zhang, Institute of Vegetables and Flowers, Chinese Academy of Agricultural Science, China

**Funding:** This research was supported by Conselho Nacional de Desenvolvimento Científico e Tecnológico - CNPq and Financiadora de Estudos e Projetos - FINEP. The funders had no role in study design, data collection and analysis, decision to publish, or preparation of the manuscript.

**Competing Interests:** The authors have declared that no competing interests exist.

\* Email: ramalhohvv@globo.com

## Introduction

Cotton leafworm *Alabama argillacea* (Hübner) (Lepidoptera: Noctuidae) is a species native to Southern and Central America, found in almost all cotton-growing regions, extending from southern Canada to northern Argentina [1].

In Brazilian cotton-growing regions, this pest can infest the crop at any stage in its phonological development [2]. In Southern–Central Brazil, it is considered a late pest [3], but in the Northeast, with the exception of Bahia State, it attacks in the initial stages and can occur sporadically when the crop has matured [2]. The cotton leafworm, is highly destructive as one of the main defoliating pests of cotton (*Gossypium hirsutum Linnaeus*) in Brazil [4]. The most severe attacks occur after the cotton flowering period and are characterized by the destruction of the leaves on the plant's main stem, which reduces plant growth at any further stage of its development by affecting the height and the stem diameter, and

production is damaged [5]. A high-density infestation can adversely affect the cotton yield. In Brazil, losses caused by *A. argillacea* vary from 21 to 35% of the cotton lint yield [1]. Chemicals are the main method for control; however, insect resistance to these molecules has been detected [6].

As an alternative to the currently used chemical control methods, cotton cultivars resistant to the cotton leafworm can be used to minimize the damage caused by this pest in cotton growing areas. The bacteria *Bacillus thuringiensis* subsp. *kurstaki* expresses protein crystals (Cry) during sporulation, and genetically modified plants with genes from the bacteria (Bt) express these protein crystals (Cry) that are deadly when ingested by lepidopteran larvae [7,8].

The high efficiency of Bt cotton against lepidopterans has contributed to this technology being adopted quickly by cotton farmers in Brazil's Cerrado region, making it a specific tactic to protect cotton from *A. argillacea* damage [9]. However, the monophagous diet of *A. argillacea* on cotton and its high degree of susceptibility to Cry toxins may place selection pressure on populations of this insect. Consequently, the insect could alter its host selection behavior [10,11] and accelerate the development of resistance, effectively jeopardizing the control of the pest and the income of the farm [12].

Insect resistance management (IRM) refers to a set of practices applied to prevent or delay the development of pest resistance to the methods used for its control [13] in agricultural areas. In view of IRM, the best way to preserve the benefits of transgenic plants is to establish refuge areas, and configure planting so blocks of non-Bt cotton are in areas adjacent to Bt cotton planting areas [9]. However, the use of this tactic can be inconvenient for the producer because it involves an additional cost in time, labor and seeds, which may contribute to its failure being accepted by cotton growers [14].

In an attempt to slow down the development of insect resistance to Bt toxins [13,15], the main multinational seed-producing companies have considered the possibility of mixing a percentage of nontransgenic seeds directly into bags with transgenic seeds [13], to facilitate compliance with the technical standards to implement refuges. However, a concern about the use of this tactic is that the larvae, with some level of tolerance to Bt cotton, begins to feed on non-Bt cotton plants, but then disperses and feeds on Bt cotton plants, and survives, reaches adulthood and produces offspring with partial resistance to Bt toxins [16]. Another concern is that some newly hatched larvae are able to feed on Bt cotton plants, but then migrate to a nearby non-Bt cotton plant, thereby surviving to adulthood to produce offspring with partial resistance to Bt toxins [17]. Both scenarios increase the likelihood that heterozygous individuals will survive and potentially accelerate the development of resistance [18].

High relative humidity and temperatures decreased the effects of the Bt toxin, which does not occur at low temperatures, where there is a significant increase in the production of this toxin [19]. Therefore, in areas where the relative humidity and temperature are high, the efficiency of the Cry protein is possibly reduced, affecting the feeding behavior of the target pest, and consequently, resistance development. Thus, information on the feeding behavior and on the host acceptance of newly hatched cotton leafworm larvae for Bt cotton plants is essential. Information on *A. argillacea* populations and their potential dispersal and development of resistance to Bt cotton is necessary to determine whether a mixture of seeds is a viable management option compared with other tactical refuges to delay the evolution of resistance in *A. argillacea* populations. Thus, the aim of this study was to characterize and compare the degree of host acceptance of newly hatched *A. argillacea* larvae on Bt and non-Bt cotton plants at different temperatures during the first 24 h after hatching. Neonate larvae were used in these bioassays because they are very mobile and accept the host plant better than the later stages [20]. We proposed two hypotheses: a₁) a higher percentage of neonate *A. argillacea* larvae would feed on non-Bt cotton plants than on Bt cotton plants and a₂) after feeding started on the cotton plants, a higher percentage of neonate *A. argillacea* larvae would disperse from Bt cotton plants than from non-Bt cotton plants. The knowledge generated in this study will be important to develop a more effective management tactic to prevent *A. argillacea* population resistance to Bt cotton.

## Materials and Methods

### Insects and cotton cultivars

*Alabama argillacea* was grown at the Embrapa Cotton Entomology Laboratory, Biological Control Unit (UCB), Campina Grande, PB, Brazil. Larvae rearing stock were kept in a climate-controlled chamber at 25°C, with a relative humidity of 70±10% and 12-h photoperiod.

Two cotton cultivars were used in the study, one a Bt DP 404 BG (Bollgard) cultivar, and the other, an untransformed isoline, DP 4049 cultivar. The Bt and non-Bt cotton cultivars were planted separately in plastic pots (20 cm in diameter and 30 cm in height) and kept in a greenhouse.

### Bioassays

**Dispersion behavior of neonate larvae at different temperatures and over time.** To assess the effect of Bt and non-Bt cotton cultivars on the dispersion behavior of *A. argillacea* larvae at different temperatures and over time, plants from each (Bt and non-Bt) cotton cultivar that reached the eight-leaf stage received a newly hatched larva (0–24 hours old) released on a leaf in the plant's apical region. Then, each plant was covered with an organza bag and placed randomly in climate-controlled chambers at 22, 25, 28, 31 and 34°C, with a relative humidity of 70±10% and a photoperiod of 12 h. Daily the plants were moved randomly to minimize position effects within the chamber.

The experiment used was a 2×5×4 factorial in randomized blocks, where each block was divided into five parts (temperatures) and subdivided into four time intervals for assessment at 6, 12, 18 and 24 hours after plant infestation. Each subpart consisted of five replications, each with eight plants (four Bt cotton and four non-Bt cotton). After each time interval, the cotton plants were inspected and the larvae were removed with a brush. The larvae were categorized according to their location: found on the plant or on the organza bag.

### Neonate larvae feeding behavior over time

This test was conducted to quantify the percentages of neonate *A. argillacea* larvae that fed on Bt and non-Bt cotton plants. The experiment used was a 2×4 factorial in randomized blocks, with two cotton cultivars (Bt and non-Bt cotton) and four periods of plant exposure to larvae, i.e., 6, 12, 18 and 24 h after plant infestation. The experimental unit consisted of a Bt or non-Bt cotton plant cultivar that reached the eight-leaf stage and received 30 neonate *A. argillacea* larvae (0-24-h-old) released onto a leaf at the apical region of the plant. Then, each plant was covered with an organza bag and placed randomly in climate-controlled chambers at 28°C, with a relative humidity of 70±10% and photoperiod of 12 h. Daily the plants were moved randomly to minimize position effects within the chamber. After each time interval, the cotton plants and the inside of the organza bags were inspected and the larvae were removed with a brush. Then, the larvae were grouped into two categories: found on the plants or on the inside of the organza bags. To check that the larvae had fed, they were mounted on microscope slides in a solution of Karo honey diluted in water [21] and examined under a stereomicroscope (10×) according to the method adopted by Razze et al. [22]. The amount of plant material found in the gut of each caterpillar was measured by using an ocular micrometer attached to the microscope's eyepiece with phase contrast.

### Data analyses

The data from the bioassays were subjected to an analysis of variance (PROC GLM) [23] to determine whether there were effects of the cultivar (C), temperature (T), or time of exposure to cotton plants (t) on neonate *A. argillacea* larvae and to determine if the relationships between cultivar versus temperature, cultivar versus time and cultivar versus temperature versus time affected the percentage of neonate *A. argillacea* larvae recovered from the cotton plants. Additionally, the analysis of variance examined whether there was an interaction effect of cultivar (C) versus time of exposure to the cotton plants (t) on the percentage of neonate *A. argillacea* larvae that fed on Bt and non-Bt cotton cultivars at 28°C. An analysis of variance analysis was also performed (PROC GLM) [23] on the data collected for the percentage of larvae recovered at each location (plant or organza bag), on the amount of plant tissue recorded in the larvae's gut and to determine if there was an interaction between the cultivar and exposure time of the cotton plants to the larvae. The comparison of treatment means was performed using the Student-Newman-Keuls test (P = 0.05). A linear model (PROC CATMOD) [23] was used to estimate the recovery of *A. argillacea* larvae in Bt and non-Bt cotton plants depending on the temperature.

## Results

### Dispersion behavior at different time intervals and temperatures

There was a significant interaction between cotton cultivar (C) and the length of time *A. argillacea* larvae were exposed to the cotton plants (t) ($F_{(C \text{ versus } t)3, 196} = 10.29$, P<0.0001) for the percentage of *A. argillacea* larvae recovered from the cotton plants (Table 1). Since P<0.05, there was significant interaction among cultivar (C), the length of time *A. argillacea* larvae were exposed to the cotton plants (t) and temperature (T). However, there were no interaction effects between cultivar (C) and temperature (T) ($F_{(C \text{ versus } T)4, 196} = 1.53$, P = 0.1967) or between the length of time *A. argillacea* larvae were exposed to the cotton plants (t) and temperature (T) ($F_{(t \text{ versus } T)12, 196} = 0.99$, P = 0.4595) in the percentage of recovered *A. argillacea* larvae from the cotton plants (Table 1). Therefore, the effects of cotton cultivar (Bt or non-Bt) on the percentage of recovered *A. argillacea* larvae from cotton plants depended on the length of time *A. argillacea* larvae were exposed to the cotton plants and the temperature.

According to the analysis of variance, significant differences were found between cultivars Bt and non-Bt ($F_{(C)1, 196} = 151.75$, P<0.0001) (Table 1) and among temperatures tested (T) ($F_{(T)4, 196} = 8.53$, *P*<0.0001) (Table 1), but not for the exposure times of *A. argillacea* larvae to the cotton plants (t) ($F_{(t)3, 196} = 2.12$, P = 0.0997) (Table 1) in the percentage of recovered *A. argillacea* larvae from the cotton plants. However, there was no significant

difference between the cotton cultivars (Bt or non-Bt) for the percentage of *A. argillacea* larvae recovered from the cotton plants in the first 6 h after release (Table 2). In the other time intervals, however, the Bt cotton cultivar had a significant reduction in the percentage of *A. argillacea* larvae recovered from the cotton plants when compared with the non-Bt plants, which was most evident in the time interval between 18 and 24 h (Table 2).

### Dispersion behavior after 24 h

For the live and dead *A. argillacea* larvae recovered from the Bt and non-Bt cotton plants, the percentage of live *A. argillacea* larvae recovered from the cotton plants was significantly affected by the cotton cultivar (Bt and non-Bt) ($F_{(C)1, 36} = 127.78$, P< 0.0001) (Table 3) and by the temperature ($F_{(T)4, 36} = 7.81$, P< 0.0001) (Table 3), while the percentage of dead *A. argillacea* was only affected by the cotton cultivar ($F_{1, 36} = 19.84$, P<0.0001) (Table 3). The cultivar versus temperature interaction was not significant for either the percentage of live *A. argillacea* recovered from cotton plants ($F_{(C \text{ versus } T)4, 36} = 0.98$, P = 0.4320) (Table 3) or the percentage of dead *A. argillacea* recovered ($F_{(C \text{ versus } T)4, 36} = 2.06$, P = 0.0610) (Table 3). Therefore, the effect of the cotton cultivar on the percentage of dead or alive *A. argillacea* larvae recovered from the cotton plants did not depend on temperature.

The percentage of live *A. argillacea* larvae recovered from the cotton plants after 24 h, at each of the temperatures, was significantly higher in the non-Bt cotton cultivar than in the Bt cotton cultivar (Fig. 1A). Regarding the influence of temperature, the percentage of live *A. argillacea* larvae recovered from the cotton plants was lower at 31 and 34°C than at the other temperatures, and there were no significant differences among the other temperatures (Fig. 1B). When summed across all temperatures, the percentage of live *A. argillacea* larvae recovered from the Bt cotton plants after 24 h was significantly lower than those recovered from the non-Bt cotton plants (Fig. 1C). The percentage of dead *A. argillacea* larvae recovered from the cotton plants after 24 h, summed across all temperatures, was significantly higher in the Bt cotton cultivar than in the non-Bt cotton cultivar (Fig. 1D).

A linear model best described the percentage of live *A. argillacea* larvae recovered from Bt or non-Bt cotton plants as a function of temperature (Fig. 2). The linear models showed that 81% and 87% of the variation for the average percentage of live *A.*

**Table 1.** Summarized model of the three-way analysis of variance (ANOVA) for the effects of cultivar[1], exposure time interval of neonate larvae to Bt cotton or non-Bt cotton[2], and temperature[3] on the percentage of neonate larvae of *A. argillacea* recovered from Bt cotton and non-Bt near isoline cotton plants.

| Source | Models | DF | F ratio | Prob> F |
|---|---|---|---|---|
| Percentage of cotton leafworm | Model | 43 | 6.32 | 0.0001 |
| larvae recovered from cotton plant | | | | |
| | Cultivar (C) | 1 | 151.75 | 0.0001 |
| | Time (t) | 3 | 2.12 | 0.0997 |
| | Temperature (T) | 4 | 8.53 | 0.0001 |
| | C x t | 3 | 10.29 | 0.0001 |
| | C x T | 4 | 1.53 | 0.1967 |
| | t x T | 12 | 0.99 | 0.4595 |
| | C x t x T | 12 | 2.00 | 0.0274 |

[1]Cultivars: Bt cotton and non-Bt near isoline cotton.
[2]Time intervals: 0–6 h, 6–12 h, 12–18 h, and 18–24 h.
[3]Temperatures (°C): 22, 25, 28, 31, and 34. Analysis was performed with the data transformed with arcsine square root percentage.

**Table 2.** Mean ($\pm$ SE) percentage of neonates recovered from the cotton plants in the test for abandonment of neonate larvae of *A. argillacea* from Bt and non-Bt cotton plants during four exposure time intervals ($F_{(C \times t) 3, 196} = 10.29$, $P < 0.00001$).

| Exposure time interval of neonate larvae to cotton plants (h) | Cultivar[1] | |
|---|---|---|
| | **Bt cotton** | **Non-Bt cotton** |
| 0–6 | 83.08±12.96 Aa | 89.57±7.07 Aa |
| 6–12 | 75.49±16.47 Bb | 93.60±8.48 Aa |
| 12–18 | 71.36±17.69 Bb | 93.05±8.46 Aa |
| 18–24 | 62.26±23.40 Cb | 94.07±8.09 Aa |

[1]Means within the same cultivar column with the same capital letters or means between cultivars within the same row with the same lower case letters are not significantly different (P = 0.05, Student-Newman-Keuls test). Original data.

*argillacea* larvae recovered from the non-Bt and Bt cotton plants, respectively, was explained by temperature (Fig. 2). However, the percentage of live *A. argillacea* larvae recovered from the cotton plants ranged from approximately 10.93% (34°C) to 48.60% (22°C) in Bt cotton plants, whereas it ranged from 63.80% (34°C) to 84.70% (22°C) in non-Bt cotton plants (Fig. 1A and Fig. 2).

## Feeding behavior in different time intervals

The percentages of neonate *A. argillacea* larvae that had fed and were found on the organza bag were affected significantly by the cotton cultivar ($F_{(C)1, 21} = 5.70$, P<0.0264), the exposure time ($F_{(Et)3, 21} = 13.48$, P<0.0001) and the cultivar versus exposure time interaction ($F_{(C \; versus \; Et)3, 21} = 36.54$, P<0.0001) (Table 4). However, the percentages of neonate *A. argillacea* larvae that had fed and were found on the plant were not affected by the cultivars ($F_{(C)1, 21} = 0.01$, P = 1.0000), the exposure time ($F_{(Et)3, 21} = 0.64$, P = 0.5999) or the cultivar versus exposure time interaction ($F_{(C \; versus \; Et)3, 21} = 1.27$, P = 0.3094) (Table 4). The average percentage of larvae that had fed and were on the organza bag was higher in the non-Bt cotton cultivar than in the Bt cotton cultivar (Fig. 3A). The percentages of larvae that had fed and that were found on the organza bag increased over exposure time for both cotton cultivars (Bt and non-Bt) (Fig. 3A).

The quantities of plant tissue measured in the intestines of the neonate *A. argillacea* larvae that fed and were found on the plant or in the organza bag were affected by the cotton cultivar (plant: $F_{(C)1, 21} = 95.69$, P<0.0001; organza bag: $F_{(C)1, 21} = 25.25$, P< 0.0001), the exposure time (plant: $F_{(Et)1, 21} = 16.65$, P<0.0001; organza bag: $F_{(C)1, 21} = 25.51$, P<0.0001) and the cultivar versus exposure time interaction (plant: $F_{(C \; versus \; Et)3, 21} = 2.97$, P = 0.0550; organza bag: $F_{(C \; versus \; Et)3, 21} = 30.90$, P<0.0001). The average amount of plant tissue found in the gut of the fed *A. argillacea* larvae, found on the plant or in the organza bag, was higher for non-Bt cotton plants than Bt plants (Figs. 3B and 3C), except at 24 h after infestation by the larvae found in the organza bag (Fig. 3B). These values increased with exposure time for both Bt cotton plants and non-Bt plants (Figs. 3B and 3C).

## Discussion

The dispersion of neonate lepidopteran pest larvae can be the result of genetic programming to reduce competition for resources and ensure the survival of the larvae [17,22]. The dispersal of *A. argillacea* larvae on Bt and non-Bt cotton plants could be related to host plant acceptance, suggesting that the *A. argillacea* larvae are more likely to abandon Bt plants than non-Bt plants, which would result in less feeding on Bt plants. The percentage of live

**Table 3.** Summarized model of the two-way analysis of variance (ANOVA) for the effects of cultivar[1] and temperature[2] on the percentage of neonate larvae of *A. argillacea* recovered alive or dead from Bt cotton or non-Bt near isoline plants after 24 h.

| Source | Models | DF | F ratio | Prob> F |
|---|---|---|---|---|
| Neonate larvae of *A. argillacea* | Model | 13 | 12.74 | 0.0001 |
| recovered alive from cotton plant | | | | |
| (%)[3] after 24 h | | | | |
| | Cultivar (C) | 1 | 127.78 | 0.0001 |
| | Temperature (T) | 4 | 7.81 | 0.0001 |
| | C x T | 4 | 0.98 | 0.4320 |
| Neonate larvae of *A. argillacea* | Model | 13 | 3.15 | 0.0032 |
| recovered dead from cotton plant | | | | |
| (%)[3] after 24 h | | | | |
| | Cultivar (C) | 1 | 19.84 | 0.0001 |
| | Temperature (T) | 4 | 1.94 | 0.1250 |
| | C x T | 4 | 2.06 | 0.0610 |

[1]Cultivars: Bt cotton and non-Bt near isoline cotton. [2]Temperatures (°C): 22, 25, 28, 31, and 34. [3]Data were arcsine-square root transformed prior to statistical analyses.

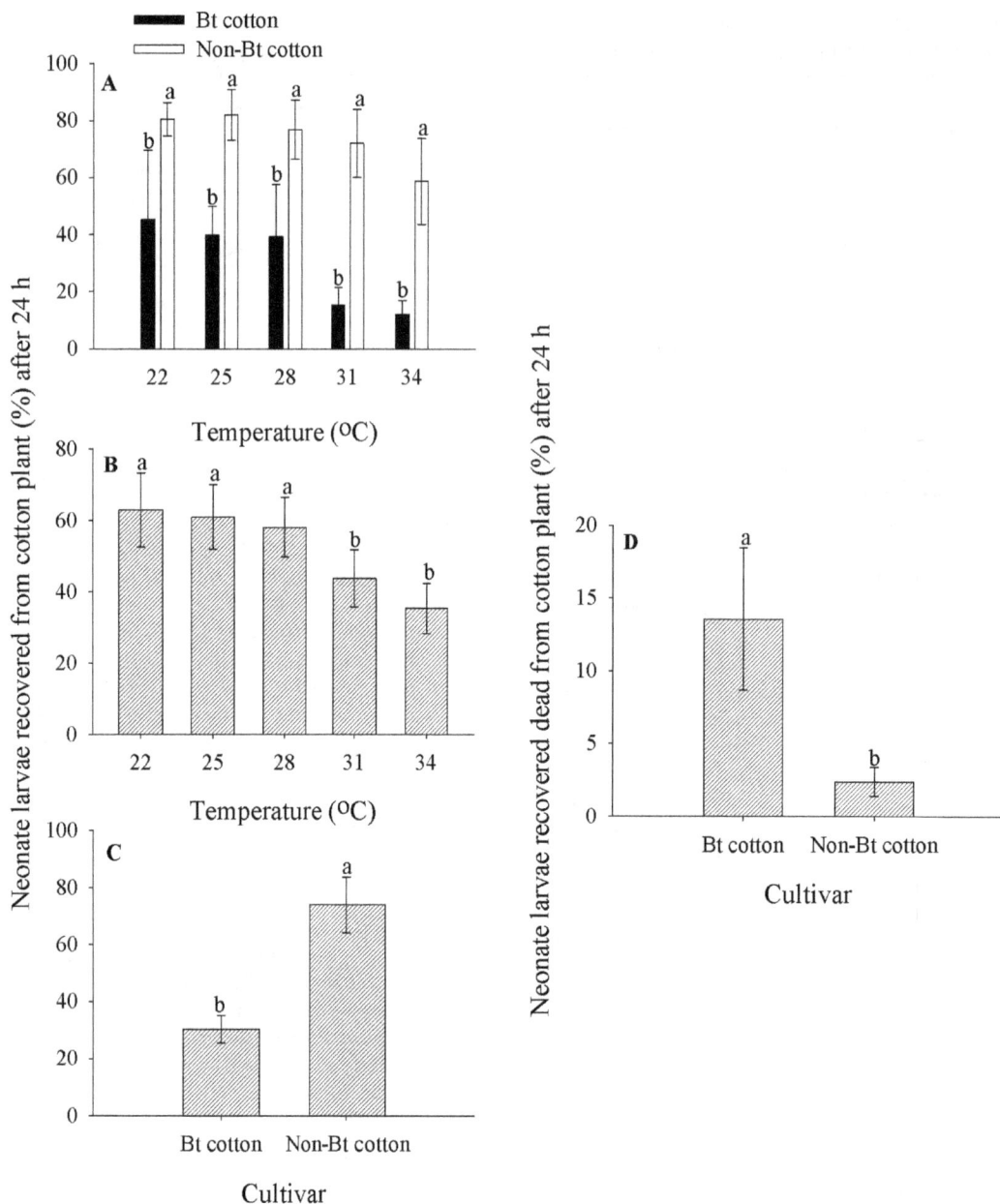

**Figure 1. Mean percentage (± SE) of neonate larvae of *A. argillacea* recovered from cotton plants after 24 h. A.** Bt and non-Bt cotton plants at each temperature (means followed by the same letter within each temperature are not significantly different by the Student-Newman-Keuls test, P = 0.05), **B.** Neonate larvae recovered alive from the cotton plants (means followed by the same letter are not significantly different by the Student-Newman-Keuls test, P = 0.05), **C.** Bt and non-Bt cotton plants ($F_{(C)1, 36}$ = 127.78, P<0.0001), and **D.** Neonate larvae recovered dead from Bt and non-Bt cotton plants after 24 h ($F_{(C)1,36}$ = 19.84, P<0.0001). Original data.

neonate *A. argillacea* larvae recovered from the non-Bt cotton cultivars was significantly higher than the percentage of *A. argillacea* larvae recovered from the Bt cotton cultivars. The one exception was in the initial interval 6 h after the neonate *A. argillacea* larvae were released onto the cotton plants, because there was no significant difference between Bt and non-Bt cotton cultivars in larvae recovered (Table 2). Razze et al. [22] reported that over 95% of neonate *Ostrinia nubilalis* (Hübner) (Lepidoptera: Cambridae) larvae who left the maize plants (*Zea mays* Linnaeus) in the first 6 h after they were exposed to the maize plants (Bt and non-Bt corn) fed very little or did not feed, regardless of whether the

plants were Bt or non-Bt corn cultivars. Goldstein et al. [17] found that significantly more neonate larvae of *O. nubilalis* present on non-Bt than on Bt corn 24 h after blackhead egg masses were placed on the plants.

We found a higher percentage of neonate *A. argillacea* larvae recovered from the organza bags with evidence that they fed on the Bt cotton cultivar compared with the non-Bt cultivar after 24 h of infestation (Fig. 3A). According to Razze et al. [22] after 48 h, there was a significantly higher percentage of *O. nubilalis* larvae that had evidence of feeding that were found on the bag on Bt corn compared with non-Bt near isoline.

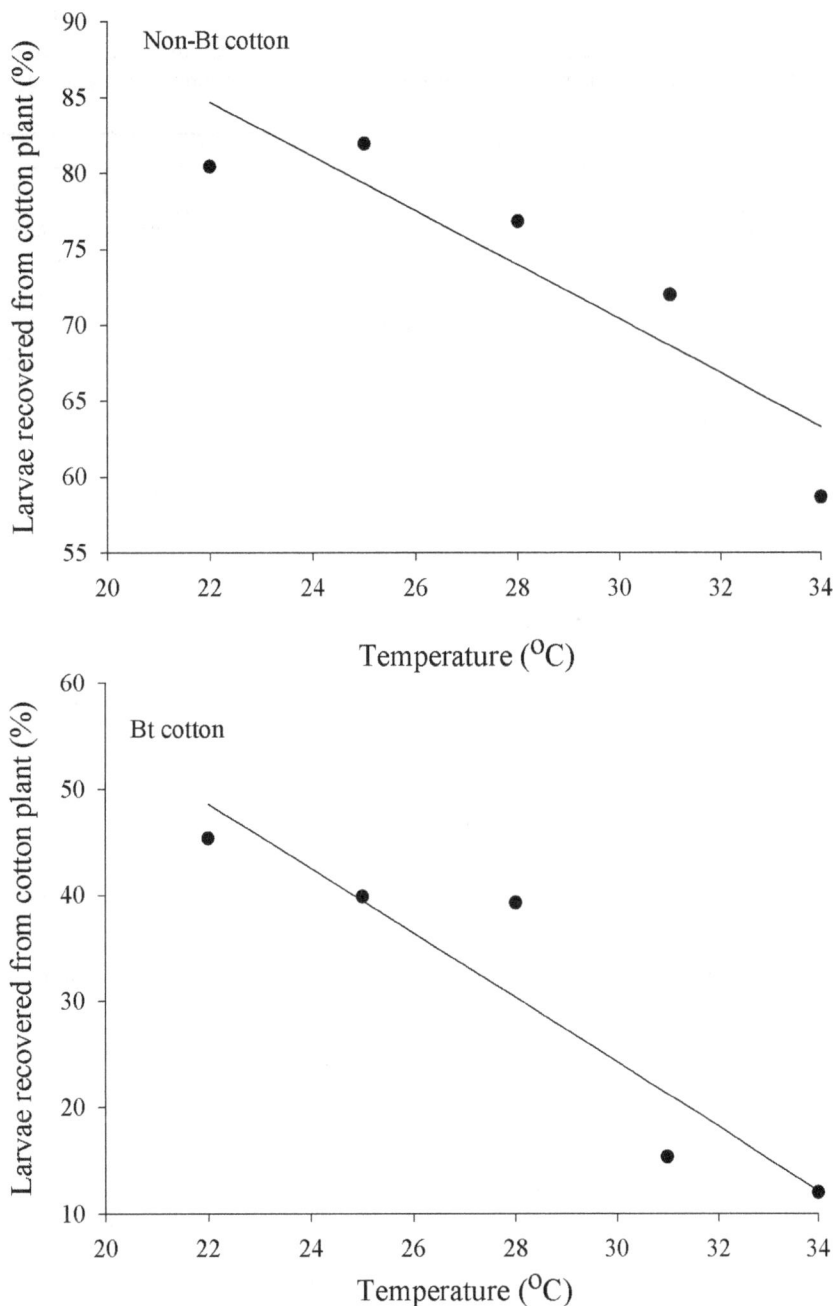

**Figure 2. Relationship between the mean percent of neonate larvae of** *A. argillacea* **recovered from plants of Bt cotton (y = 115.47 – 3.04x, $R^2$ = 0.87, $F_{1, 3}$ = 19.90, P<0.0210) and non-Bt cotton (y = 123.90 – 1.78x, $R^2$ = 0.81, $F_{1, 3}$ = 13.09, P<0.0363) and temperature after 24 h.** Original data.

This, combined with less plant material found in the intestines of larvae from the Bt plants compared with the non-Bt plants (Figs. 3B and 3C), may explain how *A. argillacea* responds to the Cry1Ac protein in cotton plants. Larvae at the start of the test fed on Bt cotton plants and realized that it was not palatable and likely sought to leave the Bt plant for a more suitable host plant. If the larvae had located another Bt plant, it would most likely try to feed and again find that the plant was not palatable and so attempt to leave the plant. This could be repeated several times. If the larva was constantly exposed to Bt cotton plants, the larva could try feeding several times; however, the larvae would not accumulate much plant tissue in its intestine. On the other hand, as the Cry1Ac protein is harmful to the insects, Bt-cotton might disturb the digestion of the plant material and contribute to the accumulation of the plant material in the gut. Our results demonstrated that after 18 h the amount of plant tissue found in the intestines of larvae exposed to Bt plants was less than that recorded in the intestines of larvae exposed to non-Bt plants, both for the larvae recovered on the plant (Fig. 3C) and for the larvae recovered on the organza bag (Fig. 3B). According to Davis and Coleman [24], *O. nubilalis* larvae will rarely stop feeding when continually exposed to a Bt plant.

**Table 4.** Summarized model of the two-way analyses of variance (ANOVA) for the effects of cultivar[1] and exposure time[2] on the percentage of neonate larvae of *A. argillacea* that had fed and were found on the plant or the bag (Bt or non-Bt near isoline)[3] and on the amount of plant tissue in the gut of neonate larvae of *A. argillacea* that were found on the plant or the bag at 6, 12, 18, and 24 h.

| Source | Models | DF | F ratio | Prob> F |
|---|---|---|---|---|
| Neonate larvae of *A. argillacea* that had fed that were found on the plant (%) | Model | 10 | 0.76 | 0.6609 |
| | Cultivar (C) | 1 | 0.01 | 1.0000 |
| | Exposure time (Et) | 3 | 0.64 | 0.5999 |
| | C x Et | 3 | 1.27 | 0.3094 |
| Neonate larvae of *A. argillacea* that had fed that were found on the bag (%) | Model | 10 | 15.96 | 0.0001 |
| | Cultivar (C) | 1 | 5.70 | 0.0264 |
| | Exposure time (Et) | 3 | 13.48 | 0.0001 |
| | C x Et | 3 | 36.54 | 0.0001 |
| Plant tissue amount in the gut of neonate larvae of *A. argillacea* that were found on the plant ($\mu m^2$) | Model | 10 | 15.79 | 0.0001 |
| | Cultivar (C) | 1 | 95.69 | 0.0001 |
| | Exposure time (Et) | 3 | 16.65 | 0.0001 |
| | C x Et | 3 | 2.97 | 0.0550 |
| Plant tissue amount in the gut of neonate larvae of *A. argillacea* that were found on the bag ($\mu m^2$) | Model | 10 | 19.77 | 0.0001 |
| | Cultivar (C) | 1 | 25.25 | 0.0001 |
| | Exposure time (Et) | 3 | 25.51 | 0.0001 |
| | C x Et | 3 | 30.90 | 0.0001 |

[1]Cultivars: Bt cotton and non-Bt near isoline cotton. [2]Time: 6, 12, 18, and 24 h. [3]Data were arcsine-square root transformed prior to statistical analyses.

It is likely that a percentage of individuals produced by *A. argillacea* moths are genetically programmed to disperse from the cotton plants without feeding on the host. One of the advantages of neonate *A. argillacea* larvae immediately migrating from the host plant is to escape interspecific competition to increase its survivability. According to Schultz and Baldwin (1982) [25], *Lymantria dispar dispar* larvae (Linnaeus) (Lepidoptera: Erebidae) can induce changes in *Quercus* spp. leaves, which may make them less palatable for subsequent herbivores. Additionally, the dispersion of some neonate *A. argillacea* larvae from cotton plants to other host plants can be a selective advantage for some larvae to avoid predators and parasitoids that are attracted to the aggregation of the neonate *A. argillacea* larvae. Goldstein et al. [17] and Razze et al. [22] suggested that various behaviors of neonate *O. nubilalis* larvae hatched in a mass of eggs were genetically controlled.

This dispersion behavior displayed by the *A. argillacea* larvae on cotton plants can have various consequences, such as contributing to the survival of the target pest (*A. argillacea*). Hence, dispersion behavior may reduce the effectiveness of strategies, such as mixing transgenic and nontransgenic seeds, to manage and delay lepidopteran pests from developing resistance to the Bt cotton cultivars. Moreover, the early migration of neonate *A. argillacea* larvae stimulated by the Bt cotton cultivars compared with non-Bt plants does not prevent larvae in more developed instars, which are more tolerant to Bt proteins, from migrating later from non-Bt cotton plants to Bt cotton plants; thus, favoring

the development of resistance in lepidopteran pests to Bt cotton cultivars [26].

One of the main environmental risks associated with Bt crops is the potential for populations of target pests to develop resistance to Bt proteins, where the pests are controlled using this technology [12]. Cases of lepidopteran pests becoming resistant to Bt cotton cultivars were detected for *Pectinophora gossypiella* (Saunders) (Lepidoptera: Gelechiidae) [27] and *Helicoverpa armigera* (Hübner) (Lepidoptera: Noctuidae) [28]. Larvae that have emerged on a Bt plant can test, stop eating, and disperse to a more acceptable host [22].

The results of our study show that 6 h after the neonate *A. argillacea* larvae were released on Bt and non-Bt cotton cultivar plants, especially between 18 and 24 h (Table 2), the *A. argillacea* larvae acceptance rate of Bt cotton cultivar plants was significantly lower than for non-Bt cotton cultivar plants. This behavior shown by the neonate *A. argillacea* larvae suggests that in the first 6 h they are exposed to the cotton plants, they are not able to identify if the cotton cultivar is Bt or non-Bt; however, identification occurs between 6-12 h after it comes in contact with the plants.

The high susceptibility of neonate *A. argillacea* larvae to the Bt toxin may have induced the dispersal of larvae in a significantly higher percentage in the Bt cotton cultivar compared with non-Bt cotton cultivar. According to Razze et al. [22] and Davis and Onstad [29], newly hatched *O. nubilalis* larvae dispersed more quickly on Bt corn plants than on non-Bt corn plants. Similar behavior was reported by López et al. [26] for *Sesamia*

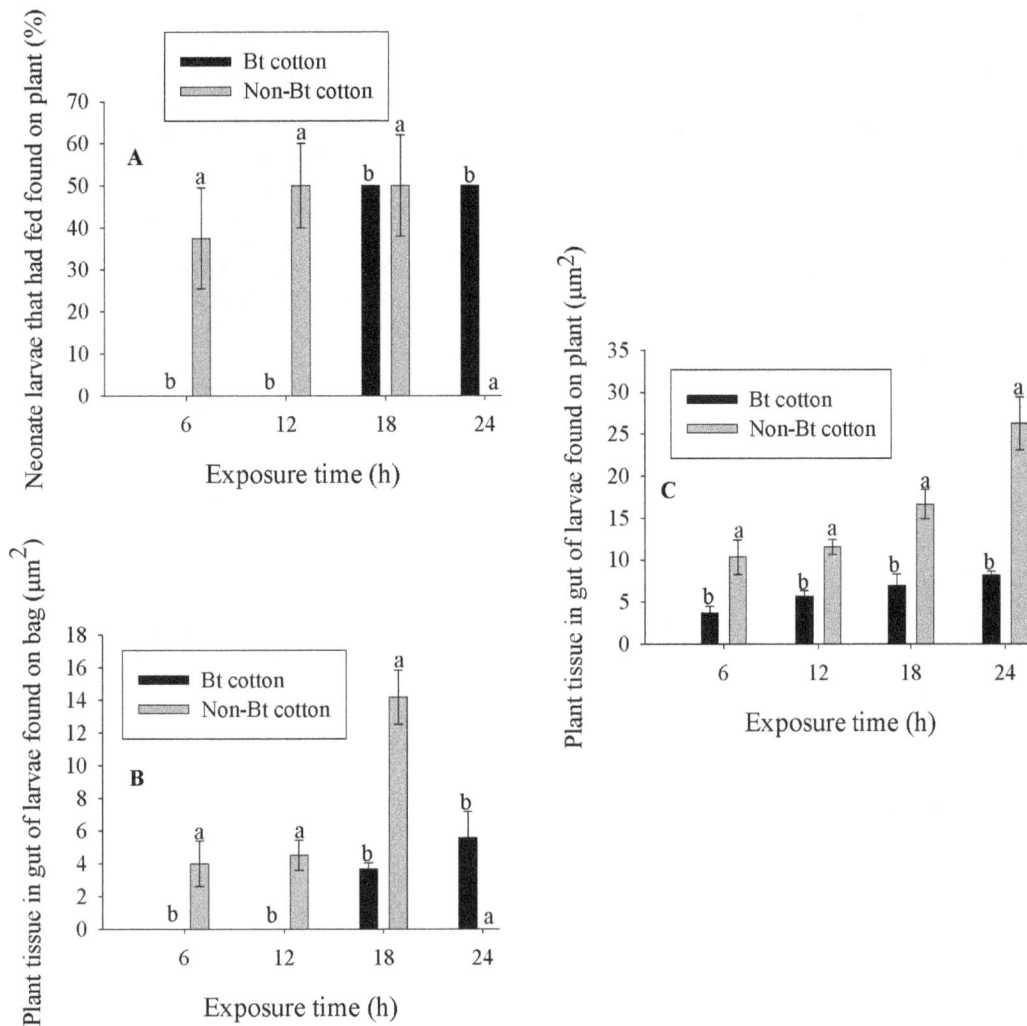

**Figure 3. Mean percentage (± SE) of neonates of** *A. argillacea* **that had fed and were found on the plant on Bt and non-Bt cotton at 6, 12, 18 and 24 h (A), (B) mean gut values for plant tissue area (± SE) for larvae that fed and were found on the bag, or (C) on the Bt or non-Bt cotton plants at 28°C.** Original data.

*nonagrioides* Lefebvre (Lepidoptera: Noctuidae) larvae on Bt corn. According to Goldstein et al. [17], the neonate *O. nubilalis* larvae were able to quickly detect Bt toxins in the leaves of Bt corn plants. Thus, the behavior exhibited by the neonate *A. argillacea* larvae to quickly disperse from Bt cotton cultivar plants is most likely because they are able to quickly detect the presence of the Bt toxin in the Bt cotton cultivar plants. However, an increase in dispersal of larvae caused by Bt plants may result in a lower efficiency for the refuge strategy [26]. Although *A. argillacea* is susceptible to the Cry1Ac toxin [30,31], the risk of developing resistance to this toxin is high because populations of this Noctuidae are exposed to high selection pressure in Bt cotton.

The *A. argillacea* larvae remained on the Bt cotton generally survived because the mean percentage of neonate larvae recovered dead from the plant was <15% (Fig. 1D), but the survival rate was about 30% (Fig. 1C); *A. argillacea* larvae mortality was significantly higher on Bt cotton cultivar plants than on non-Bt cotton cultivar plants (Fig. 1D). Similar results were found by López et al. [26] for *S. nonagrioides* larvae on Bt corn plants. Sousa et al. [30] reported a mortality of 90% of *A. argillacea* larvae that fed on Bt cotton leaves, 1 h after ingestion.

A differentiation in an *A. argillacea* larvae's acceptance of Bt cotton cultivar plants compared with non-Bt cotton cultivar plants may result in greater selection pressure on *A. argillacea* populations for resistance to Bt toxins. For example, if the mixed seed strategy was used, *A. argillacea* larvae could disperse to the nearest non-Bt cotton plant or migrate to structures with lower toxin concentrations, such as the bud bracts [32]. Therefore, the behavior of *A. argillacea* larvae not accepting the Bt cotton plants as food is important, and it should be considered in decisions on managing the resistance of *A. argillacea* to the Cry1Ac toxin.

For Brazilian cotton farmers, refuge-in-a-bag tactics can provide some advantages because there is no need to plan structured refuge areas (block or strip plantings), and the lack of adoption of standard measures in refuge areas, such as intensive use of chemicals to control target and non-target arthropods on genetically modified plants, can be avoided. Logistically, the refuge-in-a-bag tactic could be of paramount importance for Brazilian producers. This approach could make at least one resistance management tactic be adopted, the use of structured refuges, which in Brazil has received little attention for various reasons. However, the refuge-in-a-bag tactic might accelerate the development of resistance to Bt toxins in *A. argillacea*. Moreover, according to population simulations by Mallet and Porter [18], the refuge-in-a-bag tactic is not an effective measure to slow the development of resistance; paradoxically, this refuge tactic may even contribute to resistance developing, especially in situations where the level of dominance is approximately 0.01 and under conditions where the probability to select resistant individuals is high. Considering that the high dispersal of *A. argillacea* larvae within the "patch" also tends to cause an effective dominance, it is likely that an *A. argillacea* larva that feeds on a non-Bt cotton cultivar plant can grow and then migrate to and damage a Bt cotton cultivar plant and, thus, heterozygous individuals can still survive. To implement this measure, individuals must be crossed from both situations, i.e., a certain synchronicity of insect emergences from Bt and non-Bt cotton plants.

Another disadvantage of adopting the refuge-in-a-bag tactic, aimed at managing pest resistance in Brazil, is the scarcity of information on the mobility of target pests on transgenic plants, on the initial frequency of alleles that provide resistance, and on studies of the genetic structure of these populations [33–35].

High temperatures can influence the expression of the toxin in Bt crops. This can have several effects on the *A. argillacea* larvae's host acceptance behavior and the consequent dispersal and survival of *A. argillacea*. Studies have reported that high temperatures (36 to 40°C) can reduce Bt protein production in the Bt cotton plants, possibly inactivating the genes that express them, resulting in a lower efficiency of Bt plants against larvae during the open boll period [36–38]. However, this was not observed during vegetative growth or during flowering of the Bt cotton plants [36,38]. Our study showed that the percentage of *A. argillacea* larvae recovered from the cotton plants after 24 h, regardless of the temperature, was significantly higher for the non-Bt cotton cultivar than for the Bt cotton cultivar. However, the percentage of neonate *A. argillacea* larvae recovered on cotton plants was lower at 31 and 34°C than at 22, 25 and 28°C, with no differences among the other temperatures. According to Medeiros et al. [39], *A. argillacea* larvae reached thermal stress at 33°C; it was assumed that at 35°C and above, the production of enzymes in *A. argillacea* larvae was partially inhibited [39]. In addition to the *A. argillacea* larvae's low acceptance of the host plant, the heat stress, regardless of the cultivars, may have stimulated the dispersion behavior of the neonate *A. argillacea* larvae. In general, the percentage of *A. argillacea* larvae found on Bt cotton plants was less than on non-Bt cotton plants.

The eggs, larvae and pupae of *A. argillacea* are attacked by natural enemies (entomopathogens, predators and parasitoids) [39–43], which together with abiotic factors contribute to a relatively high natural mortality [1]. Therefore, the probability of an *A. argillacea* larva dispersing from a Bt cotton plant to a non-Bt cotton plant and surviving is very low with the refuge-in-a-bag strategy. The efficiency of natural enemies in reducing populations of this pest in a structured refuge may be different from the refuge-in-a-bag tactic. The structured refuge may offer a more favorable environment for the development of natural enemies than the refuge-in-a-bag, especially for a structured refuge with a high concentration of *A. argillacea* on non-Bt cotton plants. However, the effect of the refuge configuration on the abundances of the populations of the natural enemies of *A. argillacea* (entomopathogens, predators and parasitoids) needs to be explored further, and cotton growers should consider establishing a mixed-seed-refuge.

Based on the percentage of neonate *A. argillacea* larvae recovered from the organza bags for Bt and non-Bt cotton cultivars, in the first 24 h that they were exposed to the plants, the dispersal of neonate *A. argillacea* larvae was quite high. Although

the cotton plants had been protected by the organza bag and were under laboratory conditions, our results indicated that there were differences in the behavior of neonate *A. argillacea* larvae on Bt cotton plants compared with non-Bt cotton plants. After 24 h, a high percentage of *A. argillacea* larvae remained on the non-Bt cotton plants (Fig. 1A), while the dispersion of larvae from Bt cotton plants remained high (Fig. 1A). Similar results were found by Tang et al. [44] who studied the dispersion behavior of the third-instar larvae of *Plutella xylostella* (Linnaeus) (Lepidoptera: Plutellidae) on Bt and non-Bt broccoli (*Brassica oleracea* Linné) plants for a 72-h period. They found that when the release host was a Bt plant, most larval movement off the Bt plant occurred during the first 48 h with little change of movement between 48 and 72 h. When the release host was a non-Bt plant, all larval movement occurred during the first 24 h with little change of movement between 24 and 72 h. Most movement onto the second plant occurred within the first 24 h of the release with little movement occurring afterward [44]. These findings indicate that most of the cotton leafworm neonates are to detect the Bt endotoxins when exposed to the plant for 24 h and elicit behaviors leading to plant abandonment in response. If neonates abandoning Bt cotton are able to survive and find a more suitable host plant, there could be selection for behavioral resistance.

The results of our bioassays support the hypothesis that the dispersion behavior of neonate *A. argillacea* larvae is significantly different for Bt plants compared with non-Bt cotton plants. The presence of the Cry1Ac toxin in Bt cotton plants and its probable detection by the *A. argillacea* larvae tasting or feeding increases the probability of dispersion from the plant where they hatched. To understand the movement of *A. argillacea* larvae between Bt and non-Bt cotton plants and the likelihood of their survival after ingesting the Cry1Ac toxins, the relationships between their feeding behavior and their dispersion behavior need to be explored. Results obtained from the laboratory suggest that the last instar larvae of *O. nubilalis* can move from non-Bt corn plants to Bt corn plants and then survive until they reach adulthood [45,46]; however, there was little evidence that this happened in the field. Therefore, in the case of cotton, more research is needed to determine the differences in feeding behavior and dispersal of *A. argillacea* larvae on Bt cotton plants compared to non-Bt plants in the field. Once this knowledge has been obtained, the effectiveness of the refuge tactic known as a refuge-in-a-bag can be determined as well as the risk of developing resistance in *A. argillacea* populations to Bt toxins. These findings may help to understand how the *A. argillacea* larvae detect the Cry1Ab toxin in Bt cotton and how it affects the dispersion behavior of the larvae over time. Therefore, our results are extremely important for the resistance management of *A. argillacea* populations on Bt cotton cultivars.

## Author Contributions

Conceived and designed the experiments: FSR JKSP JCZ JBM. Performed the experiments: JKSP JBM FSF. Analyzed the data: FSR ACSL JKSP. Contributed reagents/materials/analysis tools: FSR JKSP JBM FSF JCZ. Wrote the paper: FSR JKSP ACSL JBM JCZ FSF.

# References

1. Carvalho SM (1981) Biologia e nutrição quantitative de *Alabama argillacea* (Hubner, 1818) (Lepidoptera: Noctuidae) em três cultivares de algodoeiro. Dissertação, ESALQ/USP, Piracicaba, SP, Brazil.
2. Nascimento ARB, Ramalho FS, Azeredo TL, Fernandes FS, Nascimento JL Jr, et al. (2011) Feeding and life history of *Alabama argillacea* (Lepidoptera: Noctuidae) on cotton cultivars producing colored fibers. Ann. Entomol. Soc. Am. 104: 613–619.
3. Oliveira JEM, Bertoli SA, Miranda JE, Torres JB, Zanuncio JC (2008) Predação por *Podisus nigrispinus* (Heteroptera: Pentatomidae) sob efeito de densidades de *Alabama argillacea* (Lepidoptera: Noctuidae) e idades do algodoiro. Cientifica 36: 1–9.
4. Ramalho FS (1994) Cotton pest management: Part 4. A Brazilian perspective. Annu. Rev. Entomol. 39: 563–578.
5. Gravena S, Cunha HF (1991) Artrópodos predadores na cultura algodoeira: atividades sobre sobre *Alabama argillacea* (Hub.) com breves referências a *Heliothis* sp. (Lepidoptera, Noctuidae). Jaboticabal: FUNEP (Boletim Técnico).
6. Silva TBM, Siqueira HAA, Oliveira AC, Torres JB, Oliveira JV, et al. (2011) Insecticide resistance in Brazilian populations of the cotton leafworm, *Alabama argillacea*. Crop Prot. 30: 1156–1161.
7. Schnepf EN, Crickmore N, Rie JV, Lereclur D, Baum J, et al. (1999) A review of *Bacillus thuringiensis* (Bt) production and use in Cuba. Biocontrol N. Informa 20: 47–48.
8. Vachon V, Laprade R, Schwartz JL (2012) Current models of the mode of action of *Bacillus thuringiensis* insecticidal crystal proteins: A critical review. J. Invertebr. Pathol. 111:1–12.
9. Hardee DD, Van Duyn JW, Layton MB, Bagwell RD (2000) Bt cotton and the tobacco budworm-bollworm complex. USDA-ARS-154. 37pp.
10. Papaj DR, Rausher MD (1983) Individual variation in host location by phytophagous insects. In: Ahmad S (ed) Herbivorous insects: host-seeking behavior and mechanisms. Academic Press, New York, pp. 77–124.
11. Jongsma MA, Gould F, Legros M, Yang L, Van Loon JJA, et al. (2010) Insect oviposition behavior affects the evolution of adaptation to Bt crops: consequences for refuge policies. Evol. Ecol. 24: 1017–1030.
12. Andow DA (2008) The risk of resistance evolution in insects to transgenic insecticidal crops. Collect. Biosaf. Rev. 4: 142–199.
13. Agi AL, Mahaffey JS, Bradley JR Jr, Van Duyn JW (2001) Efficacy of seed mixes of transgenic Bt and nontransgenic cotton against bollworm, *Helicoverpa zea* Boddie. J. Cotton Sci. 5: 74–80.
14. Bates SL, Zhao JZ, Roush RT, Shelton AM (2005) Insect resistance management in GM crops: past, present and future. Nat. Biotechnol. 23: 57–62.
15. Onstad DW, Mitchell PD, Hurley TM, Lundgren JG, Porter RP, et al. (2011) Seeds of change: corn seed mixtures for resistance management and IPM. J. Econ. Entomol. 104: 343–352.
16. Gould F (2000) Testing Bt refuge strategies in the field. Nat. Biotechnol. 18: 266–276.
17. Goldstein JA, Mason CE, Pesek J (2010) Dispersal and movement behavior of neonate European corn borer (Lepidoptera: Crambidae) on non-Bt and transgenic Bt corn. J. Econ. Entomol. 103: 331–339.
18. Mallet J, Porter P (1992) Preventing insect adaptation to insect – resistant crops: are seed mixes or refuge the best strategy? P. Roy. Soc. Lond. B Bio. 250: 165–169.
19. Chen Y, Chen Y, Wen Y, Zhang X, Chen D (2012) The effects of the relative humidity on the insecticidal expression level of Bt cotton during bolling period under high temperature. Field Crop Res. 137: 141–147.
20. Zalucki MP, Clarke AR, Malcolm SB (2002) Ecology and behavior of first instar larval Lepidoptera. Annu. Rev. Entomol. 47: 361–393.
21. Johansen DA (1940) Plant microtechnique. McGraw-Hill, Washington DC.
22. Razze JM, Mason CE, Pizzolato TD (2011) Feeding behavior of neonate *Ostrinia nubilalis* (Lepidoptera: Crambidae) on Cry1Ab Bt corn: Implications for resistance management. J. Econ. Entomol. 104: 806–813.
23. Sas Institute (2006) SAS/STAT user's guide. SAS Institute, Cary, NC.
24. Davis PM, Coleman SB (1997) European corn borer (Lepidoptera: Pyralidae) feeding behavior and survival on transgenic corn containing Cry1A(b) protein from *Bacillus thuringiensis*. J. Kans. Entomol. Soc. 70: 31–38.
25. Schultz JC, Baldwin IT (1982) Oak leaf quality declines in response to defoliation by gypsy moth larvae. Science 217: 149–151.
26. López C, Hernández-Escareño G, Eizaguirre M, Albajes R (2013) Antixenosis and larval and adult dispersal in the Mediterranean corn borer, *Sesamia nonagrioides*, in relation to Bt maize. Entomol. Exp. Appl. 149: 256–264.
27. Dhurua S, Gujar G (2011) Field-evolved resistance to *Bt* toxin Cry1Ac in the pink bollworm, *Pectinophora gossypiella* (Saunders) (Lepidoptera: Gelechiidae), from India. Pest Manag. Sci. 67: 898–903.
28. Liu F, Xu Z, Zhu YC, Huang F, Wang Y, et al. (2010) Evidence of field-evolved resistance to Cry1Ac-expressing Bt cotton in *Helicoverpa armigera* (Lepidoptera: Noctuidae) in northern China. Pest Manag. Sci. 66: 155–161.
29. Davis PM, Onstad DW (2000) Seed mixtures as a resistance management strategy for European corn borers (Lepidoptera: Crambidae) infesting transgenic corn expressing Cry1Ab protein. J. Econ. Entomol. 93: 937–948.
30. Sousa MEC, Santos FAB, Wanderley-Teixeira V, Teixeira AAC, Siqueira HAA de, et al. (2010) Histopathology and ultrastructure of midgut of *Alabama argillacea* (Hübner) (Lepidoptera: Noctuidae) fed Bt-cotton. J. Insect Physiol. 56: 1913–1919.
31. Santos RL, Torres JB (2010) Produção da proteína Cry1Ac em algodão transgênico e controle de lagartas. Rev. Bras. Ciênc. Agrár. 5: 509–517.
32. Lima MS, Torres JB (2011) Cry1Ac toxin production and feeding and oviposition preference of *Alabama argillacea* in Bt cotton under water stress. Pesqu. Agropecu. Bras. 46: 451–457.
33. Pavinato VAC, Bajay MM, Martinelli S, Monteiro M, Pinheiro JB, et al. (2011) Development and characterization of microsatellite markers for genetic studies of *Alabama argillacea* (Hüeb.) (Lepidoptera:Noctuidae): an important cotton pest in Brazil. Mol. Ecol. Resour. 11: 219–222.
34. Albernaz KC, Merlin BL, Martinelli S, Head GP, Omoto C (2012) Baseline susceptibility to Cry1Ac insecticidal protein in *Heliothis virescens* (Lepidoptera: Noctuidae) populations in Brazil. J. Econ. Entomol. 106: 1819–1824.
35. Domingues FA, Brandão KLS, Abreu AG, Pereira OP, Blanco CA, et al. (2012) Genetic structure and gene flow among Brazilian populations of *Heliothis virescens* (Lepidoptera: Noctuidae). J. Econ. Entomol. 105: 2136–2146.
36. Chen D, Ye G, Yang C, Chen Y, Wu Y (2005) The effect of high temperature on the insecticidal properties of Bt Cotton. Environ. Exp. Bot. 53: 333–342.
37. Olsen KM, Daly JC, Finnegan EJ, Mahon RJ (2005) Changes in Cry1Ac Bt transgenic cotton in response to two environmental factors: temperature and insect damage. J. Econ. Entomol. 98:1382 1390.
38. Chen Y, Wen Y, Chen Y, Zhang X, Wang Y, et al. (2013) The recovery of Bt toxin content after temperature stress termination in transgenic cotton. Span. J. Agric. Res. 11: 438–446.
39. Medeiros RS, Ramalho FS, Zanuncio JC, Serrão JE (2003) Estimate *Alabama argillacea* (Hubner) (Lepidoptera: Noctuidae) development with nonlinear models. Braz. J. Biol. 63: 589–598.37.
40. Bleicher E, Silva AL, Santos WJ, Gravena S, Nakano O, et al. (1981) Manual de manejo integrado das pragas do algodoeiro. Embrapa-CNPA, Campina Grande, PB (Documento 2). 12p.
41. Andrade CFS, Habib MEM 91984) Natural occurrence of baculoviruses in populations of some Heliconiini (Lepidoptera: Nymphalidae) with symptomatological notes. Rev. Bras. Zool. 2: 55–62.
42. Askew RR, Shaw MR (1986) Parasitoid communities: their size, structure and development. In Waage J, Greathead D (eds) Insect parasitoids. Academic Press, London, pp. 225–264.
43. Fernandes MG, Busoli AC, Degrande PE (1999) Parasitismo natural de ovos de *Alabama argillacea* Hüb. e *Heliothis virescens* Fab. (Lep.: Noctuidae) por *Trichogramma pretiosum* Riley (Hym.: Trichogrammatidae) em algodoeiros no Mato Grosso do Sul. An. Soc. Entomol. Brasil 28: 695–701.
44. Tang JD, Collins HL, Metz TD, Earle ED, Zhao JZ, et al. (2001) Greenhouse tests on resistance management of Bt transgenic plants using refuge strategies. J. Econ. Entomol. 94: 240–247.
45. Walker KA, Hellmich RL, Lewis LC (2000) Late-instar European corn borer (Lepidoptera: Crambidae) tunneling and survival in transgenic corn hybrids. J. Econ. Entomol. 93: 1276–1285.
46. Huang F, Bushman LL, Higgins RA, Li H (2002) Survival of Kansas dipel-resistant European corn borer (Lepidoptera: Crambidae) on Bt and non-Bt corn hybrids. J. Econ. Entomol. 95: 614–621.

# Identification of Top-Down Forces Regulating Cotton Aphid Population Growth in Transgenic Bt Cotton in Central China

**Peng Han[1,2], Chang-ying Niu[1]\*, Nicolas Desneux[2]\***

**1** Hubei Key Laboratory of Insect Resources Application and Sustainable Pest Control, Plant Science & Technology College, Huazhong Agricultural University, Wuhan, China, **2** French National Institute for Agricultural Research (INRA), Sophia-Antipolis, France

## Abstract

The cotton aphid *Aphis gossypii* Glover is the main aphid pest in cotton fields in the Yangtze River Valley Cotton-planting Zone (YRZ) in central China. Various natural enemies may attack the cotton aphid in Bt cotton fields but no studies have identified potential specific top-down forces that could help manage this pest in the YRZ in China. In order to identify possibilities for managing the cotton aphid, we monitored cotton aphid population dynamics and identified the effect of natural enemies on cotton aphid population growth using various exclusion cages in transgenic Cry1Ac (Bt)+CpTI (Cowpea trypsin inhibitor) cotton field in 2011. The aphid population growth in the open field (control) was significantly lower than those protected or restricted from exposure to natural enemies in the various exclusion cage types tested. The ladybird predator *Propylaea japonica* Thunberg represented 65% of Coccinellidae predators, and other predators consisted mainly of syrphids (2.1%) and spiders (1.5%). The aphid parasitoids Aphidiines represented 76.7% of the total count of the natural enemy guild (mainly *Lysiphlebia japonica* Ashmead and *Binodoxys indicus* Subba Rao & Sharma). Our results showed that *P. japonica* can effectively delay the establishment and subsequent population growth of aphids during the cotton growing season. Aphidiines could also reduce aphid density although their impact may be shadowed by the presence of coccinellids in the open field (likely both owing to resource competition and intraguild predation). The implications of these results are discussed in a framework of the compatibility of transgenic crops and top-down forces exerted by natural enemy guild.

**Editor:** Guy Smagghe, Ghent University, Belgium

**Funding:** This study was financially supported by the National Natural Science Foundation of China (Grant IBN-31071690, 31371945), and International Atomic Energy Agency (via Research Contract No. 16015, 17153 to CYN). The funders had no role in study design, data collection and analysis, decision to publish, or preparation of the manuscript.

\* Email: niuchangying88@163.com (CYN); nicolas.desneux@sophia.inra.fr (ND)

## Introduction

The widespread adoption of insect-resistant genetically modified (GM) Bt cotton has led to decreased use of chemical insecticides and enhanced biocontrol services provided by natural enemies in Northern China [1,2]. The Yangtze River Valley Cotton-planting Zone (YRZ), which located in central China, is one of the largest cotton-growing regions nationwide [3]. In this region, several insect-resistant GM cotton cultivars, notably the transgenic cotton that combines the two genes *Cry1Ac* (Bt endotoxin) and *CpTI* (Cowpea Trypsin Inhibitor), have been widely adopted during the past decade [4–6]. The cotton aphid *Aphis gossypii* Glover (Hemiptera: Aphidiae), a pest not targeted by Bt endotoxin (as is the case with other aphids, e.g. see [7,8]), is considered a secondary insect pest in the YRZ. Although cotton aphid populations have shown continuous decline in seasonal density in cotton fields in the past 15 years in Northern China [2], cotton aphid outbreaks may occur and reach economically damaging levels [1] owing to particular weather conditions (e.g. less rainfall during the aphid population-growth season) or pesticide resistance [9].

In agro-ecosystems, natural enemies play an important role in controlling arthropod pest populations [2,10]. For example, Hawkins and Marino [11] reported that insect parasitoids caused the highest mortality among the biotic factors for many pest species (mortality compiled for 78 pest species). Symondson *et al.* [10] stressed the importance of generalist predators in regulating pest populations. Various studies have documented top-down forces regulating herbivore populations and crop biomass yield [12–15] and also identified key natural enemies of predators involved in pest suppression in specific crops [14,16–18]. Indeed, it is crucial to characterize the guild of potential natural enemies capable of attacking targeted pest(s) for developing a sustainable Integrated Pest Management (IPM) program in any cropping system [19–21]. For example, identifying key natural enemies in a given ago-ecosystem may orient further research on how these natural enemies may be promoted to enhance biological control [22–25]. Therefore, studies documenting top-down forces in agro-ecosystems are crucial for developing effective IPM programs.

In Bt-cotton cropping systems in the YRZ, no systematic study has been carried out to characterize *A. gossypii* population dynamics and to identify the specific top-down forces that may help managing this pest in Bt cotton fields. In the present study, using various types of natural enemy exclusion cages and artificially released aphid populations, we aimed to (i) monitor aphid population dynamics in open field, (ii) assess specific effects of natural enemies on *A. gossypii* population dynamics, and (iii) identify the key natural enemies of *A. gossypii* in Bt cotton. The results of the present study will help optimize integrated management of *A. gossypii* in Bt-cotton cropping systems in central China.

## Materials and Methods

### Cotton field and aphid colony

Experiments were conducted during the summer of 2011 at Ezhou experimental station (Huazhong Agricultural University), Ezhou, Hubei province, China (114.7 E, 30.3 N). The GM cotton cultivar CCRI41 (Zhongmian 41) which produces insecticidal proteins Cry1Ac (*Bt* endotoxin) and CpTI (Cowpea trypsin inhibitor) [4,26] was used during the study. The CCRI41 seeds were provided by the Institute of Cotton Research of Chinese Academy of Agricultural Sciences (CAAS), Anyang, China. Seeds were sowed on April 27[th] in a 1.5-ha cotton field with 1-m spacing between rows and the cotton was cultivated using standard agronomic techniques except that no pesticides were applied. The field had been used for cotton cultivation for several years. The area surrounding the field consisted of mainly cotton (55%), rice (30%), sweet potato (10%), other minor cropping plants, and natural habitats.

Naturally occurring *A. gossypii* were collected in May from a cotton field at Huazhong Agricultural University (Wuhan, China) which had been cropped without insecticide applications. These aphids were used to establish a colony in the laboratory (on cotton) at the university and used as the source of aphids for infesting the plants during the field study.

### Experimental setup

Four different degrees of predator exclusion were tested using various exclusion designs in the Bt cotton field: (i) Exclusion cages with 530×530 μm openings in which aphids were fully protected from all insect natural enemies. (ii) Restriction cages with 3×3 mm openings in which aphids were partially protected. This size of openings restricted entry by large predators i.e. Coccinellids, but allowed small predators to enter [13,27]. (iii) Sham cages built with 530×530 μm mesh netting but included a 40 cm high opening in the middle and the bottom respectively (modified from [15]). This treatment was used to assess possible disruptive effect of caging (e.g. mesh, wood sticks, etc.) on the activity of natural enemies and aphid population growth within the plots. (iv) No cage, a completely open area (named "open field" hereafter), which used four wood sticks standing upright into the ground and a tape surrounding them as guidance for sampling range and plot size and position.

The four different treatments were established on July 28[th] (Fig. 1a) using a completely randomized block design (Fig. 1b). The distance between treatments inside each block was 3 m and between blocks was 10 m. The field cages were made of wood frames (2×2×2 m, length×width×height) covered by fine nylon mesh netting with openings of 530 μm or 3 mm according to the various designs used, see above). Four plants were enclosed in each cage with a distance of 1 m between plants. We used 2×2×2 m cages because the cotton cultivar used could grow up to 1.8 m height and 1 m width during the season [4,5]. One side of each cage was equipped with a zipper to enable sampling.

Prior to the artificial aphid infestation, any resident aphids and other insects were removed by hands, brushes and mouth aspirators in all of the cages and plants in the open field plots (20 plots total). On July 28[th], ten aphids were released on each plant of the four different treatments. Aphids were placed on the highest central leaf of the plants using a camel's hair brush. From August 4[th] to Sept 30[th], all arthropod pests and natural enemies on the four plants within each plot were recorded and identified to family or species level. In the case of aphid parasitoids, the non-emerged parasitoid mummies (pupae stage of the parasitoid) were counted (with black- and tan-colored mummies assigned to the Aphelinidae and Aphidiinae parasitoid families, respectively). The field survey was carried out on a weekly basis (every 7–8 days) from noon to 6 pm for each date of survey. Mummy samples were collected from the various plots during the course of the study (mainly from Aug 20[th] to Sept 7[th] when parasitoid densities were at high levels) for further identification of parasitoids using appropriate identification keys by [28–31]. The collected mummies (n = 119) were brought back to the laboratory and placed in Petri dishes in a Climatic Chamber (25°C, 65% RH and 16:8 h/L:D) until parasitoid adults emerged.

### Statistical analysis

We tested the effect of predator exclusion degree (factor: cage type), as well as the effect of the date (factor: date) on the aphid counts and on numbers of main natural enemies recorded (see below) using a generalized linear model based on a Poisson distribution and a log-link function (Proc Genmod in the SAS statistical package, SAS Institute, NC, USA).

## Results

Overall, three dominant arthropod guilds were identified during the surveys: (i) pest insects, (ii) natural enemies of *A. gossypii* and (iii) omnivorous insects (Table 1). *Aphis gossypii* accounted for 85.1% of total pest insects recorded; the other three main pest species were the leafhopper *Empoasca biguttula* Shiraki, the whitefly *Bemisia tabaci* Gennadius and the common cutworm *Spodoptera litura* Fabricius (Lepidoptera: Noctuidae). The natural enemy guild was largely dominated by the aphid parasitoids which accounted for 76.7% of all natural enemies recorded during the study. Aphidiines (tan-colored mummies) were most commonly observed; only 3 Aphelinidae mummies were found during the study. The parasitoids identified (i.e. those emerged from mummies brought back to the laboratory) were primarily *Lysiphlebia japonica* Ashmead and *Binodoxys (Trioxys) indicus* Subba Rao & Sharma (51.8% and 37.7% of samples collected, respectively). Two other species were also identified at lower rates: *Aphidius gifuensis* Ashmead and *B. near communis* (8.8% and 1.8%, respectively). Coccinellids represented 11.2% of all natural enemies observed, with *Propylaea japonica* Thunberg being the dominant species belonging to this group of predators (65.03%). *Harmonia axyridis* Pallas (20.04%) and *Coccinella septempunctata* Linnaeus (14.92%) were also observed as less common coccinellid species. The other natural enemies belonged to the syrphid, spider and lacewing predator groups. Omnivorous insects were also observed, mainly Hemipteran piercing-sucking bugs belonging to the Miridae, Nabidae and Anthocoridae families.

*Aphis gossypii* densities we recorded differed significantly among cage types (Fig. 2, cage type factor: $\chi^2 = 12.20$, df = 3, $P = 0.007$) and as function of the dates when the aphid populations were surveyed during the season (date factor: $\chi^2 = 17.73$, df = 7, $P = 0.013$). The two factors did not interact significantly when analyzing aphid counts ($\chi^2 = 20.07$, df = 21, $P = 0.521$). More aphids were found in exclusion cages and restriction cages than in sham cages or open field plots. There was a 180-fold aphid

(A)

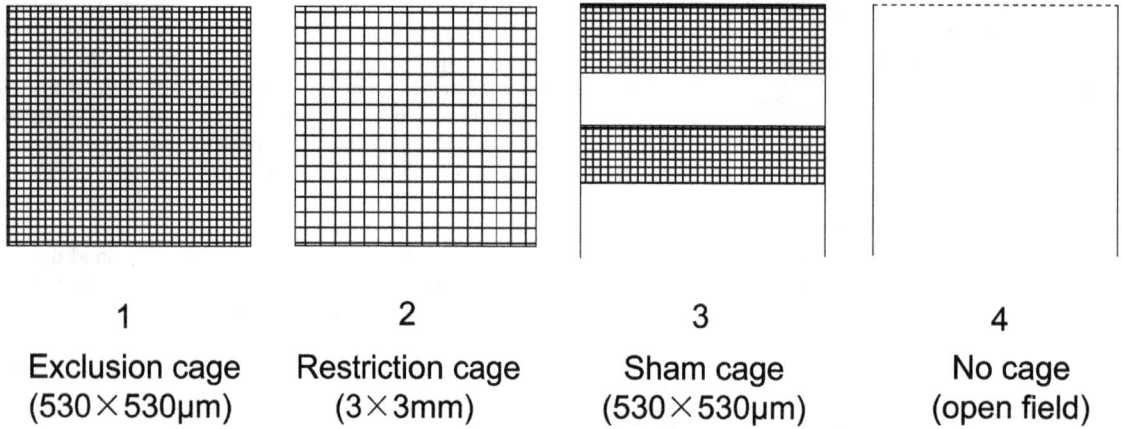

(B)

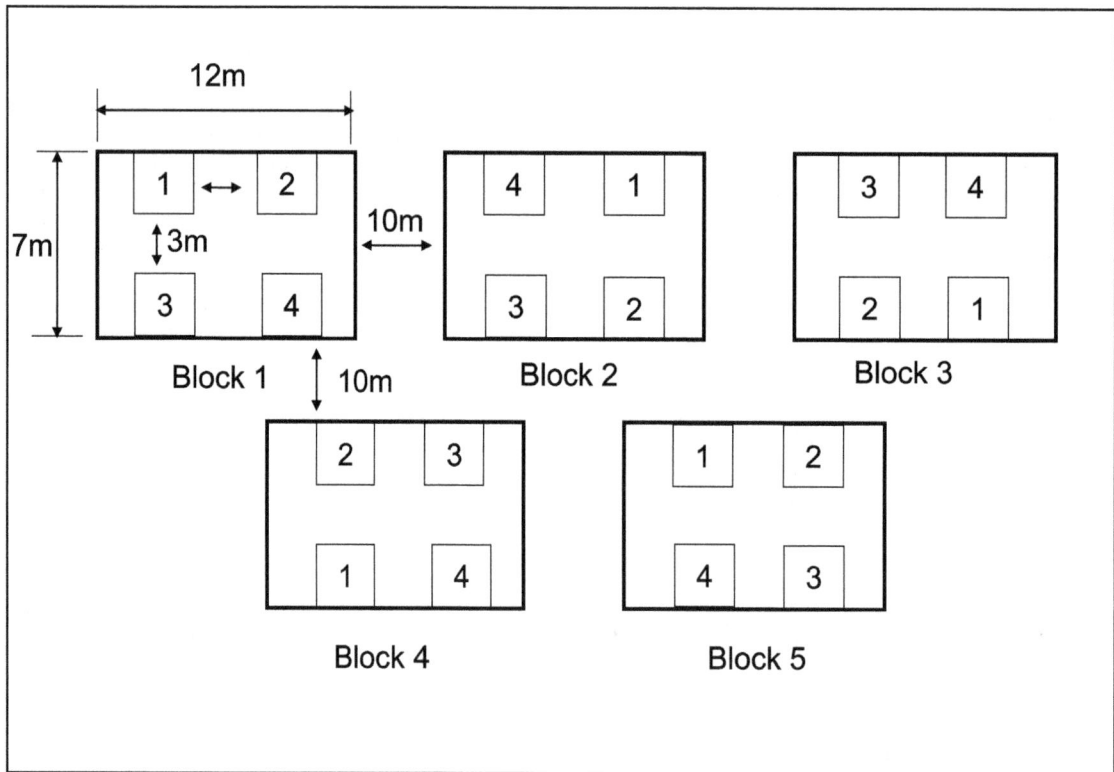

**Figure 1. Design of the field study. (A)** The four different cage treatments, i.e. natural enemy exclusion degree, in the field study; Exclusion cages: prevented natural enemy (predators and aphid parasitoids) movement, Restriction cage: prevented predator movement but allowed aphid parasitoids to colonize the plants, Sham cage and No cage: allowed free access to the plants for all natural enemies. **(B)** Within- and among block design: the distance among treatments within a block was 3 m, and among blocks was 10 m. The experimental cotton field was 70 m×30 m.

**Table 1.** Dominant arthropods, per guild, found during the surveys.

| Guild | Taxonomy | Total counts | Percentage within guild (%) |
|---|---|---|---|
| Pest insects | *Aphis gossypii* Glover | 30611 | 85.1 |
| | *Empoasca biguttula* Shiraki | 1693 | 4.7 |
| | *Bemisia tabaci* Gennadius | 2727 | 7.6 |
| | *Spodoptera litura* Fabricius | 924 | 2.6 |
| Natural enemies | Coccinellids[a] | 449 | 11.2 |
| | Aphid parasitoids (Aphidiines) | 3081 | 76.7 |
| | Syrphidae | 83 | 2.1 |
| | Araneae[b] | 60 | 1.5 |
| | Chrysopa (lacewings) | 52 | 1.3 |
| Omnivorous insects | Hemiptera (bugs)[c] | 124 | |

Total counts of dominant arthropods per guild in the experimental blocks during the field survey from August 4[th] to September 30[th], 2011, in Ezhou (China).
[a]mainly *Propylaea japonica* Thunberg (292, 65.03%), *Harmonia axyridis* Pallas (90, 20.04%) and *Coccinella septempunctata* Linnaeus (67, 14.92%).
[b]mainly *Erigonidium graminicolum* Sundevall.
[c]mainly from Miridae, Nabidae and Anthocoridae families.

population growth by August 29[th] from the initial aphid count at the July 28[th] release in the exclusion cages, whereas there was only a 10-fold aphid population increase in the sham cages and open field plots. Sham cages and open field plots showed no difference in aphid numbers during the course of this study (Fig. 2).

The numbers of coccinellids recorded also differed significantly among cage types (Fig. 3A, cage type factor: $\chi^2 = 18.20$, df = 3, $P<0.001$) and among dates of sampling (date factor: $\chi^2 = 19.52$, df = 7, $P = 0.007$). There was no significant interaction between the two factors ($\chi^2 = 19.87$, df = 21, $P = 0.134$). Many more coccinellids were recorded in sham cages and open field plots than in exclusion cages and restriction cages; however no difference in coccinellids was observed between sham cages and open field plots. *Propylaea japonica* was the dominant species

among the Coccinellidae family during the survey (Fig. 3B). The counts for this species followed the same trends as were observed for the coccinellid group as a whole: more *P. japonica* were found in sham cages and open field plots (significant cage type factor: $\chi^2 = 19.00$, df = 3, $P<0.001$, and date factor: $\chi^2 = 19.89$, df = 7, $P = 0.006$, no significant interaction: $\chi^2 = 23.12$, df = 21, $P = 0.333$).

The numbers of Aphidiine mummies differed significantly between cage types (Fig. 4, cage type factor: $\chi^2 = 8.91$, df = 3, $P = 0.031$) and dates (date factor: $\chi^2 = 19.03$, df = 7, $P = 0.008$), but the two factors did not interact significantly overall ($\chi^2 = 22.89$, df = 21, $P = 0.274$). Overall, many more Aphidiine parasitoids were found in restriction cages than in the other three cage treatments on Aug 20[th], Aug 29[th] and Sept 7[th] (Fig. 4); the

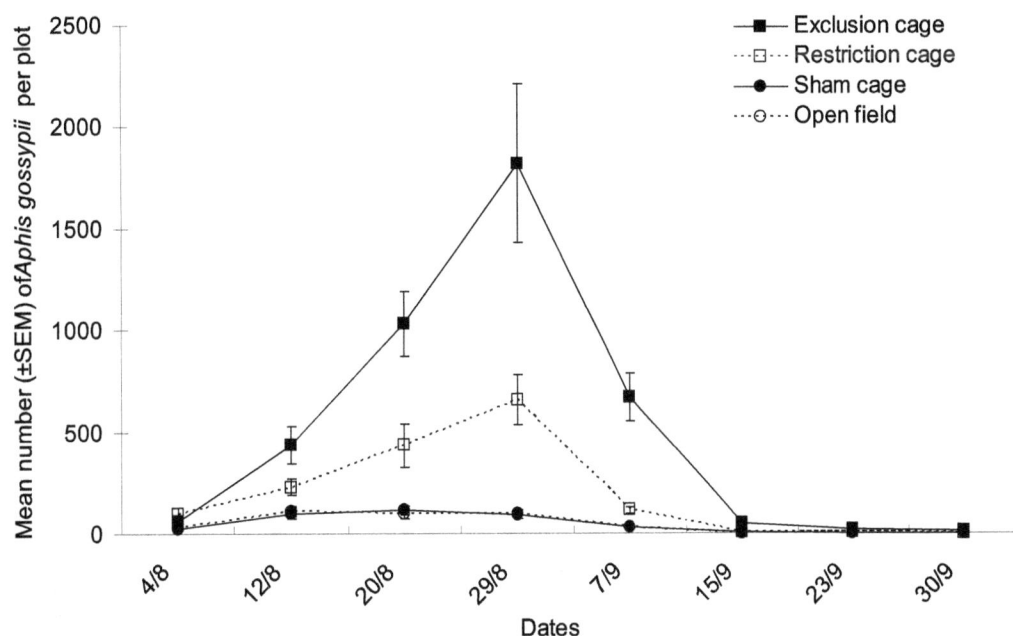

**Figure 2. Cotton aphid population dynamics.** Mean numbers (±SEM) of *A. gossypii* per plot in the various natural enemy exclusion treatments from early August to end of September in Bt cotton in Ezhou (China).

(A)

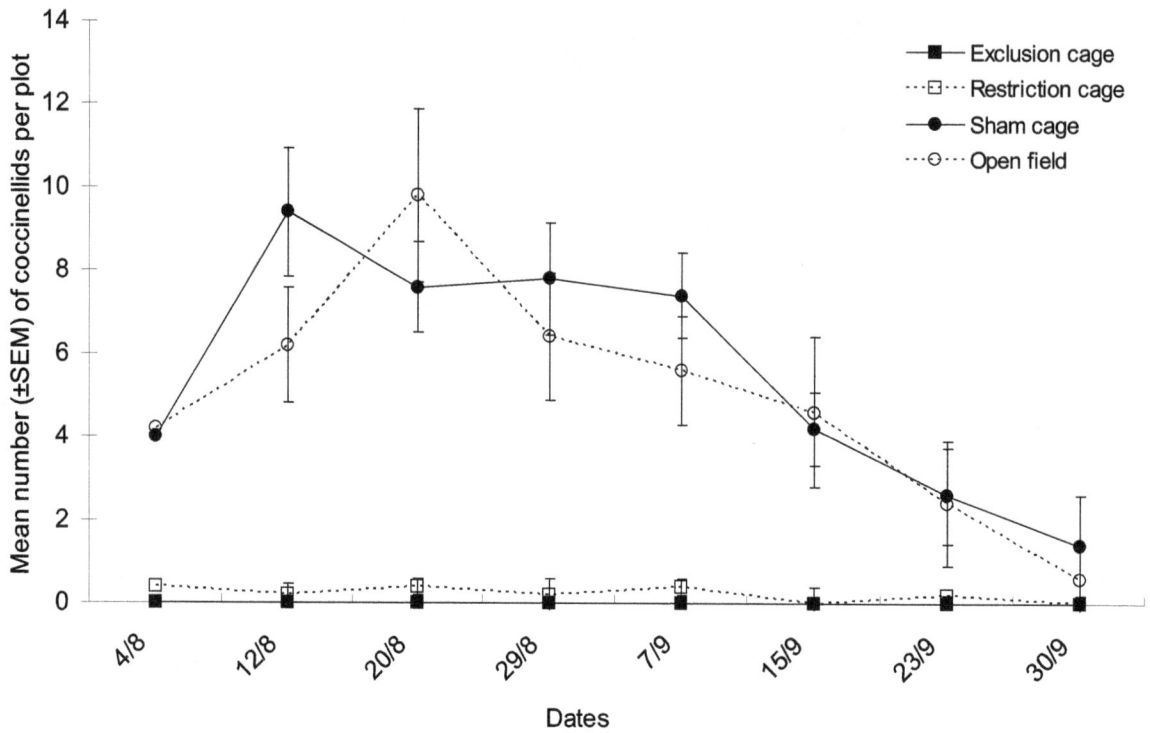

(B)

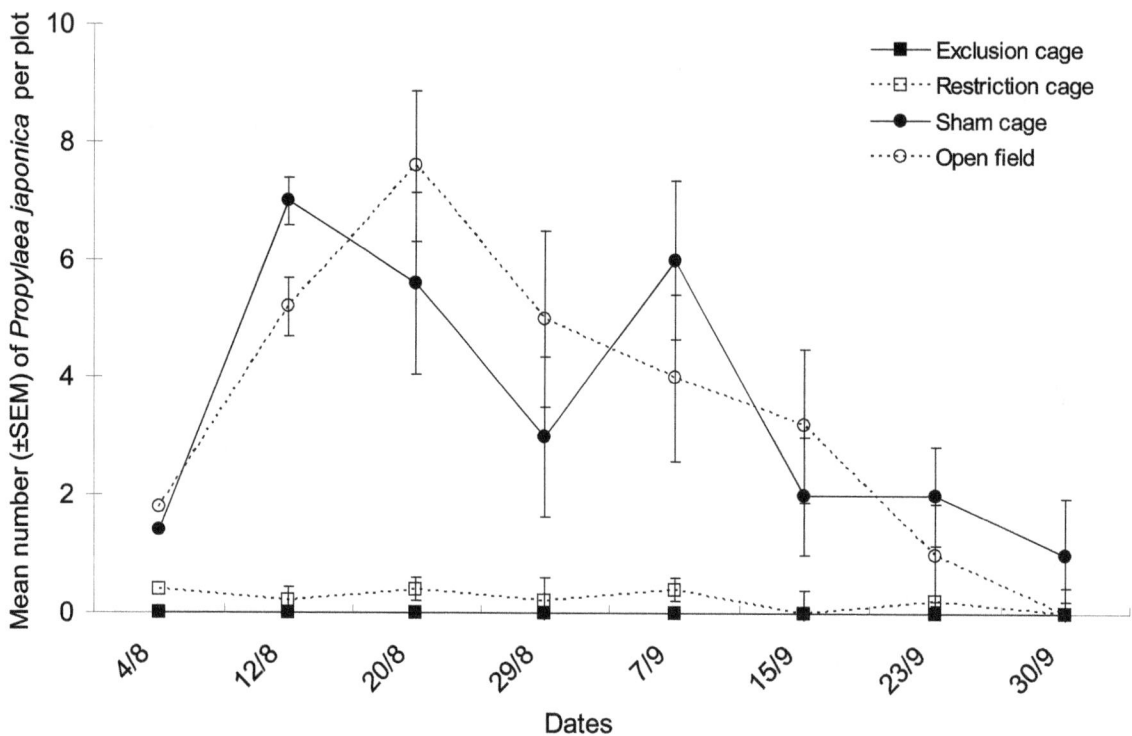

**Figure 3. Population dynamics of coccinellid predators.** Mean numbers ($\pm$SEM) of (**A**) all Coccinellids and (**B**) *P. japonica* per plot in the various natural enemy exclusion treatments from early August to end of September in Bt cotton in Ezhou (China).

parasitoid density increased markedly by 30- to 40-fold beginning Aug 12[th] and reached a peak on Aug 29[th].

The numbers of other natural enemies differed significantly among cage types as well (Fig. 5, cage type factor: $\chi^2 = 10.16$, df = 3, $P = 0.017$) and dates (date factor: $\chi^2 = 18.61$, df = 7, $P = 0.046$), no significant interaction was observed between the two factors ($\chi^2 = 25.67$, df = 21, $P = 0.458$).

## Discussion

Our study demonstrated the contribution of natural enemies (predators and parasitoids) on cotton aphid population growth in Bt+CpTI cotton field. In the absence of predators and parasitoids resulting from exclusion cages, cotton aphid populations increased up to maximum of 180-fold from aphid density at the initial release date, while in the presence of natural enemies (open field plots or sham cages) aphid populations showed a maximum 10-fold increase. These major differences in aphid population dynamics show the importance of top-down forces on this pest infesting Bt cotton. We identified the coccinellid *P. japonica* and the Aphidiine parasitoids as the predominant natural enemies in the cotton field, with distinct but additive effects on cotton aphid population growth. The best control of aphid populations was obtained when both natural enemy types had access to the aphids in open field plots or sham cages.

The coccinellid *P. japonica* proved to be an important natural enemy for suppressing cotton aphid population growth in Bt cotton fields in the YRZ in China. *Propylaea japonica* is a well-known predator of *A. gossypii* [32,33] and its life history characteristics and phenology make it a good candidate biocontrol agent for management of the aphid in Bt cotton. This predator colonizes cotton fields early in the cotton seedling stage, at the same time as the aphid population starts infesting the cotton field. Being a generalist predator, it can feed on a variety of prey including spider mites, thrips, whites flies and other small species [34,35], including those observed during our study (e.g. whiteflies

and leafhoppers, see Table 1). Alternative prey can help promote establishment of predators early in the season when the targeted pest is scarce (e.g. see [36]). Therefore, *P. japonica* can effectively delay the establishment and subsequent population growth of aphids early in the growing season. Such characteristics often make, generalist predators useful in the strategy of conservation biological control (e.g. see [10,37,38]).

The Aphidiine parasitoids, mainly *L. japonica* and *B. indicus*, were also found to suppress cotton aphid population growth. In the restriction cages, when coccinellid predators did not have access to the aphid populations, the parasitoids reduced aphid peak population by nearly 2/3 (see aphid densities in exclusion cages vs. restriction cages, Fig. 2). However, Aphidiines alone could not totally prevent aphid population growth, as aphid density reached ~600 aphids per plot by Aug 29[th]. In these restriction cages there was a rapid early season aphid population growth because predators known to limit pest population increase early in the season were excluded [10]. However, as aphid density increased in these plots, aphid parasitoid adults were attracted and this resulted in abundant parasitized mummies in the following weeks. When predators were present (in the sham cages and open field plots) the parasitoid populations remained at low densities throughout the season, either because of possible intraguild predation [39,40] of parasitoid mummies by coccinellids (e.g. see [41]), or through resource competition of aphid parasitoids (aphids) with the generalist predators in the plots [42,43]. In this instance, the aphid parasitoids may help reduce aphid densities primarily when aphid populations have already reached a certain density. Previous surveys of natural enemies of cotton aphid carried out in different regions of China produced variable collections of species records. Sun *et al.* [34] reported that the predator guild in cotton fields near Beijing (Xibeiwang) was dominated by *Chrysoperla sinica* Tjeder, *P. japonica*, various spiders and *Orius minutus* L. The same authors also reported that *Lysiphlebia japonica* was the dominant aphid parasitoid; the parasitoid guild in Xibeiwang region may be similar to the one

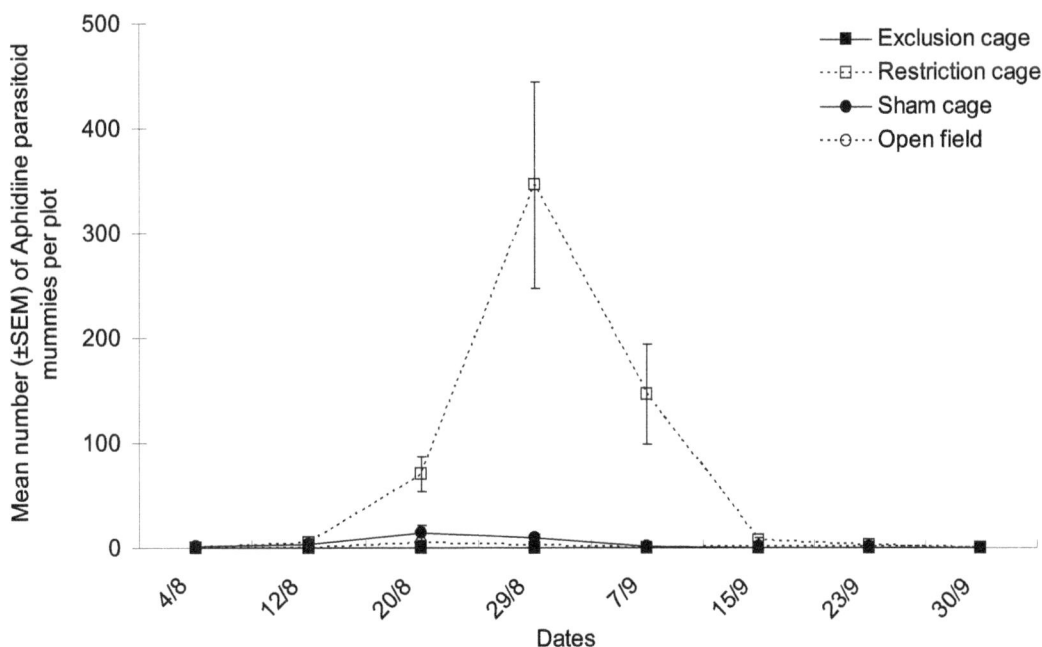

**Figure 4. Population dynamics of aphid parasitoids.** Mean numbers (±SEM) of Aphidiine per plot in the various natural enemy exclusion treatments from early August to end of September in Bt cotton in Ezhou (China).

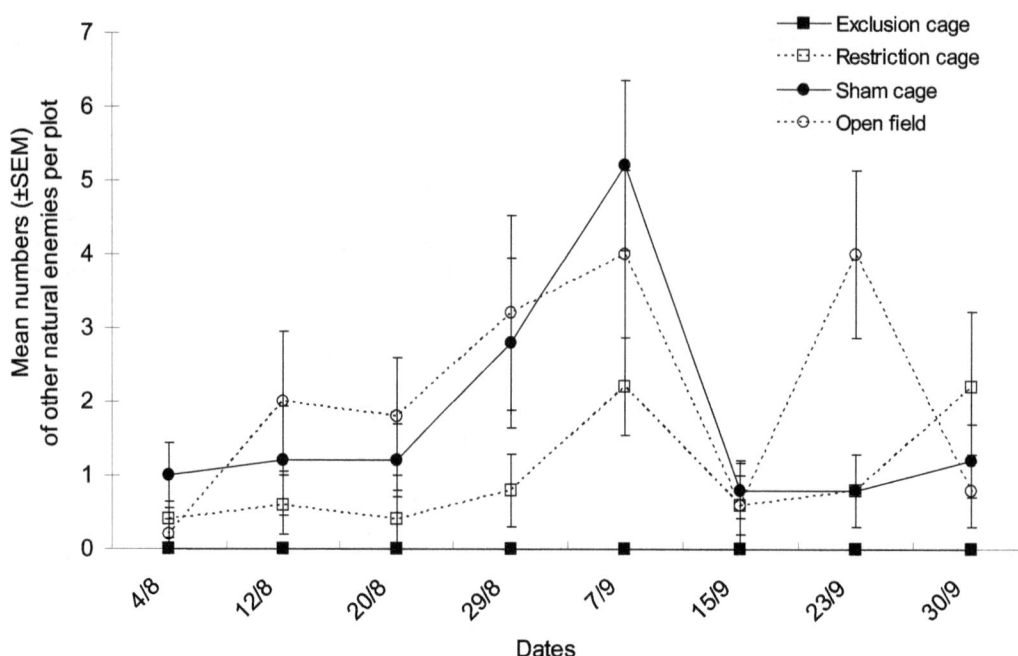

**Figure 5. Population dynamics of other natural enemies.** Mean numbers (±SEM) of other natural enemies per plot in the various natural enemy exclusion treatments from early August to end of September in Bt cotton in Ezhou (China).

recorded in YRZ. Zhou *et al.* [44] reported similar findings to ours as *L. japonica* and *P. japonica* were dominant in Hebei province. However, in contrast to our results, they reported very low biodiversity in coccinellid species (we found that *H. axyridis* and *C. septempunctata* were well represented in our plots) and that *C. sinica* and heteropteran predators (mainly *O. minutus* and the mirid predator *Campylomma diversicornis* Reuter) were quite abundant (as much as coccinellids). In the Xinjiang region, Xu *et al.* [45] conducted surveys on predators and found them, in order of importance in terms of density, coccinellids>spiders>lacewings>heteropteran predators. These contrasting results among geographic regions highlight the need to identify the specific natural enemies at play in a given region when developing conservation biological control programs.

When examining the aphid parasitoid group, it is worth mentioning that Aphelinid parasitoids were nearly absent from the field (as reported in other agro-ecosystems, e.g. in *Brassicae* crops [46,47]). When considering Aphidiine parasitoids, *L. japonica* proved to be a key natural enemy of cotton aphid in Northern China [44,48]. This species is also a natural enemy of phylogenetically closely related aphid species [49] e.g. the soybean aphid *Aphis glycines* Matsumura in Japan and Indonesia [50] and the brown citrus aphid *Toxoptera citricida* Kirkaldy [51,52]. Several species from the *Binodoxys* genus are known to efficiently attack *A. gossypii* [31,53,54] and *B. indicus* may be an important natural enemy of this aphid pest in the YRZ region as well as other regions not extensively surveyed.

Ecological compatibility of GM crops and natural enemies is a key issue for implementing biological control programs within GM cropping systems [55–57]. Previous studies suggested that Bt+ CpTI cotton might not affect population dynamics of natural enemies [34,45]. No effect was observed on the fitness of *P. japonica* when fed with *A. gossypii* on Bt cotton ([33,58], but see [59]). In addition, aphid parasitoids may not be exposed to Bt toxins [7,8]. However, they can be negatively affected by Trypsin Inhibitors [60] e.g. CpTI. Zhou *et al.* [44] reported a 44%

decrease in *L. japonica* population density in Bt-CpTI cotton fields. Although we did not carry out a formal comparison between non Bt and Bt cotton cultivar, we highlighted a strong top-down effect on cotton aphid populations. Therefore natural enemies, as a whole group, are effective in limiting aphid population growth in Bt-CpTI cotton fields.

Our study demonstrated the importance of the top-down force exerted by natural enemies, mainly coccinellids and Aphiddiine parasitoids, on cotton aphid in Bt cotton field in China. However, the relative strength of top-down vs. bottom-up forces on *A. gossypii* still needs to be studied in order to develop IPM including such forces in a sustainable and comprehensive way, especially since various studies have already identified the importance of bottom-up forces (e.g. fertilization regime) on herbivore population dynamics [61–64]. Developing such optimized IPM would help manage secondary pests that may show population outbreaks in Bt cotton since its wide spread adoption in China. For example, *S. litura* larvae were found in relatively high density during our surveys and this species can cause considerable damage to cotton crops. This finding is consistent with the reported low susceptibility of this pest species to current Bt cotton cultivars [3]. Secondary pests may promote applications of insecticides in Bt cotton with potential associated multiple negative effects on human health and non-target organisms [65–67]. Highly selective chemical pesticides may be required at times [68,69] but limiting the application of pesticides and promoting more sustainable pest management strategies should be prioritized. For example, optimized IPM may aim at combining biocontrol agents as top-down force [70,71] with bottom-up forces like fertilization regimes and/or cultural practices [13,72–75] for efficient management of pests. In addition, the sustainable use of GM crops can lead to drastic reduction in pesticide usage at the wide scale [2]; developing optimized IPM in Bt crops such Bt cotton would help capitalize on the benefits provided by transgenic methods in cropping systems.

## Acknowledgments

We thank Professor Kongming Wu (CAAS) and Tim Oppenheim for comments on the manuscript and Jintao Wang, Zhenzhong Chen, Jian Zhu, Xuejian Huang for support during the field experiments.

## Author Contributions

Conceived and designed the experiments: PH CYN ND. Performed the experiments: PH. Analyzed the data: PH ND. Contributed reagents/materials/analysis tools: CYN. Wrote the paper: PH CYN ND.

## References

1. Wu KM, Guo YY (2005) The evolution of cotton pest management practices in China. Annu Rev Entomol 50: 31–52.
2. Lu YH, Wu KM, Jiang YY, Guo YY, Desneux N (2012) Widespread adoption of Bt cotton and insecticide decrease promotes biocontrol services. Nature 487: 362–365.
3. Wan P, Wu KM, Huang MS, Yu DZ, Wu JP (2008) Population dynamics of Spodoptera litura (Lepidoptera : Noctuidae) on Bt cotton in the Yangtze River Valley of China. Environ Entomol 37:1043–1048.
4. Han P, Niu CY, Lei CL, Cui JJ, Desneux N (2010) Quantification of toxins in a Cry1Ac+CpTI cotton cultivar and its potential effects on the honey bee Apis mellifera L. Ecotoxicology 19: 1452–1459.
5. Han P, Niu CY, Lei CL, Cui JJ, Desneux N (2010) Use of an innovative T-tube maze assay and the Proboscis Extension Response assay to assess sublethal effects of GM products and pesticides on learning capacity of the honey bee Apis mellifera L. Ecotoxicology 19: 1612–1619.
6. Chen LZ, Cui JJ, Ma WH, Niu CY, Lei CL (2011) Pollen from Cry1Ac/CpTI-transgenic cotton does not affect the pollinating beetle Haptoncus luteolus. J Pest Sci 84: 9–14.
7. Ramirez-Romero R, Desneux N, Chaufaux J, Kaiser L (2008) Bt-maize effects on biological parameters of the non-target aphid Sitobion avenae (Homoptera : Aphididae) and Cry1Ab toxin detection. Pestic Biochem Phys 91: 110–115.
8. Romeis J, Meissle M (2011) Non-target risk assessment of Bt crops - Cry protein uptake by aphids. J Appl Entomol 135: 1–6.
9. Yi F, Zou CH, Hu QB, Hu MY (2012) The Joint Action of Destruxins and Botanical Insecticides (Rotenone, Azadirachtin and Paeonolum) Against the Cotton Aphid Aphis gossypii Glover. Molecules 17: 7533–7542.
10. Symondson WOC, Sunderland KD, Greenstone MH (2002) Can generalist predators be effective biocontrol agents? Annu Rev Entomol 47: 561–594.
11. Hawkins BA, Marino PC (1997) The colonization of native phytophagous insects in North America by exotic parasitoids. Oecologia 112: 566–571.
12. Liu J, Wu KM, Hopper R, Zhao K (2004) Population dynamics of Aphis glycines (Homoptera: Aphididae) and its natural enemies in soybean in northern China. Ann Entomol Soc Am 97: 235–239.
13. Costamagna AC, Landis DA (2006) Predators exert top-down control of soybean aphid across a gradient of agricultural management systems. Ecol Appl 16: 1619–1628.
14. Desneux N, O'Neil RJ, Yoo HJS (2006) Suppression of population growth of the soybean aphid, Aphis glycines Matsumura, by predators: the identification of a key predator, and the effects of prey dispersal, predator density and temperature. Environ Entomol 35: 1342–1349.
15. Costamagna AC, Landis DA, Difonzo AC (2007) Suppression of soybean aphid by generalist predators results in a trophic cascade in soybeans. Ecol Appl 17: 441–451.
16. Naranjo SE (2001) Conservation and evaluation of natural enemies in IPM systems for Bemisia tabaci. Crop Prot 20: 835–852.
17. Carrero DA, Melo D, Uribe S, Wyckhuys KAG (2013) Population dynamics of Dasiops inedulis (Diptera: Lonchaeidae) and its biotic and abiotic mortality factors in Colombian sweet passionfruit orchards. J Pest Sci 86:437–447.
18. Straub CS, Simasek NP, Gapinski MR, Dohm R, Aikens EO, et al. (2013) Influence of non host plant diversity and natural enemies on the potato leafhopper, Empoasca fabae, and pea aphid, Acyrthosiphon pisum, in alfalfa. J Pest Sci 86:235–244.
19. Norris RF, Caswell-Chen EP, Kogan M (2003) Concepts in Integrated Pest Management. Prentice Hall, New Jersey, 586 p.
20. Desneux N, Wajnberg E, Wyckhuys KAG, Burgio G, Arpaia S, et al. (2010) Biological invasion of European tomato crops by Tuta absoluta: Ecology, history of invasion and prospects for biological control. J Pest Sci 83: 197–215.
21. Ragsdale DW, Landis DA, Brodeur J, Heimpel GE, Desneux N (2011) Ecology and Management of the Soybean Aphid in North America. Annu Rev Entomol 56: 375–399.
22. Landis DA, Wratten SD, Gurr GM (2000) Habitat management to conserve natural enemies of arthropod pests in agriculture. Annu Rev Entomol 45: 175–201.
23. Zappalà L, Biondi A, Alma A, Al-Jboory IJ, Arnò J, et al. (2013) Natural enemies of the South American moth, Tuta absoluta, in Europe, North Africa and Middle-East, and their potential use in pest control strategies. J Pest Sci 86: 635–647.
24. Lundgren JG, Wyckhuys KAG, Desneux N (2009) Population responses by Orius insidiosus to vegetational diversity. Biocontrol 54: 135–142.
25. Wratten SD, Gillespie M, Decourtye A, Mader E, Desneux N (2012) Pollinator habitat enhancement: benefits to other ecosystem services. Agr Ecosyst Environ 159: 112–122.
26. Han P, Niu CY, Biondi A, Desneux N (2012) Does transgenic Cry1Ac+CpTI cotton pollen affect hypopharyngeal gland development and midgut proteolytic

27. Fox TB, Landis DA, Cardoso FF, Difonzo CD (2005) Impact of predation on establishment of the soybean aphid, Aphis glycines, in soybean Glycine max. BioControl 50: 545–563.
28. Stary P (1966) Aphid parasites of Czechoslovakia, Junk, The Hague.
29. Stary P (1979) Aphid parasites (Hymenoptera, Aphidiidae) of the central Asian area, Junk, The Hague, The Netherlands.
30. Stary P, Schlinger EI (1967) A revision of the Far East Asian Aphidiidae (Hymenoptera). Junk, Dan Haag, The Netherlands.
31. Desneux N, Stary P, Delebecque CJ, Gariepy TD, Barta RJ, et al. (2009) Cryptic species of parasitoids attacking the soybean aphid, Aphis glycines Matsumura (Hemiptera: Aphididae), in Asia: Binodoxys communis Gahan and Binodoxyx koreanus Stary sp. n. (Hymenoptera: Braconidae: Aphidiinae). Ann Entomol Soc Am 102: 925–936.
32. Zhu SZ, Su JW, Liu XH, Du L, Yardim EN, et al. (2006) Development and Reproduction of Propylaea japonica (Coleoptera: Coccinellidae) Raised on Aphis gossypii (Homoptera: Aphididae) Fed Transgenic Cotton. Zool Stud 45: 98–103.
33. Zhang GF, Wan FH, Liu WX, Guo HY (2006) Early instar response to plant-delivered Bt-toxin in a herbivore (Spodoptera litura) and a predator (Propylaea japonica). Crop Prot 25: 527–533.
34. Sun CG, Zhang QW, Xu J, Wang YX, Liu JL (2003) Efects of transgenic Bt cotton and transgenic Bt+CpTI cotton on population dynamics of main cotton pests and their natural enemies. Acta Entomologica Sinica 46: 705–712.
35. Zhang GF, Lu ZC, Wan FH (2007) Detection of Bemisia tabaci remains in predator guts using a sequence-characterized amplified region marker. Entomol Exp Appl 123: 81–90.
36. Harwood JD, Desneux N, Yoo HYS, Rowley D, Greenstone MH, et al. (2007) Tracking the role of alternative prey in soybean aphid predation by Orius insidiosus: A molecular approach. Mol Ecol 16: 4390–4400.
37. Bompard A, Jaworski CC, Bearez P, Desneux N (2013) Sharing a predator: can an invasive alien pest affect the predation on a local pest? Population Ecology, 55:433–440.
38. Juen A, Hogendoorn K, Ma G, Schmidt O, Keller MA. (2012) Analysing the diets of invertebrate predators using terminal restriction fragments. J Pest Sci 85:89–100.
39. Chailleux A, Bearez P, Pizzol J, Amiens-Desneux E, Ramirez-Romero R, et al. (2013) Potential for combined use of parasitoids and generalist predators for biological control of the key invasive tomato pest, Tuta absoluta. J Pest Sci 86: 533–541.
40. Velasco-Hernandez MC, Ramirez-Romero R, Cicero L, Michel C, Desneux N (2013) Intraguild predation on the whitefly parasitoid Eretmocerus eremicus by the generalist predator Geocoris punctipes: a behavioral approach. PLoS ONE doi:10.1371/journal.pone.0080679
41. Chacón JM, Heimpel GE (2010) Density-dependent intraguild predation of an aphid parasitoid. Oecologia 164: 213–220.
42. Elliott N, Kieckhefer R, Kauffman W (1996) Effects of an invading coccinellid on native coccinellids in an agricultural landscape. Oecologia 105: 537–544.
43. Bogran CE, Heinz KM, Ciomperlik MA (2002) Interspecific competition among insect parasitoids: Field experiments with whiteflies as hosts in cotton. Ecology 83: 653–668.
44. Zhou HX, Guo JY, Wan FH (2004) Effect of transgenic Cry1Ac+CpTI cotton (SGK321) on population dynamics of pests and their natural enemies. Acta Entomologica Sinica 47: 538–542.
45. Xu Y, Wu KM, Li HB, Liu J, Ding RF, et al. (2012) Effects of Transgenic Bt+CpTI Cotton on Field Abundance of Non-Target Pests and Predators in Xinjiang, China. J Integr Agric 11: 1493–1499.
46. Desneux N, Rabasse JM, Ballanger Y, Kaiser L (2006) Parasitism of canola aphids in France in autumn. J Pest Sci 79: 95–102.
47. Amini B, Madadi H, Desneux N, Lotfalizadehb HA (2012) Impact of irrigation systems on seasonal occurrence of Brevicoryne brassicae and its parasitism by Diaeretiella rapae on canola. J Entomol Res Soc 14: 15–26.
48. Hou ZY, Chen X, Zhang Y, Guo BQ, Yan FS (1997) EAG and orientation tests on the parasitoid Lysiphlebia japonica (Hym., Aphidiidae) to volatile chemicals extracted from host plants of cotton aphid Aphis gossypii (Hom., Aphidae). J Appl Entomol 121: 9–10.
49. Desneux N, Blahnik R, Delebecque CJ, Heimpel GE (2012) Host phylogeny and host specialization in parasitoids. Ecol Lett 15: 453–460.
50. Takada H, Chifumi N, Miyazaki M (2011) Parasitoid spectrum (Hymenoptera: Braconidae; Aphelinidae) of the soybean aphid Aphis glycines (Homoptera: Aphididae) in Japan and Indonesia (Java and Bali). Entomol Sci 14: 216–219.
51. Deng YX, Tsai JH (1998) Development of Lysiphlebia japonica (Hymenoptera : Aphidiidae), a parasitoid of Toxoptera citricida (Homoptera : Aphididae) at five temperatures. Florida Entomologist 81: 415–423.

52. Michaud JP (2002) Classical biological control: A critical review of recent programs against citrus pests in Florida. Ann Entomol Soc Am 95: 531–540.

53. Desneux N, Barta RJ, Hoelmer KA, Hopper KR, Heimpel GE (2009) Multifaceted determinants of host specificity in an aphid parasitoid. Oecologia 160: 387–398.

54. Desneux N, Barta RJ, Delebecque CJ, Heimpel GE (2009) Transient host paralysis as a means of reducing self-superparasitism in koinobiont endoparasitoids. J Insect Physiol 55: 321–327.

55. Lundgren JG, Gassmann AJ, Bernal J, Duan JJ, Ruberson J (2009) Ecological compatibility of GM crops and biological control. Crop Prot 28: 1017–1030.

56. Desneux N, Bernal JS (2010) Genetically modified crops deserve greater ecotoxicological scrutiny. Ecotoxicology 19: 1642–1644.

57. Desneux N, Ramirez-Romero R, Bokonon-Ganta AH, Bernal JS (2010) Attraction of the parasitoid Cotesia marginiventris to host frass is affected by transgenic maize. Ecotoxicology 19: 1183–1192.

58. Zhang SY, Li DM, Cui J, Xie BY (2006) Effects of Bt-toxin Cry1Ac on Propylaea japonica Thunberg (Col., Coccinellidae) by feeding on Bt-treated Bt-resistant Helicoverpa armigera (Hubner) (Lep., Noctuidae) larvae. L Appl Entomol 130: 206–212.

59. Zhang GF, Wan FH, Lovei GL, Liu WX, Guo JY (2006) Transmission of Bt toxin to the predator Propylaea japonica (Coleoptera : Coccinellidae) through its aphid prey feeding on transgenic Bt cotton. Environ Entomol 35: 143–150.

60. Azzouz H, Campan EDM, Cherqui A, Saguez J, Couty A, et al. (2005) Potential effects of plant protease inhibitors, oryzacystatin I and soybean Bowman-Birk inhibitor, on the aphid parasitoid Aphidius ervi Haliday (Hymenoptera, Braconidae). J Insect Physiol 51: 941–951.

61. Stiling P, Rossi AM (1997) Experimental manipulations of top-down and bottom-up factors in a tri-trophic system. Ecology 78: 1602–1606.

62. Denno RF, Gratton C, Peterson MA, Langellotto GA, Finke DL (2002) Bottom-up forces mediate natural-enemy impact in a phytophagous insect community. Ecology 83: 1443–1458.

63. Huberty AF, Denno RF (2004) Plant water stress and its consequences for herbivorous insects: A new synthesis. Ecology 85: 1383–1389.

64. Ai TC, Liu ZY, Li CR, Luo P, Zhu JQ, et al. (2011) Impact of fertilization on cotton aphid population in Bt-cotton production system. Ecol Complex 8: 9–14.

65. Weisenburger DD (1993) Human health - effects of agrichemicals use. Hum Pathol 24: 571–576.

66. Desneux N, Decourtye A, Delpuech JM (2007) The sublethal effects of pesticides on beneficial arthropods. Ann Rev Entomol 52: 81–106.

67. Biondi A, Mommaerts V, Smagghe G, Vinuela E, Zappalà L, et al. (2012) The non-target impact of spinosyns on beneficial arthropods. Pest Manag Sci 68: 1523–1536.

68. Marcic D (2012) Acaricides in modern management of plant-feeding mites. J Pest Sci 85: 395–408.

69. Shad SA, Sayyed AH, Fazal S, Saleem MA, Zaka SM, et al. (2012) Field evolved resistance to carbamates, organophosphates, pyrethroids, and new chemistry insecticides in Spodoptera litura Fab. (Lepidoptera: Noctuidae). J Pest Sci 85: 153–162.

70. Kuusk AK, Ekbom B (2012) Feeding habits of lycosid spiders in field habitats. J Pest Sci 85: 253–260.

71. Yuan XH, Song LW, Zhang JJ, Zang LS, Zhu L, et al. (2012) Performance of four Chinese Trichogramma species as biocontrol agents of the rice striped stem borer, Chilo suppressalis, under various temperature and humidity regimes. J Pest Sci 85: 497–504.

72. Hejcman M, Strnad L, Hejcmanova P, Pavlu V (2012) Effects of nutrient availability on performance and mortality of Rumex obtusifolius and R. crispus in unmanaged grassland. J Pest Sci 85: 191–198.

73. Han P, Lavoir AV, Le Bot J, Amiens-Desneux E, Desneux N (2014) Nitrogen and water availability to tomato plants triggers bottom-up effects on the leafminer Tuta absoluta. Sci Rep 4:4455. doi:10.1038/srep04455

74. Caballero-Lopez B, Blanco-Moreno JM, Perez-Hidalgo N, Michelena-Saval JM, Pujade-Villar J, et al. (2012) Weeds, aphids, and specialist parasitoids and predators benefit differently from organic and conventional cropping of winter cereals. J Pest Sci 85: 81–88.

75. Lu ZZ, Zalucki MP, Perkins LE, Wang DY, Wu LL (2013) Towards a resistance management strategy for Helicoverpa armigera in Bt-cotton in northwestern China: an assessment of potential refuge crops. J Pest Sci 86: 695–703.

# Competitive Ability and Fitness Differences between Two Introduced Populations of the Invasive Whitefly *Bemisia tabaci* Q in China

**Yi-Wei Fang, Ling-Yun Liu, Hua-Li Zhang, De-Feng Jiang, Dong Chu***

Key Lab of Integrated Crop Pest Management of Shandong Province, College of Agronomy and Plant Protection, Qingdao Agricultural University, Qingdao, China

## Abstract

***Background:*** Our long-term field survey revealed that the *Cardinium* infection rate in *Bemisia tabaci* Q (also known as biotype Q) population was low in Shandong, China over the past few years. We hypothesize that (1) the *Cardinium*-infected ($C^+$) *B. tabaci* Q population cannot efficiently compete with the *Cardinium*-uninfected ($C^-$) *B. tabaci* Q population; (2) no reproductive isolation may have occurred between $C^+$ and $C^-$; and (3) the $C^-$ population has higher fitness than the $C^+$ population.

***Methodology and Results:*** To reveal the differences in competitive ability and fitness between the two introduced populations ($C^+$ and $C^-$), competition between $C^+$ and $C^-$ was examined over several generations. Subsequently, the reproductive isolation between $C^+$ and $C^-$ was studied by crossing $C^+$ with $C^-$ individuals, and the fitnesses of $C^+$ and $C^-$ populations were compared using a two-sex life table method. Our results demonstrate that the competitive ability of the $C^+$ whiteflies was weaker than that of $C^-$. There is that no reproductive isolation occurred between the two populations and the $C^-$ population had higher fitness than the $C^+$ population.

***Conclusion:*** The competitive ability and fitness differences of two populations may explain why $C^-$ whitefly populations have been dominant during the past few years in Shandong, China. However, the potential role *Cardinium* plays in whitefly should be further explored.

**Editor:** Murad Ghanim, Volcani Center, Israel

**Funding:** This research was supported by the High-Level Talents Fund of Qingdao Agricultural University (631212), the National Natural Science Foundation of China (31272105), the Science and Technology Development Planning Program of Qingdao (13-1-3-108-nsh), and the Taishan Mountain Scholar Constructive Engineering Foundation of Shandong to D. Chu. The funders had no role in study design, data collection and analysis, decision to publish, or preparation of the manuscript.

**Competing Interests:** The authors have declared that no competing interests exist.

* Email: chinachudong@qau.edu.cn

## Introduction

The sweet potato whitefly, *Bemisia tabaci* (Gennadius) (Hemiptera: Aleyrodoidea), is a major crop pest [1,2] and this species complex contains at least 31 cryptic species identified based on mitochondrial cytochrome oxidase I (mtCOI) sequences and crossing experiments [3,4]. The best known species are MEAM1 (commonly known as biotype B, hereafter referred to as *B. tabaci* B or B) and MED (commonly known as biotype Q, hereafter referred to as *B. tabaci* Q or Q) because both are extremely invasive and globally distributed [3]. *B. tabaci* Q was first detected on the ornamental poinsettia (*Euphorbia pulcherrima* Willd.) in China in 2003 [5]. Since then, it has gradually displaced *B. tabaci* B which was introduced in the mid-1990s. Since 2008, *B. tabaci* Q has become the dominant whitefly species in most regions of China [6–9].

The symbiont *Cardinium* was first found in cell cultures established from the tick *Ixodes scapularis* Say [10] and was named *Candidatus* Cardinium hertigii by Zchori-Fein *et al.* [11]. *Cardinium* can alter the reproduction of its hosts by feminization [12,13], parthenogenesis [14], or cytoplasmic incompatibility of infected

hosts [15–17]. Prior studies showed that infection with this symbiont can improve the fitness of its host [18–20]. Harris *et al.* [21] reported that *Cardinium* infection frequency can increase greatly within the wasp *Encarsia pergandiella* population after nine generations. However, our long-term field survey revealed that the *Cardinium* infection rate in *B. tabaci* Q populations was low (7.6% to17.3%) in Shandong, China, during 2006–2009 [22], though the whitefly may have 10–12 generations per year in this area. Pan *et al.* [23] also reported that the infection frequency of *Cardinium* in *B. tabaci* Q was not very high (16%) in 61 localities in 19 provinces of China in 2009.

On the basis of these data, we hypothesize that (1) the *Cardinium*-infected ($C^+$) *B. tabaci* Q population cannot efficiently compete with the *Cardinium*-uninfected ($C^-$) *B. tabaci* Q population; (2) no reproductive isolation may have occurred between $C^+$ and $C^-$, and if there had been reproductive isolation between them, $C^+$ would have been completely displaced by $C^-$ after long-term coexistence in the field; and (3) the $C^+$ population has higher fitness than the $C^-$ population.

To reveal the differences in competitive ability and fitness between the two introduced populations ($C^+$ and $C^-$), competition between $C^+$ and $C^-$ was examined over several generations. Subsequently, the reproductive isolation between $C^+$ and $C^-$ was studied by crossing $C^+$ with $C^-$ individuals, and the fitnesses of $C^+$ and $C^-$ populations were compared using a two-sex life table method [24].

## Materials and Methods

### Ethics Statement

The research complies with all laws of the country (China) in which it was performed and was approved by the Department of Science and Technology of the Qingdao Agricultural University, China (permit number: 20110712).

### *Bemisia Tabaci* Laboratory Population

The stock population of *B. tabaci* was obtained from laboratory colonies established from prior field collections. The $C^+$ and $C^-$ populations were provided cotton plants and maintained in isolated whitefly-proof screen cages in a greenhouse under controlled lighting and constant temperature ($27 \pm 1°C$) for about ten generations. The primary symbiont, *Porteria*, as well as secondary symbionts belonging to the genera *Arsenophonus*, *Cardinium*, *Hamiltonella*, *Rickttisia*, and *Wolbachia*, have been detected in *B. tabaci* Q [22,25–27]. Using the specific primers of the primary symbiont, *Portiera*, and the secondary symbionts, *Arsenophonus*, *Cardinium*, *Fritschea*, *Hamiltonella*, *Rickettsia*, and *Wolbachia*, we found that the $C^+$ and $C^-$ populations were also infected with *Portiera* and *Hamiltonella*. Both $C^+$ and $C^-$ populations were maintained in separate cultures on potted cotton plants, Lu-Mian-Yan 21 cultivar. The purity of each of the cultures was monitored every 30 days by sampling 20 adults using PCR. Cotton plants were cultivated in 1.5 L plastic pots with nutritional soil and enclosed in whitefly-proof screen cages under controlled light and temperature in a screen house. Three pesticide-free, insect-free, young potted cotton plants were used in the large cages (40 cm×25 cm×50 cm). Plants were at the five to seven fully expanded true leaf stage. Plants were watered and replaced as necessary. All experiments were conducted in controlled climate chambers ($27 \pm 1°C$, 16L: 8D, and $60 \pm 5\%$ RH).

### *Bemisia Tabaci* Species Determination and Detection of *Cardinium*

Adult whiteflies were collected with a hand-held aspirator, preserved immediately in 95% ethanol, and stored at $-20°C$ until processing. Genomic DNA was extracted from individual adult whiteflies of each collection using the DNAzol kit (Molecular Research Center, Inc., Cincinnati, OH) and stored at $-20°C$ for subsequent use. The cleavage amplified polymorphic sequence (CAPS) of the mtCOI gene was used to determine the species of *B. tabaci*. The primers used for detection of the species were C1-J-2195 (5′-TTGATTTTTTGGTCATCCAGAAGT-3′) [28] and R-BQ-2819 (5′-CTGAATATCGRCGAGGCATTCC −3′) [29]. All PCR reactions were performed in 20 µl buffer containing 2 µl 10× buffer, 1.5 mM MgCl$_2$, 0.2 µM of each primer, 0.2 µM dNTPs, 1 unit Taq DNA polymerase, and 2 µl template DNA. Reaction conditions were as follows: 1 cycle of 94°C for 5 min, 35 cycles of 94°C for 1 min, 52°C for 1 min, 72°C for 1 min, and final extension at 72°C for 10 min. The presence of mtCOI amplicons was visualized by electrophoresis in 1.0% agarose gel electrophoresis and ethidium bromide staining. The mtCOI fragment (623 bp) was cleaved using the restriction endonuclease *Vsp*I [30]. Aliquots of the PCR products (13 µl) were each digested with 5 U *Vsp*I (in 20 µl total reaction volume) at 37°C for 2 h.

Specimens whose mtCOI fragments were cut by *Vsp*I were identified as *B. tabaci* Q.

PCR detection of *Cardinium* was performed using the primers Car-sp-F (5′-CGG CTT ATT AAG TCA GTT GTG AAA TCC TAG-3′) and Car-sp-R (5′-TCC TTC CTC CCG CTT ACA CG-3′). All PCR reactions were performed in 20 µl of reaction buffer containing 1×buffer, 0.16 mM of each dNTP, 0.5 mM of each primer, 0.5 unit *Taq* DNA polymerase, and 2 µl template DNA. Reaction conditions were as follows: 1 cycle of 95°C for 1 min, 35 cycles of 95°C for 30s, 57°C for 30 s, 72°C for 1 min, and final extension at 72°C for 5 min. All PCR reactions included a negative control (sterile water instead of DNA) to detect DNA contamination, and a positive control (DNA from previous sequencing) to prevent false negatives. The PCR products (544 bp) were electrophoresed in a 1.0% agarose gel in TAE [31].

### Identification and Analysis of Orthologous Genes between $C^+$ and $C^-$ Populations

To reveal genetic divergence between $C^+$ and $C^-$ populations, orthologs of *cytochrome P450* genes in the transcriptomes of these populations were analyzed. About 20 ug of total RNA ($\geq 300$ ng/ ul) from each population ($C^+$ or $C^-$) was sent to the Shanghai Sangon Institute for library preparation and sequencing on an Illumina HiSeq2000. Raw reads was obtained and *de novo* transcriptome assembly was done with the short-read assembly program trinity [32]. Pairs of sequences longer than 1000 bp that mapped unambiguously to *cytochrome P450* in the Swissprot database (E value$<1e^{-5}$) were selected as *cytochrome P450* genes. Some *P450* genes such as *P450 4C1* gene has high variation in *B. tabaci* cryptic species [33].

### Competition between $C^+$ and $C^-$ Populations

To compare the competitive abilities of $C^+$ and $C^-$ populations, we conducted a cage experiment and followed the frequencies of *Cardinium* infection in whiteflies raised on cotton over ten generations. To observe changes in the relative proportion of $C^+$ and $C^-$ individuals, 30 pairs of $C^+$ and 30 pairs of $C^-$ newly emerged adults whitefly cultures were released into a cage and the infection frequency of *Cardinium* was monitored every 25 days (approximately one generation) by sampling 50 adults using PCR. Infection frequency monitoring of *Cardinium* started from the 75th day. Three replicates were carried out for this study.

### Crossing Experiments

To examine the possibility of reproductive isolation between $C^+$ and $C^-$, we carried out crossing experiments between *Cardinium*-infected and uninfected populations, ♀$C^-$×♂$C^-$, ♀$C^-$×♂$C^+$, ♀$C^+$×♂$C^-$, and ♀$C^+$×♂$C^+$, using virgin whiteflies. To obtain newly emerged unmated adults for experiments, adults were allowed to emerge in isolation and were kept individually before crossing. In the evening, cotton leaves with whitefly pupae (late 4th instar nymphs with red eyes) were cut from plants, and individual pupae with the attached portion of the leaf were placed into a Petri dish. To maintain humidity, a moist filter paper was put on the bottom of the Petri dish. The next morning (at 7:00 am), the newly emerged adults were collected and sexed using a stereomicroscope [34].

For each cross, newly emerged females were individually transferred onto a cotton seedling as described by Li *et al.* [35] together with three adult males. Each female was allowed to lay eggs for 72 h. Adults were then collected using a small aspirator and stored at $-20°C$ for later PCR confirmation of identity. To determine the sex ratio, the eggs laid by each female were observed daily until adult emergence and the newly emerged adults were collected and sexed using a stereomicroscope. Data were analyzed using a one-way analysis of variance (ANOVA) and means were compared using a least significant difference (LSD) test. To normalize the data, an arcsine transformation was used for sex ratio.

## Fitness Assessment of $C^+$ and $C^-$ Population using the Two-Sex Life Table Method

To reveal differences in the fitnesses of $C^+$ and $C^-$, we analyzed the demographic parameters of the population using the two-sex life table method. For the life table study, the rearing containers were made of plastic pots (11.5 cm top diameter, 7.8 cm bottom diameter, and 15.5 cm height), with inverted plastic cups (11.5 cm top diameter, 7.8 cm bottom diameter, and 15.5 cm height) used as covers. The bottom of the plastic cup was cut out and covered with fine mesh cloth for ventilation. A cotton seedling at the two-true-leaves stage was used in this study, only one true leaf was kept on the seedling and the other leaf was removed.

For the life table study, approximately 15 pairs of whiteflies were transferred into rearing containers with a cotton seedling. Eggs laid within 24 h were collected and 15 eggs were used for the life table study. Other eggs were removed. For the $C^+$ or $C^-$ populations, 15 replicates were carried out. The developmental stage and survival status of individual eggs were recorded daily starting at day 14, which corresponded to the day before adult emergence. Using this method, the survival rate of eggs was determined.

To determine fecundity, newly emerged whiteflies were collected and paired in egg-laying units [35]. The whiteflies were checked daily for survival and fecundity until the death of all individuals. Cotton seedlings were replaced daily.

For each population, the life history raw data of all individuals were analyzed based on the age-stage, two-sex life table theory [24] and the method described by Chi [36]. The software TWOSEX-MSChart [36] (available at http://140.120.197.173/Ecology/download/Twosex-MSChart.rar) was used for raw data analysis. The age-stage specific survival rate ($s_{xj}$ where $x$ = age and $j$ = stage), age-stage specific fecundity ($f_{xj}$), age-specific survival rate ($l_x$), age-special fecundity ($m_x$), age-stage life expectancy ($e_{xj}$), reproductive value ($v_{xj}$), preoviposition period of the adult female (APOP), and total preoviposition period of the female from birth (TPOP) were calculated. Among these parameters, $s_{xj}$ represents the probability that a newborn will survive to age $x$ and stage $j$, and $f_{xj}$ represents the mean number of offspring produced by a female of age $x$. In the age-stage, two-sex life table, according to Chi and Liu [37], $l_x$ is estimated as $l_x = \sum_j^k s_{xj}$, and $m_x$ is estimated as $m_x = \frac{\sum_{j=1}^k s_{xj}f_{xj}}{\sum_{j=1}^k s_{xj}}$, where $k$ is the number of stages. The age-stage life expectancy ($e_{xj}$) is the length of time that an individual of age $x$ and stage $j$ is expected to live. The life expectancy for an individual of age $x$ and stage $y$ was calculated as $e_{xy} = \sum_{i=x}^n \sum_{j=y}^m s'_{ij}$, where $n$ is the number of age groups, $m$ is the number of stages, and $s'_{ij}$ is the probability that an individual of age $x$ and stage $y$ will survive to age $i$ and stage $j$ and is calculated by assuming $s'_{xy} = 1$ [37].

Population parameters, including the intrinsic rate of increase ($r$), finite rate of increase ($\lambda$), net reproductive rate ($R_0$), and mean generation time ($T$), were calculated as well. A bootstrap method [38] was used to estimate the standard errors of the population parameters. Among these parameters, the intrinsic rate of increase was estimated using the iterative bisection method from the Euler-Lotka equation ($\sum_{x=0}^\infty e^{-r(x+1)}l_xm_x = 1$) with age indexed from 0

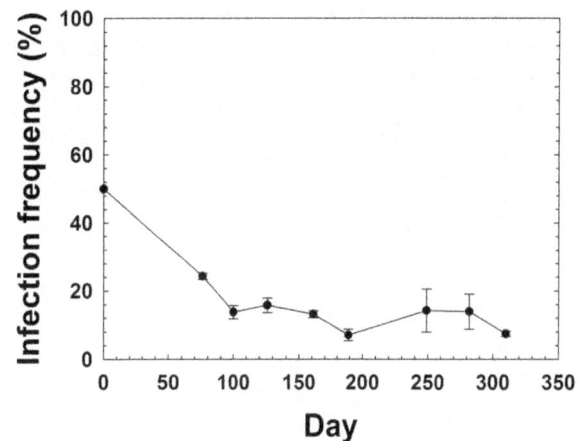

**Figure 1. Percentages of *Cardinium*-infected *B. tabaci* Q ($C^i$) in a mixed cohort.**

[39]. The net reproduction rate ($R_0$) is calculated as $R_0 = \sum_{x=0}^\infty l_xm_x$. The mean generation time is defined as the length of time that a population needs to increase $R_0$ fold of its size ($e^{rT} = R_0$ or $\lambda^T = R_0$) at a stable age distribution, and the mean generation time was calculated as $T = \frac{\ln R_0}{r}$. Student's *t-test* was used to determine differences in developmental times, fecundity, and population parameters between the $C^+$ and $C^-$ populations.

## Results

### Identification and Analysis of Orthologous Genes between $C^+$ and $C^-$ Populations

In the preliminary analysis, 40 *cytochrome P450* genes were identified (Table S1). Based on the *cytochrome P450* genes identified in this study, the average identity between $C^+$ and $C^-$ populations was 99.7% (range, 98.4% to 100.0%) (Table S1), indicating that the genetic backgrounds of these two populations are highly similar.

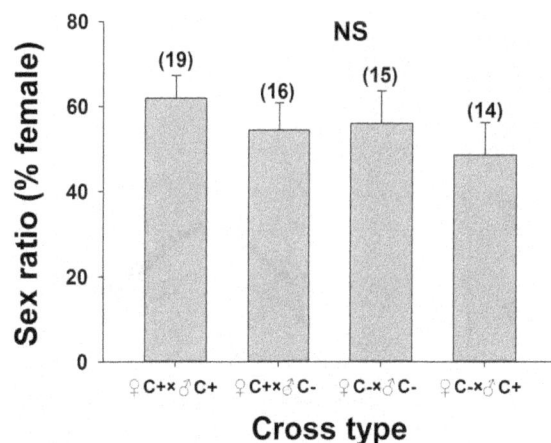

**Figure 2. Sex ratio (proportion of females) in F1 offspring produced by crosses between *Cardinium*-infected ($C^i$) and uninfected ($C^-$) *B. tabaci* Q.** Results are mean percent sex ratio ± SE. Number of each of the cross types are shown in parentheses. NS represent not significant at the 5% level (LSD test, $P > 0.05$).

## Competition between $C^+$ and $C^-$ Populations

The changes in the relative ratio of $C^+$ and $C^-$ individuals within 310 days (approximately 11 generations) are shown in Fig. 1. The mixed cohort began with 50% $C^+$ and 50% $C^-$ individuals. The relative ratio of $C^-$ individuals reached 76% after 75 days and increased steadily over time, reaching 93% after 310 days. By contrast, the relative ratio of $C^+$ individuals decreased from the initial 50% to 24% after 75 days and remained between 7% and 20% thereafter.

## Sex Ratio among Crosses between $C^+$ and $C^-$ Populations

No significant differences were observed in the sex ratio of the F1 generation among the four possible crosses between $C^+$ and $C^-$ populations ($P>0.05$, ANOVA) (Fig. 2). These results suggest that

no reproductive isolation or reproductive abnormalities occurred between $C^+$ and $C^-$ populations.

## $C^+$ and $C^-$ Population Parameters based on the Two-sex Life Table Method

The age-stage survival rate ($s_{xj}$) showed the probability that a newly deposited egg of *B. tabaci* Q would survive to age $x$ and stage $j$ (Fig. 3). The mean number of offspring produced by a female adult at age $x$ relative to the age-stage specific fecundity ($f_{xj}$) is shown in Fig. 4. The maximum lifelong fecundity was found to be 152 eggs per female for the $C^+$ population and 192 eggs per female for the $C^-$ population.

The age-specific survival rate ($l_x$) curve is the age-specific survival rate, including all individuals of the cohort and ignoring the stage of differentiation, while the age-specific fecundity ($m_x$) is the mean fecundity of all individuals in the total population. The product of $l_x$ and $m_x$ is the age-specific maternity ($l_x m_x$) of the $C^+$

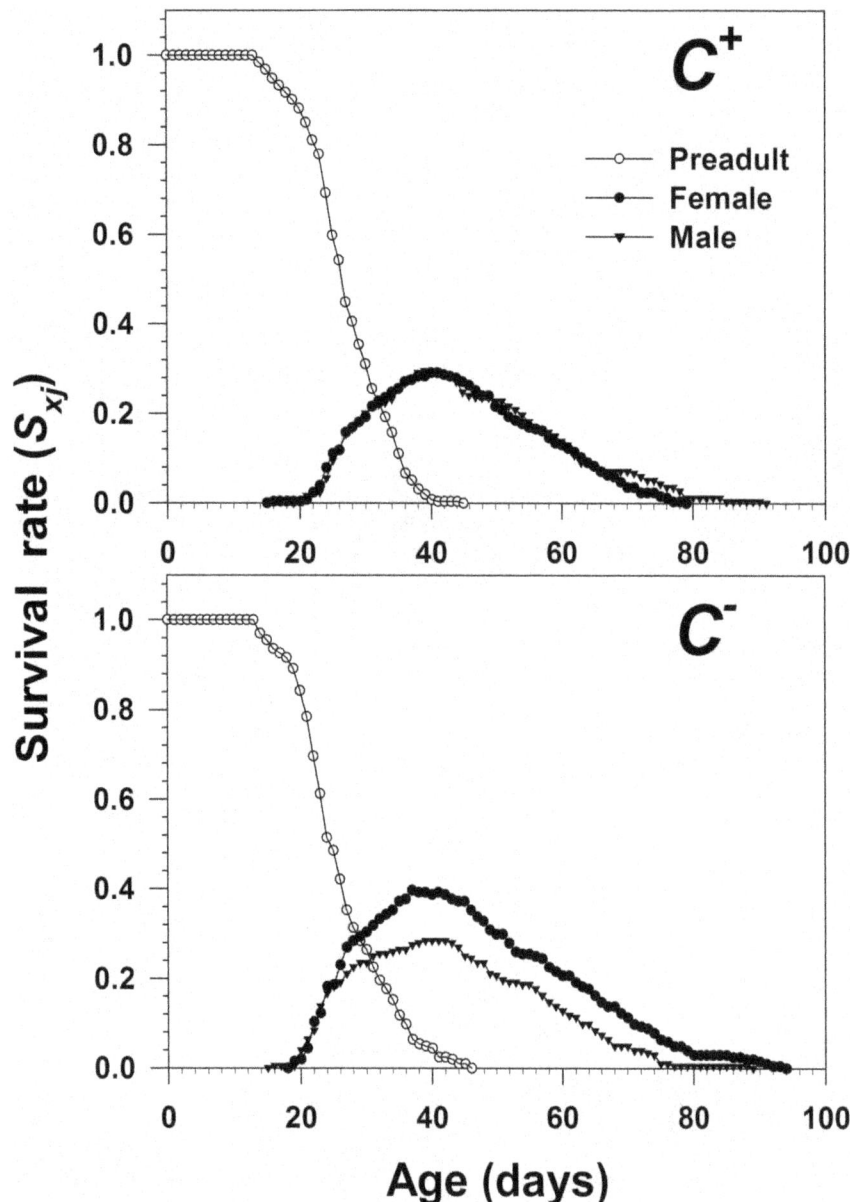

Figure 3. Age-stage specific survival rate ($s_{xj}$) of *Cardinium*-infected ($C^+$) and uninfected ($C^-$) *B. tabaci* Q.

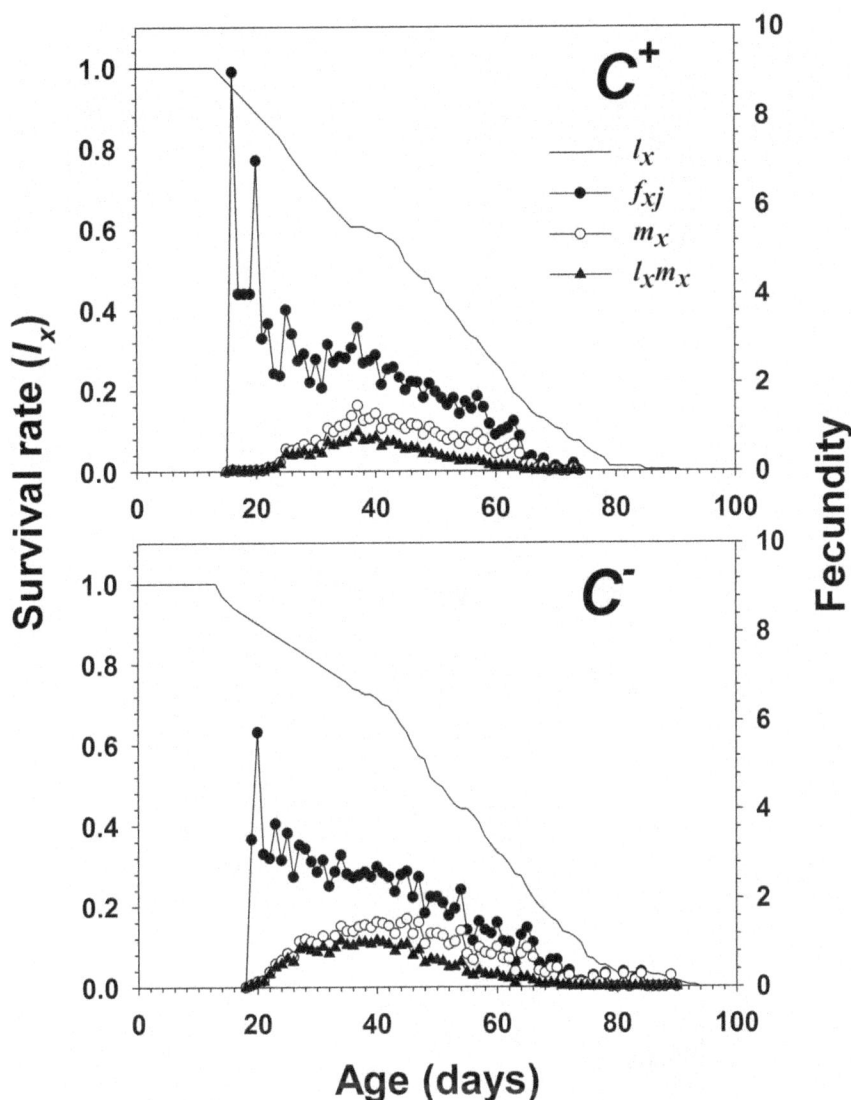

**Figure 4. Age-specific survival rate ($l_x$), female age-specific fecundity ($f_{xj}$), age-specific fecundity of the total population ($m_x$), and age-specific maternity ($l_x m_x$) of *Cardinium*-infected ($C^+$) and uninfected ($C^-$) *B. tabaci* Q.**

and $C^-$ populations. Higher peaks of $m_x$ and $l_x m_x$ were observed in the $C^-$ population. The age-stage specific life expectancy ($e_{xj}$) of *B. tabaci* is shown in Fig. 5. The maximum life expectancy was 45.75 days for the $C^+$ population and 50.48 days for the $C^-$ population. The age-stage life expectancy of the $C^+$ population was significantly shorter than the $C^-$ population.

The age-stage reproductive value ($v_{xj}$) (Fig. 6) delineates the contribution of an individual to age $x$ and stage $j$ of the future population [40]. In our study, the reproductive values increased sharply to 47.23 in the $C^+$ population when females started to emerge at day 19. The corresponding reproductive values increased sharply to 33.96 in the $C^-$ population when females started to emerge at day 16, 3 days earlier than the $C^+$ population.

The fecundity and mean developmental time of each life stage, including pre-adult duration, adult (female, male) longevity, APOP (adult preoviposition period), TPOP (total preoviposition period), and oviposition period, are given in Table 1. The pre-adult duration of the $C^-$ population was significantly shorter than that of the $C^+$ population ($P<0.05$). However, differences in the other

parameters between the two populations were not significant ($P>0.05$).

Based on the two-sex life table method (Table 2), the intrinsic rate of increase ($r$), net reproductive rate ($R_0$), and finite rate of increase ($\lambda$) of the $C^-$ population (0.09192 d$^{-1}$, 29.06 offspring, and 1.096 d$^{-1}$, respectively) were significantly higher than that of the $C^+$ population (0.07561 d$^{-1}$, 17.82 offspring, and 1.078 d$^{-1}$, respectively). The mean generation time ($T$) of the $C^+$ population (38.09 d) was significantly longer than that of $C^-$ population (36.67 d).

## Discussion

Our results showed that the percentage of $C^+$ whitefly decreased after three generations, indicating that the $C^+$ whitefly has weak competitive ability. The results proved the initial hypothesis that the competitive ability of the infected host whiteflies was weaker than that of uninfected ones, which is in agreement with results of previous field surveys in China [8,9]. Our results also revealed that no reproductive isolation occurred between the two populations

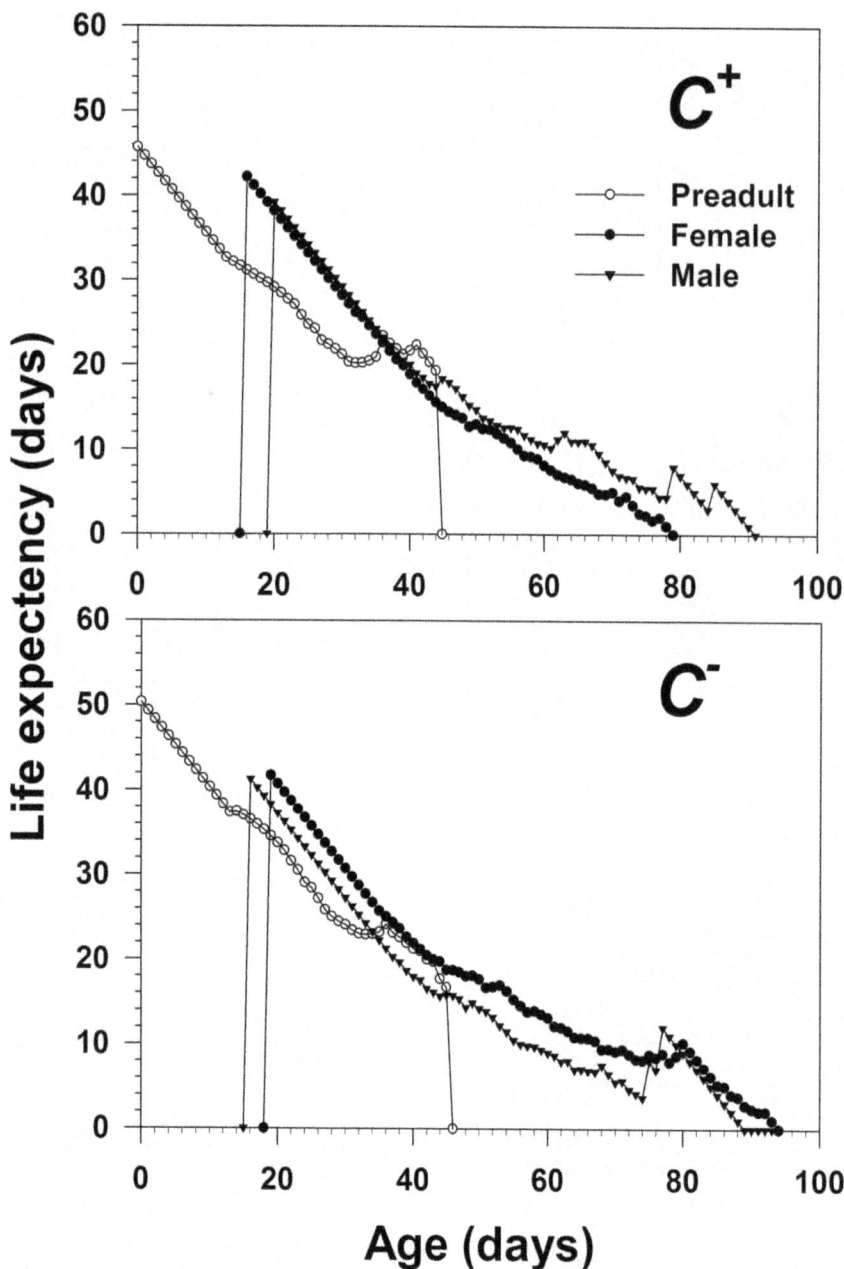

**Figure 5. Age-stage specific life expectancies ($e_{xj}$) of *Cardinium*-infected ($C^+$) and uninfected ($C^-$) *B. tabaci* Q.**

and the $C^-$ population had higher fitnesses than the $C^+$ population. The $C^-$ whiteflies had higher $r$ values (because they developed faster), higher survivorship of immature stages, and a higher net reproductive rate than those of $C^+$ whiteflies at 27°C. These differences may explain why $C^-$ whitefly populations have been dominant during the past few years in Shandong province, China.

For whiteflies, they possess a haplo-diploid sex determination system, in which unfertilized eggs developed into males and fertilized eggs develop into females [41]. In our study, we did not observe parthenogenesis or feminization in the experimental crosses. Although an analysis of *cytochrome P450* genes in the two whitefly populations revealed that they were highly similar, the effect of *Cardinium* on whitefly reproduction should be examined in more detail in future studies. In recent decades, multiple roles of

bacterial symbionts in arthropods have been revealed [42,43]. Bacterial symbionts can manipulate the reproductive biology of hosts or affect the host fitness to increase transmission of the symbiont [18,21,43–54]. Most studies have shown that *Cardinium* can alter the reproduction of its hosts, which, in turn, is helpful to the spread of the symbiont within the population [12–20,41]. For instance, embryonic mortality resulting from cytoplasmic incompatibility is the most common effect associated with endosymbiont infection [50]; consequently, the symbionts can maximize their spread. Cytoplasmic incompatibility induced by *Cardinium* has been widely reported in arthropods, such as the parasitoid wasp *Encarsia pergandiella* [15], spider mite *Eotetranychus suginamensis* [16], sexual spider mite *Bryobia sarothamni* [55], and whitebacked planthopper *Sogatella furcifera* [17]. If *Cardinium* does not alter the reproduction of the whitefly host, then this symbiont may play a

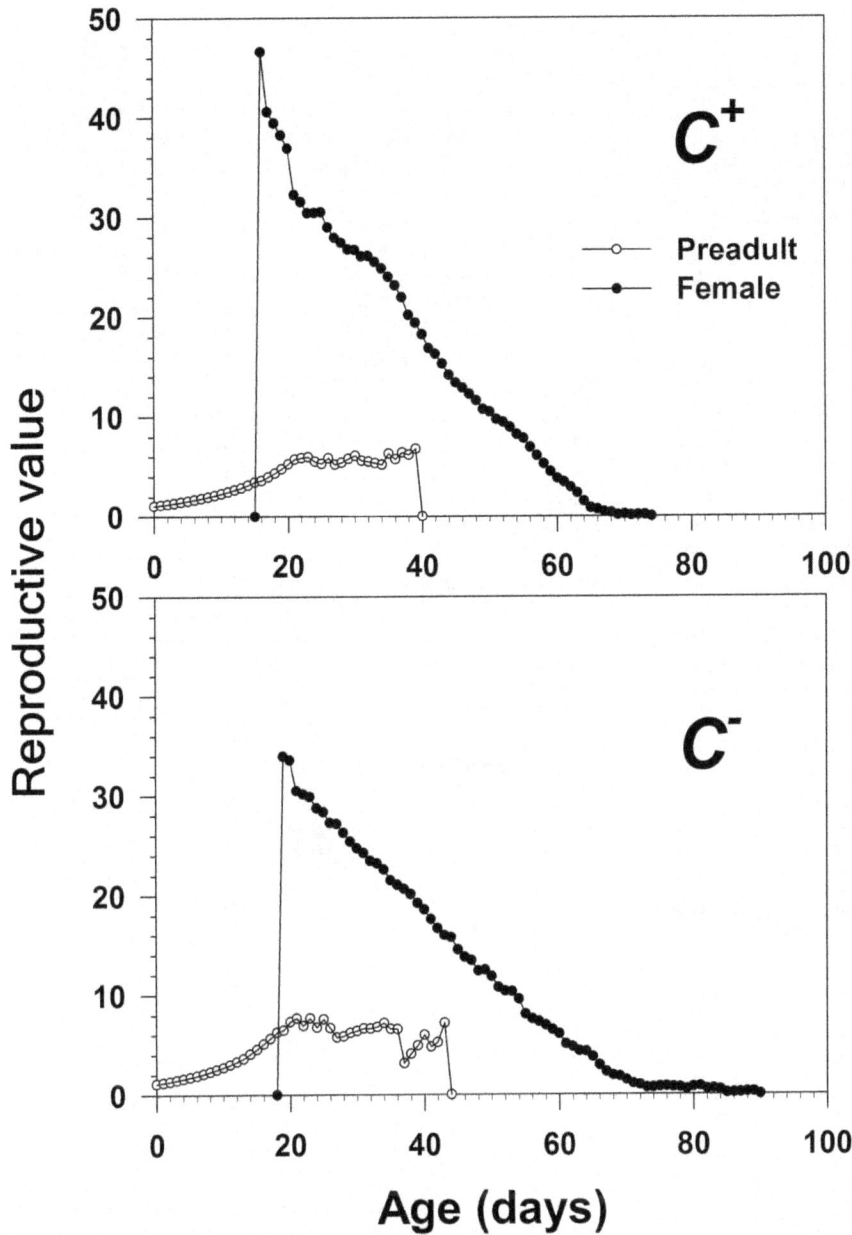

**Figure 6.** Age-stage specific reproductive value ($v_{xj}$) of *Cardinium*-infected ($C^+$) and uninfected ($C^-$) *B. tabaci* Q.

**Table 1.** Basic statistics of the life history of *Cardinium*-infected and uninfected *Bemisia tabaci* Q.

| Statistic | Stage or sex | $C^+$ | | $C^-$ | | P |
|---|---|---|---|---|---|---|
| | | n | Mean ± SE | n | Mean ± SE | |
| Pre-adult duration (d) | Pre-adult | 155 | 28.97±0.43* | 152 | 26.30±0.53 | 0.017 |
| Adult longevity (d) | Female | 75 | 29.79±1.06 | 87 | 33.37±1.51 | 0.05 |
| | Male | 80 | 29.94±1.24 | 65 | 30.37±1.29 | 0.81 |
| TPOP (d) | Female | 75 | 29.23±0.65 | 87 | 28.06±0.69 | 0.23 |
| APOP (d) | Female | 75 | 0.68±0.10 | 87 | 0.56±0.10 | 0.44 |
| Oviposition (d) | Female | 75 | 18.25±0.85 | 87 | 20.19±0.98 | 0.18 |
| Fecundity (eggs per female) | Female | 75 | 60.36±3.37 | 87 | 68.12±4.43 | 0.15 |

All P values are calculated using Student's t-test. APOP (adult preoviposition period) and TPOP (total preoviposition period) were calculated using females that produced fertile eggs.
*Significant difference (Student's t-test, $P<0.05$).

**Table 2.** Population parameters for *Cardinium*-infected and uninfected *Bemisia tabaci* Q.

| Population parameters | $C^+$ (SE) | $C^-$ (SE) |
|---|---|---|
| r (d−1) | 0.07561 (0.0035) | 0.09192 (0.0036)* |
| λ | 1.078 (0.0037) | 1.096 (0.012)* |
| R0 | 17.82 (2.04) | 29.06 (3.04)* |
| T (d) | 38.09 (0.82)* | 36.67 (0.81) |

*Significant difference (Student's *t*-test, $P<0.05$)

different role in the whitefly than observed in previous studies that showed it induces feminization [12,13], parthenogenesis [14], and cytoplasmic incompatibility of infected hosts [15–17]. Our results are similar to the observations of White *et al.* [56,57] and Stefanini & Duron [58], which showed there was no cytoplasmic incompatibility or progeny sex ratio distortion in *Cardinium*-infected individuals of the parasitic wasp *Encarsia inaron* [56,57] or the spider *Holocnemus pluchei* [58].

Based on the two-sex life table method in this study, we revealed that the $C^-$ population had higher fitness than the $C^+$ population (Tables 1 and 2), which might indicate that *Cardinium* may have cryptic negative effects on the fitness of the host. The role *Cardinium* plays in changing the fitness cost of the whitefly should be examined further in future studies. Many studies have suggested that *Cardinium* can increase the fitness of the hosts [13,18–20]. For example, Weeks & Stouthamer [18] found that the fecundity advantage of infected females was approximately 1.6 fold higher than that of uninfected females over a 6-day egg-laying period in the predatory mite, *Metaseiulus occidentailis*, and Giorgini *et*

*al.* [13] showed that the antibiotic removal of *Cardinium* reduced offspring production by adult *Encarsia hispida* females. If it is the case that *Cardinium* has cryptic negative effects on the fitness of the whitefly host, then this would be in disagreement with previous studies that showed that *Cardinium* has no detectable effect on either reproduction or development of the host [56–59]. Similar negative effects of symbiont on host insect have been revealed for *Wolbachia* in *Drosophila*, in which the infection drastically reduces life span [60], and causes widespread degeneration of tissues, culminating in early death of *Drosophila melanogaster* [61]. On the other hand, because *Cardinium* still exists in the field, there might be a benefit to the whitefly. The evolution of this symbiont strain needs to be further explored.

Finally, the two-sex life table gives the most comprehensive description and analysis of the survival and reproduction of a population and, thus, this method may be highly beneficial in revealing the difference in the two populations. This method takes into account the male population and the variable developmental rate occurring among individuals and can overcome the short-coming of the traditional female-based, age-specific life table method, which ignores the male individuals, the stage of differentiation, and variable developmental rates among individuals.

## Author Contributions

Conceived and designed the experiments: DC. Performed the experiments: YWF. Analyzed the data: YWF. Contributed reagents/materials/analysis tools: DFJ. Wrote the paper: LYL HLZ DFJ.

## References

1. Oliveira M, Henneberry T, Anderson P (2001) History, current status, and collaborative research projects for *Bemisia tabaci*. Crop Prot 20: 709–723.
2. Czosnek H, Ghanim M (2011) *Bemisia tabaci-Tomato yellow leaf curl virus* interaction causing worldwide epidemics. In: Winston M.O. Thompson. The whitefly, *Bemisia tabaci* (Homoptera: Aleyrodidae) interaction with geminivirus-infected host plants. Springer 51–67.
3. De Barro PJ, Liu SS, Boykin LM, Dinsdale AB (2011) *Bemisia tabaci*: a statement of species status. Annu Rev Entomol 56: 1–19.
4. Wang HL, Yang J, Boykin LM, Zhao QY, Li Q, et al. (2013) The characteristics and expression profiles of the mitochondrial genome for the Mediterranean species of the *Bemisia tabaci* complex. BMC Genomics 14: 401.
5. Chu D, Zhang YJ, Brown JK, Cong B, Xu BY, et al. (2006) The introduction of the exotic Q biotype of *Bemisia tabaci* from the Mediterranean region into China on ornamental crops. Fla Entomol 89: 168–174.
6. Chu D, Wan FH, Zhang YJ, Brown JK (2010) Change in the biotype composition of *Bemisia tabaci* in Shandong Province of China from 2005 to 2008. Environ Entomol 39: 1028–1036.
7. Teng X, Wan FH, Chu D (2010) *Bemisia tabaci* biotype Q dominates other biotypes across China. Fla Entomol 93: 363–368.
8. Pan HP, Chu D, Ge DQ, Wang SL, Wu QJ, et al. (2011) Further spread of and domination by *Bemisia tabaci* biotype Q on field crops in China. J Econ Entomol 20: 978–985.
9. Chu D, Gao CS, De Barro P, Zhang YJ, Wan FH (2011) Investigation of the genetic diversity of an invasive whitefly in China using both mitochondrial and nuclear DNA markers. Bull Entomol Res 101: 467–475.
10. Kurtti TJ, Munderloh UG, Andreadis TG, Magnarelli LA, Mather TN (1996) Tick cell culture isolation of an intracellular prokaryote from the tick *Ixodes scapularis*. J Invertebrate Pathol 67: 318–321.
11. Zchori-Fein E, Perlman SJ, Kelly SE, Katzir N, Hunter MS (2004) Characterization of a '*Bacteroidetes*' symbiont in *Encarsia* wasps (Hymenoptera: Aphelinidae): proposal of 'Candidatus Cardinium hertigii'. Int J Syst Evol Microbiol 54: 961–968.
12. Weeks AR, Marec F, Breeuwer JA (2001) A mite species that consists entirely of haploid females. Science 292: 2479–2482.
13. Giorgini M, Monti MM, Caprio E, Stouthamer R, Hunter MS (2009) Feminization and the collapse of haplodiploidy in an asexual parasitoid wasp harboring the bacterial symbiont *Cardinium*. Heredity 102: 365–371.
14. Provencher LM, Morse GE, Weeks AR, Normark BB (2005) Parthenogenesis in the *Aspidiotus nerii* complex (Hemiptera: Diaspididae): a single origin of a worldwide, polyphagous lineage associated with *Cardinium* bacteria. Ann Entomol Soc Amer 98: 629–635.
15. Hunter MS, Perlman SJ, Kelly SE (2003) A bacterial symbiont in the Bacteroidetes induces cytoplasmic incompatibility in the parasitoid wasp *Encarsia pergandiella*. Proc R Soc Lond Ser B-Biol Sci 270: 2185–2190.
16. Gotoh T, Noda H, Ito S (2007) *Cardinium* symbionts cause cytoplasmic incompatibility in spider mites. Heredity 98: 13–20.
17. Zhang XF, Zhao DX, Hong XY (2012) *Cardinium*-the leading factor of cytoplasmic incompatibility in the planthopper *Sogatella furcifera* doubly infected with *Wolbachia* and *Cardinium*. Environ Entomol 41: 833–840.
18. Weeks AR, Stouthamer R (2004) Increased fecundity associated with infection by a cytophaga-like intracellular bacterium in the predatory mite, *Metaseiulus occidentalis*. Proc R Soc Lond Ser B 271: S193.
19. Zchori-Fein E, Gottlieb Y, Kelly S, Brown J, Wilson J, et al. (2001) A newly discovered bacterium associated with parthenogenesis and a change in host selection behavior in parasitoid wasps. Proc Natl Acad Sci USA 98: 12555–12560.
20. Kenyon S, Hunter M (2007) Manipulation of oviposition choice of the parasitoid wasp, *Encarsia pergandiella*, by the endosymbiotic bacterium *Cardinium*. J Evol Biol 20: 707–716.
21. Harris L, Kelly S, Hunter M, Perlman S (2010) Population dynamics and rapid spread of *Cardinium*, a bacterial endosymbiont causing cytoplasmic incompatibility in *Encarsia pergandiella* (Hymenoptera: Aphelinidae). Heredity 104: 239–246.
22. Chu D, Gao C, De Barro P, Zhang Y, Wan F, et al. (2011) Further insights into the strange role of bacterial endosymbionts in whitefly, *Bemisia tabaci*: Comparison of secondary symbionts from biotypes B and Q in China. Bull Entomol Res 101: 477–486.
23. Pan H, Li X, Ge D, Wang S, Wu Q, et al. (2012) Factors affecting population dynamics of maternally transmitted endosymbionts in *Bemisia tabaci*. PLoS ONE 7: e30760.
24. Chi H (1988) Life-table analysis incorporating both sexes and variable development rates among individuals. Environ Entomol 17: 26–34.
25. Nirgianaki A, Banks GK, Frohlich DR, Veneti Z, Braig HR, et al. (2003) *Wolbachia* infections of the whitefly *Bemisia tabaci*. Curr Microbiol 47: 93–101.

26. Zchori-Fein E, Brown J (2002) Diversity of prokaryotes associated with *Bemisia tabaci* (Gennadius) (Hemiptera: Aleyrodidae). Ann Entomol Soc Am 95: 711–718.

27. Chiel E, Gottlieb Y, Zchori-Fein E, Mozes-Daube N, Katzir N, et al. (2007) Biotype-dependent secondary symbiont communities in sympatric populations of *Bemisia tabaci*. Bull Entomol Res 97: 407–413.

28. Simon C, Frati F, Beckenbach A, Crespi B, Liu H, et al. (1994) Evolution, weighting, and phylogenetic utility of mitochondrial gene sequences and a compilation of conserved polymerase chain reaction primers. Ann Entomol Soc Am 87: 651–701.

29. Chu D, Hu XS, Gao CS, Zhao HY, Nichols RL, et al. (2012) Use of mtCOI PCR-RFLP for identifying subclades of *Bemisia tabaci* Mediterranean group. J Econ Entomol 105: 242–251.

30. Khasdan V, Levin I, Rosner A, Morin S, Kontsedalov S, et al. (2005) DNA markers for identifying biotypes B and Q of *Bemisia tabaci* (Aleyrodidae) and studying population dynamics. Bull Entomol Res 95: 605–613.

31. Nakamura Y, Kawai S, Yukuhiro F, Ito S, Gotoh T, et al. (2009) Prevalence of *Cardinium* bacteria in planthoppers and spider mites and taxonomic revision of "*Candidatus* Cardinium hertigii" based on detection of a new *Cardinium* group from biting midges. Appl Environ Microbiol 75: 6757–6763.

32. Grabherr MG, Haas BJ, Yassour M, Levin JZ, Thompson DA, et al. (2011) Full-length transcriptome assembly from RNA-Seq data without a reference genome. Nat Biotechnol 29: 644–652.

33. Wang XW, Luan JB, Li JM, Su YL, Xia J, et al. (2011) Transcriptome analysis and comparison reveal divergence between two invasive whitefly cryptic species. BMC Genomics 12: 458.

34. Luan JB, Ruan YM, Zhang L, Liu SS (2008) Pre-copulation intervals, copulation frequencies, and initial progeny sex ratios in two biotypes of whitefly, *Bemisia tabaci*. Entomol Exp Appl 129: 316–324.

35. Li X, Degain BA, Harpold VS, Marçon PG, Nichols RL, et al. (2012). Baseline susceptibilities of B-and Q-biotype *Bemisia tabaci* to anthranilic diamides in Arizona. Pest Manag Sci 68: 83–91.

36. Chi H (2013) TWOSEX-MsChart: a computer program for the age-stage, two-sex life table analysis. http://140.120.197.173/Ecology/download/Twosex-MSChart. rar.

37. Chi H, Su HY (2006) Age-stage, two-sex life tables of *Aphidius gifuensis* (Ashmead) (Hymenoptera: Braconidae) and its host *Myzus persicae* (Sulzer) (Homoptera: Aphididae) with mathematical proof of the relationship between female fecundity and the net reproductive rate. Environ Entomol 35: 10–21.

38. Efron B, Tibshirani R (1993) An introduction to the bootstrap. London: Chapman & hall.

39. Goodman D (1982) Optimal life histories, optimal notation, and the value of reproductive value. Am Nat 803–823.

40. Fisher RA (1930) The genetical theory of natural selection. Clarendon Press, Oxford, United Kingdom.

41. Blackman RL, Cahill M (1998) The karyotype of *Bemisia tabaci* (Hemiptera: Aleyrodidae). Bull Entomol Res 88: 213–215.

42. Werren JH (1997) Biology of *Wolbachia*. Annu Rev Entomol 42: 587–609.

43. Werren JH, Baldo L, Clark ME (2008) *Wolbachia*: master manipulators of invertebrate biology. Nat Rev Microbiol 6: 741–751.

44. Hagimori T, Abe Y, Date S, Miura K (2006) The first finding of a *Rickettsia* bacterium associated with parthenogenesis induction among insects. Curr Microbiol 52: 97–101.

45. Haine ER (2008) Symbiont-mediated protection. Proc R Soc B 275: 353–361.

46. Wang JJ, Dong P, Xiao LS, Dou W (2008) Effects of removal of *Cardinium* infection on fitness of the stored-product pest *Liposcelis bostrychophila* (Psocoptera: Liposcelididae). J Econ Entomol 101: 1711–1717.

47. Jaenike J, Unckless R, Cockburn SN, Boelio LM, Perlman SJ (2010) Adaptation via symbiosis: recent spread of a *Drosophila* defensive symbiont. Science 329: 212–215.

48. Himler AG, Adachi-Hagimori T, Bergen JE, Kozuch A, Kelly SE, et al. (2011) Rapid spread of a bacterial symbiont in an invasive whitefly is driven by fitness benefits and female bias. Science 332: 254–256.

49. Vanthournout B, Swaegers J, Hendrickx F (2011) Spiders do not escape reproductive manipulations by *Wolbachia*. BMC Evol Biol 11: 15.

50. Rousset F, Bouchon D, Pintureau B, Juchault P, Solignac M (1992) *Wolbachia* endosymbionts responsible for various alterations of sexuality in arthropods. Proc R Soc Lond Ser B-Biol Sci 250: 91–98.

51. Stouthamer R, Breeuwer JA, Hurst GD (1999) *Wolbachia pipientis*: microbial manipulator of arthropod reproduction. Annu Rev Microbiol 53: 71–102.

52. Stouthamer R, Breeuwer J, Luck R, Werren J (1993) Molecular identification of microorganisms associated with parthenogenesis. Nature 361: 66–68.

53. Hurst GD, Jiggins FM, von der Schulenburg JHG, Bertrand D, West SA, et al. (1999) Male-killing *Wolbachia* in two species of insect. Proc R Soc Lond Ser B 266: 735–740.

54. Breeuwer J, Stouthamer R, Barns S (1992) Phylogeny of cytoplasmic incompatibility micro-organisms in the parasitoid wasp genus *Nasonia* (Hymenoptera: Pteromalidae) based on 16S ribosomal DNA sequences. Insect Mol Biol 1: 25–36.

55. Ros VID, Breeuwer JAJ (2009) The effects of, and interactions between, *Cardinium* and *Wolbachia* in the doubly infected spider mite *Bryobia sarothamni*. Heredity 102: 413–422.

56. White JA, Kelly SE, Perlman SJ, Hunter MS (2009) Cytoplasmic incompatibility in the parasitic wasp *Encarsia inaron*: disentangling the roles of *Cardinium* and *Wolbachia* symbionts. Heredity 102: 483–489.

57. White JA, Kelly SE, Cockburn SN, Perlman SJ, Hunter MS (2011) Endosymbiont costs and benefits in a parasitoid infected with both *Wolbachia* and *Cardinium*. Heredity 106: 585–591.

58. Stefanini A, Duron O (2012) Exploring the effect of the *Cardinium* endosymbiont on spiders. J Evol Biol 25: 1521–1530.

59. Bull JJ (1983) Evolution of sex determining mechanisms. Menlo Park (California): Benjamin/Cummings.

60. Brownstein J, Hett E, O'Neill S (2003) The potential of virulent *Wolbachia* to modulate disease transmission by insects. J Invertebrate Pathol 84: 24–29.

61. Min KT, Benzer S (1997) Genetics *Wolbachia*, normally a symbiont of *Drosophila*, can be virulent, causing degeneration and early death. Proc Natl Acad Sci 94: 10792–107961.

# A Meta-Analysis of the Impacts of Genetically Modified Crops

**Wilhelm Klümper, Matin Qaim\***

Department of Agricultural Economics and Rural Development, Georg-August-University of Goettingen, Goettingen, Germany

## Abstract

*Background:* Despite the rapid adoption of genetically modified (GM) crops by farmers in many countries, controversies about this technology continue. Uncertainty about GM crop impacts is one reason for widespread public suspicion.

*Objective:* We carry out a meta-analysis of the agronomic and economic impacts of GM crops to consolidate the evidence.

*Data Sources:* Original studies for inclusion were identified through keyword searches in ISI Web of Knowledge, Google Scholar, EconLit, and AgEcon Search.

*Study Eligibility Criteria:* Studies were included when they build on primary data from farm surveys or field trials anywhere in the world, and when they report impacts of GM soybean, maize, or cotton on crop yields, pesticide use, and/or farmer profits. In total, 147 original studies were included.

*Synthesis Methods:* Analysis of mean impacts and meta-regressions to examine factors that influence outcomes.

*Results:* On average, GM technology adoption has reduced chemical pesticide use by 37%, increased crop yields by 22%, and increased farmer profits by 68%. Yield gains and pesticide reductions are larger for insect-resistant crops than for herbicide-tolerant crops. Yield and profit gains are higher in developing countries than in developed countries.

*Limitations:* Several of the original studies did not report sample sizes and measures of variance.

*Conclusion:* The meta-analysis reveals robust evidence of GM crop benefits for farmers in developed and developing countries. Such evidence may help to gradually increase public trust in this technology.

**Editor:** emidio albertini, University of Perugia, Italy

**Funding:** This research was financially supported by the German Federal Ministry of Economic Cooperation and Development (BMZ) and the European Union's Seventh Framework Programme (FP7/2007-2011) under Grant Agreement 290693 FOODSECURE. The funders had no role in study design, data collection and analysis, decision to publish, or preparation of the manuscript. Neither BMZ nor FOODSECURE and any of its partner organizations, any organization of the European Union or the European Commission are accountable for the content of this article.

**Competing Interests:** The authors have declared that no competing interests exist.

\* Email: mqaim@uni-goettingen.de

## Introduction

Despite the rapid adoption of genetically modified (GM) crops by farmers in many countries, public controversies about the risks and benefits continue [1–4]. Numerous independent science academies and regulatory bodies have reviewed the evidence about risks, concluding that commercialized GM crops are safe for human consumption and the environment [5–7]. There are also plenty of studies showing that GM crops cause benefits in terms of higher yields and cost savings in agricultural production [8–12], and welfare gains among adopting farm households [13–15]. However, some argue that the evidence about impacts is mixed and that studies showing large benefits may have problems with the data and methods used [16–18]. Uncertainty about GM crop impacts is one reason for the widespread public suspicion towards this technology. We have carried out a meta-analysis that may help to consolidate the evidence.

While earlier reviews of GM crop impacts exist [19–22], our approach adds to the knowledge in two important ways. First, we include more recent studies into the meta-analysis. In the emerging literature on GM crop impacts, new studies are published continuously, broadening the geographical area covered, the methods used, and the type of outcome variables considered. For instance, in addition to other impacts we analyze effects of GM crop adoption on pesticide quantity, which previous meta-analyses could not because of the limited number of observations for this particular outcome variable. Second, we go beyond average impacts and use meta-regressions to explain impact heterogeneity and test for possible biases.

Our meta-analysis concentrates on the most important GM crops, including herbicide-tolerant (HT) soybean, maize, and cotton, as well as insect-resistant (IR) maize and cotton. For these crops, a sufficiently large number of original impact studies have been published to estimate meaningful average effect sizes. We estimate mean impacts of GM crop adoption on crop yield, pesticide quantity, pesticide cost, total production cost, and farmer profit. Furthermore, we analyze several factors that may influence outcomes, such as geographic location, modified crop trait, and type of data and methods used in the original studies.

## Materials and Methods

### Literature search

Original studies for inclusion in this meta-analysis were identified through keyword searches in relevant literature databanks. Studies were searched in the ISI Web of Knowledge, Google Scholar, EconLit, and AgEcon Search. We searched for studies in the English language that were published after 1995. We did not extend the review to earlier years, because the commercial adoption of GM crops started only in the mid-1990s [23]. The search was performed for combinations of keywords related to GM technology and related to the outcome of interest. Concrete keywords used related to GM technology were (an asterisk is a replacement for any ending of the respective term; quotation marks indicate that the term was used as a whole, not each word alone): GM*, "genetically engineered", "genetically modified", transgenic, "agricultural biotechnology", HT, "herbicide tolerant", Roundup, Bt, "insect resistant". Concrete keywords used related to outcome variables were: impact*, effect*, benefit*, yield*, economic*, income*, cost*, soci*, pesticide*, herbicide*, insecticide*, productivity*, margin*, profit*. The search was completed in March 2014.

Most of the publications in the ISI Web of Knowledge are articles in academic journals, while Google Scholar, EconLit, and AgEcon Search also comprise book chapters and grey literature such as conference papers, working papers, and reports in institutional series. Articles published in academic journals have usually passed a rigorous peer-review process. Most papers presented at academic conferences have also passed a peer-review process, which is often less strict than that of good journals though. Some of the other publications are peer reviewed, while many are not. Some of the working papers and reports are published by research institutes or government organizations, while others are NGO publications. Unlike previous reviews of GM crop impacts, we did not limit the sample to peer-reviewed studies but included all publications for two reasons. First, a clear-cut distinction between studies with and without peer review is not always possible, especially when dealing with papers that were not published in a journal or presented at an academic conference [24]. Second, studies without peer review also influence the public and policy debate on GM crops; ignoring them completely would be short-sighted.

Of the studies identified through the keyword searches, not all reported original impact results. We classified studies by screening titles, abstracts, and full texts. Studies had to fulfill the following criteria to be included:

- The study is an empirical investigation of the agronomic and/ or economic impacts of GM soybean, GM maize, or GM cotton using micro-level data from individual plots and/or farms. Other GM crops such as GM rapeseed, GM sugarbeet, and GM papaya were commercialized in selected countries [23], but the number of impact studies available for these other crops is very small.

- The study reports GM crop impacts in terms of one or more of the following outcome variables: yield, pesticide quantity (especially insecticides and herbicides), pesticide costs, total variable costs, gross margins, farmer profits. If only the number of pesticide sprays was reported, this was used as a proxy for pesticide quantity.

- The study analyzes the performance of GM crops by either reporting mean outcomes for GM and non-GM, absolute or percentage differences, or estimated coefficients of regression models that can be used to calculate percentage differences between GM and non-GM crops.

- The study contains original results and is not only a review of previous studies.

In some cases, the same results were reported in different publications; in these cases, only one of the publications was included to avoid double counting. On the other hand, several publications involve more than one impact observation, even for a single outcome variable, for instance when reporting results for different geographical regions or derived with different methods (e.g., comparison of mean outcomes of GM and non-GM crops plus regression model estimates). In those cases, all observations were included. Moreover, the same primary dataset was sometimes used for different publications without reporting identical results (e.g., analysis of different outcome variables, different waves of panel data, use of different methods). Hence, the number of impact observations in our sample is larger than the number of publications and primary datasets (Data S1). The number of studies selected at various stages is shown in the flow diagram in Figure 1. The number of publications finally included in the meta-analysis is 147 (Table S1).

### Effect sizes and influencing factors

Effect sizes are measures of outcome variables. We chose the percentage difference between GM and non-GM crops for five different outcome variables, namely yield, pesticide quantity, pesticide cost, total production cost, and farmer profits per unit area. Most studies that analyze production costs focus on variable costs, which are the costs primarily affected through GM technology adoption. Accordingly, profits are calculated as revenues minus variable production costs (profits calculated in this way are also referred to as gross margins). These production costs also take into account the higher prices charged by private companies for GM seeds. Hence, the percentage differences in profits considered here are net economic benefits for farmers using GM technology. Percentage differences, when not reported in the original studies, were calculated from mean value comparisons between GM and non-GM or from estimated regression coefficients.

Since we look at different types of GM technologies (different modified traits) that are used in different countries and regions, we do not expect that effect sizes are homogenous across studies. Hence, our approach of combining effect sizes corresponds to a random-effects model in meta-analysis [25]. To explain impact heterogeneity and test for possible biases, we also compiled data on a number of study descriptors that may influence the reported effect sizes. These influencing factors include information on the type of GM technology (modified trait), the region studied, the type of data and method used, the source of funding, and the type of publication. All influencing factors are defined as dummy variables. The exact definition of these dummy variables is given in Table 1. Variable distributions of the study descriptors are shown in Table S2.

### Statistical analysis

In a first step, we estimate average effect sizes for each outcome variable. To test whether these mean impacts are significantly

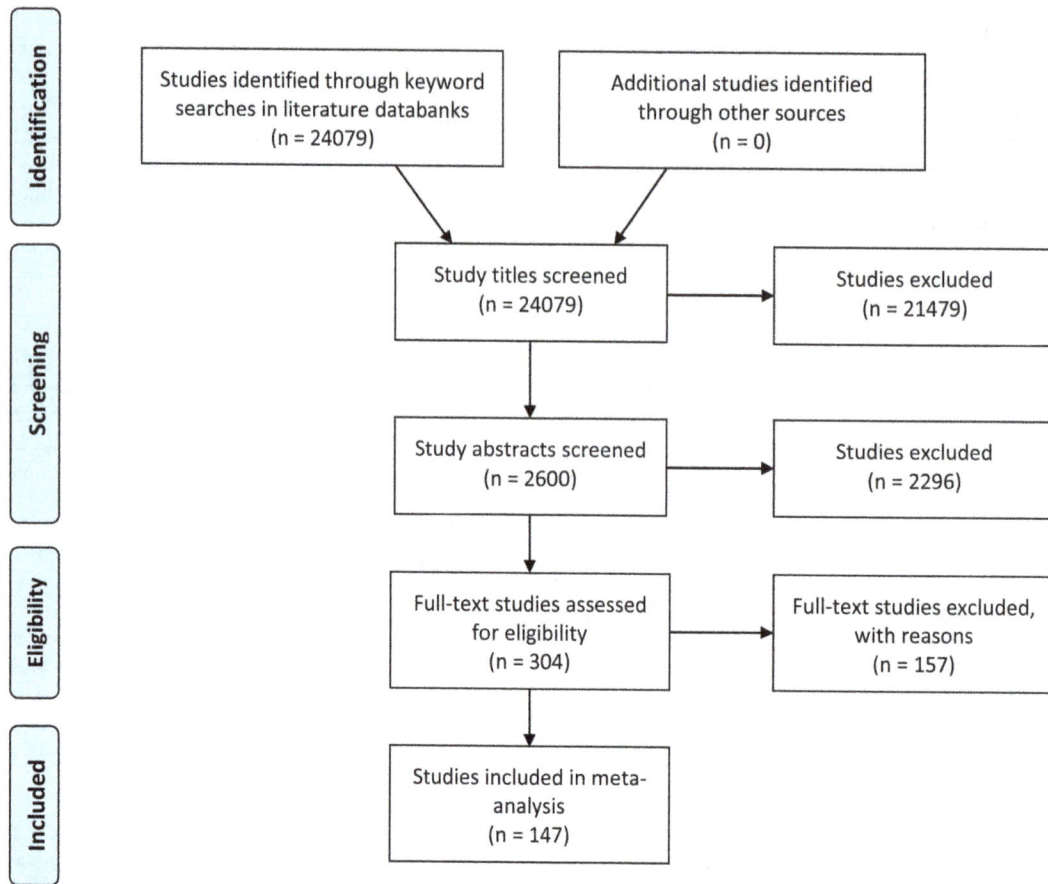

**Figure 1. Selection of studies for inclusion in the meta-analysis.**

different from zero, we regress each outcome variable on a constant with cluster correction of standard errors by primary dataset. Thus, the test for significance is valid also when observations from the same dataset are correlated. We estimate average effect sizes for all GM crops combined. However, we expect that the results may differ by modified trait, so that we also analyze mean effects for HT crops and IR crops separately.

Meta-analyses often weight impact estimates by their variances; estimates with low variance are considered more reliable and receive a higher weight [26]. In our case, several of the original studies do not report measures of variance, so that weighting by variance is not possible. Alternatively, weighting by sample size is common, but sample sizes are also not reported in all studies considered, especially not in some of the grey literature publications. To test the robustness of the results, we employ a

**Table 1.** Variables used to analyze influencing factors of GM crop impacts.

| Variable name | Variable definition |
|---|---|
| Insect resistance (IR) | Dummy that takes a value of one for all observations referring to insect-resistant GM crops with genes from *Bacillus thuringiensis* (Bt), and zero for all herbicide-tolerant (HT) GM crops. |
| Developing country | Dummy that takes a value of one for all GM crop applications in a developing country according to the World Bank classification of countries, and zero for all applications in a developed country. |
| Field-trial data | Dummy that takes a value of one for all observations building on field-trial data (on-station and on-farm experiments), and zero for all observations building on farm survey data. |
| Industry-funded study | Dummy that takes a value of one for all studies that mention industry (private sector companies) as source of funding, and zero otherwise. |
| Regression model result | Dummy that takes a value of one for all impact observations that are derived from regression model estimates, and zero for observations derived from mean value comparisons between GM and non-GM. |
| Journal publication | Dummy that takes a value of one for all studies published in a peer-reviewed journal, and zero otherwise. |
| Journal/academic conference | Dummy that takes a value of one for all studies published in a peer-reviewed journal or presented at an academic conference, and zero otherwise. |

different weighting procedure, using the inverse of the number of impact observations per dataset as weights. This procedure avoids that individual datasets that were used in several publications dominate the calculation of average effect sizes.

In a second step, we use meta-regressions to explain impact heterogeneity and test for possible biases. Linear regression models are estimated separately for all of the five outcome variables:

$$\%\Delta Y_{hij} = \alpha_h + X_{hij}\beta_h + \varepsilon_{hij}$$

$\%\Delta Y_{hij}$ is the effect size (percentage difference between GM and non-GM) of each outcome variable $h$ for observation $i$ in publication $j$, and $X_{hij}$ is a vector of influencing factors. $\alpha_h$ is a coefficient and $\beta_h$ a vector of coefficients to be estimated; $\varepsilon_{hij}$ is a random error term. Influencing factors used in the regressions are defined in Table 1.

## Results and Discussion

### Average effect sizes

Distributions of all five outcome variables are shown in Figure S1. Table 2 presents unweighted mean impacts. As a robustness check, we weighted by the inverse of the number of impact observations per dataset. Comparing unweighted results (Table 2) with weighted results (Table S3) we find only very small differences. This comparison suggests that the unweighted results are robust.

On average, GM technology has increased crop yields by 21% (Figure 2). These yield increases are not due to higher genetic yield potential, but to more effective pest control and thus lower crop damage [27]. At the same time, GM crops have reduced pesticide quantity by 37% and pesticide cost by 39%. The effect on the cost of production is not significant. GM seeds are more expensive than non-GM seeds, but the additional seed costs are compensated through savings in chemical and mechanical pest control. Average profit gains for GM-adopting farmers are 69%.

Results of Cochran's test [25], which are reported in Figure S1, confirm that there is significant heterogeneity across study observations for all five outcome variables. Hence it is useful to

further disaggregate the results. Table 2 shows a breakdown by modified crop trait. While significant reductions in pesticide costs are observed for both HT and IR crops, only IR crops cause a consistent reduction in pesticide quantity. Such disparities are expected, because the two technologies are quite different. IR crops protect themselves against certain insect pests, so that spraying can be reduced. HT crops, on the other hand, are not protected against pests but against a broad-spectrum chemical herbicide (mostly glyphosate), use of which facilitates weed control. While HT crops have reduced herbicide quantity in some situations, they have contributed to increases in the use of broad-spectrum herbicides elsewhere [2,11,19]. The savings in pesticide costs for HT crops in spite of higher quantities can be explained by the fact that broad-spectrum herbicides are often much cheaper than the selective herbicides that were used before. The average farmer profit effect for HT crops is large and positive, but not statistically significant because of considerable variation and a relatively small number of observations for this outcome variable.

### Impact heterogeneity and possible biases

Table 3 shows the estimation results from the meta-regressions that explain how different factors influence impact heterogeneity. Controlling for other factors, yield gains of IR crops are almost 7 percentage points higher than those of HT crops (column 1). Furthermore, yield gains of GM crops are 14 percentage points higher in developing countries than in developed countries. Especially smallholder farmers in the tropics and subtropics suffer from considerable pest damage that can be reduced through GM crop adoption [27].

Most original studies in this meta-analysis build on farm surveys, although some are based on field-trial data. Field-trial results are often criticized to overestimate impacts, because farmers may not be able to replicate experimental conditions. However, results in Table 3 (column 1) show that field-trial data do not overestimate the yield effects of GM crops. Reported yield gains from field trials are even lower than those from farm surveys. This is plausible, because pest damage in non-GM crops is often more severe in farmers' fields than on well-managed experimental plots.

**Table 2.** Impacts of GM crop adoption by modified trait.

| Outcome variable | All GM crops | Insect resistance | Herbicide tolerance |
|---|---|---|---|
| Yield | 21.57*** | 24.85*** | 9.29** |
| | (15.65; 27.48) | (18.49; 31.22) | (1.78; 16.80) |
| *n/m* | 451/100 | 353/83 | 94/25 |
| Pesticide quantity | −36.93*** | −41.67*** | 2.43 |
| | (−48.01; −25.86) | (−51.99; −31.36) | (−20.26; 25.12) |
| *n/m* | 121/37 | 108/31 | 13/7 |
| Pesticide cost | −39.15*** | −43.43*** | −25.29*** |
| | (−46.96; −31.33) | (−51.64; −35.22) | (−33.84; −16.74) |
| *n/m* | 193/57 | 145/45 | 48/15 |
| Total production cost | 3.25 | 5.24** | −6.83 |
| | (−1.76; 8.25) | (0.25; 10.73) | (−16.43; 2.77) |
| *n/m* | 115/46 | 96/38 | 19/10 |
| Farmer profit | 68.21*** | 68.78*** | 64.29 |
| | (46.31; 90.12) | (46.45; 91.11) | (−24.73; 153.31) |
| *n/m* | 136/42 | 119/36 | 17/9 |

Average percentage differences between GM and non-GM crops are shown with 95% confidence intervals in parentheses. *, **, *** indicate statistical significance at the 10%, 5%, and 1% level, respectively. *n* is the number of observations, *m* the number of different primary datasets from which these observations are derived.

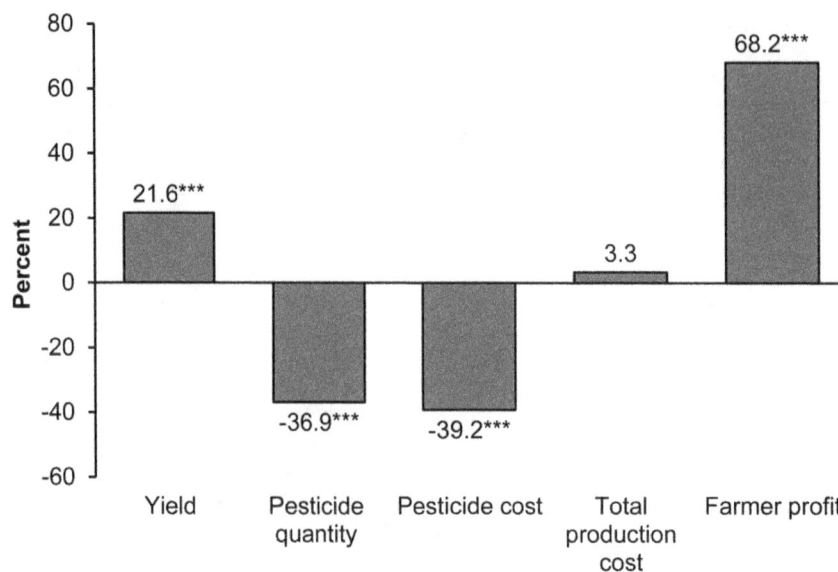

**Figure 2. Impacts of GM crop adoption.** Average percentage differences between GM and non-GM crops are shown. Results refer to all GM crops, including herbicide-tolerant and insect-resistant traits. The number of observations varies by outcome variable; yield: 451; pesticide quantity: 121; pesticide cost: 193; total production cost: 115; farmer profit: 136. *** indicates statistical significance at the 1% level.

Another concern often voiced in the public debate is that studies funded by industry money might report inflated benefits. Our results show that the source of funding does not significantly influence the impact estimates. We also analyzed whether the statistical method plays a role. Many of the earlier studies just compared yields of GM and non-GM crops without considering possible differences in other inputs and conditions that may also affect the outcome. Net impacts of GM technology can be estimated with regression-based production function models that control for other factors. Interestingly, results derived from regression analysis report higher average yield effects.

Finally, we examined whether the type of publication matters. Controlling for other factors, the regression coefficient for journal publications in column (1) of Table 3 implies that studies published in peer-reviewed journals show 12 percentage points higher yield gains than studies published elsewhere. Indeed, when only including observations from studies that were published in journals, the mean effect size is larger than if all observations are included (Figure S2). On first sight, one might suspect publication bias, meaning that only studies that report substantial effects are accepted for publication in a journal. A common way to assess possible publication bias in meta-analysis is through funnel plots [25], which we show in Figure S3. However, in our case these funnel plots should not be over-interpreted. First, only studies that report variance measures can be included in the funnel plots, which holds true only for a subset of the original studies used here. Second, even if there were publication bias, our mean results would be estimated correctly, because we do include studies that were not published in peer-reviewed journals.

Further analysis suggests that the journal review process does not systematically filter out studies with small effect sizes. The journal articles in the sample report a wide range of yield effects, even including negative estimates in some cases. Moreover, when combining journal articles with papers presented at academic conferences, average yield gains are even higher (Table 3, column 2). Studies that were neither published in a journal nor presented at an academic conference encompass a diverse set of papers, including reports by NGOs and outspoken biotechnology critics.

These reports show lower GM yield effects on average, but not all meet common scientific standards. Hence, rather than indicating publication bias, the positive and significant journal coefficient may be the result of a negative NGO bias in some of the grey literature.

Concerning other outcome variables, IR crops have much stronger reducing effects on pesticide quantity than HT crops (Table 3, column 3), as already discussed above. In terms of pesticide costs, the difference between IR and HT is less pronounced and not statistically significant (column 4). The profit gains of GM crops are 60 percentage points higher in developing countries than in developed countries (column 6). This large difference is due to higher GM yield gains and stronger pesticide cost savings in developing countries. Moreover, most GM crops are not patented in developing countries, so that GM seed prices are lower [19]. Like for yields, studies published in peer-reviewed journals report higher profit gains than studies published elsewhere, but again we do not find evidence of publication bias (column 7).

## Conclusion

This meta-analysis confirms that – in spite of impact heterogeneity – the average agronomic and economic benefits of GM crops are large and significant. Impacts vary especially by modified crop trait and geographic region. Yield gains and pesticide reductions are larger for IR crops than for HT crops. Yield and farmer profit gains are higher in developing countries than in developed countries. Recent impact studies used better data and methods than earlier studies, but these improvements in study design did not reduce the estimates of GM crop advantages. Rather, NGO reports and other publications without scientific peer review seem to bias the impact estimates downward. But even with such biased estimates included, mean effects remain sizeable.

One limitation is that not all of the original studies included in this meta-analysis reported sample sizes and measures of variance. This is not untypical for analyses in the social sciences, especially when studies from the grey literature are also included. Future

**Table 3.** Factors influencing results on GM crop impacts (%).

| Variables | (1) Yield | (2) Yield | (3) Pesticide quantity | (4) Pesticide cost | (5) Total cost | (6) Farmer profit | (7) Farmer profit |
|---|---|---|---|---|---|---|---|
| Insect resistance (IR) | 6.58** (2.85) | 5.25* (2.82) | -37.38*** (11.81) | -7.28 (5.44) | 5.63 (5.60) | -22.33 (21.62) | -33.41 (21.94) |
| Developing country | 14.17*** (2.72) | 13.32*** (2.65) | -10.23 (8.99) | -19.16*** (5.35) | 3.43 (4.78) | 59.52*** (18.02) | 60.58*** (17.67) |
| Field-trial data | -7.14** (3.19) | -7.81** (3.08) | –# | -17.56 (11.45) | -10.69* (5.79) | –# | –# |
| Industry-funded study | 1.68 (5.30) | 1.05 (5.21) | 37.04 (23.08) | -7.77 (10.22) | –# | –# | –# |
| Regression model result | 7.38* (3.90) | 7.29* (3.83) | 9.67 (10.40) | –# | –# | -11.44 (24.33) | -9.85 (24.03) |
| Journal publication | 12.00*** (2.52) | - | 9.95 (6.79) | -3.71 (4.09) | -3.08 (3.30) | 48.27*** (15.48) | - |
| Journal/academic conference | - | 16.48*** (2.64) | - | - | - | - | 65.29*** (17.75) |
| Constant | -0.22 (2.84) | -2.64 (2.86) | -4.44 (10.33) | -16.13 (4.88) | -1.02 (4.86) | 8.57 (24.33) | -1.19 (24.53) |
| Observations | 451 | 451 | 121 | 193 | 115 | 136 | 136 |
| $R^2$ | 0.23 | 0.25 | 0.20 | 0.14 | 0.12 | 0.12 | 0.14 |

Coefficient estimates from linear regression models are shown with standard errors in parentheses. Dependent variables are GM crop impacts measured as percentage differences between GM and non-GM. All explanatory variables are 0/1 dummies (for variable definitions see Table 1). The yield models in columns (1) and (2) and the farmer profit models in columns (6) and (7) have the same dependent variables, but they differ in terms of the explanatory variables, as shown. *, **, *** indicate statistical significance at the 10%, 5%, and 1% level, respectively. # indicates that the variable was dropped because the number of observations with a value of one was smaller than 5.

impact studies with primary data should follow more standardized reporting procedures. Nevertheless, our findings reveal that there is robust evidence of GM crop benefits. Such evidence may help to gradually increase public trust in this promising technology.

## Supporting Information

**Figure S1 Histograms of effect sizes for the five outcome variables.**

**Figure S2 Impacts of GM crop adoption including only studies published in journals.**

**Figure S3 Funnel plots for the five outcome variables.**

**Table S1 List of publications included in the meta-analysis.**

**Table S2 Distribution of study descriptor dummy variables for different outcomes.**

**Table S3 Weighted mean impacts of GM crop adoption.**

**Data S1 Data used for the meta-analysis.**

## Acknowledgments

We thank Sinja Buri and Tingting Xu for assistance in compiling the dataset. We also thank Joachim von Braun and three reviewers of this journal for useful comments.

## Author Contributions

Conceived and designed the research: WK MQ. Analyzed the data: WK MQ. Contributed to the writing of the manuscript: WK MQ. Compiled the data: WK.

## References

1. Gilbert N (2013) A hard look at GM crops. Nature 497: 24–26.
2. Fernandez-Cornejo J, Wechsler JJ, Livingston M, Mitchell L (2014) Genetically Engineered Crops in the United States. Economic Research Report ERR-162 (United Sates Department of Agriculture, Washington, DC).
3. Anonymous (2013) Contrary to popular belief. Nature Biotechnology 31: 767.
4. Andreasen M (2014) GM food in the public mind–facts are not what they used to be. Nature Biotechnology 32: 25.
5. DeFrancesco L (2013) How safe does transgenic food need to be? Nature Biotechnology 31: 794–802.
6. European Academies Science Advisory Council (2013) Planting the Future: Opportunities and Challenges for Using Crop Genetic Improvement Technologies for Sustainable Agriculture (EASAC, Halle, Germany).
7. European Commission (2010) A Decade of EU-Funded GMO Research 2001–2010 (European Commission, Brussels).
8. Pray CE, Huang J, Hu R, Rozelle S (2002) Five years of Bt cotton in China - the benefits continue. The Plant Journal 31: 423–430.
9. Huang J, Hu R, Rozelle S, Pray C (2008) Genetically modified rice, yields and pesticides: assessing farm-level productivity effects in China. Economic Development and Cultural Change 56: 241–263.
10. Morse S, Bennett R, Ismael Y (2004) Why Bt cotton pays for small-scale producers in South Africa. Nature Biotechnology 22: 379–380.
11. Qaim M, Traxler G (2005) Roundup Ready soybeans in Argentina: farm level and aggregate welfare effects. Agricultural Economics 32: 73–86.
12. Sexton S, Zilberman D (2012) Land for food and fuel production: the role of agricultural biotechnology. In: The Intended and Unintended Effects of US Agricultural and Biotechnology Policies (eds. Zivin, G. & Perloff, J.M.), 269–288 (University of Chicago Press, Chicago).
13. Ali A, Abdulai A (2010) The adoption of genetically modified cotton and poverty reduction in Pakistan. Journal of Agricultural Economics 61, 175–192.
14. Kathage J, Qaim M (2012) Economic impacts and impact dynamics of Bt (*Bacillus thuringiensis*) cotton in India. Proceedings of the National Academy of Sciences USA 109: 11652–11656.
15. Qaim M, Kouser S (2013) Genetically modified crops and food security. PLOS ONE 8: e64879.
16. Stone GD (2012) Constructing facts: Bt cotton narratives in India. Economic & Political Weekly 47(38): 62–70.
17. Smale M, Zambrano P, Gruere G, Falck-Zepeda J, Matuschke I, et al. (2009) Measuring the Economic Impacts of Transgenic Crops in Developing Agriculture During the First Decade: Approaches, Findings, and Future Directions (International Food Policy Research Institute, Washington, DC).
18. Glover D (2010) Is Bt cotton a pro-poor technology? A review and critique of the empirical record. Journal of Agrarian Change 10: 482–509.
19. Qaim M (2009) The economics of genetically modified crops. Annual Review of Resource Economics 1: 665–693.
20. Carpenter JE (2010) Peer-reviewed surveys indicate positive impact of commercialized GM crops. Nature Biotechnology 28: 319–321.
21. Finger R, El Benni N, Kaphengst T, Evans C, Herbert S, et al. (2011) A meta analysis on farm-level costs and benefits of GM crops. Sustainability 3: 743–762.
22. Areal FJ, Riesgo L, Rodríguez-Cerezo E (2013) Economic and agronomic impact of commercialized GM crops: a meta-analysis. Journal of Agricultural Science 151: 7–33.
23. James C (2013) Global Status of Commercialized Biotech/GM Crops: 2013. ISAAA Briefs No.46 (International Service for the Acquisition of Agri-biotech Applications, Ithaca, NY).
24. Rothstein HR, Hopewell S (2009) Grey literature. In: Handbook of Research Synthesis and Meta-Analysis, Second Edition (eds. Cooper, H., Hegdes, L.V. & Valentine, J.C.), 103–125 (Russell Sage Foundation, New York).
25. Borenstein M, Hedges LV, Higgins JPT, Rothstein HR (2009) Introduction to Meta-Analysis (John Wiley and Sons, Chichester, UK).
26. Shadish WR, Haddock CK (2009) Combining estimates of effect size. In: Handbook of Research Synthesis and Meta-Analysis, Second Edition (eds. Cooper, H., Hegdes, L.V. & Valentine, J.C.), 257–277 (Russell Sage Foundation, New York).
27. Qaim M, Zilberman D (2003) Yield effects of genetically modified crops in developing countries. Science 299: 900–902.

# The Entomopathogenic Fungal Endophytes *Purpureocillium lilacinum* (Formerly *Paecilomyces lilacinus*) and *Beauveria bassiana* Negatively Affect Cotton Aphid Reproduction under Both Greenhouse and Field Conditions

**Diana Castillo Lopez[1]\*, Keyan Zhu-Salzman[1], Maria Julissa Ek-Ramos[2], Gregory A. Sword[1]**

**1** Department of Entomology, Texas A&M University, College Station, Texas, United States of America, **2** Department of Immunology and Microbiology, Autonomous University of Nuevo Leon, San Nicolás de los Garza, Nuevo Leon, Mexico

## Abstract

The effects of two entomopathogenic fungal endophytes, *Beauveria bassiana* and *Purpureocillium lilacinum* (formerly *Paecilomyces lilacinus*), were assessed on the reproduction of cotton aphid, *Aphis gossypii* Glover (Homoptera:Aphididae), through *in planta* feeding trials. In replicate greenhouse and field trials, cotton plants (*Gossypium hirsutum*) were inoculated as seed treatments with two concentrations of *B. bassiana* or *P. lilacinum* conidia. Positive colonization of cotton by the endophytes was confirmed through potato dextrose agar (PDA) media plating and PCR analysis. Inoculation and colonization of cotton by either *B. bassiana* or *P. lilacinum* negatively affected aphid reproduction over periods of seven and 14 days in a series of greenhouse trials. Field trials were conducted in the summers of 2012 and 2013 in which cotton plants inoculated as seed treatments with *B. bassiana* and *P. lilacinum* were exposed to cotton aphids for 14 days. There was a significant overall effect of endophyte treatment on the number of cotton aphids per plant. Plants inoculated with *B. bassiana* had significantly lower numbers of aphids across both years. The number of aphids on plants inoculated with *P. lilacinum* exhibited a similar, but non-significant, reduction in numbers relative to control plants. We also tested the pathogenicity of both *P. lilacinum* and *B. bassiana* strains used in the experiments against cotton aphids in a survival experiment where 60% and 57% of treated aphids, respectively, died from infection over seven days versus 10% mortality among control insects. Our results demonstrate (i) the successful establishment of *P. lilacinum* and *B. bassiana* as endophytes in cotton via seed inoculation, (ii) subsequent negative effects of the presence of both target endophytes on cotton aphid reproduction using whole plant assays, and (iii) that the *P. lilacinum* strain used is both endophytic and pathogenic to cotton aphids. Our results illustrate the potential of using these endophytes for the biological control of aphids and other herbivores under greenhouse and field conditions.

**Editor:** Thomas L. Wilkinson, University College Dublin, Ireland

**Funding:** The work was supported by Cotton Incorporated Grant and Good Neighbor Scholarship. The funders had no role in study design, data collection and analysis, decision to publish, or preparation of the manuscript.

**Competing Interests:** This study was funded in part by a grant from the Cotton Incorporated Core Program (#12-387) to GAS.

\* Email: dianacastillo8@tamu.edu

## Introduction

Fungal endophytes can protect plants from a wide range of stressors including insect pests [1]. In this study we refer to an endophyte as defined by Schulz (2005) [2] to be microorganisms (fungi or bacteria) found in asymptomatic plant tissues for all or part of their life cycle without causing detectable damage to the host. The need for the development of new strategies for the control of agricultural insect pests continues to increase due to factors such as development of insecticide resistance [3–5]. Here we focus on entomopathogenic fungal endophytes [6] and the ecological role these fungi can play in agricultural systems.

Entomopathogenic fungal endophytes have been isolated from a variety of different plant species and tissues, and can be inoculated to establish endophytically in a range of other plants to test for adverse effects, if any, on different insect herbivores [1] [6–7]. These entomopathogenic fungal endophytes are classified as non-clavicipitaceous [8]; referring to fungal endophytes that are usually horizontally transmitted. Clavicipitaceous endophytes, on the other hand, are found in grasses and are typically vertically transmitted, potentially leading to an obligate relationship and higher infection rates with their hosts [8–9]. Clavicipitaceous endophytes, named true endophytes, have been studied more extensively than non-clavicipitaceous species and are generally considered mutualistic. Evidence suggests that these fungal endophytes can significantly improve host plant tolerance to drought, insects, diseases, and nematodes, and in exchange, plants provide protection, nutrition and dissemination of the fungi [10].

A number of benefits to plants are also conferred by non-clavicipitaceous endophytes [9] [11–14]. As endophytes, several non-clavicipitaceous entomopathogens including *Beauveria bassiana*, *Lecanicillium lecanii*, *Metharizium anisoplae* and *Isaria spp.* can have negative effects on insect pests when *in planta*, antagonize plant pathogens and promote plant growth [6] [15]. The activity of *B. bassiana* has received particular attention due to its negative effects on a variety of insect herbivores including the cotton aphid [7] [16–22].

The fungus *P. lilacinum* is more widely known as *Paecilomyces lilacinus*, having undergone a recent taxonomic revision [23]. To our knowledge there are no studies demonstrating *P. lilacinum* as an endophytic fungus causing negative effects on insect herbivores, but there are reports of it being pathogenic to a number of insects including *Ceratitis capitata*, *Setora nitens*, *A. gossypii*, and *Triatoma infestans* [24–28]. Both *B. bassiana* and *P. lilacinum* are commercially available for use as biocontrol agents, but *P. lilacinum* is mainly considered to be a nematophagous, egg-parasitizing fungus, specifically against root-knot nematode, *Meloidogyne incognita*, and several other nematode species including *Radopholus similis*, *Heterodera spp*, *Globodeera spp* [29–32].

Cotton aphids, *A. gossypii*, have a broad range of host plants including cultivated cotton, causing damage directly by plant feeding and indirectly through virus transmission and physical contamination of cotton by honeydew production [33]. Most commonly, *A. gossypii* is considered a mid- to late-season pest in cotton. However, extensive use of insecticides such as pyrethroids can decrease its natural enemy community, thereby contributing to the establishment of the aphid as a season-long pest across cotton production areas [34–35]. Chronic insecticide use for aphid control has also increased its resistance to several classes of insecticides [36–38]. Considering the increasing need for alternative insect management strategies in agricultural systems, we investigated the effects of two entomopathogens, *B. bassiana* and *P. lilacinum*, on the cotton aphid when present endophytically in cotton. Specifically, we tested: 1) the ability of *B. bassiana* and *P. lilacinum* to establish as endophytes in cotton seedlings when inoculated at the seed stage, and 2) the effects of these endophytes on cotton aphid reproduction using *in planta* feeding trials in both greenhouse and field environments.

## Materials and Methods

### Plants and endophytic fungi strains

The cotton seeds used for all experiments were variety LA122 (All-Tex Seed, Inc.). The *P. lilacinum* strain was isolated from a field survey of naturally-occurring fungal endophytes in cotton [39]. This strain was confirmed to be *P. lilacinum* (formerly *P. lilacinus*) by diagnostic PCR and subsequent sequencing of the ribosomal ITS region using specific species primers [40]. The *B. bassiana* was cultured from a commercially obtained strain (Botanigard, BioWorks Inc, Victor, NY). Stock spore solutions of each fungus were made by adding 10 ml of sterile water to the fungi cultured on potato dextrose agar (PDA) in 10 cm diameter petri dish plates and scraping them with a sterile scalpel. The resulting mycelia and spores were then filtered through cheese cloth into a sterile beaker. A haemocytometer was used to calculate the conidia concentrations of the resulting stock solutions. Final treatment concentrations were reached by dilution using sterile water.

### Cotton seed inoculation

Seeds were surfaced sterilized prior to soaking in different spore concentrations by immersion in 70% ethanol for 3 minutes with constant shaking, then 3 minutes in 2% sodium hypochlorite (NaOCl) followed by three washes in sterile water, based on Posada et al. [18]. The third wash was plated on PDA media to confirm surface sterilization efficiency. Seeds were then soaked for 24 hours in two different spore concentrations of the two fungi and sterile water was used as control. Spore concentrations for each fungus were zero (control), $1 \times 10^6$ spores/ml (treatment 1) and $1 \times 10^7$ spores/ml (treatment 2) based on inoculum concentrations used in previous studies of endophytic entomopathogens [7] [17–18] [22] [70]. Beakers containing the seeds were placed in a dark environment chamber at 28°C until the next day for planting. Soaked seeds were planted in individual pots (15 cm diameter) containing unsterilized Metro mix 900 soil consisting of 40–50% composted pine bark, peat moss, vermiculite, perlite and dolomitic limestone (Borlaug Institute, Texas A&M). All plants were grown in a greenhouse at ~25°C with natural photoperiod for the duration of the experiment. Pots were placed in a complete randomized design, watered as needed, and no fertilizer was applied throughout the experiments.

### Confirmation of plant colonization by endophytic fungi

We have no reason to assume that 100% of the endophyte-treated plants are always colonized by the endophytes when inoculated as seed treatments. Given this constraint, we decided to use two detection methods simultaneously, PDA culturing and diagnostic PCR analysis, to positively confirm the presence of the target endophytes in the experimental plants from the greenhouse experiments, but not for our field experiments. At the end of each greenhouse trial, all treated and control plants were harvested, and each plant was cut in half longitudinally using a sterile scalpel. Fragments of leaves of 1 cm$^2$, stems and roots of 1 cm length were plated on PDA media and placed in growth chamber at 28°C to check for presence of the endophytes. The other half of the plant was freeze dried and DNA was extracted utilizing the CTAB protocol [41]. Species specific oligonucleotide primers for *B. bassiana* 5′CGGCGGACTCGCCCCAGCCCG 3′, 3′ CCGCGTCGGGGGTTCCGGTGCG 5′ [39] and *P. lilacinum* 5′ CTCAGTTGCCTCGGCGGGAA 3′, 3′ GTGCAACTCAGAGAAGAAATTCCG 5′ [40] (Sigma-Aldrich, Inc St Louis, MO) were used for diagnostic PCR assays. PCR products were visualized on a 2% agarose gel to determine the presence of the inoculated fungal endophytes based on amplification of a DNA fragment of the expected size (positive control). Given the larger size of the plants utilized in our field trials and the impracticality of PDA plating and extracting genomic DNA from entire large plants, we did not test for the presence of the target endophytes in the experimental plants. Instead, we analyzed our data as treatment groups [control, *B. bassiana* ($10^6$), *B. bassiana* ($10^7$), *P. lilacinum* ($10^6$) and *P. lilcainum* ($10^7$)] with concentration effects nested within endophyte treatment and present our results as such.

### Cotton aphid reproduction tests

A colony of *A. gossypii* was maintained on caged cotton plants in the same greenhouse as the experimental plants as described above. For all endophyte-aphid greenhouse trials, second instar nymphs were placed directly on to the experimental control and endophyte-treated cotton plants. Experimental and control plants with aphids were placed in individual clear plastic cages of 45 cm height and 20 cm diameter, then sealed on top with no-see-um mesh (Eastex products, NJ) to avoid aphid escape or movement between plants.

## B. bassiana cotton aphid greenhouse experiments

Greenhouse assays of the effects of endophytic *B. bassiana* on cotton aphid reproduction consisted of three independent tests, each utilizing slightly different protocols. The first was initiated when plants were 13 days old ($1^{st}$ true leaf stage) with aphids allowed to feed for seven days on 10 plants per treatment group. For the second trial, we used older plants (20 days old/third true leaf stage) and aphids were left to reproduce for a longer period of time (14 days) on 10 plants per treatment. At the end of each trial, total aphid numbers were recorded on each individual plant. The third independent test consisted of only a single reproduction trial in which ten $2^{nd}$ instar aphids were placed on 15 day old plants (second true leaf stage) and left to reproduce 14 days on 15 plants per treatment group, but the cohorts of aphids on each plant were sampled twice at 7 and then again at 14 days.

## P. lilacinum cotton aphid greenhouse experiments

We conducted two replicate experiments testing for effects of endophytic *P. lilacinum* on cotton aphid reproduction utilizing the same reproduction test protocol for each trial. In these trials, ten $2^{nd}$ instar aphids were left to reproduce on the same plants for 14 days consecutively and sampled twice at 7 and then again at 14 days. Ten $1^{st}$ true leaf stage plants per treatment group were utilized for the first trial; 15 plants per treatment group were used for the second trial.

## Cotton aphid field trials for both B. bassiana and P. lilacinum

During the summers of 2012 and 2013, experimental field trials were conducted at the Texas A&M University Field Station located near College Station in Burleson, Co., TX (N 30° 26′ 48′′ W 96° 24′ 05.12′′) at an elevation of 68.8 m. We utilized a randomized block design with five seed inoculation treatments (T1: Control, T2: *B. bassiana* $1 \times 10^6$, T3: *B. bassiana* $1 \times 10^7$, T4: *P. lilacinum* $1 \times 10^6$ and T5: *P. lilacinum* $1 \times 10^7$). Surface sterilized seeds were inoculated with the different treatments as described in our greenhouse assay protocol. Treatments were replicated six times, making a total of 30 plots in the field. Each plot was comprised of 4 rows of 16.6 m length and planted with 15 seeds per meter. For the aphid reproduction experiments, we utilized the same protocol during both field seasons whereby a total of 75 cone shaped metal framed cages (0.35 m of height) were randomly assigned to be placed over endophyte-inoculated and control plants (15 cages/treatment) and set up on May 17, 2012 and June 24, 2013, respectively (delayed experiment due to rain in 2013). Predators were eliminated if found prior to enclosing the caged plants with no-see-um mesh (Eastex products, NJ) to prevent aphid escapes and entrance of predators. Ten second instar aphid nymphs from the laboratory colony were placed on each plant and left to reproduce for 14 days. At the end of the experiment, cages were removed, the entire plant was bagged and brought back to laboratory for total aphid number counts.

## Fungal pathogenicity experiment

To assess pathogenicity of both the *P. lilacinum* strain recovered in our endophyte survey of cotton [39], and the commercial *B. bassiana* strain utilized in our endophyte trials, we performed a cotton aphid survival experiment as per Gurunlin-gappa et al. [22] and Vega et al. 2008 [70] with slight modification. The same spore concentrations used in our endophyte *in planta* experiment were used for this test for both endophytes (0, $1 \times 10^6$ and $1 \times 10^7$ spores/ml). Thirty $2^{nd}$ instar aphids per treatment were dipped in spore solutions for 5 seconds, and then placed on fresh cotton leaves kept on moistened filter paper (to prevent drying out) inside 10 cm diameter petri dishes sealed with parafilm (Bemis flexible packaging, Neenah, WI). Ten aphids per petri dish were placed in three replicate petri dishes per treatment. Aphids were checked daily for mortality and dead aphids were removed, plated and incubated on PDA media to confirm emergence of the entomopathogens from aphid cadavers.

## Statistical analyses

All data were tested for normality assumptions using a qqplot, Levene's homogeneity test and the Shapiro-Wilk normality test at alpha = 0.05 significance level. For the first independent *B. bassiana* greenhouse experiment, ANOVA and t-tests were performed to compare aphid reproduction differences among plants after 7 days of feeding. In the second and third *B. bassiana* tests, the data were non-normal and nonparametric Kruskal-Wallis and Mann-Whitney U tests were used. For both *P. lilacinum* greenhouse trials, a repeated measures ANOVA was performed with time as a repeated factor to test for differences in aphid numbers between plants after 7 and 14 days of reproduction because aphids on the same plants were sampled sequentially. Aphid field trials for both 2012 and 2013 were analyzed using ANOVA followed by pairwise comparisons (control vs. treatment). We conducted a combined ANOVA analysis of the field data across both 2012 and 2013 to test for year, treatment, and year by treatment effects. For the cotton aphid pathogenicity experiment, a Kaplan-Meier survival analysis was performed to compare the cumulative survival of treated vs. untreated control aphids. All analyses were conducted using SPSS 22 (IBM SPSS, Armonk NY).

# Results

## Plant colonization by endophytic fungi

Our culturing results showed no fungal growth on the PDA plating of the third sterile water wash of either the surface sterilized seeds or plant samples, indicating the efficacy of our surface sterilization. Thus, we assume that the fungi growing in the media from surface-sterilized plant materials were endophytes that came from within plant tissues and not epiphytes from the plant surface. Utilizing combined PDA plating and diagnostic PCR detection methods revealed 30–45% more instances of positive endophytic colonization relative to PDA plating alone. *B. bassiana* was detected in 35% and 55% of the treated plants in the first (7 day) and second (14 day) greenhouse trials, respectively. For the third *B. bassiana* trial which consisted of using the same plants for both measurements of aphid reproduction at 7 and 14 days, *B. bassiana* was detected in 53.3% of the treated plants. In the *P. lilacinum* experiments, the target endophyte was detected in 55% and 45% of plants in the first and second trials, respectively.

## B. bassiana cotton aphid greenhouse experiments

Our results were analyzed both as treatments (control, low and high concentration) and by confirmed positive colonization of plants by the target endophyte (colonized vs. uncolonized). In the first test, the mean number of cotton aphids per plant on *B. bassiana* treated plants was not significantly different from those on control plants after 7 days of reproduction when analyzed by treatment groups (F = 2.07; df = 2,29; P = 0.145), but was significantly different when analyzed by positive colonization of the endophyte (t-test; P = 0.014) (Fig 1a). In the second test, we observed a significant negative effect on reproduction of cotton aphids after 14 days when analyzed by treatment groups (Kruskal-Wallis = 6.744; P = 0.034) as well as by positive colonization of the endophyte (Mann Whitney U = 44; P = 0.004) (Fig 1b). In our third *B. bassiana* trial, there was no significant effect on the number of aphids per plant after 7 days when analyzed by treatment (Kruskal-Wallis = 4.74; P = 0.093), but there was a significant effect on aphids when analyzed by positive colonization by the endophyte (Mann-Whitney U = 60.50; P = <0.0001) (Fig 1c). Similarly at the end of the 14 days in the same experiment, there were no significant effects on the number of aphids when the data were analyzed by treatment (Kruskal Wallis = 3.069; P = 0.216), but a significant effect was observed when the data were analyzed by plant positive colonization by the endophyte (Mann Whitney U = 58; P<0.0001) (Fig 1d).

### P. lilacinum cotton aphid greenhouse experiments

As with the *B. bassiana* trials above, we present the results of analyses categorizing the data as both treatment groups and positive versus negative colonization. In the first *P. lilacinum* trial, aphid numbers varied significantly with time (Repeated Measures ANOVA F = 60.40; df = 1,28; P = 0.0001), but no significant endophyte treatment effect was observed when data were analyzed by plant positive colonization (F = 0.026; df = 1,28; P = 0.873). However, when analyzed based on treatment groups, there was a significant effect of time (F = 69.56; df = 1,27; P<0.0001) as well as endophyte treatment (F = 140.48; df = 2,27; P = 0.049) (Fig 2a). After increasing our sample size in the second trial, we observed a significant effect of both time (F = 53.73; df = 1,42;P = 0.0001) and treatment when analyzed based on plant positive colonization by the endophyte (F = 8.05; df = 1,42; P = 0.007) (Fig 2c). Although there was a significant effect of time (F = 52.52; df = 1,41; P< 0.000) on the number of aphids when we analyzed our data by treatment groups (control, low or high concentration), the effect of endophyte treatment was not significant (F = 0.546; df = 241; P = 0.583).

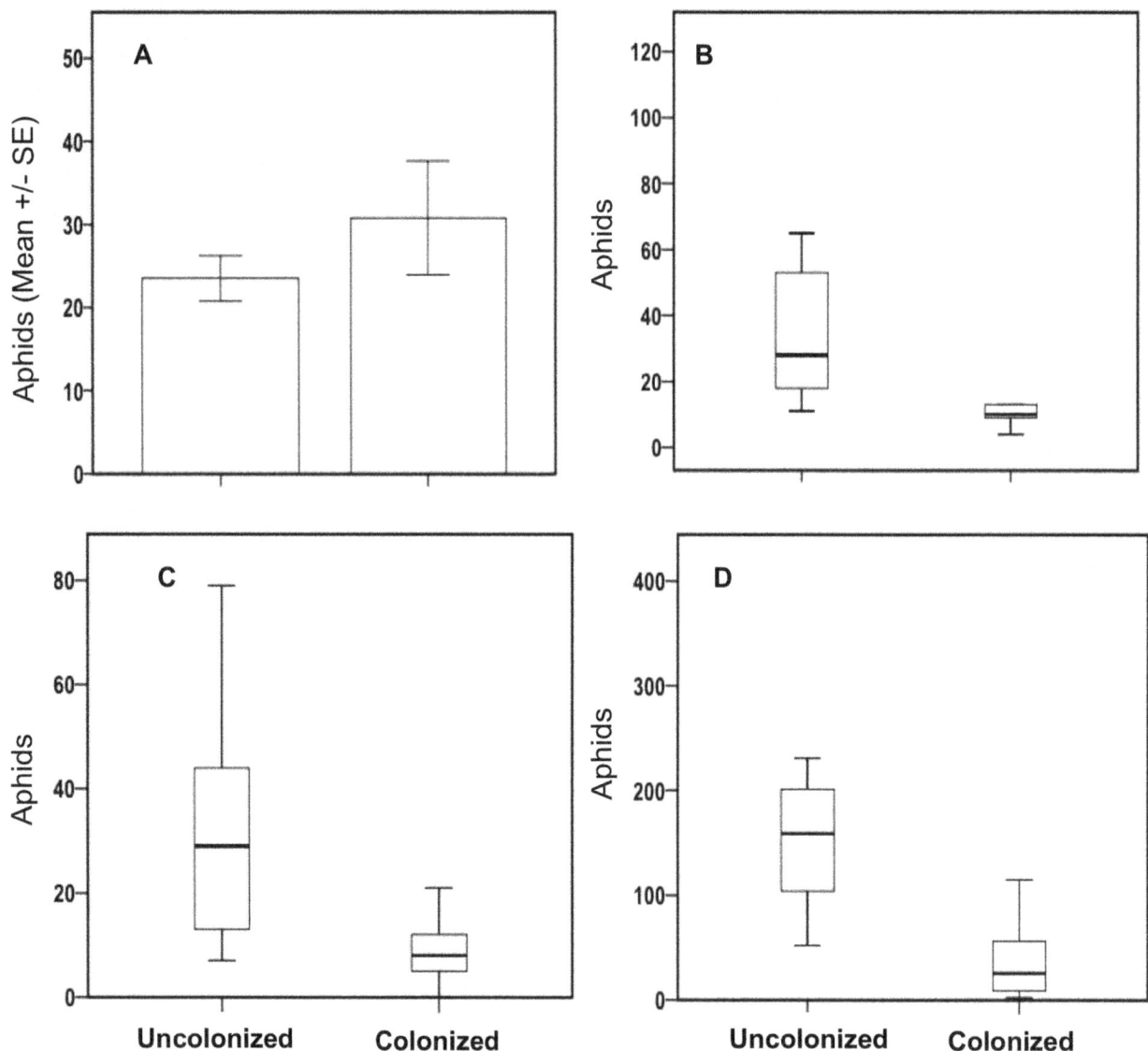

**Figure 1. Effects of endophytic *B. bassiana* on cotton aphid reproduction in three independent greenhouse assays.** Cotton aphid reproduction on plants positively colonized by endophytic *B. bassiana* versus uncolonized plants after (a) 7 days in the first trial, (b) 14 days in the second trial, and (c) 7 and (d) 14 days successively in the third trial.

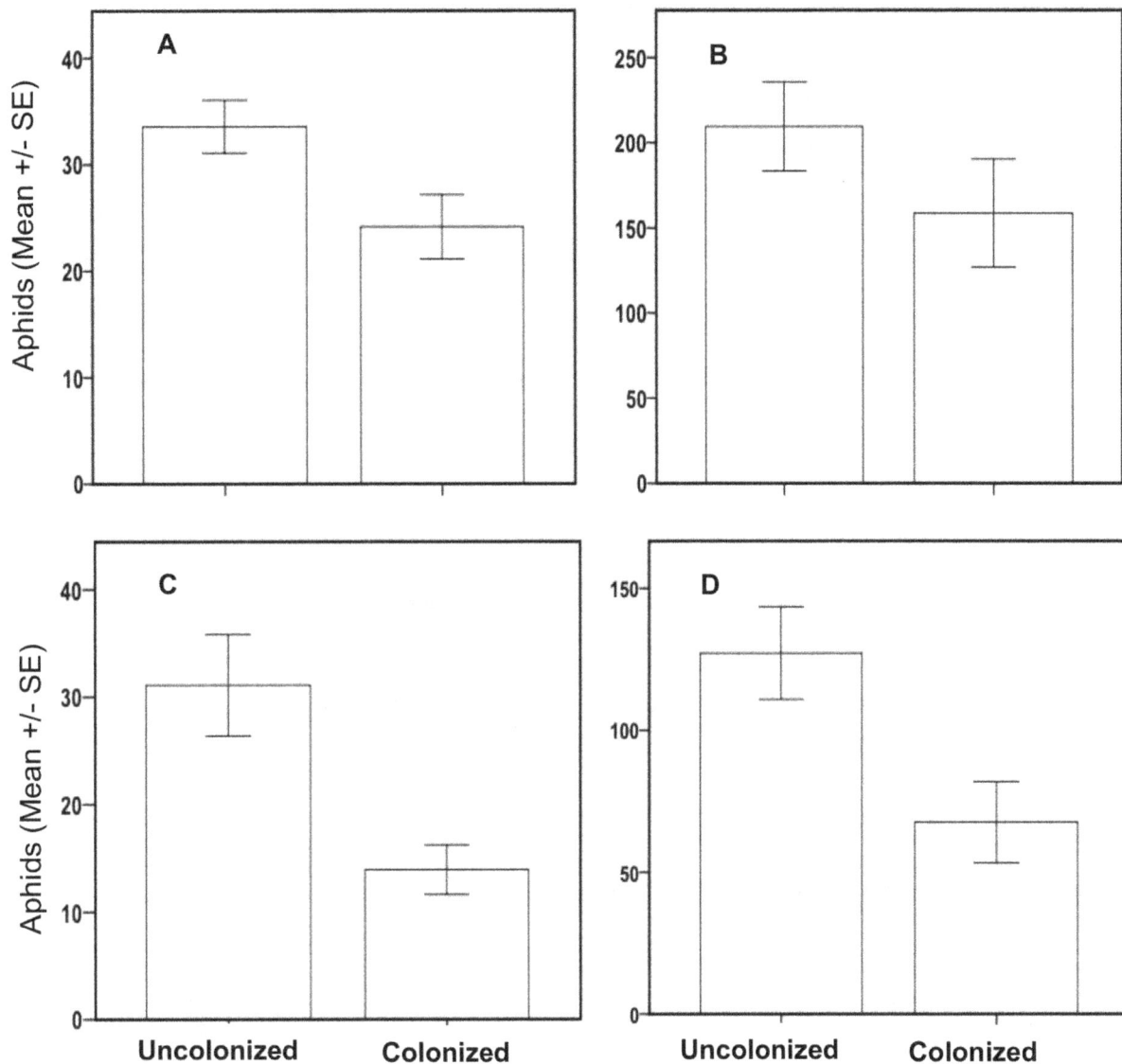

**Figure 2. Effects of endophytic _P. lilacinum_ on cotton aphid reproduction in two replicate greenhouse assays.** Cotton aphid reproduction on plants positively colonized by endophytic _P. lilacinum_ versus control plants after 7 days in the first (a) and second (c) trials, followed by 14 days in the same trials (b & d, respectively).

## Cotton aphid field trials of both _B. bassiana_ and _P. lilacinum_

In both 2012 and 2013 there was no effect of seed treatment spore concentration within each endophyte treatment (2012 Nested ANOVA, F = 1.95; df = 2,77; P = 0.149 and 2013 Nested ANOVA F = .935; df = 2,67; P = 0.398), therefore data from both concentrations were grouped for each endophyte in subsequent analyses. Across both years of the field trial, there was a significant effect of endophyte treatment (ANOVA, F = 7.31; df = 5,132; P = 0.001) and also a significant year effect (ANOVA, F = 17.43; df = 5,132; P<0.0001), but no endophyte by year interaction (ANOVA, F = 0.547; df = 5,132; P = 0.580). During the summer of 2012, there was a significant overall effect of endophyte treatment on the number of cotton aphids per plant at the end of 14 days of reproduction (ANOVA, F = 4.12; df = 2,73; P = 0.02). Follow-up pairwise comparisons revealed that there were significantly fewer aphids on cotton plants from _B. bassiana_-treated vs. control plots (P = 0.006). The difference in aphid numbers on plants in _P._

_lilacinum_-treated vs. control plots exhibited a similar but non-significant reduction (P = 0.085) (Fig 3a). Similarly in 2013, there was a significant overall effect of endophyte treatment on aphid reproduction at the end of 14 days (ANOVA, F = 3.13; df = 2,59; P = 0.05). Pairwise comparisons indicated that inoculation of plants with _B. bassiana_ had a significant negative effect on aphid reproduction vs. control (P = 0.016), but only a non-significant trend was observed with _P. lilacinum_ vs. the control (P = 0.086) (Fig 3b).

## Cotton aphid survival experiment

There was no significant difference in aphid mortality between those treated with two different concentrations ($1 \times 10^6$ or $1 \times 10^7$) of conidia solutions of each fungus. Thus, the data from both concentrations were pooled and analyzed together for each fungus. There was a highly significant increase in mortality between aphids treated with either _P. lilacinum_ (60%) or _B._

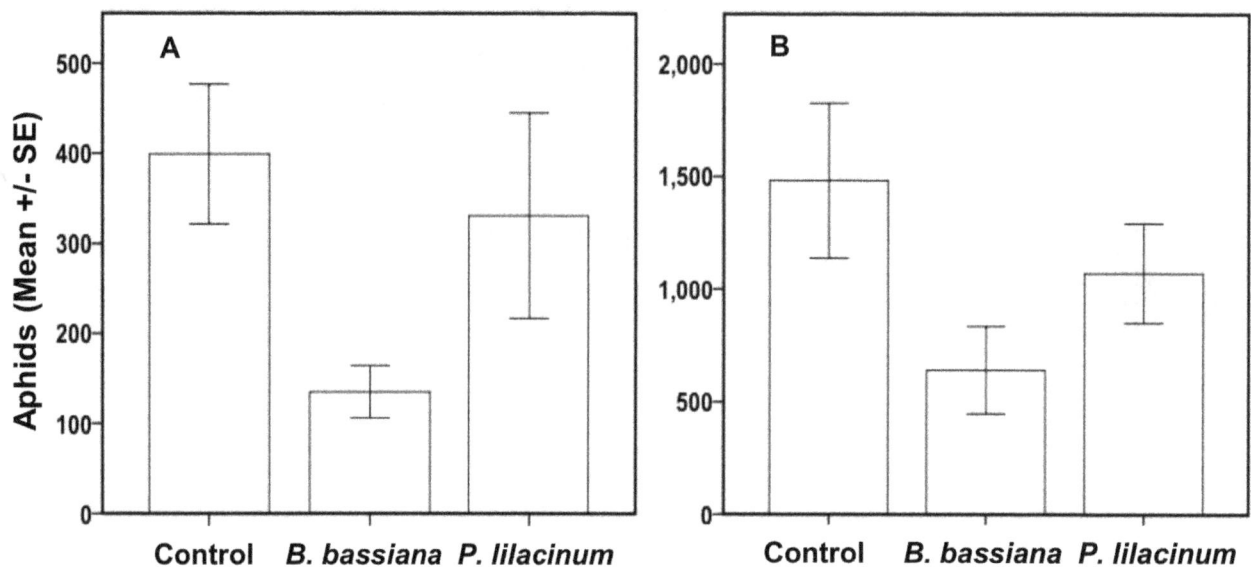

**Figure 3. Effects of endophytic** *B. bassiana* **and** *P. lilacinum* **on cotton aphid reproduction under field conditions.** Cotton aphid reproduction after 14 days on plants inoculated as seeds with either *B. bassiana* or *P. lilacinum* versus uninoculated control plants under field conditions in (a) 2012 and (b) 2013.

*bassiana* (57%) vs. the controls (10%) (Kaplan-Meier, P<0.0001 for both fungi).

## Discussion

Our results provide the first report of the negative effects of two endophytic entomopathogenic fungi, *B. bassiana* and *P. lilacinum*, on cotton aphid reproduction when feeding on whole intact cotton plants inoculated as seed treatments. Importantly, we observed negative effects under both greenhouse and field conditions. We also provide the first evidence for an endophytic effect of *P. lilacinum* on herbivorous insect performance.

After analyzing our data based on positive plant colonization by the target endophyte, we found that aphid reproduction on cotton plants positively colonized by *B. bassiana* was reduced in three independent greenhouse trials. Although the results of our first trial testing the effects of *P. lilacinum* as an endophyte on aphid reproduction revealed only a significant effect of time but not treatment, we attributed this to a small sample size for the given effect size based on the results of power analysis (Power = 0.175) (Fig. 2b). After increasing the sample size in the second *P. lilacinum* trial, we observed a significant effect of both time and treatment on the reproduction of cotton aphid with lower aphid numbers on endophyte-colonized plants (Figs. 2c & 2d). Our greenhouse endophyte trial results using *A. gossypii* are similar to those of Martinuz et al. [42] in which whole squash plants were inoculated with *Fusarium oxysporum* as an endophyte via soil drench, resulting in negative effects on *A. gossypii* choice and performance. Similarly, Akello et al. [43] showed that *Aphis fabae* feeding on bean plants colonized independently by strains of either *B. bassiana*, *Trichoderma asperellum* or *Gibberella moniliformis* reproduced poorly compared to those on control plants. Both Martinuz et al. [42] and Akello et al [43] attribute the negative effects on aphid fitness to be due to chemical changes in the plant that were systemically induced by the presence of the endophyte, though the specific mechanism by which these fungi activated a systemic response within the plants was not investigated.

The ability of *B. bassiana* to establish as an endophyte across a range of plants has been well established [e.g., cotton, corn, bean, wheat, pumpkin, tomato [7]; coffee [18]); sorghum [44]; banana [19]; tomato [20]; jute [21] and pine [45]. A number of plant-endophyte-insect interaction experiments, including a cotton aphid study by Gurunlingappa et al. [7] have been performed using cut leaf bioassays rather than whole intact plant experiments [25] [46–48]. Utilizing leaf cuts rather than whole intact plants can potentially cause release of allelochemicals due to direct plant damage that may have negative effects on insects that could obscure those caused by the presence of an endophyte [49]. Alternatively, cutting plants and abscising leaves may induce changes in plant chemistry that alter the interaction between the endophyte and host in ways not observed in intact plants [49]. Demonstrations of negative effects of endophytic entomopathogens including *B. bassiana* on herbivores in more natural whole plant feeding assays are relatively rare, but have been shown for a few species including aphids [42–43]. Similarly, there are only a few examples of negative effects on lepidopteran species caused by endophytic colonization by *B. bassiana* using whole plant assays including *Ostrinia nubilalis* and *Helicoverpa zea* [16] [20].

To our knowledge, there are no reports in the literature of negative endophytic effects of *P. lilacinum* on herbivorous insects. This is not surprising since this fungus was until recently thought to mainly have pathogenic properties against nematodes and not insects. Historically, *P. lilacinum* has been considered largely as a soil-born nematode egg parasite and used as a biocontrol agent against nematode pests such as root-knot, *Meloidogyne incognita*, and reniform, *Rotylenchulus reniformis*, nematodes [50–52]. However, recent evidence indicates that *P. lilacinum* can also be an entomopathogen [24–28]. Our results indicate that the *P. lilacinum* strain isolated from cotton by Ek-Ramos et al. [39] can negatively affect insect herbivores when present as an endophyte and that it is also pathogenic to insects. Interestingly, the same strain has also been observed to parasitize root-knot nematode eggs in simple lab bioassays and negatively affect nematode reproduction when present as an endophyte in *in planta* assays (W. Zhou, J.T. Starr and G.A. Sword, unpublished results).

The mechanisms by which herbivores can be negatively affected by clavicipitaceous obligate endophytes have been studied in a few different grass species and can vary from antixenosis and/or antibiosis mediated by constitutive production and or induction of secondary compounds produced by the plant [53–55] or secondary metabolites produced by the endophytes themselves [13] [22] [56–61][64]. It is important to mention that infection rates of natural populations of grasses by these endophytes can vary depending on the genetic and environmental background of the population and these factors can determine if this symbiosis goes from mutualistic to antagonistic [63–67]. Another hypothesis for the mechanism by which endophytes can negatively affect herbivores is based on the idea that endophytes can alter the phytosterol profiles of plants and compete with insects for these compounds which are essential for their development [46] [62]. The mechanisms by which entomopathogenic endophytic fungi may protect plants from insect herbivores are unknown. Although these endophytes do produce secondary metabolites [22] [68], we do not know if this is the main cause for the negative effects on aphids when feeding on endophytically-colonized plants observed in our study. The literature also suggests a systemic response in the plant can be induced by the presence of some entomopathogenic endophytes including *B. bassiana* that confers resistance against plant pathogens [68–69]. Whether an induced systemic response accounts for the negative effects on insects observed in our study remains to be determined.

The mode of establishment and duration of presence of endophytic fungi in plants varies among the different plant-endophyte combinations tested to date [7] [17–21] [44–45]. In some cases, intentionally inoculated endophytes can be retained within plants for considerable amounts of time, including *B. bassiana* found for as long as eight months in coffee [18] or nine months in *Pinus radiata* [45]. Our study indicates that *B. bassiana* and *P. lilacinum* were still present in cotton plants up to 34 days following inoculation as a seed treatment. This duration does not necessarily indicate that *B. bassiana* and *P. lilacinum* can only be present in cotton as endophytes for this period of time, but rather that we did not test for the presence/absence of the endophytes beyond 34 days. The average recovery success of the target endophytes used in our studies ranged from 35–55%. Though not a high colonization frequency, we were still able to detect negative effects on aphids feeding on plants colonized by the endophytes. We have not yet rigorously studied the endophytic colonization of cotton by *P. lilacinum* and *B. bassiana*, but *P. lilacinum* was primarily detected in the root tissues whereas *B. bassiana* was found mostly in the above ground tissues. Fungal endophytes are known to occur throughout an entire plant including leaves, stems, roots and reproductive parts, however, tissue specific presence in plants is not required for negative effects on target herbivores. For example, endophytic fungi inhabiting roots can negatively affect the performance and fitness of caterpillars feeding on above ground tissues [13,71]. Our results support this scenario given that *P. lilacinum* negatively affects aphids feeding on cotton leaves above ground, but is recovered more commonly from below ground root tissues.

The manipulation of endophytic fungi, many of which are completely unstudied, has the potential to protect plants from insect herbivores and other stress factors [1]. We have provided novel evidence showing that the endophytic establishment in cotton of the entomopathogens *B. bassiana* and *P. lilacinum* when inoculated as seeds can adversely affect cotton aphid reproduction not only in greenhouse assays, but also under field conditions. Although we observed a significant year effect, this was due to differences in the total aphid numbers across years (Fig. 3a&b). Importantly, there was no year by endophyte treatment interaction effect. Our field results exhibited the same pattern of negative effects of endophytes on cotton aphids across years in both 2012 and 2013. The consistency of results across years under field conditions that can vary in variety of uncontrolled environmental variables (e.g. precipitation and temperature regimes) is particularly encouraging for the potential reliability of incorporating fungal endophyte manipulations into IPM strategies. Future directions of our work include testing these entomopathogenic endophytes against other insect and nematode herbivores along with phytohormone and transcriptomic analysis to investigate the mechanisms by which these endophytes confer protection to their plant hosts.

## Acknowledgments

We would like to thank Cesar Valencia and Josephine Antwi for help provided during greenhouse and field trials both in 2012 and 2013. Dr. Steve Hague from the Soil and Crop Department at Texas A&M University provided critical logistical support for the field trials. Charlie Cook from All-Tex Seed, Inc. generously provided seeds for these experiments.

## Author Contributions

Conceived and designed the experiments: DCL KZS GAS. Performed the experiments: DCL. Analyzed the data: DCL GAS. Contributed reagents/materials/analysis tools: DCL GAS KZS MJER. Wrote the paper: DCL GAS.

## References

1. Porras-Alfaro A, Bayman P (2011) Hidden fungi, emergent properties: endophytes and microbes. Annu Rev Phytopathol 49: 291–315.

2. Schulz B, Boyle C (2005) The endophytic continuum. Mycol Res 109: 661–686.

3. Gould F (1995) Comparison between resistance management strategies for insects and weeds. Weed Technology 9(4): 830–839.

4. Gassmann AJ (2009) Evolutionary analysis of herbivorous insects in natural and agricultural environments. Pest Manag Sci 65: 1174–1181.

5. Silva AX, Jander G, SamadiegoH, Ramsey JS, Figueroa CC (2012) Insecticide resistance mechanisms in the green peach aphid *Myzus persicae* (Hemiptera:Aphididae): A transcriptomic survey. PLOS ONE 7(6): 363–366.

6. Vega FE, Mark S, Goettel MS, Blackwell M, Chandler D, et al. (2009) Fungal entomopathogens: new insights in their ecology. Fungal Ecology 2: 149–159.

7. Gurulingappa P, Sword GA, Murdoch G, McGee PA (2010) Colonization of crop plants by fungal entomopathogens and their effects on two insect pests when *in planta*. BioControl 55: 34–41.

8. Rodriguez RJ, White JF Jr, Arnold AE, Redman RS (2009) Fungal endophytes: diversity and functional roles. New Pathologist 182(2): 314–330.

9. Hartley SE, Gange AC (2009) The impacts of symbiotic fungi on insect herbivores: mutualism in a multitrophic context. Ann Rev Entomol 54: 323–342.

10. Schardl CL, Leuchtmann A, Spiering MJ (2004) Symbioses of grasses with seedborne fungal endophytes. Annu Rev Plant Biol 55: 315–340.

11. Omacini M, Chaneton EJ, Ghersa CM, Muller CB (2001) Symbiotic fungal endophytes control insect host-parasite interaction webs. Nature 409: 78–81.

12. Jung HS, Lee HB, Kim K, Lee EY (2006) Selection of *Lecancillium* strains for aphid (*Myzus persicae*) control. Korean J Mycol 34: 112–118.

13. Jaber LR, Vidal S (2010) Fungal endophyte negative effects on herbivory are enhanced on intact plants and maintained in a subsequent generation. Ecol Entomol 35: 25–36.

14. Gange AC, Eschen R, Wearn J, Thawer A, Sutton B (2012) Differential effects of foliar endophytic fungi on insect herbivores attacking a herbaceous plant. Oecologia 168: 1023–1031.

15. Vega FE, Posada F, Aime MC, Pava-Ripoll M, Infante F, et al. (2008) Entomopathogenic fungal endophytes. Biological Control 46: 72–82.

16. Bing LA, Lewis LC (1991) Suppression of *Ostrinia nubilalis* by endophytic *Beauveria bassiana*. Environ Entomol 20: 1207–1211.

17. Posada FJ, Vega FE (2005) Establishment of the fungal entomopathogen *Beauveria bassiana* (Ascomycota: Hypocreales) as an endophyte in cocoa seedlings (*Theobroma cacao*). Mycologia 97: 1208–1213.

18. Posada F, Aime MC, Peterson SW, Rehner SA, Vega FE (2007) Inoculation of coffee plants with the fungal entomopathogen *Beauveria bassiana* (Ascomycota:Hypocreales). Mycol Res 111: 748–757.

19. Akello J, Dubois T, Coyne D, Kyamanywa S (2008) Endophytic *Beauveria bassiana* in banana (*Musa spp.*) reduces banana weevil (*Cosmopolites sordidus*) fitness and damage. Crop Prot 27: 1437–1441.

20. Powell WA, Klingeman WE, Ownley BH, Gwinn KD (2009) Evidence of Endophytic *Beauveria bassiana* in seed treated tomato plants acting as a systemic entomopathogen to larval *Helicoverpa zea* (Lepidoptera:Noctuidae). J Entomol Sci 44(4): 391–396.

21. Biswas C, Dey P, Satpathy S, Satya P (2011) Establishment of the fungal entomopathogen *Beauveria bassiana* as a season long endophyte in jute (*Corchorus olitorius*) and its rapid detection using SCAR marker. BioControl. DOI 10.1007/s10526-011-9424-0.

22. Gurunlingappa P, McGee PA, Sword GA (2011) Endophytic *Lecanicillum lecanii* and *Beauveria bassiana* reduce the survival and fecundity of *Aphis gossypii* following contact with conidia and secondary metabolites. Crop Prot 30: 349–353.

23. Luangsa-ard JJ, Houbraken J, van Doorn T, Hong SB, Borman AM, et al. (2011) *Purpureocillium*, a new genus for the medically important *Paecilomyces lilacinum*. FEMS Microbiol Lett 321: 141–149.

24. Imoulan A (2011) Natural occurrence of soil-borne entomopathogenic fungi in the Moroccan Endemic forest of *Argania spinosa* and their pathogenicity to *Ceratitis capitata*. J Microbiol & Biotech 27(11): 2619–2628.

25. Wakil W, Ashfaq M, Ghazanfar MU, Kwon YJ, Ullah E, et al. (2012) Testing *Paecilomyces lilacinus*, diatomaceous earth and *Azadirachta indica* alone and in combination against cotton aphid (*Aphis gossypii* Glover) (Insecta: Homoptera: Aphididae). African J Biotech 11(4): 821–828.

26. Rao NBVC, Snehalatharani A, Emmanuel N (2012) New record of *Paecilomyces lilacinus* (Deuteromycotina: Hyphomycetes) as an entomopathogenic fungi on slug caterpillar of coconut. Insect Environ 17(4): 151–153.

27. Marti GA, Lastra CC, Pelizza SA, García JJ (2006) Isolation of *Paecilomyces lilacinus* (Thom) Samson (Ascomycota: Hypocreales) from the Chagas disease vector, *Triatoma infestans* Klug (Hemiptera:Reduviidae) in an endemic area in Argentina. Mycopathologia 162(5): 369–72.

28. Fiedler Ž and Sosnowska D (2007) Nematophagous fungus *Paecilomyces lilacinus* (Thom) Samson is also a biological agent for control of greenhouse insects and mite pests. BioControl 52(4): 547–8.

29. Sharma S, Trivedi PC (2012) Application of *Paecilomyces lilacinus* for the control of *Meloidogyne incognita* infecting *Vigna radiate*. Indian J Nematol 42(1): 1–4.

30. Kannan R (2012) Effect of different dose and application methods of *Paecilomyces lilacinus* (Thom.) Samson against Root Knot Nematode, *Meloidogyne incognita* (Kofoidand White) Chitwood in Okra. J Ag Science 4(11): 119–127.

31. Carrion G, Desgarennes D (2012) Effect of *Paecilomyces lilacinus* in free-living nematodes to the rhizosphere associates potatoes grown in the Cofre of Perote region, Veracruz, Mexico. Revista Mexicana de Fitopatologia 30(1): 86–90.

32. Khan MR (2012) Management of root-knot disease in eggplant through the application of biocontrol fungi and dry neem leaves. Turkish J Biol 36(2): 161–169.

33. Godfrey LD, Fuson KJ, Wood JP, Wright SD (1997) Physiological and yield responses of cotton to mid-season cotton aphid infestations in California. Proc Beltwide Cotton Conferences 1048–1051.

34. King EG, Phillips JR, Head RB (1987) 40th Annual conference report on cotton insect research and control. In: JM Brown and DA Richter Proc Beltwide Cotton Prod.

35. Godfrey LD, Rosenheim JA, Goodell PB (2000) Cotton aphid emerges as major pest in SVJ cotton. California Agriculture 54(6): 32–34.

36. O'Brien PJ, Hardee DD, Grafton-Caldwell EE (1990) Screening of *Aphis gossypii* for insecticide tolerance. Insecticide and Acaracide Tests 15: 254–255.

37. Grafton-Caldwell EE (1991) Geographical and temporal variation in response to insecticides in various life stages of *Aphis gossypii* (Homoptera:Aphididae) infesting cotton in California. J Econ Entomol 84: 741–749.

38. Kerns DL, Gaylor MJ (1992) Insecticide resistance in field populations of the cotton aphid (Homoptera:Aphididae). J Econ Entomol 85: 7–8.

39. Ek-Ramos MJ, Zhou W, Valencia CU, Antwi JB, Sword GA (2013) Spatial and temporal variation in fungal endophyte communities isolated form cultivated cotton (*Gossypium hirsutum*). PLOS ONE 8: e66049.

40. Atkins SD, Clark IM, Pande S, Hirsch PR, Kerry BR (2004) The use of real time PCR and species specific primers for the identification and monitoring of *Paecilomyces lilacinus*. FEMS Microbiol Ecol 51: 257–264.

41. Doyle JJ, Doyle JL (1987) A rapid DNA isolation procedure for small quantities of fresh leaf tissue. Phytochem Bull 19: 11–15.

42. Martinuz A, Schouten A, Menjivar RD, Sikora RA (2012) Effectiveness of systemic resistance toward *Aphis gossypii* (Aphididae) as induced by combined applications of the endophytes *Fusarium oxysporum* Fo162 and *Rhizobium etli* G12. Biological Control 62: 206–212.

43. Akello J and Sikora R (2012) Systemic acropedal influence of endophyte seed treatment on *Acyrthosipon pisum* and *Aphis fabae* offspring development and reproductive fitness. Biological Control 61: 215–221.

44. Reddy N, Ali Khan AP, Devi UK, Sharma HC, Reineke A (2009) Treatment of millet crop plant (*Sorghum bicolor*) with the entomopathogen fungus (*Beauveria bassiana*) to combat infestation by the stem borer, *Chilo partellus* Swinhoe (Lepidoptera:Pyralidae). J of Asia-Pacific Entomol 12: 221–226.

45. Brownbridge M, Reay SD, Nelson TL, Glare TR (2012) Persistence of *Beauveria bassiana* (Ascomycota:Hypocreales) as an endophyte following inoculation of radiata pine seed and seedlings. Biological Control 62(3): 194–200.

46. Raps A, & Vidal S (1998) Indirect effects of an unspecialized endophytic fungus on specialized plant-herbivorous insect interactions. Oecologia 114: 541–547.

47. McGee PA (2002) Reduced growth and deterrence from feeding of the insect pest *Helicoverpa armiguera* associated with fungal endophytes from cotton. Australian J Exp Ag 42(7): 995–999.

48. Vicari M, Hatcher PE, Ayres PG (2002) Combined effect of foliar and mycorrhizal endophytes on an insect herbivore. Ecology 83: 2452–2464.

49. Price PW, Denno RF, Eubanks MD, Finke DL, Kaplan I (2011) Plant and herbivore interactions. In: Insect Ecology. pp. 121–127. Cambridge University Press.

50. Munawar F, Bhat MY, Ashaq M (2011) Combined application of *Paecilomyces lilacinus* and carbosulfan for management of *Meloidogyne incognita* and *Rotylenchulus reniformis*. Ann Plant Protection Sciences 19(1): 168–173.

51. Kiewnick S (2011) Effect of *Meloidogyne incognita* inoculum density and application rate of *Paecilomyces lilacinus* strain 251 on biocontrol efficacy and colonization of egg masses analyzed by real-time quantitative PCR. Phytopathology 101(1): 105–12.

52. Chaudhary KK, Kaul RK (2012) Compatibility of *Pausteria penetrans* with fungal parasite *Paecilomyces lilacinus* against root knot nematode on chilli: *Capsicum annuum*. South Asian J Exp Biol 1(1): 36–42.

53. Clay K, Marks S, Cheplick GP (1993) Effects of insect herbivory and fungal endophyte infection on competitive interactions among grasses. Ecology 74: 1767–77.

54. Clay K (1996) Interactions among fungal endophytes, grasses and herbivores. Res Popul Ecol 38(2): 191–201.

55. Carriere Y, Bouchard A, Bourassa S, Brodeur J (1998) Effect of endophyte incidence in perennial ryegrass on distribution, host choice, and performance of the hairy chinch bug (Hemiptera: Lygaeidae). J Econ Entomol 91: 324–328.

56. Gindin G, Barash I, Harari N, Raccah B (1994) Effect of endotoxic compounds isolated from *Verticillium lecanii* on the sweetpotato whitefly, *Bemisia tabaci*. Phytoparasitica 22: 189–196

57. Wang L, Huang J, You M, Guan X, Liu B (2007) Toxicity and feeding deterrence of crude toxin extracts of *Lecanicillium* (*Verticillium*) *lecanii* (Hyphomycetes) against sweet potato whitefly, *Bemisia tabaci* (Homoptera: Aleyrodidae). Pest Manag Sci 63: 381–387.

58. Ball OJP, Barker GM, Prestidge RA, Lauren DR (1997a) Distribution and accumulation of the alkaloid peramine in *Neotyphodium lolii* infected perennial ryegrass. J Chem Ecol 23: 1419–1434.

59. Ball OJP, Miles CO, Prestidge RA (1997b) Ergopeptine alkaloids and *Neotyphodium lolii* mediated resistance in perennial ryegrass against adult *Heteronychus arator* (Coleoptera: Scarabaeidae). J Econ Entomol 90: 1382–1391.

60. Latch GCM (1993) Physiological interactions of endophytic fungi and their hosts: biotic stress tolerance imparted to grasses by endophytes. Agriculture, Ecosystems and Environment 44: 143–156.

61. Bush LP, Wilkinson HH, Schardl CL (1997) Bioprotective alkaloids of grass-fungal endophyte symbioses. Plant Physiol 114(1): 1–7.

62. Dugassa-Gobena D, Raps A, Vidal S (1996) Einuû von *Acremonium strictum* auf den Sterolhaushalt von Panzen: ein moÈ glicher Faktor zum veraÈ nderten Verhalten von Herbivoren. Mitt Biol Bundesanst 321: 299.

63. Saikkonen K, Wali PR, Helander M (2010) Genetic compatibility determines endophyte grass combinations. PLOS ONE 5(6): e11395.

64. Saari S, Helander M, Faeth SH (2010) The effects of endophytes on seed production and seed predation of tall fescue and meadow fescue. Microb Ecol 60: 928–934.

65. Rasmussen S, Parksons AJ, Fraser K, Xue H, Newman JA (2008) Metabolic profiles of Lolium perenne are differentially affected by Nitrogen supply, Carbohydrate content and fungal endophyte infection. Plant Physiol 146(3):1440–1453.

66. Saikkonen K, Lehtonen P, Helander M, Koricheva J, Faeth SH (2006) Model systems in ecology: dissecting the endophyte-grass literature. Trends Plant Sci 11: 428–433.

67. Young C, Wilkinson H, (2010) Epichloe endophytes: models of an ecological strategy. In Cellular and Molecular Biology of Filamentous Fungi. ASM Press, Whashington DC. Pp 660–671.

68. Ownley BH, Griffin MR, Klingeman WE, Gwinn KD, Moulton JK, et al. (2008) Beauveria bassiana: endophytic colonization and plant disease control. J Invertebr Pathol 3: 267–270

69. Ownley BH, Gwinn KD, Vega FE (2010) Endophytic fungal entomopathogens with activity against plant pathogens: ecology and evolution. BioControl 55: 113–128.

70. Vega FE, Posada F, Aime MC, Pava-Ripolli M, Infante F, et al. (2008) Entomopathogenic fungal endophytes. Biological Control 46: 72–82.

71. Raps A, Vidal S. (1998) Indirect effects of an unspecialized endophytic fungus on specialized plant – herbivorous insect interactions. Oecologia 114: 541–547.

# Genetic Structure, Linkage Disequilibrium and Association Mapping of Verticillium Wilt Resistance in Elite Cotton (*Gossypium hirsutum* L.) Germplasm Population

**Yunlei Zhao, Hongmei Wang\*, Wei Chen, Yunhai Li**

State Key Laboratory of Cotton Biology, Institute of Cotton Research of Chinese Academy of Agricultural Sciences (CAAS), Anyang, People's Republic of China

## Abstract

Understanding the population structure and linkage disequilibrium in an association panel can effectively avoid spurious associations and improve the accuracy in association mapping. In this study, one hundred and fifty eight elite cotton (*Gossypium hirsutum* L.) germplasm from all over the world, which were genotyped with 212 whole genome-wide marker loci and phenotyped with an disease nursery and greenhouse screening method, were assayed for population structure, linkage disequilibrium, and association mapping of Verticillium wilt resistance. A total of 480 alleles ranging from 2 to 4 per locus were identified from all collections. Model-based analysis identified two groups (G1 and G2) and seven subgroups (G1a–c, G2a–d), and differentiation analysis showed that subgroup having a single origin or pedigree was apt to differentiate with those having a mixed origin. Only 8.12% linked marker pairs showed significant LD (P<0.001) in this association panel. The LD level for linked markers is significantly higher than that for unlinked markers, suggesting that physical linkage strongly influences LD in this panel, and LD level was elevated when the panel was classified into groups and subgroups. The LD decay analysis for several chromosomes showed that different chromosomes showed a notable change in LD decay distances for the same gene pool. Based on the disease nursery and greenhouse environment, 42 marker loci associated with Verticillium wilt resistance were identified through association mapping, which widely were distributed among 15 chromosomes. Among which 10 marker loci were found to be consistent with previously identified QTLs and 32 were new unreported marker loci, and QTL clusters for Verticillium wilt resistanc on Chr.16 were also proved in our study, which was consistent with the strong linkage in this chromosome. Our results would contribute to association mapping and supply the marker candidates for marker-assisted selection of Verticillium wilt resistance in cotton.

**Editor:** Jinfa Zhang, New Mexico State University, United States of America

**Funding:** This research was supported by the National Natural Science Foundation of China (Grant No. 31000733) and by the National Hi-Tech Research and Development Program of China (Grant No. 2012AA101108-02-02). The funders had no role in study design, data collection and analysis, decision to publish, or preparation of the manuscript.

**Competing Interests:** The authors have declared that no competing interests exist.

\* E-mail: wanghm@cricaas.com.cn

## Introduction

Cotton is an important economic crop worldwide, which provides the most important natural fiber for the textile industry. Genetic improvement of yield, fiber quality and disease resistance is the most important objectives in cotton breeding programs worldwide. However, it is a challenging task for breeders to realize the synchronous improvement of yield, fiber quality and disease resistance because of the negative genetic correlation between them [1]. The development of molecular quantitative genetics has made it possible to map the quantitative trait loci (QTLs) for yield, fiber quality and disease resistance, thus facilitating the application of marker-assisted selection (MAS) for genetic improvement. In cotton, numerous QTLs for yield, fiber quality and disease resistance were identified [2–11]. In all these studies, the QTL mapping had been performed in segregating populations derived from biparental crosses. Due to limited recombination events, it is difficult for biparental segregating populations to detect closely linked markers for marker-assisted selection. What's more, the

frequency of polymorphic loci in biparental populations is limited and some minor QTLs are not detected. An alternative approach to QTL mapping is association analysis, also known as LD mapping. In contrast to QTL mapping based on bi-parental populations, association mapping is based on linkage disequilibrium (LD) and uses a sample of lines from the broader breeding population, unrelated by any specific crossing design [12]. So, the higher number of historical recombination events can be explored in natural population than that in the biparental segregating populations, resulting in a higher resolution of QTL mapping [13]. What's more, association mapping has been used to identify causal polymorphism within a gene that is responsible for the phenotypic variations [14].

The starting point for association mapping studies is based on the non-random association of alleles at different loci (linkage disequilibrium, LD), namely between a marker locus and a phenotypic trait locus. LD can be caused by unknown population structure and several forces, including mutation, drift, genetic bottlenecks, founder effects, selection, and specifically for plants, level of inbreeding caused by their mating systems [15]. In order to appropriately apply LD mapping in crop plants, it is a prerequisite to characterize LD levels and patterns in a population analyzed. It is also important to distinguish between physical LD and the other different forces that can create LD in natural populations, to avoid the detection of spurious associations [16]. The decay or decrease of LD with increasing map distance between markers in outcrossing plants is usually faster than that in inbreeding plants [16]. For example, LD decays rapidly within 1–5 kb in maize diverse inbred lines [17], 1.1 kb in cultivated sunflower [18], 300 bp in wild grapevine [19], whereas LD decays slowly within 250 kb in Arabidopsis [20], 212 kb in elite barley cultivars [21], 100–200 kb in rice diverse lines [22,23] and 250 kb in cultivated soybean [24]. Also, The decay or decrease of LD in wild relatives is faster than that in modern varieties [25,26].

The tetraploid species, *Gossypium hirsutum* L. (n = 26, AD genome), showing obvious economic importance such as high yield and environmental suitability, have attracted considerable scientific interest for plant breeders and agricultural scientists and been planted widespreadly, and have been responsible for 95% of the annual cotton crop in the world [27]. However, due to the complexity of genome structure and the lack of high-quality molecular markers, studies on the population structure and LD in cotton (*Gossypium hirsutum* L.) is limited so far and lagged behind the other crop species. Recently, several studies have investigated the level of linkage disequilibrium among genetic markers in various cotton populations. For example, at the significance threshold ($r^2 \geq$ 0.1), LD decays up to 25 cM in 335 *G. hirsutum* germplasm [28], less than 10 cM in 208 landrace stocks and more than 30 cM in 77 photoperiodic variety accessions [29], while LD decays within 13–14 cM in 81 Upland cotton cultivars [30]. The fast LD decay of cotton cultivars illustrates the significant potential for LD-based association mapping for agronomic traits. Elite cotton (*Gossypium hirsutum* L.) germplasm is the important resources in cotton breeding, which possess the following one or more characters of high yield, good fiber quality, earliness, disease and pest resistance. Therefore, further characterization of the population structure and LD levels in elite cotton (*Gossypium hirsutum* L.) germplasm collected from all over the world will be a benefit for association mapping of complex traits in cotton.

Verticillium wilt, incited by fungal pathogen called *Verticillium dahliae*, is a serious soil-borne disease with international consequences for cotton production. To date, the most effective and feasible way to control wilt disease has been to develop new cotton varieties resistant to Verticillium wilt. Since most commercial cultivars of upland cotton are susceptible or only slightly resistant to cotton wilt disease [10], it is necessary for cotton breeder to improve Verticillium wilt resistance in cotton (*Gossypium hirsutum* L.) by conducting introgression of resistance genes in sea island cotton or gene pyramiding from different sources of resistance. The most effective selection for introgression of resistance genes or gene pyramiding was using marker-assisted selection (MAS). To date, at least 60 different Verticillium wilt resistance QTLs have been reported on 10 chromosomes or linkage groups of cotton [8–11]. However, these QTLs were mapped in four different biparental populations and poorly colocalized, thus markers linked to these QTLs are not directly used in cotton breeding. QTL effects needed to be verified in other genetic backgrounds prior to widespread application of QTL-linked markers in MAS. Against this backdrop, association mapping using a sample of lines from the broader breeding population showed great potential for QTL detection, which can explore the higher number of historical recombination events than the biparental segregating populations, resulting in a higher resolution of QTL mapping. Currently,

association mapping for yield and fiber traits in cotton has been conducted in several studies [28–30], but no report has been found for association mapping of Verticillium wilt resistance in cotton. Therefore, our study on association mapping of Verticillium wilt resistance in cotton would be a beneficial supplementary and verification for current QTL mapping of verticillium wilt resistance genes in cotton.

In this study, we genotyped a population of 158 elite cotton (*Gossypium hirsutum* L.) germplasm from all over the world using genome-wide molecular markers. The aims of this study were to assess the population structure, linkage disequilibrium (LD), and association of molecular markers with Verticillium wilt resistance in a collection of 158 elite cotton germplasm accessions.

## Materials and Methods

### Sampling of cotton accessions

A collection of 329 cotton (*Gossypium hirsutum* L.) accessions from the China cotton germplasm collection were planted in the experimental field at Cotton Research Institute of Chinese Academy of Agricultural Sciences, Anyang, China in 2007,2008 and 2009 to evaluate their agronomic traits of yields, fiber quality, growth period and disease resistance. Some of varieties with same pedigree and similar performance in agronomy traits were excluded. Finally, a panel of 158 cotton accessions (Table S1) were selected, which included 106 accessions from China, 41 accessions from America, 3 accessions from Africa, 4 accessions from Former Soviet Union, 1 accession from French, 1 accession from Pakistan, 1 accession from Australia, and 1 accession with unknown origin. These accessions have been strictly self-pollinated during the past decades for germplasm renewing and the residual heterozygosity have been decreased remarkably.

### Phenotypic evaluation

Verticillium wilt resistance of 158 cotton lines was evaluated by two methods. *Verticillium dahliae* isolate Vd080, which is a defoliating, moderate pathogenic to cotton, was used in both methods. The resistance during adult-plant stage was evaluated in the artificial Verticillium wilt nursery in Anyang, China, in 2009. The disease nursery was constructed by uniformly mixing natural soil with *Verticillium dahliae* cotton seed cultivation with an amount of 450~750 kilogram per hectare. To ensure enough and uniform infection, the disease nursery was devided into several disease pools and each pool was isolated physically from natural soil around it and below. The inoculation amount for each pool was controlled artificially to ensure the uniform and severe infection. At the same time, for each pool, the same susceptible control was used to judge the severity of infection and decrease the error of investigation in different pools (see below). Based on the experiences for many years, the disease pools in the nursery had been severely infected with *Verticillium dahliae*, which ensured the susceptible control to reach anticipative severity of infection (see below). The experimental design was a randomized block with three replications. The different cotton lines were grown in two row plots, 6.0 m long and 0.8 m row space. Jimian 11 acted as a susceptible control to estimate the severity of disease and determine the optimal time for investigation. The 0–5 scales were used for disease severity ratings based on the percentage of diseased leaves of the whole plants, where 0 = no symptoms, 1 = less than 25%, 2 = 25–50%, 3 = 50–75%, and 4 = more than 75% of leaves showing symptoms. Since the 0–5 scales were based on the investigation for a single plant and were not visual to estimate the severity of infection for a certain line, they were converted into the disease index (DI). The disease index (DI) was calculated as follows: $DI = \dfrac{\sum(d_c \times n_c)}{n_t \times 4} \times 100$, where $d_c$ was disease severity rating, $n_c$ the numbers of plants with each of the corresponding disease severity rating, and $n_i$ the total number of plants assessed. Since the disease nursery was composed of several disease pools and each pool has the same susceptible control, the DI was further adjusted into relative disease index (RDI) to decrease the error of investigation in different pools by a correction factor K. The K was defined as follow: $K = \dfrac{50.00}{DI_{CK}}$\$, where 50.00 was regarded as the standard DI of the susceptible control, which means the uniform and severe infection in the pool; $DI_{CK}$ was the factual DI of the susceptible control. The RDI was defined as follow: $RDI = DI \times K$, where $DI$ was the factual DI of the testing lines. The cotton seeds were planted at the end of April, and the plants usually showed symptoms of Verticillium wilt in June. The disease symptoms gradually increased along with the growth of cotton, and reached their peak in blossoming and boll forming stage. After disease symptoms appeared, all the cotton lines were observed for the severity of infection. The disease severity ratings of the susceptible control were investigated at regular intervals. When the DI of the susceptible control was around 50 (optimally in the range of 40 to 66.67), implying the uniform and severe infection, the disease severity ratings of each cotton line was investigated and the disease index (DI) and relative disease index (RDI) were calculated, which served as an evaluation of disease resistance for a cotton line at adult-plant stage.

The second method was to estimate disease resistance at seedling stage. The experiment was conducted in a greenhouse with 12-h photoperiod and the temperature variation of 23–30°C. The experimental design was also a randomized block with three replications. Cotton seeds were sown in paper pots (6 cm in diameter and 10 cm in height, made up of newspaper and without bottom) filled with autoclaved substrate (vol/vol, vermiculite:sand = 6:4). The paper pots were placed on plastic trays. *Verticillium dahliae* Vd080 was cultured in Czapek liquid medium for 10 days. Spores were collected by filteration with 4 lays gauzes and diluted by sterilized distilled water to approximately $1.0 \times 10^7$ spores/ml. Seedlings were inoculated with spore suspension 18 or 22 days after sowing. The seedlings were inoculated by placing the paper pots onto a plate (10 cm in diameter) containing 20 ml of a spore suspension and incubating for 40 min; the pots were then returned to the plastic trays. Seedlings dipped in sterile water were used as the control. Each treatment with three replications(n = 3) had five pots, and each pot contained three to five plants. The susceptible line-Jimian 11 acted as a susceptible control for estimation of RDI. The cotton seedlings generally showed up symptoms 7 days after inoculation. At 18 and 25 days after inoculation, the level of severity was recorded. The disease rating scale was as follows: 0 = healthy plants, no symptoms on leaves; 1 = one or two cotyledons showing symptoms and no symptoms on true leaves; 2 = both cotyledons and one true leaf showing symptoms; 3 = both cotyledons and two true leaves showing symptoms; 4 = all of the leaves showing symptoms, symptomatic leaves dropped, the apical meristem was necrotic or the plant died. The disease index (DI) and relative disease index (RDI) were calculated according to the same method as those for the disease nursery in the field above.

## SSR genotyping

Genomic DNA was isolated individually from each of the 158 cotton accessions, starting from fresh leaf tissues and using the CTAB method [31]. A total of 1482 SSRs on the AD-genome wide Reference Map (http://www.cottongen.org/tools/cmap/viewer), which evenly distributed on 26 chromosomes in cotton,

were screened and those SSRs showing polymorphism among the 158 cotton accessions were retained for genotyping. Another 17 SSRs and 2 RGAP markers showing polymorphism but without chromosome location information were also included in the analysis (Table S2).Since not all the SSRs on the AD-genome wide Reference Map showed polymorphism among the 158 cotton accessions, we only got 212 informative markers which were used for genotyping. The chromosome locations of these SSR markers and positions of each locus(Table S2)were obtained from the AD-genome wide Reference Map (http://www.cottongen.org/tools/cmap/viewer) and previous studies [32–41]. For each marker, only the clear major bands were recorded and each band was corresponding to an allele. Each SSR locus was scored with "1" for one band, "2" for another band, and "3" for the third band, etc., which distinguished different alleles. The occasional non-amplification or missing data state was scored with "−9" or "?", depending on the software requirement. To avoid assigning incorrect allelic relationships,the following criteria were used: (i) alleles were regarded as belonging to the same locus if they showed an obvious codominant relationship from their segregation patterns among different lines; (ii) when amplicons of the alleles were very close but different in molecular size, they were considered allelic; and (iii) ampicons that were not judged their allelic relationship with the above criteria were regarded as novel loci. Most SSRs were considered codominant markers. For dominant SSRs showing only one band in some lines and no band in other lines, the present state was regarded as a allele and scored with "1", and the absent state as another allele and scored with "2". Markers were analysed by PCR and 6% polyacrylamide gel electrophoresis (PAGE). PCR runs were performed 35 cycles of 45 s at 94_°C, at the annealing temperature for 45 s and 72_°C for 90 s, and a final extension step at 72_°C for 10 min. For each SSR locus, alleles were scored in ascending order according to the amplified fragment size.

## Statistical analysis

**Genetic diversity.** The cotton germplasm used in this study were strictly self-pollinated during the past decades for germplasm renewing. Preliminary analysis conducted on this panel using 83 simple sequence repeats (SSR) markers (not shown) showed only a couple of heterozygous loci in a few individuals. Therefore, the individuals in the present study were assumed to be homozygous. The number of alleles, gene diversity, and polymorphism information content (PIC) were estimated using the PowerMarker version 3.25 [42]. The differences of allele richness between different samples were compared using the rarefraction method in the HP-RARE package [43]. The significance of different statistics including gene diversity, PIC and allelic richness was assessed using Wilcoxon's paired test across loci.

**Population structure and differentiation analyses.** The model-based (Bayesian) cluster software STRUCTURE 2.2 [44] was chosen to estimate the population structure of the 158 cotton accessions and assign accessions to groups or subgroups with the 212 molecular markers which distributed across all cotton chromosomes. For structure analysis, each individual was coded using a two-row format: $(x_j^{i,\,1}, x_j^{i,\,2})$, which represents the genotype of individual i at locus j as described by Pritchard et al. [44]. We ran STRUCTURE under the 'admixture model' with a burn-in period of 10 000 followed by 100 000 replications of Markov Chain Monte Carlo. Five independent runs each were performed with the number of clusters (K) varying from 1 to 15. An ad hoc measure $\Delta K$ based on the relative rate of change in the likelihood of the data between successive K values were used to determine the optimal number of clusters [45]. That run with the maximum likelihood was adopted to divide the cotton accessions into different groups with the membership probabilities threshold of 0.60 as well as the maximum membership probability among groups. Those accessions with less than 0.60 membership probabilities were retained in the admixed group. The inferred groups were further subdivided into subgroups using a similar methodology. Because the pedigree information of many cotton accessions was unknown, the classification of the accessions was largely based on the STRUCTURE results. No a priori population information was used.

The unrooted neighbor-joining(N-J) tree was applied using the software Powermarker 3.25 under the Nei 1983 model [42] to investigate the tridimensional structure of elite cotton germplasm accessions. Using inferred groups and subgroups, genetic differentiation within and among predefined groups and pairwise Fst genetic distances were measured by molecular variance analysis (AMOVA) using ARLEQUIN2.0 [46], with 1,000 permutations and sum of squared size differences as molecular distance.

**Relative kinship.** Pairwise kinship estimates were calculated by constructing relative kinship matrix according to Hardy and Vekemans [47] using the software SPAGeDi. The kinship matrix compared the identity by descent (IBD) among all pairs of the 158 cotton accessions genotyped using 212 markers, by adjusting the probability of identity by state between two individuals with the average probability of identity by state between random individuals. All negative kinship values between individuals were set to zero [48].

**Linkage disequilibrium.** LD was estimated by calculating the square value of correlation coefficient ($r^2$) between all pairs of markers with the software package TASSEL 2.1[49]. Only marker loci with minor allele frequency values above 0.05 and having at least 80% successful calls among the sample set were included further for LD analyses. P-values for each $r^2$ estimate were obtained with a two-sided Fisher's exact test as implemented in TASSEL. Each pair of loci was categorized as unlinked (marker loci located on different chromosomes) or linked (marker loci located on the same chromosome).The LD was estimated for global, linked and unlined markers, respectively. The LD values between all pairs of marker loci were plotted as triangle LD plots using TASSEL to estimate the general view of genome-wide LD patterns and evaluate 'block-like' LD structures. To display the change in LD as a function of genetic distance, the position information of linked markers was acquired according to position references (Table S2), and only when the position information of linked markers came from the same position reference was the genetic distance calculated. The $r^2$ value corresponding to the genetic distance was acquired by running the software. LD plots against genetic map distance were generated in Microsoft Excel, where only $r^2$ values with P<0.001 were included. The $r^2$ value for marker distance of 0 cM was assumed to be 1 as previously described [50]. A curve was drawn to describe the trend of LD decay using the nonlinear regression model, which revealed an overall correlation between the genetic distance of markers on the same chromosome and LD.

**Association analysis.** Since most lines in the cotton panel have no or very weak kinship, the general linear model (GLM) was performed to calculate the marker-trait association using the TASSEL 2.1 software package [49]. In GLM model, association was estimated by using the percentages of admixture of each accession (Q matrix) as covariates to take population structure into account, thus avoided the detection of spurious associations. The Q matrix was created by programme STRUCTURE 2.2. The significant associations were compared with published literature information to judge obtained associations. The 5% 'minor alleles'

filtered SSR datasets were used for this association mapping models.

## Results

### Phenotypic analysis of Verticillium wilt resistance

Each line was rated for Verticillium wilt resistance in the disease nursery and in the greenhouse, respectively. The DI was calculated and adjusted into RDI with a correction factor K. The RDI of all the lines ranged from10.10 to 76.60, with an average of 38.23 in the Verticillium wilt nursery environment and from 17.01 to 72.63, with an average of 38.51 in the greenhouse environment, and the coefficient of variation (CV) was 0.34 and 0.29, respectively. ANOVA showed that there were not significant differences for the estimated RDI between two environments (no significant environment effect), but there were significant differences (P<0.001) for RDI among the 158 lines (Table 1). The frequency distribution of RDI showed a continuous variation (Figure 1).

### Population structure and kinship in the panel of 158 cotton accessions

Population structure inference for the panel of 158 cotton accessions was performed by using a model-based software STRUCTURE and 212 molecular markers. The structure analysis was performed by setting the number of clusters ($K$) from 1 to 15 with five replications for each $K$. The LnP(D) value increased continuously with $K$ from 1 to 15 (Figure 2A), and the highest $\Delta K$ value was observed at $K=2$ followed by a drastic decline of $\Delta K$ from $K=3$ (Figure 2B). Accordingly, the total panel could be assigned into two main groups, designated as G1 and G2, respectively. Using a probability of membership threshold of 60%, 58 lines were assigned to G1, 73 lines to G2 and the remaining 27 lines were considered as intermediates (Figure 3, Table S3). G1 is consisted of 19 American variety accessions, 34 Chinese variety accessions, 3 Former Soviet Union variety accessions and 2 Africa variety accessions. G2 is consisted of 15 American variety accessions, 55 Chinese variety accessions, 1 Africa variety accessions, 1 Pakistan variety accession and 1 Former Soviet Union variety accession. The intermediates are consisted of 8 American variety accessions, 17 Chinese variety accessions, 1 Auatralia variety accession and 1 French variety accession. The two groups inferred from structure did not show an association with the geographic origin of the materials, reflecting the probable extensive exchange of parental lines by breeders worldwide. The tree-based analyses gave very similar results as the STRUCTURE analysis (Figure S1).

**Table 1.** ANOVA of Verticillium wilt resistance ratings (indicated by RDI) in two environments, Verticillium wilt nursery environment and greenhouse environment.

| Source | DF | Sum of squares | Mean square | F value | P value |
|---|---|---|---|---|---|
| Environment | 1 | 37.722 | 37.722 | 0.439 | 0.509 |
| Genotype | 157 | 33243.084 | 211.739 | 2.463 | <0.0001 |
| Error | 157 | 13499.436 | 85.984 | | |
| Total | 315 | 46780.242 | | | |

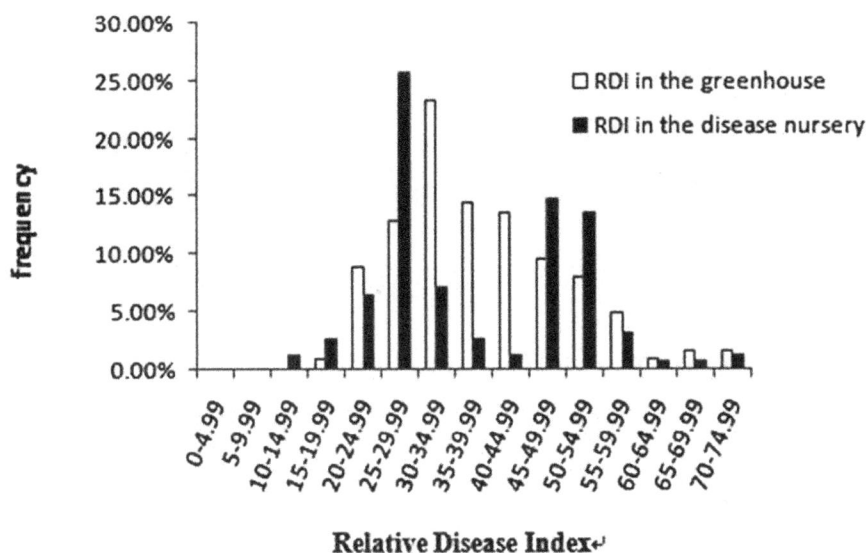

**Figure 1. The frequency distribution of Verticillium wilt resistance ratings (indicated by RDI) of 158 accessions in the disease nursery and in the greenhouse environment.**

The continuous increase of LnP(D) after $K=2$ and a small peak of $\Delta K$ at $K=7$ (Figure 2C) implied there are subtle sub-structures within the two inferred groups. Therefore, we performed further independent simulations within each of the two groups. For the G1 group, two sharp peaks of $\Delta K$ appeared at $K=3$ and $K=10$, respectively (Figure 2E), but most lines were assigned to intermediates (membership probabilities less than 60%) at $K=10$. Accordingly, $K=3$ was regarded as ideal number of cluster for the G1 group. For the G2 group, a sharp peaks of $\Delta K$ appeared at $K=4$, implying four subgroups were included (Figure 2G). The G1 group was classified into three subgroups, named as G1a, G1b and G1c, and a mixed subgroup including 13 lines. G1a contained 12 lines, 8 of which were breeding lines from China, and the remaining 4 lines were from USA. G1b contained 7 lines, which were representative of five lines collected from north early maturity cotton area in China. G1c contained 26 lines, most of which collected from abroad (America, Former Soviet Union, and Africa). The G2 group included four subgroups, named as G2a, G2b, G2c, and G2d, and a mixed subgroup including 11 lines. G2a contained 10 lines, 4 of which were lines from China, and the remaining 6 lines were from USA. G2b contained 36 lines, which originated from USA(3), Africa(1), and China(32). G2c contained 5 lines, among which there were four innovative lines created by Atomic energy mutation of one America-originated breeding line, Arcot-1. G2d contained 11 lines, which were representative of Xingtai79-11, a breeding line from Huanghe River valley area in China and Sulian8908, a breeding line from Former Soviet Union (Figure 3, Table S1, Table S4).

The pairwise kinship estimates based on 212 informative molecular markers showed that the majority of the pairs of cotton accessions(53.67%) had zero estimated kinship values, while 18.34% kinship estimates ranged from 0 to 0.05, 13.51% from 0.05 to 0.1, and 10.96% of the pairs had a value 0.1–0.20. The remaining pairs of accessions (3.52%) had >0.25 kinship values, suggesting involvement of some common parental genotypes in the breeding history of these germplasm groups (Figure 4). These results indicated that most lines in the panel have no or very weak kinship, which might be attributed to the broad range collection of genotypes and the exclusion of similar genotypes before analysis.

## Population differentiation

The genetic diversity within and among predefined groups (G1a–c,G2a–d) was estimated by AMOVA test. AMOVA results indicated that 4.46% ($P<0.05$) of the total molecular variation in the panel was attributed to the differentiation between groups, and 17.08% ($P<0.01$) was attributed to the differentiation among subgroups (Table 2). Although differentiation among subgroups was highly significant ($P<0.0001$), 76.95% of total genetic variance was attributed to the difference within subgroups and 1.51% within individuals. Pairwise $F_{st}$ of the two inferred groups was 0.10320 ($P<0.001$), suggesting that G1 is significantly divergent from G2. The levels of differentiation between subgroups were variable with $F_{st}$ ranging from 0.15315 (G2a with G2b, $P<0.001$) to 0.57518 (G1b with G2c, $P<0.001$) (Table 3). The differentiation among subgroups in the G1 group was lower than that in the G2 group. Among subgroups in the G2 subgroup, the differentiation between G2c with G2d was the highest ($F_{st}=0.40944$, $P<0.001$) while the difference between G2a and G2b was the lowest ($F_{st}=0.15315$, $P<0.001$).

## Genetic diversity within groups and subgroups

The allele number, allelic richness, gene diversity and PIC were calculated to estimate the genetic diversity in the total panel, each inferred group and subgroup. The total number of alleles in the total panel was 480, with an average of 2.26 alleles per locus. The gene diversity, PIC, and allele richness were 0.34 (0.0126–0.7159), 0.28 (0.0125–0.6643), and 2.26 (2.00–4.00), respectively (Table 4). The G1 and G2 groups contained 465 and 447 alleles, with 2.18 and 2.09 alleles per locus, respectively. Though the sample size of G2 was larger than that of G1, G1 had a higher level of gene diversity ($z=-5.260$, $P<0.001$), PIC ($z=-5.484$, $P<0.001$) and allele richness ($z=-2.910$, $P=0.004$). Within G1, subgroup G1c and G1b contained the highest (average of 2.08) and lowest (average of 1.57) number of alleles per locus, respectively. Within G2, the number of alleles in each G2 subgroup ranged from 285 (G2c, 1.34 alleles per locus) to 418 (G2b, 1.97 alleles per locus). Group-specific alleles, gene diversity, PIC and allelic richness showed a similar trend in subgroups (Table 4).

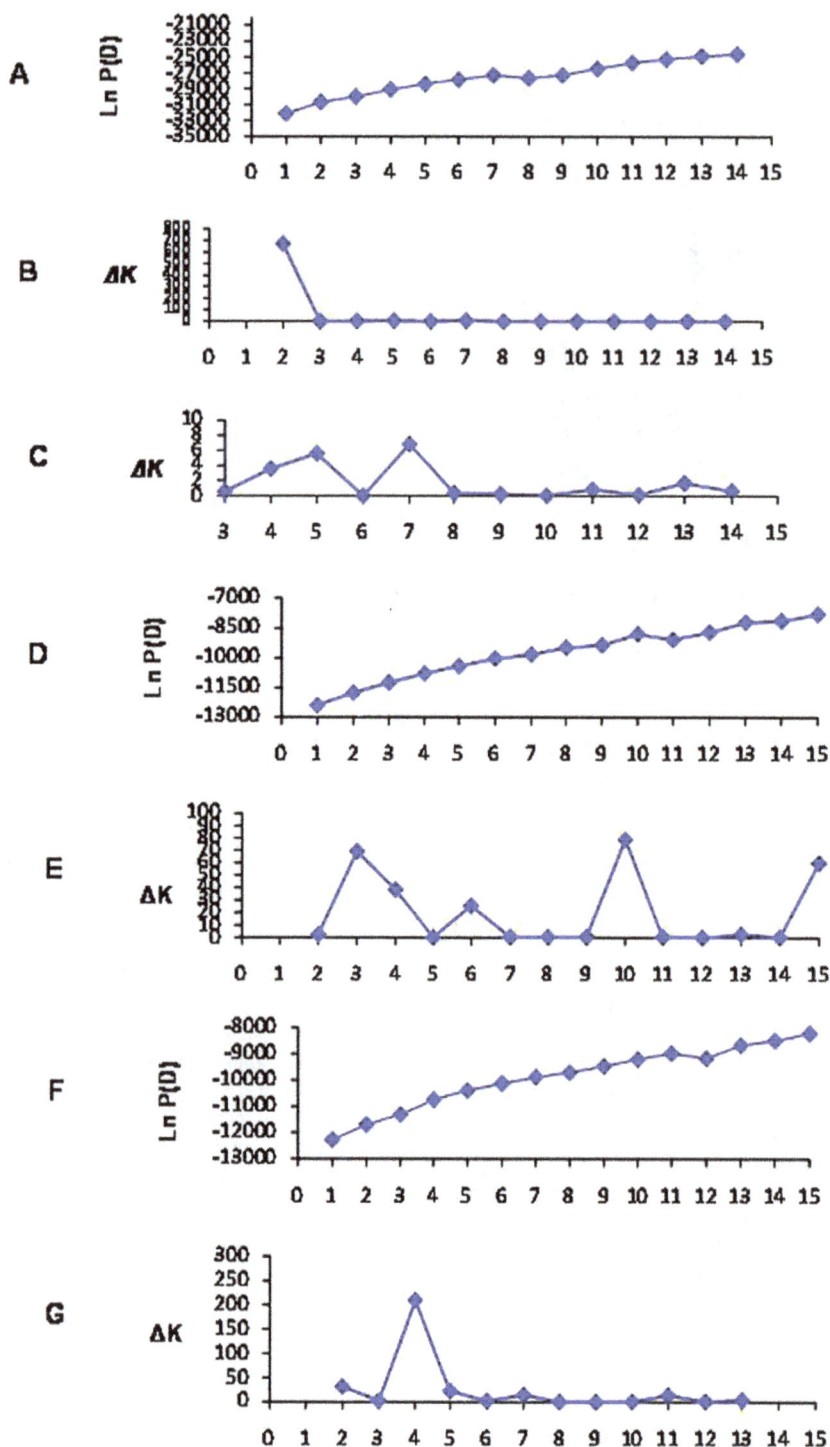

**Figure 2. The average LnP(D) and ΔK in the total panel and inferred groups. A–C** LnP(D) with k = 1–15, *ΔK* with k = 2–15, and *ΔK* with k = 3–15 for simulations using all 158 accessions; **D–E** LnP(D) with k = 1–15 and ΔK with k = 2–15 for inferred G1 group; **F–G** LnP(D) with k = 1–15 and *ΔK* with k = 2–15 for inferred G2 group.

## Pairwise linkage disequilibrium and LD decay in the whole genome level

As the 158 cotton accessions could be divided into two distinct groups or seven subgroups, pairwise LD estimates were performed in the total panel and in each group and subgroup using a total of 212 molecular markers. In the total panel, the average $r^2$ of global marker pairs was 0.0132, and only 1.83% of the total possible marker locus pairs were in significant LD ($P < 0.001$), suggesting that the LD level is very low in the panel (Table 5). Moreover, the average $r^2$ of linked marker pairs was 0.0362, and the percentage of linked marker pairs in significant LD ($P < 0.001$) was 8.33%, both of which were higher than those for unlinked marker pairs

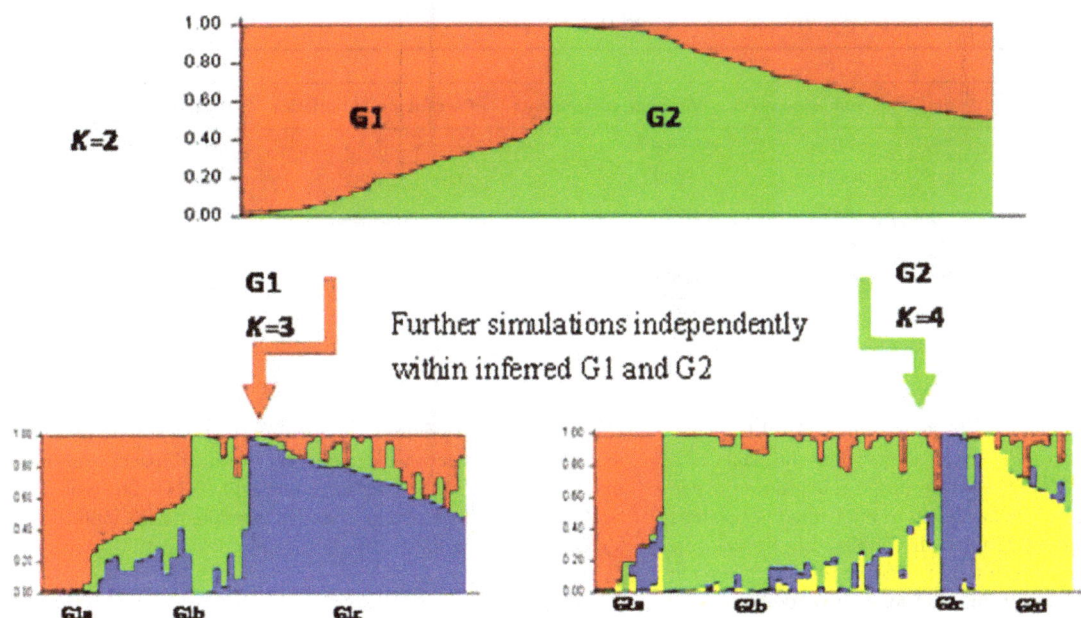

**Figure 3. Relationship between the inferred populations.** The two inferred clusters (k = 2) resulted from simulation using all 158 accessions in one and correspond to G1 and G2, respectively. Then three and four clusters (k = 3 and 4) were inferred within inferred G1 and G2 independently.

(0.0121 and 0.972%, respectively), demonstrating that physical linkage is predominant in determining LD compared with random forces [51]. For global marker pairs, the mean $r^2$ both in groups (ranging from 0.0213 to 0.0267) and in subgroups (ranging from 0.0354 to 0.1379) was larger than that in the total panel, suggesting that the LD level was elevated when the panel was classified into groups and subgroups (Table 5). Further analysis of the LD in all groups and subgroups showed that both average $r^2$ and proportion of significant LD for linked markers were still higher than those for unlinked markers, which reinforced the view that physical linkage strongly influences LD in this panel of inbred lines.

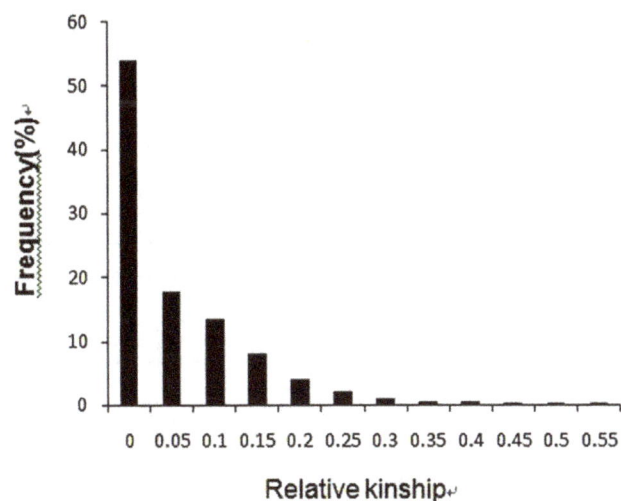

**Figure 4. Distribution of pairwise relative kinship estimates between 158 cotton accessions.** Values are from SPAGeDi estimates using 212 SSRs. For simplicity, only percentages of relative kinship estimates ranging from 0 to 0.50 are shown.

To explore the LD at the single chromosome level, we performed the same evaluations for the single chromosome as those for the whole genome, using SSR markers evenly distributing in the single chromosome. In this part, only five chromosomes (Chr.11, 16, 18, 19 and 23) were choosed for analyzing, which representing the concentrated distribution of Verticillium wilt resistance related QTLs reported by previous studies [8–11]. In order to get enough marker pairs for LD estimation at a single chromosome level, we set the significance of $r^2$ value with P<0.05. In the total panel, the percentage of mean locus pairs in significant LD was 19.25% ranging from 14.29% to 26.46%, and mean $r^2$ was 0.03165 ranging from 0.0221 to 0.0442. The Chr.16 showed a relative higher mean $r^2$ values and highest percentage of linked marker pairs in significant LD in these five chromosomes. What's more, both in the group G1 and G2, we detected the highest mean $r^2$ values (0.0494 and 0.0613, respectively) and most SSR locus pairs in significant LD (8.13% and 15.83% of marker pairs, respectively) in Chr.16 in the five chromosomes, implying the stronger linkage in this chromosome. At the individual chromosome level, for Chr.11, 18 and 23, the G1 group had more SSR locus pairs in significant LD and higher mean $r^2$ values than the G2 group. While for Chr.16, the G1 group had less SSR locus pairs in significant LD and lower mean $r^2$ values than the G2 group (Table 6).

Triangle plots for pairwise LD between SSR markers revealed significant LD block, or so called a genome-wide LD decay in the genome-wide LD analysis (Figure S2). However, we only estimated the LD decay distances on Chr.11, 16, 19 and 23 because of the similar informative marker coverage on these chromosomes (Table 7). In this study, $r^2$ threshold of 0.1 was adopted according to previous study [52]. We found that for the same gene pool, different chromosomes showed a very big change in LD decay distances. For example, in the total panel, LD decay distance was 1–2 cM for Chr.16 and 19, but 5–10 cM for Chr.23 and 15–20 cM for Chr.11. In the G1 group, LD decay distance was 10–15 cM for Chr.19 and 20–25 cM for Chr.23, but >50 cM for Chr.16 and >100 cM for Chr.11. At the same time, for the same

**Table 2.** Analysis of molecular variance (AMOVA) among inferred populations.

| Source of variation | df | Sum of squares | Variance components | Percentage of variation | P-value |
|---|---|---|---|---|---|
| Among groups[a] | 1 | 398.569 | 1.3784 | 4.46 | 0.03715±0.00557 |
| Among populations[b] | 5 | 889.309 | 5.27178 | 17.08 | <0.0001 |
| Within populations | 96 | 4605.758 | 23.75531 | 76.95 | <0.0001 |
| Within individuals | 103 | 48 | 0.46602 | 1.51 | <0.0001 |
| Total | 205 | 5941.636 | 30.87151 | | |

[a]Groups were defined by two inferred groups.
[b]Populations were defined by inferred subgroups.

chromosome, different gene pools showed different LD decay distance. For each of the four chromosomes, the total panel had faster LD decay than each of the two groups. A much slower decay of LD within groups might be attributed to the limited population size and narrow genetic background that inhibit LD decay [53]. These general descriptions of LD decay distance provide important information concerning decisions on marker densities for future association analyses at the chromosome level.

### Marker loci associated with Verticillium wilt resistance

Associations between 212 marker loci and Verticillium wilt resistance were determined by GLM method. The Significant (P< 0.05) candidate markers associated with disease resistance were detected in the two environments (Table 8). A total of 42 marker loci with the $R^2$ ranged of 2.84–10.93% were identified to be significantly associated with Verticillium wilt resistance in at least one environment, six each on chromosome (Chr.) 11 and 16, four on Chr.19, three each on Chr.8, 17 and 23, two each on Chr.5 and 20, one each on Chr.1, 3, 4, 9, 15, 21 and 26, and the other six without location and position information. Of them, 15 and 32 markers were significantly associated with the RDI from the greenhouse environment and the disease nursery environment, respectively. Five markers, including NAU3828 on Chr.5, DPL0222 on Chr.9, BNL1606 on Chr.17, NAU3574 on Chr.20 and BNL3649 on Chr.21, were significant for both environments. Ten marker loci, significantly associated with Verticillium wilt resistance, including NAU3828 on Chr.5 at 24.1 cM, NAU3201 on Chr.8 at 72 cM, BNL3255 on Chr.8 at 76.5 cM, BNL2441 on Chr.16 at 76.567 cM, NAU2627 on Chr.16 at 62.301 cM, BNL3319 on Chr.16 at 57.702 cM, NAU2887 on Chr.16 at 60.867 cM, NAU5120 on Chr.16 at 47.7 cM, JESPR274 on

Chr.23 at 52.972 cM and NAU1047 on Chr.23 at 97.1 cM, were found to be consistent with previously identified association of the marker loci from QTL mapping analyses[8,10,11]. The remaining SSRs are new unreported markers revealing associations with Verticillium wilt resistance in this set of cotton germplasm.

### Discussion

#### Genetic diversity in the cotton panel

In this study, across the entire population, we observed an average number of alleles per locus of 2.26 which ranged from 2 to 4, a gene diversity of 0.34 and a PIC of 0.28 (Table 4). These values were similar to those detected in 53 *Gossypium hirsutum* L. cotton cultivars [54] and 8 cotton (*G. hirsutum*) cultivars [55], reflecting a relatively low genetic diversity in *Gossypium hirsutum* L. cotton cultivars worldwide. However, the average number of alleles per locus, gene diversity and PIC of our study are less than those detected in 47 accessions including 38 *G. hirsutum*, 2 *G. darwinii*, 2 *G. tomentosum* and 5 *G. barbadense* [56] and 35 cultivars and eight inbred lines of *G. hirsutum* L. from Africa, United States and Brazil [57]. Although the number of alleles in this study was lower than that detected in 97 cultivars and primitive species [58], the PIC values in the two studies were very similar. This can be explained by that the level of polymorphism among races and wild species of *Gossypium* was significantly higher than that within cultivated *G.hirsutum*, and cultivars domesticated directly in a native cotton growing area usually reserved their higher level of polymorphism than those in a non-native cotton growing area.

Though the genetic diversity is relatively low in the entire panel, the group G1 and G2 showed a significant differences in gene diversity and PIC. When compared with G2, G1 had a higher

**Table 3.** Fst among seven subgroups.

| Groups | Subgroups | G1 | | | G2 | | | |
|---|---|---|---|---|---|---|---|---|
| | | G1a | G1b | G1c | G2a | G2b | G2c | G2d |
| G1 | G1a | | | | | | | |
| | G1b | 0.32395** | | | | | | |
| | G1c | 0.15490** | 0.20997** | | | | | |
| G2 | G2a | 0.24737** | 0.39020** | 0.19033** | | | | |
| | G2b | 0.17797** | 0.28545** | 0.15914** | 0.15315** | | | |
| | G2c | 0.38511** | 0.57518** | 0.33934** | 0.35653** | 0.28013** | | |
| | G2d | 0.27666** | 0.42016** | 0.25889** | 0.24459** | 0.17229** | 0.40944** | |

**Significant at P<0.001.

**Table 4.** Summary of genetic diversity for overall panel, groups and subgroups.

| Items | overall | G1 | | | | | G2 | | | | | | Mixed |
| | | G1 overall | G1a | G1b | G1c | G1 mixed | G2 overall | G2a | G2b | G2c | G2d | G2 mixed | |
|---|---|---|---|---|---|---|---|---|---|---|---|---|---|
| sample size | 158 | 58 | 12 | 7 | 26 | 13 | 73 | 10 | 36 | 5 | 11 | 11 | 27 |
| Alleles | 480 | 465 | 385 | 333 | 441 | 423 | 447 | 368 | 418 | 285 | 353 | 377 | 443 |
| Alleles per locus | 2.26 | 2.19 | 1.82 | 1.57 | 2.08 | 2 | 2.11 | 1.74 | 1.97 | 1.34 | 1.67 | 1.78 | 2.09 |
| Gene diversity | 0.34 | 0.35 | 0.27 | 0.19 | 0.34 | 0.33 | 0.3 | 0.24 | 0.28 | 0.12 | 0.19 | 0.26 | 0.33 |
| PIC | 0.28 | 0.28 | 0.22 | 0.16 | 0.28 | 0.27 | 0.24 | 0.2 | 0.23 | 0.09 | 0.16 | 0.21 | 0.27 |
| Allelic richness | 2.26 | 2.19 | 1.82 | 1.57 | 2.08 | 2 | 2.11 | 1.74 | 1.97 | 1.34 | 1.67 | 1.78 | 2.09 |
| Group-specific alleles | 480 | 18 | 4 | 1 | 12 | 5 | 6 | 1 | 2 | 1 | 0 | 3 | 4 |

**Note:** Groups G1 and G2 were classified based on the results of STRUCTURE analysis of the 158 cotton lines.
The G1 group were further partitioned into G1a, G1b and G1c subgroups, and the G2 group into G2a, G2b, G2c and G2d subgroups.
The intermediates in the total panel, G1 group and G2 group were named as "Mixed", "G1 mixed" and "G2 mixed", respectively.

level of gene diversity and PIC. Significant differentiation assessed in AMOVA was observed between the two groups. G1 group has 18 group-specific alleles while G2 group 6 group-specific alleles (Table 4). The reason might be due to that the G1 group in this study contained more lines from abroad than G2. In the subgroup level, we detected a a significant difference of allele richness between China originated lines (G1b) and abroad originated lines (G1c). This can be explained by the fact that China is not a native cotton growing area, and most cotton varieties planted in China were derived from a few sources of germplasm such as DPL, Stoneville, King, Uganda, Foster, and Trice, all of which were introduced from abroad [59].

## Population structure and differentiation in the association panel

Detailed knowledge about population structure in an association panel is important to avoid spurious associations. A model-based approach using the software STRUCTURE[44,60] might be the most frequently used method to correct spurious associations. It is computationally difficult to obtain accurate estimates of the number of populations($K$). Generally, $K$ is taken to be the value with the highest estimated LnP(D) value returned by STRUC-

TURE [44]. However, in real data the value of LnP(D) continues to increase with increasing K. In this situation, an ad hoc measure $\Delta K$ based on the relative rate of change in the likelihood of the data between successive $K$ values were used to determine the optimal number of clusters [45]. In this study, the $\Delta K$ values indicated dividing the cotton panel into two groups and seven subgroups was the most biologically meaningful population structure. Our results are very similar to a cluster analysis for 53 *Gossypium hirsutum* L. cotton cultivars, which were grouped into two large groups and seven subgroups [54]. And 35 cultivars and eight inbred lines of *G. hirsutum* L. also were identified as four groups that consisted of American cultivars and inbred lines, African and Brazilian cultivars, BRS Brazilian cultivars and FM Brazilian cultivars by a structure running [57]. Also, 285 exotic *Gossypium hirsutum* accessions were classified into three groups consisted of landrace stock germplasm group, Mexican varieties and African varieties and 334 *G. hirsutum* variety accessions were identified as three groups consisted of Uzbekistan, Latin American, and Australian cotton accessions in Uzbek cotton germplasm collection [29,28]. These results demonstrated the existence of population structure in cotton germplasm of *G. hirsutum* worldwide.

**Table 5.** LD in the entire panel, groups and subgroups at the whole genome level.

| Groups[a] | Global[c] | | Unlinked[d] | | Linked[e] | |
| | $r^2$ | Significant LD (%)[f] | $r^2$ | Significant LD (%)[f] | $r^2$ | Significant LD (%)[f] |
|---|---|---|---|---|---|---|
| G1 overall[a] | 0.0267 | 0.76 | 0.0257 | 0.58 | 0.0469 | 4.47 |
| G1a[b] | 0.1379 | 0.05 | 0.1365 | 0.03 | 0.1706 | 0.41 |
| G1c | 0.0506 | 0.27 | 0.0498 | 0.17 | 0.067 | 2.22 |
| G2 overall | 0.0213 | 0.94 | 0.0202 | 0.71 | 0.0459 | 6.47 |
| G2b | 0.0354 | 0.35 | 0.0343 | 0.21 | 0.0593 | 3.34 |
| Total | 0.0132 | 1.83 | 0.0121 | 0.972 | 0.0362 | 8.33 |

[a]Groups G1 and G2 were classified based on the results of STRUCTURE analysis of the 158 cotton lines.
[b]The G1 group were further partitioned into G1a, G1b and G1c subgroups, and the G2 group into G2a, G2b, G2c and G2d subgroups. But the G1b, G2a, G2c and G2d subgroups were not included in the analysis due to their small population size.
[c]The whole set of marker pairs, including linked and unlinked markers pairs.
[d]Pairs of markers from different chromosomes.
[e]Pairs of markers on the same chromosome.
[f]Significant threshold is set to P<0.001, which determine whether pairwise LD estimate is significant statistically.

**Table 6.** LD in the entire panel, groups and subgroups at single chromosome level.

| Chr. | No. of loci | Overall[a] | | G1[c] | | | | | | G2 | | | |
| | | r² | Significant(%)[b] | Overall | | G1a[d] | | G1c | | Overall | | G2b | |
| | | | | r² | Significant(%) | r² | Significant(%) | r² | Significant(%) | r² | Significant(%) | r² | Significant(%) |
| 11 | 21 | 0.0442 | 16.5 | 0.0461 | 12.42 | 0.1241 | 3.64 | 0.0643 | 7.5 | 0.043 | 9.09 | 0.0546 | 9.26 |
| 16 | 25 | 0.0362 | 26.46 | 0.0494 | 16.27 | 0.1582 | 5.88 | 0.0699 | 6.88 | 0.0613 | 25.83 | 0.0581 | 11.67 |
| 18 | 9 | 0.02212 | 14.29 | 0.0465 | 25 | 0.1807 | 4.76 | 0.0901 | 11.54 | 0.0208 | 20 | 0.0439 | 6.67 |
| 19 | 24 | 0.0286 | 22.5 | 0.0319 | 9.8 | 0.1613 | 12.62 | 0.0581 | 3.95 | 0.0327 | 6.67 | 0.0554 | 8.57 |
| 23 | 17 | 0.02712 | 16.48 | 0.0471 | 16.48 | 0.1277 | 3.85 | 0.0779 | 6.06 | 0.0365 | 15.38 | 0.0499 | 10 |
| Mean | 19.2 | 0.031648 | 19.246 | 0.0442 | 15.994 | 0.1504 | 6.15 | 0.07206 | 7.186 | 0.03886 | 15.394 | 0.05238 | 9.234 |

[a]The total panel for the 158 cotton lines.
[b]Significant threshold is set to P<0.05, which determine whether pairwise LD estimate is significant statistically.
[c]Groups G1 and G2 were classified based on the results of STRUCTURE analysis of the 158 cotton lines.
[d]The G1 group were further partitioned into G1a, G1b and G1c subgroups, and the G2 group into G2a, G2b, G2c and G2d subgroups. But the G1b, G2a, G2c and G2d subgroups were not included in the analysis due to their small population size.

**Table 7.** Average LD decay distance(cM) in different chromosomes in the total panel, G1 and G2 groups for locus pairs with r²>0.1 at P<0.05.

| Chr. | Overall[a] | G1 | G2 |
| --- | --- | --- | --- |
| 11 | 15–20 | >100 | —[b] |
| 16 | 1–2 | >50 | 40–50 |
| 19 | 1–2 | 10–15 | — |
| 23 | 5–10 | 20–25 | 10–15 |

[a]The total panel for the 158 cotton lines
[b]The short horizontal line means that only a few marker pairs were in significant LD that a regression curve was not created to estimate the LD decay.

Several studies had showed that the genetic structure of *G. hirsutum* L. is in accordance with their geographical origins [55,28,29,57]. But in our study, it was interesting to note that in each of the two groups there were germplasm lines from several origins (China, America, Africa and former Soviet Union), indicating the exchange and domestication of germplasm between these origins. In the level of subgroups, we only detected a limited association between the subgroup structure and the geographic origin of the materials, for example, all the lines in G1b originated from China, and most lines in G2c containing the pedigree of Arcot-1, one America-originated breeding line. But other subgroups consisted of accessions derived both China and abroad. So, most of the elite *G. hirsutum* variety accessions in China had the close pedigree relationships with some exotic variety accessions (especially accessions derived from America). This result was consistent with the report by Cheng and Du (2006), who considered that many of the Chinese breeding source germplasms had been based on the introduction, selection, and domestication of germplasms from other countries thus narrowing their genetic base and possibly making them vulnerable to the present and future diseases [59]. Iqbal et al. (2001) also pointed out that the *G. hirsutum* cultivated around the world is derived from the USA, which were exported to other countries in the 19th and early twentieth century, with most upland cotton used in early Chinese cotton breeding coming from this source [61]. Though the apparent lack of diversity in cultivated *G. hirsutum*, Van Esbroeck and Bowman (1998) have argued that there is enough allelic variation, mutation or recombination in crosses between closely related individuals to allow improvement in agronomic performance and/or that the coefficient of parentage may not reflect the real genetic distance [62]. In our study we observed that all the accessions in the subgroup G2c had good fiber qualities with a fiber length of >30 mm and a fiber strength of >30 cN/tex (unpublished), which would effectively improve the fiber quality by justifying crosses between these accessions and other related individuals in cotton cultivar breeding programs.

Our current results showed the significant differentiation among groups and subgroups in our association panel. Pairwise Fst showed that G1group is significantly divergent from G2 with the fact that G1 contained more lines from abroad and had more group-specific alleles than G2. The highest differentiation between subgroups occurred in the G1b with G2c. This can be explained by all the lines in G1b originated from China and most of lines in G2c owned the pedigree of Arcot-1, one America-originated breeding line, implying the highly differentiation between China-originated and America-originated breeding lines. In G1, pairwise Fst values indicate that G1b was strongly differentiated from G1a

**Table 8.** Marker loci significantly associated with Verticillium wilt resistance and their positions on chromosomes (Chr).

| Marker name | Chr. | Position(cM) | greenhouse P value[a] | greenhouse Rsq_Marker[b] | disease nursery P value | disease nursery Rsq_Marker |
|---|---|---|---|---|---|---|
| BNL2599 | 1 | 1.633 | 0.0221 | 0.0597 | NS | |
| NAU5233 | 3 | 108 | NS | | 0.034 | 0.0287 |
| NAU3592 | 4 | 119.269 | | | 0.0057 | 0.0488 |
| NAU3828 | 5 | 24.1 | 0.0282 | 0.0387 | 6.89E-04 | 0.0727 |
| NAU3212 | 5 | 66 | NS | | 0.0441 | 0.0531 |
| BNL3255 | 8 | 76.5 | 0.0224 | 0.0411 | NS | |
| NAU3201 | 8 | 38.4 | 0.0113 | 0.0524 | NS | |
| NAU3499 | 8 | 65.3 | 0.0037 | 0.0871 | NS | |
| DPL0222 | 9 | 137.829 | 0.0339 | 0.0371 | 0.0014 | 0.0665 |
| NAU3074 | 11 | 183.689 | NS | | 0.0308 | 0.0303 |
| CIR196 | 11 | 145.826 | NS | | 0.0064 | 0.0684 |
| NAU980 | 11 | 169.5 | NS | | 0.0078 | 0.045 |
| NAU5428 | 11 | 32.076 | NS | | 6.86E-04 | 0.1093 |
| BNL1034 | 11 | 184.577 | NS | | 0.0017 | 0.0805 |
| NAU5064 | 11 | 162.6 | NS | | 0.0143 | 0.0543 |
| BNL2646 | 15 | 48.8 | NS | | 0.0067 | 0.0463 |
| BNL2441 | 16 | 76.567 | NS | | 0.0017 | 0.067 |
| NAU2627 | 16 | 62.301 | 0.0349 | 0.037 | NS | |
| BNL3319 | 16 | 57.702 | NS | | 0.0011 | 0.067 |
| TMB1114 | 16 | 41.815 | NS | | 8.39E-04 | 0.0695 |
| NAU2887 | 16 | 60.867 | NS | | 5.66E-04 | 0.0755 |
| NAU5120 | 16 | 47.7 | NS | | 0.0209 | 0.0341 |
| BNL1606 | 17 | 51.762 | 0.0101 | 0.0526 | 0.0098 | 0.0427 |
| NAU2859 | 17 | 86.286 | NS | | 0.0184 | 0.065 |
| JESPR101 | 17 | 71.031 | NS | | 4.50E-04 | 0.0956 |
| BNL4069 | 19 | 36.8 | 0.009 | 0.0619 | NS | |
| JESPR0001 | 19 | 123.567 | 0.0477 | 0.0479 | NS | |
| CIR364 | 19 | 66.663 | NS | | 0.0037 | 0.0708 |
| NAU2894 | 19 | 26.581 | NS | | 0.0184 | 0.0357 |
| BNL3646 | 20 | 3.479 | 0.0465 | 0.0317 | NS | |
| NAU3574 | 20 | 58.598 | 0.0411 | 0.0506 | 0.0443 | 0.0399 |
| BNL3649 | 21 | 10.8 | 0.0076 | 0.0566 | 0.0338 | 0.0291 |
| NAU2954 | 23 | 114.846 | 0.0093 | 0.0542 | NS | |
| JESPR274 | 23 | 52.972 | NS | | 0.017 | 0.0874 |
| NAU1047 | 23 | 97.1 | NS | | 0.0173 | 0.0362 |
| NAU4912 | 26 | | 0.0226 | 0.0594 | NS | |
| NAU5463 | — | — | NS | | 0.009 | 0.0432 |
| Gh268 | — | — | NS | | 0.0066 | 0.0466 |
| Gh454 | — | — | NS | | 0.006 | 0.0509 |
| NAU3563 | — | — | NS | | 0.0353 | 0.0284 |
| w11330 | — | — | NS | | 0.0065 | 0.065 |
| 73686-3 | — | — | NS | | 0.0313 | 0.0461 |

[a]NS, not statistically significant;
[b]Rsq_marker, total explained phenotypic variation.

and G1c. In G2, pairwise Fst values indicated that G2c was highly unrelated with G2a, G2b and G2d. These results suggested that subgroup having a single origin or pedigree was usually apt to differentiate with those having a mixed origin. A few lines in some subgroups were not consistent with pedigree information perfectly, maybe due to the unknown pedigree information in our study.

## Patterns of linkage disequilibrium in the cotton panel

The genome-wide distribution of LD estimated with a high number of markers greatly influence the resolution of association mapping [63]. In this study, we observed that a total of 8.33%, 4.47%, and 6.47% of the linked loci pairs in the entire panel, G1 group and G2 group, respectively, showed significant LD (P< 0.001) (Table 5). The percentages observed in our study were lower than those reported earlier [28,29]. This can be explained by that different significance thresholds and different plant materials were used in these studies.

Physical linkage that determines LD between molecular marker and causative polymorphisms is the genetic basis for association mapping of genes or QTLs underlying traits of interest [16]. In this study, the extent of LD of linked markers in the entire panel, groups and subgroups is significantly higher than that of unlinked markers (Table 5), suggesting that physical linkage strongly influences LD in this cotton panel, and indicating that this cotton panel is suitable for association analysis. Triangle plots for pairwise LD between SSR markers revealed significant LD decay in the genome-wide LD analysis (Figure S2). As a supplementary, we analyzed the LD delay distance in the cotton panel in a whole genome scale (Figure S3), and found that the LD in the cotton panel decayed to the background level within 10–15 cM in a whole genome scale. If we set the $r^2$ threshold of 0.2, genome-wide LD fast decayed within 1–3 cM (Figure S3). Therefore, in our association panel, the LD decayed faster than that in 335 variety accessions of *G. hirsutum* from Uzbek cotton germplasm collection, which indicated that a genome-wide average of LD extended up to genetic distance of 25 cM at $r^2 \geq 0.1$ and reduced to ~5–6 cM at $r^2 \geq 0.2$ [28]. This can be explained by that the genetic diversity in our association panel (overall PIC for SSRs was in the range of 0.0125–0.06643 with an average of 0.28) was higher than that in 335 variety accessions of *G. hirsutum* from Uzbek cotton germplasm collection (overall PIC for SSRs was in the range of 0.006–0.50 with an average of 0.082). On the other hand, the LD in our panel delayed slower than that in 208 landrace stocks of G. *hirsutum*, which indicated that LD clearly decays within the genetic distance of <10 cM with $r^2 \geq 0.1$ and reduced to ~1–2 cM at r2≥0.2 [29]. This difference can be explained by that landraces, usually having a higher genetic diversity, often showed faster LD decay than modern varieties [64].

Population structure is one of several important factors that have strong influences on LD [16]. In our LD estimations, we took into account the effect of population structure by subdividing the total panel into different groups and subgroups. Various levels of LD in groups and subgroups were observed, indicating that population structure has significant impact on LD (Table 5). Based on LD analyses both in the whole genome level and at the individual chromosome level, the LD level was elevated when the panel was classified into groups and subgroups (Table 5 and Table 6), implying that variable extents of LD are expected within the different genetic groups and highlight the fact that different marker densities will be required if association studies are planned in the different genetic groups.

So far, there is no report about the LD at the single chromosome level in cotton. Our results showed that for the same gene pool, different chromosomes showed a notable change in LD decay distances. In the total panel, Chr.11 and 23 showed wider LD decay distance than Chr.16 and 19, indicating that Chr.11 and 23 may carry more QTLs or genes related to important agronomic traits that were strongly selected in breeding [65]. In G1 and G2 gene pools, the Chr.16 showed higher mean $r^2$ values and wider LD decay distance than other chromosomes, implying the stronger linkage in this chromosome. In fact, both our study and previous study [11] proved the existence of QTL clusters for Verticillium wilt resistanc on Chr.16, which was consistent with the strong linkage in this chromosome.

## Verticillium wilt resistance associated markers and QTL identification

In this study, two hundred and twelve genome-wide distributed markers were employed in the association study (Table S2). Of more than 60 previously identified Verticillium wilt resistance QTLs [8–11], only ten were confirmed to be consistent with them. Unlike previous study in bi-parental populations [11], which showed that 41QTLs related to Verticillium wilt resistance intensively distributed on chromosomes D9(Chr.23) and D7(Chr.16), our study identified 42 associations widely distributed on 15 chromosomes(Table 8).This implied that association mapping can locate many QTLs over the entire genome since the mapping population includes a large number of diversified entries of germplasm, while conventional QTL mapping based on bi-parental populations only identified fewer QTLs which be located in a limited area in the genome where the two parents differ, thus causing QTL clustering [11].

On Chr.1, 3, 4, 9, 15, 21and 26, we identified one Verticillium wilt resistance associated marker from each chromosome. These marker loci were regarded as novel Verticillium wilt resistance QTLs that have not been reported, and it was the first findings for Chr.1, 3, 4, 9, 15 and 21 to exist Verticillium wilt resistance related QTLs on them. What's more, DPL0222 on Chr.9 and BNL3649 on Chr.21 were considered stable Verticillium wilt resistance QTLs that showed significant association with Verticillium wilt resistance both in the greenhouse environment and in the disease nursery environment.

On Chr.5 we identified two markers, NAU3212 and NAU3828. NAU3212 was located at 66 cM according to Guo et al.[33] and was regarded as a novel Verticillium wilt resistance QTL that has not been reported. NAU3828 was located at 24.1 cM and overlapped with the Verticillium wilt resistance QTL $qVL$-$A5$-$2BC_1S_2592$ found by Yang et al. [8]. What's more, this marker showed significant association at P<0.05 level in the greenhouse environment and strong association at P<0.0001 level in the disease nursery environment. So, the marker NAU3828 was regarded as a stable QTL in different environments and co-localized with previously identified Verticillium wilt resistance QTLs.

On Chr.8 our study identified NAU3201, NAU3499 and BNL3255. NAU3201 located at 38.4 cM and overlapped with the Verticillium wilt resistance QTL $qVL$-$A8$-$1F_2$ found by Yang et al. [8]. BNL3255 at 76.5 cM had been reported near the QTL $qVV$-$A8$-$1BC_1S_2BP2$ between NAU3964 at 70.7 cM and NAU920 at 129.1 cM [8]. NAU3499 at 65.3 cM was a novel Verticillium wilt resistance QTL that has not been reported.

On Chr.11 we identified five significant marker-trait associations, which were different from three loci having large effect on resistance to Verticillium wilt reported by Bolek et al.[9]. Of them, NAU5428 showed strong association with Verticillium wilt resistance at P<0.0001 level and explained the most phenotypic variation (10.93%), thus might be a new major QTL for Verticillium wilt resistance that need to be further identified.

On Chr.16, a report indicated that there existed QTL clusters with high contribution rates for Verticillium wilt resistance on this chromosome [11]. Of six identified markers in our study, TMB1114 was a novel Verticillium wilt resistance QTL that has not been reported, and other five were deduced to be co-localized with previously identified Verticillium wilt resistance QTLs. BNL2441 at 76.567 cM was overlapped with q8.24-2 reported by Wang et al.[10] and close to the QTL $qV$-$BP2M$-$D7$-$1$ (58.5~72.2) by Jiang et al.[11]. NAU2627 at 62.301 cM was within the QTL $qV$-$BP2M$-$D7$-$1$ (58.5~72.2 cM) and $qV$-$VD8M$-$D7$-$2$ (60.9~67.2 cM) reported by Jiang et al. [11]. BNL3319 at 57.702 cM was within the QTL $qV$-$VD8M$-$D7$-$1$ (52.9~60.9 cM) and close to the QTL $qV$-$BP2M$-$D7$-$1$ (58.5~72.2 cM) reported by Jiang et al. [11]. NAU2887 at 60.867 cM was within the QTL $qV$-$BP2M$-$D7$-$1$ (58.5~72.2 cM) and $qV$-$VD8M$-$D7$-$1$ (52.9~60.9 cM), and close to the QTL $qV$-$VD8M$-$D7$-$2$ (60.9~67.2 cM). NAU5120 at 47.7 was within the QTL $qV$-$BP2S1$-$D7$-$1$ (39.9~50.1) and $qV$-$T9M$-$D7$-$1$ (42.5~71.9) reported by Jiang et al. [11], and close to the $qV$-$VD8M$-$D7$-$1$ (52.9~60.9 cM). Therefore, our study further proved the QTL clusters on Chr.16. These QTL clusters confirmed by multiple studies strongly suggests a reliable location harboring Verticillium wilt resistance QTL.

On Chr.23, three markers NAU2954, JESPR274 and NAU1047 were identified. NAU2954 was regarded as a novel Verticillium wilt resistance QTL that has not been reported. JESPR274 at 52.972 cM, explaining relatively high phenotypic variation (8.74%), was close to the QTL $qV$-$VD8S2$-$D9$-$3$ (47.5~51.5) reported by Jiang et al. [11]. NAU1047 was one of the flanking markers for each of $qV$-$T9S1$-$D9$-$1$, $qV$-$MIXS2$-$D9$-$1$ and $qV$-$MIXM$-$D9$-$3$ reported by Jiang et al. [11].

The remain identified markers in our study included three on Chr.17, four on Chr.19, two on Chr.20, and six without location and position information. All these marker loci were considered novel Verticillium wilt resistance QTLs that have not been reported. Of them, BNL1606 on Chr.17 and NAU3574 on Chr.20 were considered stable QTLs that showed significant association with Verticillium wilt resistance both in the greenhouse environment and in the disease nursery environment.

Totally, in our study, ten SSR markers were colocalized with or close to previously identified Verticillium wilt resistance QTLs using conventional QTL mapping approaches. This suggests that association mapping using natural population can effectively detect major QTLs. Moreover, most SSR loci (32 of 42) were considered novel Verticillium wilt resistance QTLs that have not been reported, implying that association mapping has the advantage of being able to work with a higher number of polymorphic markers than conventional QTL mapping and locates many QTLs over the entire genome.

## Conclusion

Two groups and seven subgroups were identified in the cotton panel, demonstrating the existence of population structure in cotton germplasm of G. hirsutum worldwide. The two subgroups inferred from structure did not show an association with the geographic origin of the materials, reflecting the probable extensive exchange of parental lines by breeders worldwide. In the subgroup level, subgroup having a single origin or pedigree was usually apt to differentiate with those having a mixed origin. This fact suggested that it is a prerequisite to perform structure analysis before association mapping. Both in the whole genome level and at the individual chromosome level, the LD level was elevated when the panel was classified into groups and subgroups, highlighting the fact that different marker densities will be required if association studies are planned in the different genetic groups. For the same gene pool, different chromosomes showed a very big change in LD decay distances, indicating that different chromosomes may carry different QTLs or genes related to important agronomic traits that were strongly selected in breeding. Association mapping based on the disease nursery and greenhouse environment identified 42 marker loci associated with Verticillium wilt resistance, which widely distributed on 15 chromosomes, implying that association mapping can locate many QTLs over the entire genome. 10 marker loci were found to be consistent with previously identified QTLs, which suggests that association mapping using natural population can effectively detect major QTLs. 32 loci were new unreported markers related with Verticillium wilt resistance, implying that association mapping has the advantage of being able to work with a higher number of polymorphic markers than conventional QTL mapping and locates many QTLs over the entire genome. QTL clusters for Verticillium wilt resistanc on Chr.16 were proved by our study, which was consistant with the strong linkage in this chromosome.

## Supporting Information

**Figure S1**  Unrooted neighbor-joining tree for 158 accessions. The ancestries of the accessions in inferred populations are represented by different colours.

**Figure S2**  The triangle LD plot for a pairwise genome-wide LD between SSR loci (with a 5% minor allele filtered datasets). Polymorphic SSR sites are plotted on both X-axis and Y-axis. Each cell represents the comparison of two pairs of SSR sites with the color codes for the presence of significant LD. Colored bare code for the significance threshold levels is given.

**Figure S3**  LD decays ($r^2$) in the association panel consisting 158 cotton lines. The $r^2$ value for marker distance of 0 cM is defined as 1. The dots are $r^2$ values for linked marker pairs in significant LD ($P<0.001$). The curve was drawn across the dots using the nonlinear regression model. The horizontal line indicates the thresholds of $r^2 = 0.1$ and $r^2 = 0.2$, respectively.

**Table S1**  Accesion or cultivar, origin, subspecies, type and usage of the materials, and groups of 158 accessions. The varieties were sorted according to their STRUCTURE membership probability as Figure 3. [a] means that the innovation line was created by interspecific hybridization; [b] means that the innovation line was created by induced mutagenesis; [c] Subgroups defined by STRUCTURE.

**Table S2**  Microsatellite markers used in the study, including the location of the markers in the cotton reference map and position based on the position references. [a] The chromosome locations were based on the the AD-genome wide Reference Map. [b] The positon information was based on position references. [c] Position references for the information of the SSR: A. Yu JZ; Kohel RJ; Fang DD; Cho J; Van Deynze A; Ulloa M; Hoffman SM; Pepper E; Stelly DM; Jenkins JN; Saha S; Kumpatla SP; Shah MR; Hugie WV; Percy RG. A High-Density Simple Sequence Repeat and

Single Nucleotide Polymorphism Genetic Map of the Tetraploid Cotton Genome. G3: Genes, Genomes, Genetics Mission 2012 2(3): 43–58. B. Guo W; Cai C; Wang C; Han Z; Song X; Wang K; Niu X; Lu K; Shi B; Zhang T. A microsatellite-based, gene-rich linkage map reveals genome structure, function and evolution in gossypium. Genetics 2007 176(1): 527–541. C. Yu Y; Yuan D; Liang S; Li X; Wang X; Lin Z; Zhang X. Genome structure of cotton revealed by a genome-wide SSR genetic map constructed from a BC1 population between gossypium hirsutum and G. barbadense. BMC Genomics 2011 12:15. D. Shen X; Guo W; Lu Q; Zhu X; Yuan Y; Zhang T. Genetic mapping of quantitative trait loci for fiber quality and yield trait by RIL approach in Upland cotton. Euphytica 2007 155 371–380. E. Lacape JM, Jacobs J, Arioli T, Derijcker R, Forestier-Chiron N, Llewellyn D, Jean J, Thomas E, Viot C. A new interspecific, Gossypium hirsutum x G. barbadense, RIL population: towards a unified consensus linkage map of tetraploid cotton.Theor Appl Genet. 2009 119(2):281–92. F. Yu JW; Yu SX; Lu CR; Wang W; Fan SL; Song MZ; Lin ZX; Zhang XL; Zhang JF; Wu W. High-density Linkage Map of Cultivated Allotetraploid Cotton Based on SSR, TRAP, SRAP and AFLP Markers. Journal of Integrative Plant Biology. 2007 49: 716–724. H. Qin H; Guo W; Zhang YM; Zhang T. QTL mapping of yield and fiber traits based on a four-way cross population in Gossypium hirsutum L. Theor Appl Genet 2008 117:883–894. N. Ma XX; Zhou BL; L YH; Guo WZ; Zhang TZ. Simple Sequence Repeat Genetic Linkage Maps of A-genome Diploid Cotton (Gossypium arboreum). Journal of Integrative Plant Biology 2008 50: 491–502. P. Wang HM; Lin ZX; Zhang XL; Chen W; Guo XP; Nie YC; Li YH. Mapping and Quantitative Trait Loci Analysis of Verticillium Wilt Resistance Genes in Cotton. Journal of Integrative Plant Biology. 2007 50:174–182. R. Guo WZ; Cai CP; Wang CB; Zhao L; Wang L; Zhang TZ. A preliminary analysis of genome structure and composition in Gossypium hirsutum. BMC Genomics 2008 9: 314. — The locus was not assigned to the chromosomes of known maps. The SSR marker E38644, W11330, 73686–3 and 6738–1 were EST-SSRs developed from sequences by RNA-Seq for a cotton sample innoculated with Verticillium dahlia (unpublished). [e] The marker 1DF and 1EF are RGAP markers.

**Table S3**  Proportional memberships in groups as defined by Structure.

**Table S4**  List of accessions with their proportional memberships in model-based subgroups.

## Acknowledgments

The authors thank the research group led by Dr. Heqin Zhu for supplying the technical support in identifying disease resistance.

## Author Contributions

Conceived and designed the experiments: HMW YLZ. Performed the experiments: YLZ. Analyzed the data: YLZ. Contributed reagents/materials/analysis tools: WC YHL. Wrote the paper: YLZ.

## References

1. Zhang JF, Percy RG (2007) Improving Upland cotton by introducing desirable genes from Pima cotton. World Cotton Res Conf. http://wcrc.confex.com/wcrc/2007/techprogram/P1901.HTM.

2. He DH, Lin ZX, Zhang XL, Nie YZ, Guo XP, et al. (2005) Mapping QTLs of traits contributing to yield and analysis of genetic effects in tetraploid cotton. Euphytica 144(1-2): 141–149.

3. Wang B, Guo W, Zhu X, Wu Y, Huang N, et al. (2007) QTL mapping of yield and yield components for elite hybrid derived-RILs in upland cotton. J Genet Genomics 34: 35–45.

4. Qin H, Guo W, Zhang Y, Zhang T (2008) QTL mapping of yield and fiber traits based on a four-way cross population in Gossypium hirsutum L. Theor Appl Genet117: 883–894.

5. Yu JW, Zhang K, Li SY, Yu SX, Zhai HH, et al. (2013) Mapping quantitative trait loci for lint yield and fiber quality across environments in a *Gossypium hirsutum* × *Gossypium barbadense* backcross inbred line population. Theor Appl Genet 126:275–287.

6. Zhang T, Yuan Y, Yu J, Guo W, Kohel RJ (2003) Molecular tagging of a major QTL for fiber strength in Upland cotton and its marker-assisted selection. Theor Appl Genet 106: 262–268.

7. Lacape JM, Llewellyn D, Jacobs J, Arioli T, Becker D, et al. (2010) Meta-analysis of cotton fiber quality QTLs across diverse environments in a *Gossypium hirsutum* × *G. barbadense* RIL population. BMC Plant Biol 10:132

8. Yang C, Guo WZ, Li GY, Gao F, Lin SS, et al. (2008) QTLs mapping for Verticillium wilt resistance at seedling and maturity stages in *Gossypium barbadense* L. Plant Science174:290-298.

9. Bolek Y, El-Zik KM, Pepper AE, Bell AA, Magill CW, et al. (2005) Mapping of verticillium wilt resistance genes in cotton. Plant Science 168:1581–1590.

10. Wang HM, Lin ZX, Zhang XL, Chen W, Guo XP, et al. (2008) Mapping and Quantitative Trait Loci Analysis of Wilt Resistance Genes in Cotton. Journal of Integrative Plant Biology,50(2): 174–182.

11. Jiang F, Zhao J, Zhou L, Guo WZ, Zhang TZ (2009) Molecular mapping of Verticillium wilt resistance QTL clustered on chromosomes D7 and D9 in upland cotton. Science in China Series C: Life Sciences 52(9): 872–884.

12. Zhu C, Gore M, Buckler ES, Yu J (2008) Status and prospects of association mapping in plants. Plant Genome 1:5–20.

13. Ersoz ES, Yu J, Buckler ES (2007) Applications of linkage disequilibrium and association mapping in crop plants. Genomics-assisted crop improvement Springer Dordrecht 97–120

14. Yan J, Kandianis CB, Harjes CE, Bai L, Kim E-H, et al. (2010) Rare genetic variation at Zea mays crtRB1 increases b-carotene in maize grain. Nat Genet 42:322–327.

15. Hartl DL, Clark AG (1997) Principles of population genetics. Sinauer Associates Sunderland, MA.

16. Flint-Garcia SA, Thornsberry JM, Buckler SE (2003) Structure of linkage disequilibrium in plants. Ann Rev Plant Biol 54:357–374.

17. Yan J, Shah T, Warburton ML, Buckler ES, McMullen MD, et al. (2009) Genetic characterization and linkage disequilibrium estimation of a global maize collection using SNP markers. PLoS One 4:e8451.

18. Liu A, Burke JM (2006) Patterns of nucleotide diversity in wild and cultivated sunflower. Genetics 173:321–330.

19. Lijavetzky D, Cabezas JA, Ibanez A, Rodriguez V, Martinez-Zapater JM (2007) High throughput SNP discovery and genotyping in grapevine (Vitis vinifera L.) by combining a re-sequencing approach and SNPlex technology. BMC Genomics 8:424

20. Nordborg M, Borevitz JO, Bergelson J, Berry CC, Chory J, et al. (2002) The extent of linkage disequilibrium in Arabidopsis thaliana. Nat Genet 30:190–193.

21. Caldwell KS, Russell J, Langridge P, Powell W (2006) Extreme population dependent linkage disequilibrium detected in an inbreeding plant species, Hordeum vulgare. Genetics 172:557–567.

22. McNally KL, Childs KL, Bohnert R, Davidson RM, Zhao K, et al. (2009) Genomewide SNP variation reveals relationships among landraces and modern varieties of rice. Proc Natl Acad Sci USA 106:12273–12278.

23. Huang X, Wei X, Sang T, Zhao Q, Feng Q, et al. (2010) Genomewide association studies of 14 agronomic traits in rice landraces. Nat Genet 42:961–967.

24. Lam H, Xu X, Liu X, Chen W, Yang G, et al. (2010) Resequencing of 31 wild and cultivated soybean genomes identifies patterns of genetic diversity and selection. Nat Genet 42:1053–1059.

25. Morrell PL, Toleno DM, Lundy KE, Clegg MT (2005) Low levels of linkage disequilibrium in wild barley (Hordeum vulgare ssp. spontaneum) despite high rates of self-fertilization. Proc Natl Acad Sci USA 102:2442–2447.

26. Song B-H, Windsor AJ, Schmid KJ, Ramos-Onsins S, Schranz ME, et al. (2009) Multilocus patterns of nucleotide diversity, population structure and linkage disequilibrium in Boechera stricta, a wild relative of Arabidopsis. Genetics 181:1021–1033.

27. Chen ZJ, Scheffler BE, Dennis E, Triplett BA, Zhang TZ, et al. (2007) Toward sequencing cotton (*Gossypium*) genomes. Plant Physiol 145:1303–1310.

28. Abdurakhmonov IY, Saha S, Jenkins JN, Buriev ZT, Shermatov SE, et al. (2009) Linkage disequilibrium based association mapping of fiber quality traits in *G. hirsutum* L. variety germplasm. Genetica 136:401–417.

29. Abdurakhmonov IY, Kohel RJ, Yu JZ, Pepper AE, Abdullaev AA, et al. (2008) Molecular diversity and association mapping of fiber quality traits in exotic

30. *G. hirsutum* L. germplasm. Genomics 92: 478–487.

31. Zhang TZ, Qian N, Zhu XF, Chen H, Wang S, et al. (2013) Variations and Transmission of QTL Alleles for Yield and Fiber Qualities in Upland Cotton Cultivars Developed in China. PLoS One 8(2) : e57220.

32. Paterson AH, Brubaker CL, Wendel JF (1993) A rapid method for extraction of cotton (Gossypium spp.) genomic DNA suitable for RFLP or PCR analysis. Plant Molecular Biology Reporter 11:122–127.

33. Yu JZ, Kohel RJ, Fang DD, Cho J, Van Deynze A, et al. (2012) A High-Density Simple Sequence Repeat and Single Nucleotide Polymorphism Genetic Map of the Tetraploid Cotton Genome. G3: Genes, Genomes, Genetics Mission 2(3): 43–58.

34. Guo W, Cai C, Wang C, Han Z, Song X, et al. (2007) A microsatellite-based, gene-rich linkage map reveals genome structure, function and evolution in gossypium. Genetics 176(1): 527–541.

35. Yu Y, Yuan D, Liang S, Li X, Wang X, et al. (2011) Genome structure of cotton revealed by a genome-wide SSR genetic map constructed from a BC1 population between gossypium hirsutum and G. barbadense. BMC Genomics 12:15.

36. Shen X, Guo W, Lu Q, Zhu X, Yuan Y, et al. (2007) Genetic mapping of quantitative trait loci for fiber quality and yield trait by RIL approach in Upland cotton. Euphytica 155: 371–380.

37. Lacape JM, Jacobs J, Arioli T, Derijcker R, Forestier-Chiron N, et al. (2009) A new interspecific, *Gossypium hirsutum* × *G. barbadense*, RIL population: towards a unified consensus linkage map of tetraploid cotton. Theor Appl Genet 119(2):281–92.

38. Yu JW, Yu SX, Lu CR, Wang W, Fan SL, et al. (2007) High-density Linkage Map of Cultivated Allotetraploid Cotton Based on SSR, TRAP, SRAP and AFLP Markers. Journal of Integrative Plant Biology 49: 716–724.

39. Qin H, Guo W, Zhang YM, Zhang T (2008) QTL mapping of yield and fiber traits based on a four-way cross population in *Gossypium hirsutum* L 117:883–894.

40. Ma XX, Zhou BL, Lü YH, Guo WZ, Zhang TZ (2008) Simple Sequence Repeat Genetic Linkage Maps of A-genome Diploid Cotton (Gossypium arboreum). Journal of Integrative Plant Biology 50: 491–502.

41. Wang HM, Lin ZX, Zhang XL, Chen W, Guo XP, et al. (2007) Mapping and Quantitative Trait Loci Analysis of Verticillium Wilt Resistance Genes in Cotton. Journal of Integrative Plant Biology 50:174–182.

42. Guo W, Cai C, Wang C, Han Z, Song X, et al. (2007) A microsatellite-based, gene-rich linkage map reveals genome structure, function and evolution in gossypium. Genetics 176: 527–541.

43. Liu KJ, Muse SV (2005) PowerMarker: an integrated analysis environment for genetic marker analysis. Bioinformatics 21:2128–2129.

44. Kalinowski S (2005) HP-RARE 1.0: a computer program for performing rarefaction on measures of allelic richness. Mol Ecol Notes 5:187–189.

45. Pritchard JK, Stephens M, Donnelly P (2000) Inference of population structure using multi-locus genotype data. Genetics 155:945–959.

46. Evanno G, Regnaut S, Goudet J (2005) Detecting the number of clusters of individuals using the software STRUCTURE: a simulation study. Molecular Ecology 14: 2611–2620.

47. Schneider S, Roessll D, Excoffier L (2000) ARLEQUIN: a software for population genetics data analysis, version 2.0. Genetics and Biometry Laboratory, Department of Anthropology Geneva, Switzerland University of Geneva.

48. Hardy OJ, Vekemans X (2002) SpaGeDi: a versatile computer program to analyze spatial genetic structure at the individual or population levels. Mol Ecol Notes 2: 618–620.

49. Yu J, Pressoir G, Briggs WH, Bi IV, Yamasaki M, et al. (2006) A unified mixed-model method for association mapping that accounts for multiple levels of relatedness. Nat Genet 38:203–208.

50. Bradbury PJ, Zhang Z, Kroon DE, Casstevens TM, Ramdoss Y, et al. (2007) TASSEL: software for association mapping of complex traits in diverse samples. Bioinformatics 23:2633–2635.

51. Yan J, Shah T, Warburton ML, Buckler ES, McMullen MD, et al. (2009) Genetic characterization and linkage disequilibrium estimation of a global maize collection using SNP markers. PLoS One 4:e8451

52. Flint-Garcia SA, Thornsberry JM, Buckler ES (2003) Structure of linkage disequilibrium in plants. Ann Rev Plant Biol 54:357–374.

53. Witt SR, Buckler ES (2003) Using natural allelic diversity to evaluate gene function. Methods Mol Biol 236:123–139.

54. Ersoz ES, Yu J, Buckler ES (2007) Applications of linkage disequilibrium and association mapping in crop plants. Genomics-assisted crop improvement-SpringerDordrecht97–120.

55. Bertini CHCM, Schuster I, Sediyama T, Barros EG, Moreira MA (2006) Characterization and genetic diversity analysis of cotton cultivars using microsatellites. Genet Mol Biol 2: 321–329.

56. Rungis D, Llewellyn D, Dennis ES, Lyon BR (2005) Simple sequence repeat (SSR) markers reveal low levels of polymorphism between cotton (*Gossypium hirsutum* L.) cultivars. Australian Journal of Agricultural Research 56: 301–307.

57. Lacape JM, Dessauw D, Rajab M, Noyer JL, Hau B (2007) Microsatellite diversity in tetraploid *Gossypium* germplasm: assembling a highly informative genotyping set of cotton SSRs. Mol Breed 19:45–58.

58. Moiana LD, Filho PSV, Goncalves-Vidigal MC, Lacanallo GF, Galvan MC, et al. (2012) Genetic diversity and population structure of cotton (*Gossypium hirsutum* L. race *latifolium* H.) using microsatellite markers. African Journal of Biotechnology 11(54): 11640–11647.

59. Liu S, Cantrell RG, Mcarty JCJ, Stewart JM (2000). Simple sequence repeat-based assessment of genetic diversity in cotton race stock accessions. Crop Sci 40:1459–1469.

60. Chen G, Du XM (2006) Genetic Diversity of Source Germplasm of Upland Cotton in China as Determined by SSR Marker Analysis. Acta Genetica Sinica 33 (8): 733–745.

61. Falush D, Stephens M, Pritchard J (2003) Inference of Population Structure Using Multilocus Genotype Data: Linked Loci and Correlated Allele Frequencies. Genetics 164: 1567–1587.

62. Iqbal J, Reddy O, El-Zik Km, Pepper AEA (2001) Genetic bottleneck in the evolution under domestication of upland cotton *Gossypium hirsutum* L. examined using DNA fingerprinting. Theor Appl Genet 103:547–554.

63. Van-Esbroeck G, Bownam DT (1998). Cotton germplasm diversity and its importance to cultivar development. J Cotton Sci 2:121–129.

64. Stich B, Melchinger AE, Frisch M, Maurer HP, Heckenberger M, et al. (2005) Linkage disequilibrium in European elite maize germplasm investigated with SSRs. Theoretical and Applied Genetics 111:723–730.

65. Ha C, Wang L, Ge H, Dong Y, Zhang X (2011) Genetic diversity and linkage disequilibrium in Chinese bread wheat (*Triticum aestivum* L.) revealed by SSR markers. PLoS One6(2):e17279.

66. Zhang XY, Tong YP, You GX, Hao CY, Ge HM, et al. (2007) Hitchhiking effect mapping: A new approach for discovering agronomic important genes. Agricultural Sciences in China 6: 255–264.

# Quantification of the Pirimicarb Resistance Allele Frequency in Pooled Cotton Aphid (*Aphis gossypii* Glover) Samples by TaqMan SNP Genotyping Assay

**Yizhou Chen\*, Daniel R. Bogema, Idris M. Barchia, Grant A. Herron**

Elizabeth Macarthur Agricultural Institute, NSW Department of Primary Industries, Menangle, New South Wales, Australia

## Abstract

***Background:*** Pesticide resistance monitoring is a crucial part to achieving sustainable integrated pest management (IPM) in agricultural production systems. Monitoring of resistance in arthropod populations is initially performed by bioassay, a method that detects a phenotypic response to pesticides. Molecular diagnostic assays, offering speed and cost improvements, can be developed when the causative mutation for resistance has been identified. However, improvements to throughput are limited as genotyping methods cannot be accurately applied to pooled DNA. Quantifying an allele frequency from pooled DNA would allow faster and cheaper monitoring of pesticide resistance.

***Methodology/Principal Findings:*** We demonstrate a new method to quantify a resistance allele frequency (RAF) from pooled insects via TaqMan assay by using raw fluorescence data to calculate the transformed fluorescence ratio $k'$ at the inflexion point based on a four parameter sigmoid curve. Our results show that $k'$ is reproducible and highly correlated with RAF ($r > 0.99$). We also demonstrate that $k'$ has a non-linear relationship with RAF and that five standard points are sufficient to build a prediction model. Additionally, we identified a non-linear relationship between runs for $k'$, allowing the combination of samples across multiple runs in a single analysis.

***Conclusions/Significance:*** The transformed fluorescence ratio ($k'$) method can be used to monitor pesticide resistance in IPM and to accurately quantify allele frequency from pooled samples. We have determined that five standards (0.0, 0.2, 0.5, 0.8, and 1.0) are sufficient for accurate prediction and are statistically-equivalent to the 13 standard points used experimentally

**Editor:** Shu-Biao Wu, University of New England, Australia

**Funding:** This study is funded by the CRDC (DAN1203). The funders had no role in study design, data collection and analysis, decision to publish, or preparation of the manuscript.

**Competing Interests:** The authors have declared that no competing interests exist.

\* E-mail: yizhou.chen@dpi.nsw.gov.au

## Introduction

Insecticide resistance has long been a problem of agriculture but has risen in prominence since the introduction of synthetic organic insecticides in the 1950's [1]. While the use of toxins remains a fundamental method of pest control, resistance will continue to threaten sustainable agriculture. The threat remains despite the introduction of new transgenic cotton varieties and Integrated Pest Management (IPM). This is because transgenics, such as *Bt*-cotton, also rely on toxins and so expose pests to high selection for resistance. Furthermore, IPM systems favour the use of more selective compounds, thereby narrowing the range of chemicals used. One such compound is pirimicarb (Pirimor), an insecticide that is highly effective at killing aphids but not the desirable and beneficial predatory species associated with aphids [2].

Traditional monitoring of pesticide resistance in arthropods is performed by bioassay in which insects are exposed to insecticide and mortality is recorded at specific post-exposure interval(s) [3]. Resistance levels are determined from dose-response mortality data and expressed as $LC_{50}$ values, which are an estimate of the lethal concentration required to cause 50% mortality in the target

population tested [3]. Additionally, resistance can be monitored via a single diagnostic or discriminating dose but these are difficult to accurately set [4] and require the generation of significant base line data which, due to tedious laboratory process, can take weeks or months to produce.

Insecticide resistance can be behaviourally- or physiologically-based with the latter involving three distinct mechanisms: target site insensitivity, enhanced detoxification and reduced pesticide penetration [5]. With the recent advance of genomics it has been possible to study many possible target resistance genes often associated with the insect nervous system.

Examples of this include the point mutation in the *GABA* receptor conferring insecticide resistance in *Drosophila melanogaster* [6] and a mutation in the acetylcholinesterase gene causing pesticide resistance in a variety of insect species [7]. In other cases resistance is caused by detoxification linked to a single nucleotide mutation [8] or even single gene duplication or deletions [9]. However, only when these molecular mechanisms are identified can rapid molecular methods be developed, allowing more effective monitoring of pesticide resistance.

The cotton or melon aphid, *Aphis gossypii* Glover is a serious pest of many crop species including cotton, pumpkin, citrus and melons [10]. This species has developed resistance to multiple insecticides including the carbamate pirimicarb (Pirimor) and some specific organophosphates that has led to chemical control failures in Australian cotton production regions [11]. The causal mechanism of pirimicarb resistance in *A. gossypii* has been identified as target site mutation in the acetylcholinesterase gene [12,13]. A double nucleotide substitution (TCA → TT[T/C]) in *ACE1* causes the replacement of a serine with a phenylalanine (S431F) and has been confirmed to be the cause of the pirimicarb resistance seen in Australian field collections of *A. gossypii* associated with control failure [14]. In Australian cotton IPM, a PCR-RFLP assay has been used to monitor pirimicarb resistance in the field by individually genotyping 20–50 individual aphids [2]. However, individual genotyping by PCR-RFLP limits the number of sites that can be monitored as it is labour intensive and offers limited benefits over the traditional bioassay. It is critical to have cost effective methods to monitor resistance allele frequencies (RAF) in field populations to maintain successful IPM strategies.

An alternative method to individual aphid genotyping is to estimate allele frequency from pooled DNA using real-time PCR technology with allele-specific probes or allele-specific primers [15–18]. However these pooled DNA approaches are often designed for specific assays and, due to the complexity of non-specific binding or amplification, are not widely used.

Currently, the most widely used qPCR platform for the estimation of allele frequency from pooled DNA is the 5' nuclease assay. It utilizes TaqMan probes that possess a minor-groove binding (MGB) molecule and a fluorescent dye attached to the 3' and 5' ends, respectively. The 'gold standard' for this technique uses two probes with different reporter dyes, allowing the detection of both alleles. Quantification of allele frequency is achieved by using the threshold cycle ($C_t$) or crossing point (CP) to calculate allele ratios based on $2^{-\Delta Ct}$ [19,20]. However, significant variation can arise if the fluorescent probes differ significantly in their binding efficiency or if amplification efficiency varies between resistant and susceptible alleles. Yu *et al* [18] have used the normalized fluorescence ratio in the exponential phase of PCR with known premixed allele ratios and generated a linear regression from which an allele ratio can be estimated. However, this method suffers as the exponential phase of PCR is selected arbitrarily.

Here we have developed a simple method to estimate allele frequency using TaqMan assays. We show that by selecting a single, standard reference point RAF can be predicted from the ratio of the two fluorescence intensities. Additionally, we demonstrate that RAF is a function of the transformed fluorescence ratio (k') and that five standard-points are sufficient to develop the equation of prediction.

## Materials and Methods

### PCR Assay and Probe Design for S431F

The TaqMan SNP assay was designed based on the Genbank sequence (AF502802) using RealTimeDesign Software (Biosearch Technologies) with forward primer 5'-AACCAATATACT-CATGGGTAGTAACTC-3' and the reverse primer 5'-AACC-GCCGCATCTGCATT-3'. A dual-labeled probe, 5'-Quasar 670-CGAAGAGGGTTACTATTCAA-3'- BHQ2 for the susceptible allele was designed based on a known susceptible *A. gossypii* sequence for a strain known as 'Sonya'. Two dual-labeled probes were designed for previously-identified resistance alleles, probe 5'-Fam- CGAAGAGGGTTACTATTTTA-3'-BHQ1 matching the

allele identified in pirimicarb-resistant strain Adam and probe 5'-Fam-CGAAGAGGGTTACTAYTTCA-3'-BHQ1 for the allele identified in pirimicarb-resistant strain Togo. All primers and probes were synthesized by Biosearch Technologies Inc (Biosearch Technologies Inc, Novato USA).

### Predefined RAF with Plasmid DNA and Pooled Cotton Aphids

Fragments, 667 bp in size and containing the S431F mutation site, were amplified from the susceptible Sonya, and resistant Adam and Togo strains and cloned into the pCR4 vector (Invitrogen, USA) using RFLP genotyping primers. Plasmid DNA concentration was then measured by a Nanodrop 2000 (Nanodrop Technologies). To create a standard curve, a series of standards (T/S) with predefined RAF of 1.0, 0.95, 0.9, 0.8, 0.7, 0.6, 0.5, 0.4, 0.3, 0.2, 0.1, 0.05 and 0.0 were constructed by mixing plasmids containing the resistant Togo and susceptible Sonya alleles. A duplicate standard series (A/S) was made by mixing plasmids containing the Adam and Sonya alleles.

In addition to plasmid standards, a series of standards was prepared using susceptible and resistant aphids. Thirteen pools of 20 aphids were prepared with RAF of 1.0, 0.95, 0.9, 0.8, 0.7, 0.6, 0.5, 0.4, 0.3, 0.2, 0.1, 0.05 and 0.0. As an example, the pool for RAF 0.95 was constructed by extracting a tube containing 19 aphids from the resistant strain and 1 aphid from the susceptible strain.

### 2011/2012 Aphis Gossypii Field Collection

Methods for the collection, transport, culture and bioassay of *A. gossypii* samples have been described previously [11,14]. A total of 35 *A. gossypii* samples (or strains) collected from cotton producing farms across eastern Australia during the 2011/2012 season were genotyped individually by PCR-RFLP. Resistance allele frequencies were estimated by genotyping 20 individual aphids from each sample. Samples were further confirmed susceptible or resistant via bioassay using methods outlined in detail by Herron *et al* [11].

### DNA Extraction

*Aphis gossypii* DNA was extracted from pooled or individual aphids using Chelex –100 resin (BioRad, USA) as described in [14]. Briefly, individual or 200 pooled aphids were placed inside a 1.5 mL microcentrifuge tube containing 80 μL of 5% Chelex – 100 resin. The sample was thoroughly homogenized with a sterile micropestle and incubated first at 56°C for 30 min, then at 100°C for 5 min. The crude DNA sample was then used for real-time PCR or PCR-RFLP or stored at −20°C for future use.

### Individual Genotyping of S431F by PCR-RFLP

RFLP genotyping of the S431F mutation has been described previously [14]. Briefly, a 667 bp fragment containing the mutation was amplified with forward primer 5'- CAAGCCAT-CATGGAATCAGG-3' and reverse primer 5'-TCATCAC-CATGCATCACACC-3'. The PCR product was digested by restriction endonuclease *SspI* by adding 5 units of enzyme and *SspI* buffer (1×) to a completed PCR for 3 hours at 37°C. The resultant PCR-RFLP profile was visualized by agarose gel electrophoresis. The pirimicarb-susceptible allele shows a single intense band at 336 bp (digested by *SspI*), whereas the pirimicarb-resistant allele shows a single intense band at 667 bp.

### Real-time PCR with TaqMan Assay

PCRs contained 400 nmol forward primer and reverse primer, 200 nmol susceptible and resistant probe, in a 1×TaqMan

Universal PCR Master Mix (Applied Biosystems, USA) comprising a total 25 µl reaction volume. Each sample was set up in triplicate and one negative control sample was included in each run. Real-time PCR was performed in an ABI7500 Real-Time PCR System (Applied Biosystems, Foster City, CA, USA) with 10 min at 95°C followed by 47 cycles of 15 s at 95°C and 1 min at 60°C.

## Data Analysis

Sigmoid 4 parameter curve fitting statistical analysis was carried out with GENSTAT release 10 software [21] using nonlinear regression and linear regression functions.

## Principle of Quantification

Real-time PCR quantification is measured as the incremental change in signal ($\Delta Rn$) that is directly proportional to the amount of amplicons produced at any cycle [18,22] and is defined as follows:

$$[amplicon]_{synthesized} = \Delta R_n / \Delta \varphi \qquad (1)$$

Where $\Delta \varphi$ represents the difference between the specific fluorescence of the free fluorophore and the specific fluorescence of the probe-bound fluorophore.

The synthesized amplicon is determined by the initial template number copy ($N_0$), the number of cycles ($n$) and the amplification efficiency ($E$).

$$[amplicon]_{synthesized} = N_0 * E^n \qquad (2)$$

Combining equation 1 with 2 yields:

$$N_0 = \Delta R_n / \Delta \varphi / E^n \qquad (3)$$

Quantification of the two alleles (susceptible and resistant) was achieved with TaqMan real-time SNP assays (Figure 1). Allele R (resistance allele) and allele S (susceptible) were detected by dual-labelled probes, 5′ FAM and 3′ BHQ and 5′ Quasar 670 and 3′ BHQ, respectively.

For the resistant allele R;

$$A_0 = \Delta R_n A / \Delta \varphi A / E_A^n \qquad (4)$$

where $A_0$ is the initial copy number of allele R
$\Delta R_n A$ is the fluorescence intensity of Fam at cycle n.
$\Delta \varphi A$ is the parameter for fluorescence Fam.
$E_A^n$ is the compound amplification efficiency of allele R.
For the susceptible allele S:

$$B_0 = \Delta R_n B / \Delta \varphi B / E_B^n \qquad (5)$$

Where $\Delta R_n B$ is the fluorescence increment of Quasar 670 at cycle n.

The initial allele ratio:

$$\frac{A_0}{B_0} = \frac{\Delta R_n A}{\Delta R_n B} \times \frac{\Delta \varphi B}{\Delta \varphi A} \times \frac{E_B^n}{E_A^n} \qquad (6)$$

While for any given assay, the ratio of parameter $\Delta \varphi B / \Delta \varphi A$ and $E_B^n / E_A^n$ will be a relative constant, there will be constant relationship between $R = A_0 / B_0$ and $R' = \Delta R_n A / \Delta R_n B$.

Therefore $R$ can be predicted by the ratio $R' = \Delta R_n A / \Delta R_n B$.

## Estimation of $\Delta R_n$

Real-time PCR is modelled via a four parametric sigmoid function [23,24]:

$$y(x) = \frac{a}{1 + \exp^{-(\frac{x-b}{c})}} + y_0 \qquad (7)$$

Where:
$x$ is cycle number,
$y(x)$ is raw fluorescence of cycle x,
$y_0$ is the background fluorescence,
$a$ is the maximal height of the curve (the difference between the maximal fluorescence and background fluorescence).
$b$ is the first derivative maximum of the function (the inflexion point of the curve) and $c$ describes the slope of the curve.

If you subtract the background fluorescence the equation above can be rewritten as:

$$f(x) = \frac{a}{1 + \exp^{-(\frac{x-b}{c})}} \qquad (8)$$

Where $f(x)$ is the fluorescence minus the background which is equivalent to $\Delta R_n$ at cycle $n$.

## Selecting a Single Point in the Exponential Phase

For allele R, with fluorescence Fam,

$$f_{fam}(x) = \frac{a_{fam}}{1 + \exp^{-(\frac{x-b_{fam}}{c_{fam}})}} \qquad (9)$$

Where:
$a_{fam}$ is the maximal height of the curve for fluorescence Fam.
$b_{fam}$ is the inflexion point of the curve of allele R,
$c_{fam}$ is the slope of the curve of allele R.
For allele S with fluorescence Quasar.

$$f_{qua}(x) = \frac{a_{qua}}{1 + \exp^{-(\frac{x-b_{qua}}{c_{qua}})}} \qquad (10)$$

Where:
$a_{qua}$ is the maximal height of the curve for fluorescence Quasar.
$b_{qua}$ is the inflexion point of the curve of allele S,
$c_{qua}$ is the slope of the curve of allele S.
Use the ratio $f_{fam}(x)/f_{qua}(x)$ when one of the alleles is at its maximum speed, for example, if $b_{fam} < b_{qua}$ where the Fam reaches to its maximum speed first (Figure 2).

$$R' = f_{fam}(b_{fam})/f_{qua}(b_{fam})$$

$$R' = \frac{0.5 a_{fam}}{f(b_{fam})} = 0.5 \times \frac{a_{fam}}{\dfrac{a_{qua}}{1 + \exp(\frac{-(b_{fam} - b_{qua})}{c_{qua}})}}$$

**A.**

**B.**

**C.**

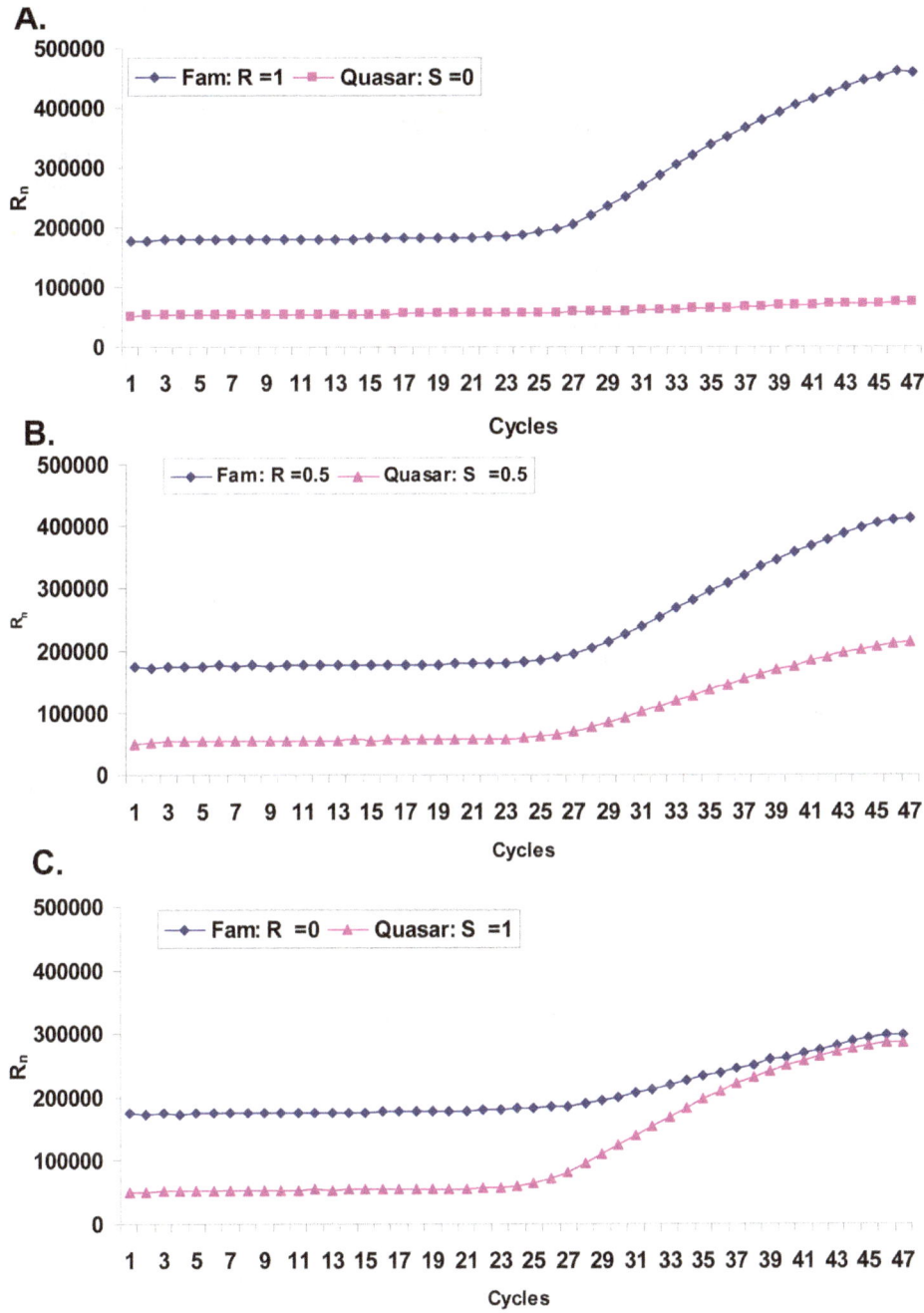

**Figure 1. Raw fluorescence plot of TaqMan assay with two probes.** Fam probe (blue) was from resistance allele and Quasar probe (red) was from susceptible allele.

$$= 0.5 \times \frac{a_{fam}}{a_{qua}} \times (1 + \exp(\frac{-(b_{fam} - b_{qua})}{c_{qua}})) \qquad (11)$$

$$R' = \frac{f_{fam}(b_{qua})}{0.5 a_{qua}} = \frac{\dfrac{a_{fam}}{1 + \exp(\frac{-(b_{qua} - b_{fam})}{c_{fam}})}}{0.5 a_{qua}}$$

However If $b_{fam} > b_{qua}$ where Quasar reaches to its maximum speed first.

$$= 2 \times \frac{a_{fam}}{a_{qua}} \times \frac{1}{1 + \exp\left(\frac{-(b_{qua} - b_{fam})}{c_{fam}}\right)} \quad (12)$$

The frequency of allele can be expressed as $k$.

$$k = \frac{A_0}{A_0 + B_0} = \frac{A_0/B_0}{A_0/B_0 + 1} = \frac{R}{R+1} \quad (13)$$

The ratio of fluorescence $R'$ can be expressed as $k'$.

$$k' = \frac{R'}{R'+1} \quad (14)$$

## A Two-step Sigmoid Curve Fitting for Standardized Parameters

To reduce parameter estimation bias caused by variable number of cycles in PCR plateau phase, the raw fluorescence data was fitted to a sigmoid curve twice. Fitting was performed first with all data points (in our case 47 cycles), and second fitting used data points with one slope after the inflexion point (cycle b+c) (see Table S1). An example of this calculation of k' is demonstrated in Table S2.

## Results

### Transformed Fluorescence Ratio k'

Essentially the transformed fluorescence ratio $k'$ is the transformation of the ratio of two fluorescence intensities when one of these intensities reaches its inflexion point (Figure 2). The transformed fluorescence ratio $k'$ comprising 4 runs of plasmid mix and 3 runs of pooled aphids with predefined RAF is summarized in Table 1 with original data included in Table S3 and Table S9.

The transformed fluorescence ratio $k'$ is highly consistent between triplicates from each standard. The average coefficient of variation within runs (intra-run) is between 1–2%. However the variation of $k'$ between runs (inter-run) for the same standard mixes is more variable and range from 1–18%. The value of $k'$ within runs for each predefined standard, follow the trends of the initial RAF. As RAF becomes higher, k' also becomes higher.

### Testing the Relationship between RAF and k'

Using both plasmid and aphid standards, we first tested the relationship between RAF and $k'$ by linear regression (as equation 6 predicts a linear relationship between these variables). In four runs using purified plasmids, a strong linear relationship was demonstrated with a high coefficient of determination ($R^2 > 0.99$). However, the linear model did not fit as well for standards made from Chelex-extracted aphids, where the three runs produced a coefficient of determination for aphid standards ranging from 0.93 to 0.95 (also see Table S4).

In addition to linear regression, we attempted a non-linear, 4 parameter sigmoid curve fitting model. Using this non-linear model, we found that the relationship between RAF and $k'$ for the three runs that used extracted aphids produced a higher coefficient of determination ($R^2 > 0.98$), which was comparable to the purified

plasmid samples. Interestingly, the purified plasmid standards also show improved coefficient of determinations using this sigmoid curve fitting model (Table S4).

### Inter-run Correlation of k'

Further analyses were performed to determine if there was correlation in the values of $k'$ between runs. Table 2 summarizes the inter-run coefficient of determination for $k'$ by linear and non-linear (sigmoid) regression. When compared like-for-like, the four plasmid and three aphid standard runs generally demonstrated a strong linear relationship ($R^2 > 0.98$). However, in contrast to above, the linear relationship was poorer when plasmid standard runs were compared to the standards made from Chelex-extracted aphids ($R^2 = 0.89$–$0.97$). When standard reactions were analyzed with the non-linear model a high correlation was observed between all runs. Coefficients of determination were higher ($R^2 > 0.99$) when plasmid and Chelex-extracted aphid runs were examined with non-linear regression and compared like-for-like. When plasmid standard runs were compared to the standards made from Chelex-extracted aphids using non-linear regression, coefficients of determination were also higher ($R^2 > 0.97$) than those generated from linear regression analysis (Table 2).

### The Number of Standards Required for Accurate Prediction

The sigmoid relationship allows for a reduction in the number of data-points required to build the standard curve and hence increases the number of samples that can be examined per run. We have determined that five standards (0.0, 0.2, 0.5, 0.8, and 1.0) are sufficient for accurate prediction and are statistically-equivalent to the 13 standard points used experimentally (see Table S5, Table S6 and Table S7). We have examined RAFs using a full 13-standard model and a reduced 5-standard model for both the purified plasmid (Table S6) and Chelex-extracted aphid standard runs (Table S7). If the predefined RAF standards (0.05, 0.10, 0.30, 0.40, 0.60, 0.70, 0.90 and 0.95) were treated as unknowns, the RAFs predicted using the reduced model are highly accurate for all runs (see Table S6 and Table S7). The correlation between actual RAFs and those predicted using the reduced model standard curve is very high ($R^2 > 0.999$).

### Combined Analysis of Multiple Runs

The sigmoid relationship between runs allows the analysis of multiple runs by normalizing all samples into a single run. The transformed fluorescence ratio $k'$ for all runs was adjusted by sigmoid function using five standards shared between each run (Table S8). By normalizing to Run 1 T/S, allele frequency can now be predicted for all runs using the equation derived from this run (Table 3). The accuracy of prediction is statistically-equivalent to intra-run prediction with 13-standard-points (Figure 3).

### Testing Pirimicarb Resistance Allele Frequency in Aphids Collected during the 2011/2012 Australian Cotton Season

To demonstrate the principle, we used qPCR to examine 35 *A. gossypii* samples collected from cotton producing farms across eastern Australia during the 2011/2012 season. Premixed DNA standards of known RAF were run simultaneously with DNAs extracted from a pool of 200 adult aphids. Table 4 lists the predicted resistance allele frequency based on $k'$. These results are consistent with resistance allele frequency obtained by individual genotyping.

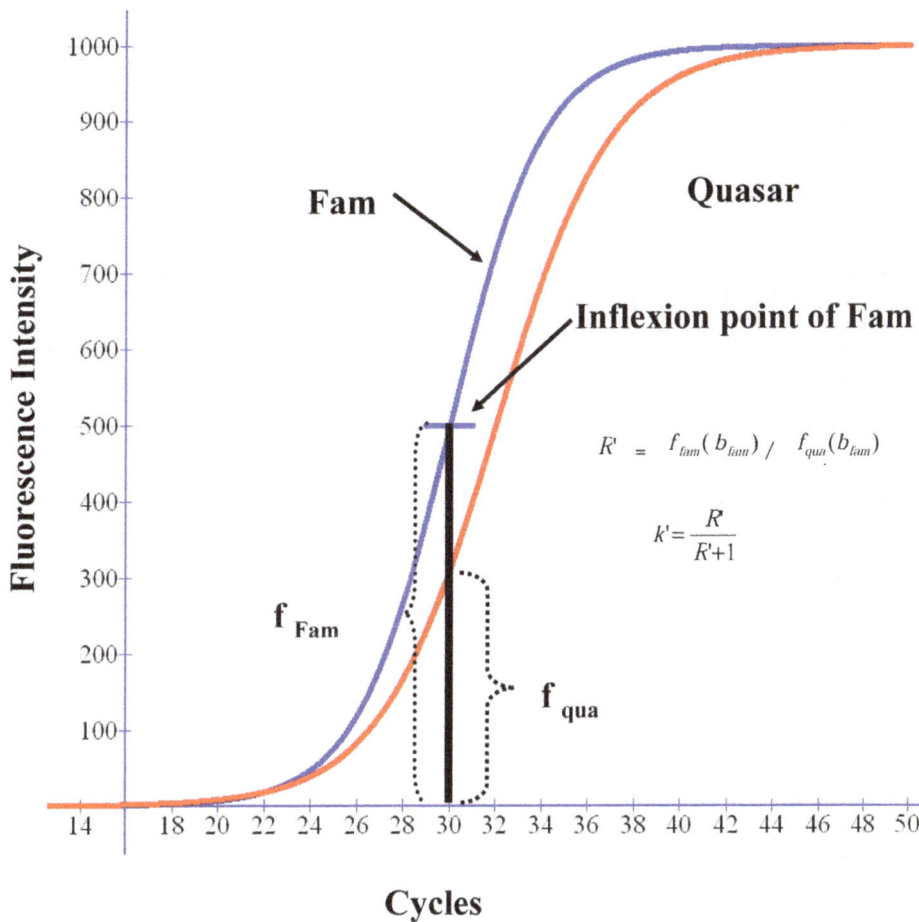

**Figure 2. Schematic of the calculation of transformed fluorescence ratio $k'$.**

## Discussion

The principle of quantification (Equation 6) states that the ratio of fluorescence from the two allele specific reporter dyes is a function of the initial allele ratio. Our method using the transformed fluorescence ratio at a single, standard time point is able to accurately quantify the allele frequency from pooled DNA samples and fully complies with the principle of PCR quantification. The method is less affected by background variation and so has the potential to overcome the intra-run and inter-run variation.

### Independence from Background Signal

A major contributor to observed variance in qPCR data outputs is baseline assignment and significant variation in the baseline fluorescence is often observed in replicate qPCR experiments [18,25]. Baseline variation affects the determination of the reaction threshold yet this parameter is often set automatically by the instrument software at 10 times the standard deviation of baseline. The fluorescence baseline commonly fluctuates between wells, runs and specific instrument being used [18]. Therefore, normalizing background fluorescence often reduces the well-to-well variation [25].

The transformed fluorescence ratio $k'$ uses raw fluorescence data points modeled by a four-parametric sigmoid function [23,24]. By using the transformation given in equation 8, the parameters; (a) the maximal height of the curve, (b) the first derivative maximum

of the function and (c) the slope of the curve are less dependent on background fluorescence and the estimation of ΔRn is standardized across different wells and runs.

### Single Time Point (Inflexion Point) from Consistent Parameter Estimates

In the past decade, 'assumption free' quantification methods of PCR based on non-linear regression (NLR) have been developed to fit observed parameters and calculate the initial number of target molecules at cycle 0 [24,26–28]. Although these models are mathematically sound and have been reported to contain less well-to-well variation, independent studies show that quantification based on these NLR methods do not outperform the conventional cycle of quantification ($C_t$) method due to the increased random error of qPCR [29,30].

One factor often unnoticed when using these models is that parameter estimates are significantly influenced by the number of cycles in the plateau phase of PCR. Sigmoid fitting methods are often not reproducible when replicate samples reach the plateau phase at slightly different cycle numbers. Our two-step sigmoid curve fitting method enables a more consistent sigmoid parameter estimate. In undertaking this method, we first fitted a sigmoid curve with all data points to obtain the proximal inflexion point (b) and the slope of the curve (c). Next the sigmoid curve was refitted with only data points from the b+c cycles. By doing that we standardized the data points so that a similar data range exists after the inflexion point for all datasets. Having an equal number

**Table 1.** The transformed fluorescence ratio k′ comprising 4 runs of plasmid mix (run1–4) and 3 runs of pooled aphids (run5–7) with predefined RAF.

| RAF | Run1 T/S K′ | CV (%) | Run2 A/S K′ | CV (%) | Run3 A/S K′ | CV (%) | Run4 T/S K′ | CV (%) | Run5 MP/S K′ | CV (%) | Run6 MP/S K′ | CV (%) | Run7 MP/S K′ | CV (%) | Inter run CV (%) |
|---|---|---|---|---|---|---|---|---|---|---|---|---|---|---|---|
| 100 | 0.924 | 0.083 | 0.908 | 0.091 | 0.917 | 0.177 | 0.916 | 0.211 | 0.917 | 2.790 | 0.888 | 0.259 | 0.925 | 0.459 | 1.358 |
| 95 | 0.887 | 0.115 | 0.862 | 0.771 | 0.869 | 0.544 | 0.890 | 0.121 | 0.886 | 4.036 | 0.871 | 0.042 | 0.916 | 0.362 | 1.923 |
| 90 | 0.851 | 0.334 | 0.823 | 1.344 | 0.828 | 0.506 | 0.861 | 0.155 | 0.881 | 0.259 | 0.862 | 2.041 | 0.885 | 0.352 | 2.574 |
| 80 | 0.791 | 0.485 | 0.752 | * | 0.755 | 0.499 | 0.802 | 0.145 | 0.821 | 1.872 | 0.826 | 0.263 | 0.849 | 0.639 | 4.045 |
| 70 | 0.710 | 1.095 | 0.683 | 2.293 | 0.679 | 2.077 | 0.738 | 1.666 | 0.730 | 0.113 | 0.757 | 4.986 | 0.819 | 0.440 | 6.487 |
| 60 | 0.654 | 1.490 | 0.622 | 2.788 | 0.621 | 1.254 | 0.672 | 0.440 | 0.721 | 1.014 | 0.738 | 0.492 | 0.778 | 1.165 | 8.421 |
| 50 | 0.588 | 0.442 | 0.564 | 3.107 | 0.559 | 1.953 | 0.615 | 0.764 | 0.665 | 0.855 | 0.687 | 3.880 | 0.705 | 0.399 | 9.271 |
| 40 | 0.531 | 1.410 | 0.498 | 0.874 | 0.491 | 1.177 | 0.554 | 2.422 | 0.632 | 0.697 | 0.632 | 1.940 | 0.651 | 1.817 | 11.404 |
| 30 | 0.444 | 0.146 | 0.395 | 1.279 | 0.386 | 3.190 | 0.496 | 1.287 | 0.551 | 1.623 | 0.564 | 1.068 | 0.528 | 1.210 | 14.763 |
| 20 | 0.392 | 1.547 | 0.359 | 1.052 | 0.347 | 2.974 | 0.444 | 2.894 | 0.453 | 0.419 | 0.446 | 2.304 | 0.508 | 0.051 | 12.900 |
| 10 | 0.321 | 2.345 | 0.308 | 1.258 | 0.295 | 3.603 | 0.382 | 1.324 | 0.424 | 1.176 | 0.421 | 1.422 | 0.383 | 0.948 | 14.500 |
| 5 | 0.286 | 2.290 | 0.280 | * | 0.272 | 8.648 | 0.310 | 5.008 | 0.262 | 0.187 | 0.254 | 4.973 | 0.281 | 7.650 | 8.021 |
| 0 | 0.255 | 3.547 | 0.246 | 9.724 | 0.233 | 2.572 | 0.248 | 2.798 | 0.184 | 0.145 | 0.165 | 3.689 | 0.161 | 10.051 | 18.964 |
| Intra run CV (%) | | 1.179 | | 2.235 | | 2.244 | | 1.479 | | 1.168 | | 2.105 | | 1.965 | |

The transformed fluorescence ratio k′ is the transformation of the ratio of two fluorescence intensities when one fluorescence reaches its inflexion point.
RAF: Predefined Resistance allele frequency (RAF) expressed as percentage.
K′: Average of transformed fluorescence ratio (equation 14) from 3 replicates.
Inter run CV (%): the average coefficient of variation for a standard among 7 runs.
Intra run CV (%): the average coefficient of variation for 13 standards within a run.
*: No replicate.

**Table 2.** Coefficient of determination $R^2$ of inter-run $k'$ with linear and 4 parameter sigmoid curve fitting.

| $R^2$ | Run2 A/S | Run3 A/S | Run4 T/S | Run5 MP/S | Run6 MP/S | Run7 MP/S |
|---|---|---|---|---|---|---|
| **Run1 T/S - linear** | 0.9975 | 0.997 | 0.9931 | 0.948 | 0.9262 | 0.9364 |
| **Run1 T/S - sigmoid** | 0.9989 | 0.9988 | 0.9953 | 0.9747 | 0.9855 | 0.9906 |
| **Run2 A/S – linear** | | 0.9996 | 0.9859 | 0.9246 | 0.9057 | 0.9184 |
| **Run2 A/S - sigmoid** | | 0.9999 | 0.9895 | 0.9675 | 0.982 | 0.9922 |
| **Run3 A/S - linear** | | | 0.9801 | 0.9229 | 0.8869 | 0.9091 |
| **Run3 A/S - sigmoid** | | | 0.9916 | 0.9743 | 0.9805 | 0.9884 |
| **Run4 T/S - linear** | | | | 0.9722 | 0.9557 | 0.9582 |
| **Run4 T/S - sigmoid** | | | | 0.9918 | 0.9946 | 0.9968 |
| **Run5 MP/S - linear** | | | | | 0.9948 | 0.9822 |
| **Run5 MP/S - sigmoid** | | | | | 0.9991 | 0.986 |
| **Run6 MP/S - linear** | | | | | | 0.9869 |
| **Run6 MP/S - sigmoid** | | | | | | 0.9874 |

Run1–4 are plasmid mix and run5–7 are pooled aphids.

of cycles after the infection point enables a more robust estimation of the parameters.

## A Single Reference Point for Fluorescence Ratio Determination

Earlier work by Oliver et al [31] to quantify the initial allele ratio by examining the qPCR end point fluorescence ratio is not ideal for accurate quantification due to the dramatic decrease of amplification efficiency in late PCR cycles. To more accurately predict the initial allele ratio, Yu et al [18] used background-normalized fluorescence from both fluorophores in exponential phase. However, the selection of the exponential cycles in this method was arbitrary, particularly when one fluorescence signal reaches the exponential phase much earlier than the other which is

often the case when one or the other allele frequency is quite low. In our two-step sigmoid curve fitting method, the fluorescence ratio is measured when a fluorescence signal first reaches the inflexion point (equation 11 and 12) and allows for a standard method of identifying the exponential phase. As the inflexion point is always in the middle of the exponential phase it shows very similar kinetics between replicate samples and so has the potential to be more accurate.

## Transformed Fluorescence Ratio $k'$

The transformed fluorescence ratio $k'$ (Equation 13) permits the development of a standard curve with allele frequency ranging from 0 to 1. The inclusion of a zero allele frequency is critical as a control to assess the sensitivity of the assay. In TaqMan assays,

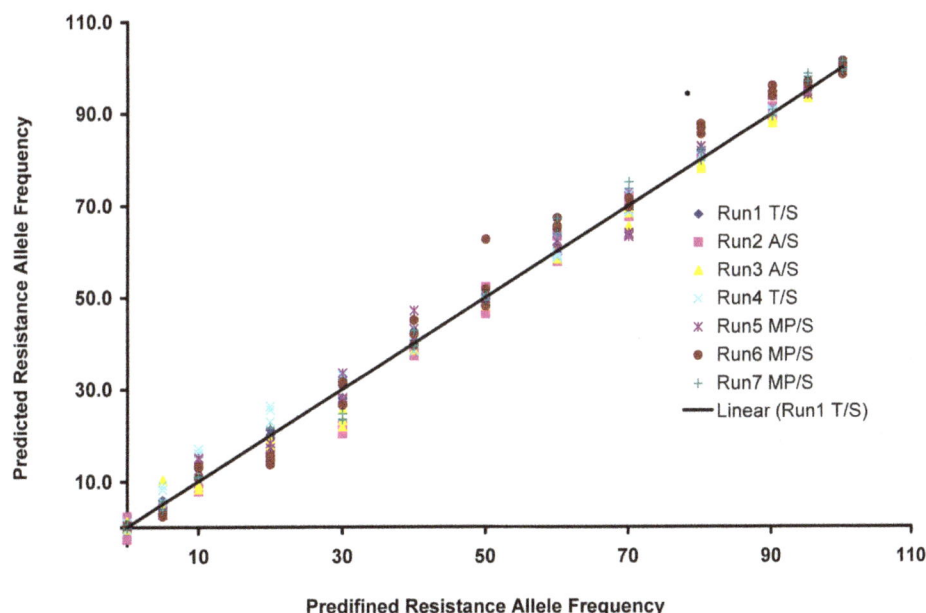

**Figure 3. Predicted resistance allele frequency (RAF) for standards based on Run1 T/S.** Prediction were based on five standards from Run1 T/S and all calculated transformed fluorescence ratio $k'$ were adjusted to Run1 T/S by sigmoid function.

**Table 3.** Prediction of RAF by normalizing $k'$ to Run1 T/S.

| RAF Standard | Run1 T/S | | Run2 A/S | | Run3 A/S | | Run4 T/S | | Run5 MP/S | | Run6 MP/S | | Run7 MP/S | |
|---|---|---|---|---|---|---|---|---|---|---|---|---|---|---|
| | $k'$ | RAF* | $k'$ adj** | RAF* | $k'$ adj | RAF* | $k'$ adj | RAF* | $k'$ adj | RAF* | $k'$ adj | RAF* | $k'$ adj | RAF* |
| 100 | 0.924 | 99.9 | 0.927 | 100.3 | 0.926 | 100.2 | 0.921 | 99.5 | 0.925 | 100.0 | 0.918 | 99.0 | 0.923 | 99.7 |
| 95 | 0.887 | 94.5 | 0.885 | 94.2 | 0.885 | 94.2 | 0.895 | 95.6 | 0.881 | 93.5 | 0.883 | 93.9 | 0.907 | 97.4 |
| 90 | 0.851 | 89.2 | 0.849 | 88.9 | 0.849 | 88.8 | 0.864 | 91.1 | 0.873 | 92.5 | 0.865 | 91.2 | 0.854 | 89.6 |
| 80 | 0.791 | 80.2 | 0.783 | 79.0 | 0.784 | 79.1 | 0.798 | 81.3 | 0.789 | 80.0 | 0.798 | 81.2 | 0.794 | 80.7 |
| 70 | 0.71 | 68.1 | 0.717 | 69.1 | 0.713 | 68.6 | 0.724 | 70.2 | 0.668 | 61.8 | 0.686 | 64.5 | 0.746 | 73.5 |
| 60 | 0.654 | 59.7 | 0.657 | 60.1 | 0.659 | 60.4 | 0.646 | 58.4 | 0.657 | 60.1 | 0.659 | 60.4 | 0.684 | 64.2 |
| 50 | 0.588 | 49.7 | 0.599 | 51.3 | 0.599 | 51.3 | 0.579 | 48.3 | 0.590 | 50.0 | 0.593 | 50.4 | 0.585 | 49.2 |
| 40 | 0.531 | 41.1 | 0.531 | 41.1 | 0.531 | 41.2 | 0.510 | 38.0 | 0.553 | 44.4 | 0.531 | 41.2 | 0.522 | 39.7 |
| 30 | 0.444 | 28.0 | 0.422 | 24.8 | 0.424 | 25.0 | 0.449 | 28.8 | 0.471 | 32.1 | 0.467 | 31.5 | 0.409 | 22.7 |
| 20 | 0.392 | 20.2 | 0.384 | 19.0 | 0.383 | 18.8 | 0.399 | 21.3 | 0.391 | 20.1 | 0.381 | 18.6 | 0.394 | 20.5 |
| 10 | 0.321 | 9.7 | 0.328 | 10.6 | 0.327 | 10.5 | 0.345 | 13.3 | 0.371 | 17.0 | 0.366 | 16.3 | 0.322 | 9.8 |
| 5 | 0.286 | 4.5 | 0.297 | 6.0 | 0.302 | 6.8 | 0.292 | 5.3 | 0.284 | 4.2 | 0.289 | 4.9 | 0.284 | 4.2 |
| 0 | 0.255 | -0.1 | 0.258 | 0.4 | 0.259 | 0.5 | 0.253 | -0.4 | 0.255 | 0.0 | 0.260 | 0.7 | 0.254 | -0.2 |
| $R^2$ | | 0.999 | | 0.998 | | 0.998 | | 0.998 | | 0.989 | | 0.995 | | 0.995 |

RAF*: Predicted resistance allele frequency from Run1 T/S.
$k'$ adj**: normalized transformed fluorescence ratio $k'$ based on Run1 T/S.

**Table 4.** Field isolates of *Aphis gossypii* collected during the 2011/2012 season showing pirimicarb resistance status determined by individual PCR-RFLP.

| Strain | Region | RAF by qPCR with pooled DNA | RAF by RFLP genotyping of 20 individual aphids | Bioassay |
|--------|--------|------|------|------|
| Alch | Darling Downs, QLD | −2.4 | 0 | S |
| And | Fitzroy, QLD | −1.6 | 0 | S |
| Aral | Darling Downs, QLD | −0.4 | 0 | S |
| Arra | Darling Downs, QLD | −1.0 | 0 | S |
| Bal F3 | S. West QLD | −2.3 | 0 | S |
| Bal Vol | S. West QLD | −2.6 | 0 | S |
| Boo Dry | Darling Downs, QLD | −1.7 | 0 | S |
| Boo Irr | Darling Downs, QLD | 0.4 | 0 | S |
| Bor P | S. West QLD | 0.6 | 0 | S |
| Both | Kimberley, WA | 0.9 | 0 | S |
| Both B | Kimberley, WA | 102.0 | 100 | R |
| Bro Cle | S. West QLD | −1.1 | 0 | S |
| Bro Tre | S. West QLD | 1.0 | 0 | S |
| Bud | Darling Downs, QLD | 0.9 | 0 | S |
| Bur Dry | S. West QLD | 102.0 | 100 | R |
| Car F3 | N. Inland, NSW | −1.1 | 0 | S |
| Car Vol | N. Inland, NSW | 1.0 | 0 | S |
| Carring | N. Inland, NSW | 0.9 | 0 | S |
| Cly | S. West QLD | −1.4 | 0 | S |
| Doo 1 | S. West QLD | −1.4 | 0 | S |
| Doo 2 | S. West QLD | −0.3 | 0 | S |
| Eum | Darling Downs, QLD | −0.6 | 0 | S |
| Fair | Darling Downs, QLD | 0.0 | 0 | S |
| Gra 148 | Fitzroy, QLD | −0.1 | 0 | S |
| Mon P | Northern QLD | 104.6 | 100 | R |
| Over | Darling Downs, QLD | −1.2 | 0 | S |
| P Seed | Kimberley, WA | −0.7 | 0 | S |
| Spri | N. Inland, NSW | −0.4 | 0 | S |
| Terr | Darling Downs, QLD | −1.3 | 0 | S |
| T Sand | Kimberley, WA | 104.1 | 100 | R |
| Walt | Darling Downs, QLD | −1.7 | 0 | S |
| Wanh F | Kimberley, WA | 103.9 | 100 | R |
| Wise | N. Inland, NSW | −1.3 | 0 | S |
| Wyad | N. Inland, NSW | −1.0 | 0 | S |
| Zig | S. West QLD | −1.7 | 0 | S |

R: Resistant to Pirimicarb.
S: Susceptible to Pirimicar.

potential errors occur due to significant cross-binding of probes. Even when one allele is absent its corresponding florescence signal can still be observed due to cross binding. Including an allele frequency of 0 and 1 makes it possible to accurately estimate unknown samples with allele frequency <0.05 or <0.95.

## Non-linear Relationship between Transformed Fluorescence Ratio *k'* and RAF

Our results demonstrate a sigmoid relationship between RAF and transformed fluorescence ratio *k'*. The predicted linear relationship between the initial allele frequency and transformed

fluorescence ratio *k'* can be achieved for PCR in optimal conditions. However, a non-linear model is more universal given most PCRs are performed in conditions that are not optimal, particularly when unknown PCR inhibitors are present.

The sigmoid function between RAF and transformed fluorescence ratio *k'* theoretically enables the construction of a standard curve using only 4 standard points and our results demonstrate that the prediction model obtained from 5 rather than 4 standard points was as accurate as the model base on 13 standard points. This reduction in the number of standards required for each run allows for a considerable increase in the number of wells that can be used for samples rather than standards.

## Inter-run Correlation

Additionally, we have found that there is a sigmoid relationship between the transformed fluorescence ratios $k'$ across multiple runs. This enables the normalization of samples from multiple experiments into a single run if at least 4 samples are shared in each. Therefore, a single analysis can be performed for all samples across all runs.

## Practical Implementation of the Method

Our two-step sigmoid curve fitting method has the potential to be used broadly for high throughput/low cost genotyping. Although this method was developed using a TaqMan assay on an ABI 7500 real-time thermocycler, the principle can theoretically be applied to other fluorescent dye and instrument platforms (such as SYBR green). In some cases, when one allele is absent, there is no PCR amplification or irregular amplification, it is possible to manually estimate fluorescence height above the background at the approximate inflexion point of the other allele for the $k'$ calculation. Alternatively, a predefined allele frequency 0.01 and 0.99 can be used as standard points.

## Diagnostic Testing the Pirimicarb Resistance Allele Frequency in Aphids Collected during the 2011/2012 Cotton Season

To examine the effectiveness of our two-step sigmoid curve fitting method we used field samples to predict the pirimicarb-RAF in 35 field isolates of *A. gossypii* and compared those estimated allele frequencies with individual genotyping of 20 aphids from each isolate. A remarkable consistency was observed between the RAF predicted by qPCR and allele frequency predicted by individual genotyping. Unfortunately, the pirimicarb-RAF observed in the 2011/2012 season where either 0% or 100%, making it difficult to statistically assess the precision of the prediction. While this data limitation could not be overcome, the method demonstrated good sensitivity when RAF is low.

This method allows for a dramatic decrease in the amount of labor required for the high-throughput monitoring of RAF in insects of agricultural importance, so aiding sustainable IPM systems. Interestingly, we found that a similar amount of time was required for an experienced worker to extract DNA from 20 aphids individually or 200 aphids combined in one tube. However, there was a great difference in the amount of time required to genotype these samples. Genotyping of $35 \times 20$ aphids individually required almost three weeks of work while genotyping of $35 \times 200$ aphids using pooled DNA, a TaqMan assay and our two-step sigmoid curve fitting method could be performed in as little as three days.

## Conclusion

We have developed a method using a TaqMan SNP assay to accurately estimate the allele frequencies from pooled DNA samples. The method uses the transformed fluorescence ratio based on a single reference point and has proven precise at predicting unknown allele frequencies. The prediction model can be built using five standard points and results can be normalized across multiple runs. The method can dramatically reduce time and labour required for insecticide resistance monitoring and has the potential for broad applications in high throughput genotyping such as genome –wide association studies, population studies even the quantitative assessment of post transplant chimeras in medicine.

## Supporting Information

**Table S1  An example of two-step four-parameter sigmoid curve fitting.**

**Table S2  Example of Calculation of $k'$.**

**Table S3  The transformed fluorescence ratio $k'$ comprising 4 runs of plasmid mix (run1-4) and 3 runs of pooled aphids (run5-7) with predefined resistance allele frequency (RAF).** The transformed fluorescence ratio $k'$ is the transformation of the ratio of two fluorescence intensity when one fluorescence reaches its inflexion point.

**Table S4  Test of the linear and non-linear (sigmoid) relationship between RAF and transformed fluorescence ratio $k'$. Run1-4 are plasmid mix and run5-7 are pooled aphids.**

**Table S5  Five standard points used for the reduced prediction model.**

**Table S6  Resistance allele frequencies (RAF) predicted from full and reduced prediction models in plasmid mix runs.**

**Table S7  Resistance allele frequencies (RAF) predicted from full and reduced prediction models in aphid mix runs.**

**Table S8  Transformed fluorescence ratio $k'$ between runs can be normalized to Run1 T/S by using five standards based on sigmoid function.**

**Table S9  Raw fluorescence data for 7 runs.**

## Author Contributions

Conceived and designed the experiments: YC GH. Performed the experiments: DB. Analyzed the data: IB YC. Contributed reagents/materials/analysis tools: GH. Wrote the paper: YC GH.

## References

1. Georguiou GP, Mellon RB (1983) Pesticide resistance in time and space. In: Georguiou GP, Saito T, editors. Pest resistance to pesticides. New York: Plenum Press. 1–46.
2. Mass S (2012) Cotton pest management guide 2012–13. Toowoomba: Greenmount Press.
3. Busvine JR (1971) A critical review of the techniques for testing insecticides. the University of Wisconsin - Madison: Commonwealth Agricultural Bureaux. 345 p.
4. Ffrench-Constant R, Roush R (1991) Resistance detection and documentation: The relative roles of pesticidal and biochemical assays. In: Roush R, Tabashnik B, editors. Pesticide resistance in arthropods: Springer US. 4–38.
5. Yu SJ (2008) The toxicology and biochemistry of insecticides. Boca Raton: CRC Press. 296 p.
6. Ffrenchconstant RH, Rocheleau TA, Steichen JC, Chalmers AE (1993) A point mutation in a drosophila gaba receptor confers insecticide resistance. Nature 363: 449–451.

7. Weill M, Fort P, Berthomieu A, Dubois MP, Pasteur N, et al. (2002) A novel acetylcholinesterase gene in mosquitoes codes for the insecticide target and is non-homologous to the ace gene Drosophila. Proc Biol Sci 269: 2007–2016.

8. Newcomb RD, Campbell PM, Ollis DL, Cheah E, Russell RJ, et al. (1997) A single amino acid substitution converts a carboxylesterase to an organophosphorus hydrolase and confers insecticide resistance on a blowfly. Proc Natl Acad Sci USA 94: 7464–7468.

9. Daborn PJ, Yen JL, Bogwitz MR, Le Goff G, Feil E, et al. (2002) A single p450 allele associated with insecticide resistance in drosophila. Science 297: 2253–2256.

10. Barbagallo S, Cravedi P, Pasqualini E, Patti I (1997) Aphids of the principal fruit bearing crops. Milan: Bayer. 123 p.

11. Herron GA, Powis K, Rophail J (2001) Insecticide resistance in *Aphis gossypii* Glover (Hemiptera : Aphididae), a serious threat to Australian cotton. Aust J Entomol 40: 85–89.

12. Andrews MC, Callaghan A, Field LM, Williamson MS, Moores GD (2004) Identification of mutations conferring insecticide-insensitive AChE in the cotton-melon aphid, Aphis gossypii Glover. Insect Mol Biol 13: 555–561.

13. Toda S, Komazaki S, Tomita T, Kono Y (2004) Two amino acid substitutions in acetylcholinesterase associated with pirimicarb and organophosphorous insecticide resistance in the cotton aphid, Aphis gossypii Glover (Homoptera: Aphididae). Insect Biochem Mol Biol 13: 549–553.

14. McLoon MO, Herron GA (2009) PCR detection of pirimicarb resistance in Australian field isolates of *Aphis gossypii* Glover (Aphididae: Hemiptera). Aust J Entomol 48: 65–72.

15. Billard A, Laval V, Fillinger S, Leroux P, Lachaise H, et al. (2012) The allele-specific probe and primer amplification assay, a new real-time pcr method for fine quantification of single-nucleotide polymorphisms in pooled DNA. Appl Environ Microbiol 78: 1063–1068.

16. Breen G, Harold D, Ralston S, Shaw D, St Clair D (2000) Determining SNP allele frequencies in DNA pools. Biotechniques 28: 464–466, 468, 470.

17. Psifidi A, Dovas C, Banos G (2011) Novel Quantitative Real-Time LCR for the Sensitive Detection of SNP Frequencies in Pooled DNA: Method Development, Evaluation and Application. PLoS One 6: e14560.

18. Yu A, Geng H, Zhou X (2006) Quantify single nucleotide polymorphism (SNP) ratio in pooled DNA based on normalized fluorescence real-time PCR. BMC Genomics 7: 143.

19. Chen J, Germer S, Higuchi R, Berkowitz G, Godbold J, et al. (2002) Kinetic polymerase chain reaction on pooled DNA: a high-throughput, high-efficiency alternative in genetic epidemiological studies. Cancer Epidemiol Biomarkers Prev 11: 131–136.

20. Germer S, Holland MJ, Higuchi R (2000) High-throughput snp allele-frequency determination in pooled DNA samples by kinetic pcr. Genome Res 10: 258–266.

21. Payne RW, Harding SA, Murray DA, Soutar DM, Baird DB, et al. (2007) Genstat release 10 reference manual. Hemel Hempstead: VSN International.

22. Swillens S, Goffard J, Marechal Y, de Kerchove d'Exaerde A, El Housni H (2004) Instant evaluation of the absolute initial number of cDNA copies from a single real-time PCR curve. Nucleic Acids Res 32: e56.

23. Johnson M, Haupt L, Griffiths L (2004) Locked nucleic acid (LNA) single nucleotide polymorphism (SNP) genotype analysis and validation using real-time PCR. Nucleic Acids Res 32: e55.

24. Liu W, Saint D (2002) Validation of a quantitative method for real time PCR kinetics. Biochem Biophys Res Commun 294: 347–353.

25. Carr AC, Moore SD (2012) Robust quantification of polymerase chain reactions using global fitting. PLoS One 7: e37640.

26. Ramakers C, Ruijter JM, Deprez RHL, Moorman AFM (2003) Assumption-free analysis of quantitative real-time polymerase chain reaction (PCR) data. Neurosci Lett 339: 62–66.

27. Rutledge RG (2004) Sigmoidal curve-fitting redefines quantitative real-time PCR with the prospective of developing automated high-throughput applications. Nucleic Acids Res 32: e178.

28. Spiess AN, Feig C, Ritz C (2008) Highly accurate sigmoidal fitting of real-time PCR data by introducing a parameter for asymmetry. BMC Bioinformatics 9: 221.

29. Karlen Y, McNair A, Perseguers S, Mazza C, Mermod N (2007) Statistical significance of quantitative PCR. BMC Bioinformatics 8: 131.

30. Bar T, Kubista M, Tichopad A (2012) Validation of kinetics similarity in qPCR. Nucleic Acids Res 40: 1395–1406.

31. Oliver DH, Thompson RE, Griffin CA, Eshleman JR (2000) Use of single nucleotide polymorphisms (snp) and real-time polymerase chain reaction for bone marrow engraftment analysis. J Mol Diagn 2: 202–208.

# Molecular Mapping and Validation of a Major QTL Conferring Resistance to a Defoliating Isolate of Verticillium Wilt in Cotton (*Gossypium hirsutum* L.)

Xingju Zhang[1,ϑ], Yanchao Yuan[1,ϑ], Ze Wei[1], Xian Guo[1], Yuping Guo[1], Suqing Zhang[1], Junsheng Zhao[2], Guihua Zhang[3], Xianliang Song[1]*, Xuezhen Sun[1]*

1 State Key Laboratory of Crop Biology/Agronomy College, Shandong Agricultural University, Taian, Shandong, China, 2 Cotton Research Center, Shandong Academy of Agricultural Sciences, Jinan, Shandong, China, 3 Heze Academy of Agricultual Sciences, Heze, Shandong, China

## Abstract

Verticillium wilt (VW) caused by *Verticillium dahliae* Kleb is one of the most destructive diseases of cotton. Development and use of a VW resistant variety is the most practical and effective way to manage this disease. Identification of highly resistant genes/QTL and the underlining genetic architecture is a prerequisite for developing a VW resistant variety. A major QTL *qVW-c6-1* conferring resistance to the defoliating isolate V991 was identified on chromosome 6 in LHB22×JM11 $F_{2:3}$ population inoculated and grown in a greenhouse. This QTL was further validated in the LHB22×NNG $F_{2:3}$ population that was evaluated in an artificial disease nursery of V991 for two years and in its subsequent $F_4$ population grown in a field severely infested by V991. The allele conferring resistance within the QTL *qVW-c6-1* region originated from parent LHB22 and could explain 23.1–27.1% of phenotypic variation. Another resistance QTL *qVW-c21-1* originated from the susceptible parent JM11 was mapped on chromosome 21, explaining 14.44% of phenotypic variation. The resistance QTL reported herein provides a useful tool for breeding a cotton variety with enhanced resistance to VW.

**Editor:** David D. Fang, USDA-ARS-SRRC, United States of America

**Funding:** This research was financially supported in part by grants from the System of Modern Agriculture Industrial Technology (SDAIT-07-011-02), the Science and Technology Development Project (2012GGB01026), the Natural Science foundation (ZR2013CM005) and the Agricultural Seed Project (cotton variety development, 2011-2013; cotton germplasm innovation, 2013) of Shandong Province. The funders had no role in study design, data collection and analysis, decision to publish, or preparation of the manuscript.

**Competing Interests:** The authors declare that they have no competing interests.

\* E-mail: songxl999@163.com (XLS); sunxz@sdau.edu.cn (XZS)

ϑ These authors contributed equally to this work.

## Introduction

Cotton (*Gossypium* spp.) is the most important fiber crop, and the second most important source for edible oil and protein in the world [1]. Cotton production is affected by both biotic and abiotic stresses, of which Verticillium wilt (VW) caused by soil-borne *Verticillium dahliae* Kleb is one of the most destructive diseases in cotton. VW was first reported in Virginia, US in 1914 [2] and is found in almost all cotton growing areas worldwide. The infected plants usually exhibit symptoms of marginal chlorosis or necrosis in leaves, discoloration of the stem vascular bundles, even full defolation and plant death. Severe infection results in significant reduction in plant mass, lint yield and fiber quality [3,4]. There were two major VW outbreaks in China, one was in 1993 and the other in 2002–2003, each resulting in about 100,000 tons of lint yield loss [5]. There are no effective solutions to control this disease once a plant is infected [6]. Thus, VW is called cancer of cotton in China.

The VW has not been effectively controlled mainly due to its biology characteristics, the indeterminacy of the genetic mechanism of resistance to VW and the lack of highly resistant commercial Upland cotton varieties, except for some modern Acala cotton cultivars developed in California and New Mexico

[7]. *V. dahliae* has a wide host range of over 400 plant species including herbaceous annuals and perennials and woody perennials [8], and can survive in soil for many years in the dormant form of microsclerotia or as mycelium or conidia in the vascular system of perennial plants. These properties make crop rotation, even field fallowing less effective. Although application of soil fumigants, such as methyl bromide, is an effective control strategy, but expensive and harmful to both environment and human health [9]. Planting a resistant or tolerant cultivar has long been considered as the most practical, economic and effective means of decreasing lose from VW.

As for VW resistance heredity, two different genetic models, qualitative trait model and quantitative trait model, were reported by researches using different materials in traditional genetic studies [10]. With the development and utilization of molecular marker technologies, much progress in QTL mapping for VW resistance in cotton has been achieved which provides more information about the genetics of VW resistance at molecular level. So far, more than 100 QTL conferring VW resistance have been mapped on twenty-two chromosomes (Chr.) of tetraploid cotton except for Chr.6, 10, 12 and 18 from interspecific populations of *G. hirsutum*×*G. barbadense* [3,11–17] and from *G. hirsutum* intraspecific populations [18–21] evaluated at different growing stages with different *V. dahliae* isolates. However, comparison of these results indicated that VW resistance was characterized as specific to both *V. dahliae* isolates and plant growth stages [15,16,21], with some exceptions where chromosomal regions with broad-spectrum VW resistance on Chr.16, 23 in *G. hirsutum* germplasm 60182 [18] and on Chr.23 in *G. hirsutum* varieties Prema [19] were identified, respectively. Previous studies also indicated that the resistance QTL number varied largely from 1~3 [13,14,21] to 14 [18] with cotton varieties/lines studied even for one *V. hahliae isolate*, implying there be differences in genetic basis for VW resistance in different materials. More recently, a large number of genes were identified to express differentially in defense responses to different races (V991 and D07038) of *V. dahliae* through cDNA-AFLP [22], RNA-sequencing [23,24] and proteomic and virus-induced gene silencing [25]. These researches indicated that genes that involved in the lignin, gossypol, brassinosteroids, jasmonic acid and the phenylalanine metabolism play important role in cotton defense responses to different races of *V. dahliae*. Several important clues regarding VW resistance have been obtained, including Bet v1 and UbI gene family in *G. hirsutum* cv. Zhongzhimian KV1 [22], lignin-synthesis related genes in *G. barbadense* cv. Hai7124 [24], the phenylalanine metabolism related genes in Zhongzhimian KV1 and *G. barbadense* cv. Xinhai 15 [23], and genes involved in gossypol, brassinosteroids and jasmonic acid in Hai7124 [25].

The defense responses to VW infection are also associated with some other factors, such as the phonological stage of the plant, environmental conditions, level of inoculum, and the virulence of *V. dahliae* strains [26]. So integrated resistance evaluation using local isolates both in greenhouse for seedling stage resistance and in disease nursery and/or naturally infected field for adult stage resistance was suggested to get reliable disease evaluation results in cotton breeding.

Our previous studies indicated LHB22 possessed high resistance to VW at seedling stage under greenhouse conditions [27] and at mature stage under artificially infected nursery conditions [28]. The objective of this study was to reassess the VW resistance of LHB22 at different growth stages under controlled (greenhouse and artificially infected disease nursery) and naturally infected field environments, and to identify QTL of VW resistance from LHB22, so as to research the genetic architecture of VW resistance and provide molecular tool for cotton breeding program.

## Materials and Methods

### Plant materials

Three variety/germplasm lines were chosen for this research. LHB22 (Lu Hirsutum-Bickii 22), characterized by pink petals and filaments, with a large purplish red spot in the petal base introgressed from *G. bickii* [27,29], exhibits high resistance to VW [27,28], and has been widely used as a parent in hybrid development and breeding programs in China. Jimian 11 (JM11), a susceptible control variety in national VW resistance evaluation test of China was provided by the Cotton Research Institute, Chinese Academy of Agricultural Sciences (CAAS). Nannonggan (NNG) introduced from Nanjing Agricultural University was also used as a susceptible parent in this study to map VW resistance in different genetic backgrounds.

In winter of 2008, LHB22 and JM11 were crossed at Sanya breeding station of Shandong province (SYBS/SDP), Hainan, China. In 2009, the subsequent $F_1$ seeds were planted and self-pollinated to produce $F_2$ progeny in the breeding field of Boyang Seed Company, Huimin, Shandong, China. The parental and $F_2$ seeds were then planted at SYBS/SDS in winter of 2009 and self-pollinated to produce 243 $F_{2:3}$ families (designated as

LHB22×JM11 population) used for VW resistance evaluation. In order to validate the QTL located on Chr.6 detected in LHB22×JM11 population, another $F_{2:3}$ population of 226 families (named as LHB22×NNG) using LHB22 and NNG as parents was constructed with the same methodology in 2009 and 2010.

### VW resistance assay under greenhouse conditions

A defoliating *V. dahliae* isolate, V991, which is prevailing in the Yellow and the Yangtze River cotton growing regions [19] was used for disease infection. This isolate was kindly provided by the Institute of Plant Protection, CAAS, Beijing, China. It was grown on agar petri dish plates for 7–10 days. Then, the isolate mass was increased by growing the isolate on boiled and sterilized cotton seeds in conical flasks for 12–15 days at 25°C. After adding 200 ml of sterile water, the flasks were shaken at 60–100 rpm for 10 minutes to produce a high concentrated conidial suspension which was filtered through 4-layers of cheesecloth, and the conidial concentration was determined with a hemacytometer and adjusted to $1×10^7$ conidia ml$^{-1}$. Conidia suspension was prepared immediately before being used for inoculation.

VW resistance evaluation of LHB22×JM11 population was conducted in a greenhouse at Crop Research Station of Shandong Agricultural University (CRS/SDAU), Taian, China. On September 18 2010, acid-delinted seeds were planted in 10-cm-diameter pots with 2 seeds per pot and 14 pots per family in each replication. The experiment was arranged in a randomized completed block design with two replications for $F_{2:3}$ families and three replications for two parents. The pots were filled with sterilized potting mixture (sand: peat: clay loam = 1:1:1, vol: vol). At two-true leaf stage, the pot bottom was gently torn off with scissors, each pot was placed in one petri dish containing 10 ml of *V. dahliae* conidia suspension. After the conidia suspension was sucked dry, the pots/plants were planted in a seedling bed in the greenhouse. The plants were irrigated once a week. The average temperature was 24–27°C in the daytime and 18–20°C at nighttime.

Twenty-five days after inoculation, phenotyping of VW resistance was conducted using a severity rating system ranging from 0 to 4 for leaf disease [19], where 0 = healthy plant without disease symptom; 1 = one to two yellowish cotyledons; 2 = two yellowish cotyledons and one symptomatic true leaf or one cotyledon abscised; 3 = yellowish cotyledons and two symptomatic true leaves, or two cotyledon abscised and one symptomatic true leaf; 4 = complete defoliation or dead plant. Mean value of all plants was calculated as the score of each $F_{2:3}$ family in a replicate and average values of two replicates were used in analysis.

### VW resistance assay under disease nursery conditions

VW resistance assays of LHB22×NNG population was performed in an artificially infected disease nursery at the CRS/SDAU in 2011 and 2012. The VW disease nursery was inoculated with cotton seed cultured with isolate V991 for three years since 2007 and once every two years since 2010, resulting in a relatively severe and uniform soil infection by VW. Seeds of plant materials were acid-delinted before use. All $F_{2:3}$ families were planted in two fully randomized replications with three replications of two parents evenly distributed among them. Each plot had two rows of 5.0 m long in 2011 and one row of 8.0 m long in 2012. The row spacing and plant spacing were 50 cm and 15 cm respectively in two years. The planting date was April 27 in 2011 and April 30 in 2012.

VW resistance was evaluated at early flowering stage on July 15 only in 2011 and at mature stage on October 6–8 in both 2011 and 2012. The disease scores at early flowering stage in 2012 were not collected due to relatively severe drought unsuitable for VW development. Leaf disease symptoms at early flowering stage were scored with a 5-score system ranging from 0 to 4 used as national standard for screening cotton VW resistance in China [16,18,19], where score 0 represents healthy plant without symptoms, score 1 represents <25% chloropic/necrotic leaves, score 2 represents 25–50% chloropic/necrotic leaves, score 3 represents 50–75% chloropic/necrotic leaves, and score 4 represents >75% chloropic/necrotic leaves or complete defoliation and plant death. The vascular tissue symptoms were evaluated by the ratio of the length of symptomatic vascular tissue to the entire plant length [16] with a 5-score system ranging from 0 to 4, where score 0 represents healthy vascular without symptoms, score 1 represents <25% symptomatic vascular area, score 2 represents 25–50% symptomatic vascular area, score 3 represents 50–75% symptomatic vascular area, and score 4 represents >75% symptomatic vascular area. All plants (55–60 plants in 2011 and 45–50 plants in 2012 respectively) in a spot were scored to get spot value and averages of two replications were used in QTL analysis.

## VW resistance assay under naturally infected field conditions

In order to further test the effect of the major resistance QTL $qVW-c6-1$ which was detected on Chr.6 at more generations, four bulked $F_4$ of the LHB22×NNG population were made through the Graphical GenoTypes software GGT32 (http://www.dpw.wau.nl/pv/pub/ggt/) based on the molecular marker data, including two resistant bulks (BulkR1, BulkR2) and two susceptible bulks (BulkS1, BulkS2). The two resistant bulks contained a homozygous chromosome segment between markers GH433 and DPL0665 from LHB22 (containing the resistance allele of $qVW-c6-1$), while the two susceptible bulks possessed homozygous chromosome segments of the same marker interval from NNG (not containing the resistance allele of $qVW-c6-1$). Each bulk was comprised of equal volume of seeds from 8 different $F_{2:3}$ families. This experiment was conducted in 2012 in a field severely infected by mixed isolates including V991 of $V. dahliae$ in Dezhou, Shandong province, which belongs to the Yellow River cotton growing region of China. The four $F_4$ bulks and the two parents (LHB22, NNG), together with another cotton ($G. hirsutum$) variety Yu2067 which is widely used as a resistant control in VW resistance evaluation test in China, were planted on April 26 in a randomized complete block design with three replications. Each replicate had two rows of 10 m long. The row space was 80 cm and plant space was 33 cm. Twenty plants in the middle of each plot were used for VW resistance evaluation. VW resistance was scored at flowering stage (July 8) and at mature stage (October 7) with the same methodology as described above.

## DNA isolation and SSR assays

Genomic DNA was isolated from young leaves from each $F_2$ plant of the two populations and the three parental varieties using the CTAB method [30]. The SSR technique was used to identify polymorphic markers for construction of a genetic linkage map. Mainly based on two previously reported maps [31,32], a total of 5400 SSR primers were used to screen polymorphisms between LHB22 and JM11. PCR amplifications were conducted on a Peltier Thermal Cycler (M J Research) using the PCR reaction mixture of a previous work [33] following the reaction program: 95°C for 2 min; 30 cycles of 94°C for 45 s, 57°C for 45 s and 72°C for 60 s; and a final extension at 72°C for 10 min. SSR primers were obtained from the Cotton Marker Database (CMD) (http://www.cottonmarker.org). PCR products were separated

and visualized using sodium dodecyl sulfate polyacrylamide gel electrophoresis and silver staining method [34].

## Map construction and QTL mapping

Genetic maps were constructed using JoinMap 3.0 [35]. A minimum LOD value of 4.0 and a recombination frequency of 0.40 were used as thresholds to identify linkage groups. Marker distances in centi Morgan (cM) were estimated with the Kosambi mapping function [36]. Linkage groups were assigned to chromosomes by bridge loci with existing consensus maps of tetraploid cotton [31,32,37]. QTL detection was performed by composite interval mapping (CIM) using Windows QTL Cartographer 2.5 (http://statgen.ncsu.edu/qtlcart/WQTLCart.htm). A stringent LOD threshold through 1000-permutation test was utilized to determine significant QTL. The percentage of the phenotypic variation explained by a QTL was estimated at the highest probability peak.

## Data analysis

Data analysis was performed with SPSS18 software. For the bulked $F_4$ assay, a one-way ANOVA was conducted and the least significant difference (LSD) was used to verify the differences between the varieties or genotypes at p = 0.05. Pearson correlation coefficients between disease severity scores at two stages were also calculated.

## Results

### VW resistance evaluation

Plant disease scores of both parents and hybrid progeny ($F_1$ and $F_{2:3}$) families were assayed for leaf symptoms at seedling stage and vascular symptoms at mature stage (Table 1). In greenhouse, both LHB22 and LHB2×JM11 $F_1$ showed high resistance (low disease scores) to the isolate V991 at seedling stage, whereas the susceptible JM11 exhibited significant higher disease scores. In the disease nursery, LHB22 and LHB2×JM11 $F_1$ showed higher resistance (0.89–1.09; 1.03–1.25) than the susceptible parent NNG (3.41–3.56). Although the disease scores of the two $F_1$s were higher than LHB22 both in greenhouse and disease nursery evaluations, the differences were not statistically significant. These data suggested a possible dominant inheritance pattern of LHB22's VW resistance. Continuous and normal distribution of disease scores was observed in both $F_{2:3}$ populations grown in a greenhouse and disease nursery at different stages (Figure 1). A significant positive correlation (r = 0.867, $p$ = 0.0078) was observed between early flowering stage resistance and mature stage resistance in LHB22×NNG $F_{2:3}$ population in 2011.

### Linkage map construction

Of the 5400 SSR primer pairs screened, 186 (3.4%) revealed polymorphism between parents LHB22 and JM11. These 186 polymorphic markers were used to genotype the $F_2$ population in order to construct a genetic map. Nine markers did not produce clear and stable PCR products among the population, and were discarded. Of the 177 markers analyzed, a total of 182 segregating loci were scored. Of them, 13 exhibited significant deviation from expected Mendelian segregation ratios for 1:2:1 or 3:1 in chi-square tests, and were excluded when mapping. Finally 141 SSR marker loci were mapped on 26 linkage groups, covering 1143.1 cM with an average distance of 8.11 cM between adjacent markers. According to the consensus maps [31,32,37], these 26 linkage groups were assigned to 23 chromosomes. No markers were mapped on Chr.2, 20 and 25. The whole linkage map is available in Figure S1.

**Table 1.** Average disease severity scores of parents and populations.

| Parents and populations | Average disease severity scores | | | |
| | Seedling stage | Flowering stage | Mature stage | |
| | Greenhouse (2010) | Nursery (2011) | Nursery (2011) | Nursery (2012) |
|---|---|---|---|---|
| LHB22 | 0.65 [aA‡] | | | |
| JM11 | 3.38 [bB] | | | |
| F₁ | 0.77 [aA] | | | |
| F₂:₃ | 2.45 | | | |
| LHB22 | | 0.89 [aA] | 1.04 [aA] | 1.09 [aA] |
| NNG | | 3.41 [bB] | 3.56 [bB] | 3.47 [bB] |
| F₁ | | 1.03 [aA] | 1.25 [aA] | 1.21 [aA] |
| F₂:₃ | | 2.65 | 2.77 | 2.58 |

[‡]Differences among parents and F₁ of a population were verified by the LSD analysis, and different lowercase letter and capital letter in the same column show difference significant at 0.05 and 0.01 level respectively.

## QTL mapping of VW resistance at seedling stage under greenhouse conditions

Based on a Composite Interval Mapping (CIM) analysis of the disease score data of the LHB22×JM11 population grown in a greenhouse, two VW resistance QTL were identified on Chr.6 and 21. The nearest marker, LOD value, peak position, marker interval, genetic effects and the percent PV explained are summarized in Table 2 and Figure 2, Figure 3). A major resistance QTL, *qVW-c6-1*, explaining 25.90% of the phenotypic variance, was detected tightly linked to MGHES18 on Chr. 6. The additive effect of this QTL could decrease the VW disease score by 0.66. The resistance allele originated from the resistant parent LHB22. The other QTL *qVW-c21-1* was mapped on Chr.21 with a LOD value of 3.4 and could explain 14.44% of the phenotypic variation, whose resistant allele was donated by the susceptible parent JM11. These results also indicated both resistant and susceptible parents may provide resistance QTL for VW resistance improvement in breeding.

## Verification of the QTL *qVW-c6-1*

In order to verify the major resistance QTL *qVW-c6-1* originated from LHB22 in a different genetic background at different growth stages, another F₂:₃ population (LHB22×NNG) was developed and evaluated for VW resistance in an artificially infected disease nursery of V991 at early flowering stage and mature stage in two years. As the *qVW-c6-1* was mapped on Chr.6, the seven SSR markers, mapped on Chr.6 in the LHB22×JM11 population, were first used to screen polymorphism between LHB22 and NNG. Only two markers MGHES18 and GH433 were polymorphic. Then 217 SSR primers previously mapped on Chr.6 available at CMD (http://www.cottonmarker.org) and by other works [16,31,32,37–39], were selected to screen polymorphism between the two parents. Six polymorphic markers were identified. Finally, a linkage map covering 84.3 cM with 8 SSR markers was constructed (Figure 2) and utilized in QTL analysis with the same methodology mentioned above. One major resistance QTL each for leaf symptom in 2011 and vascular symptom in 2011 and 2012, respectively, was detected on Chr.6

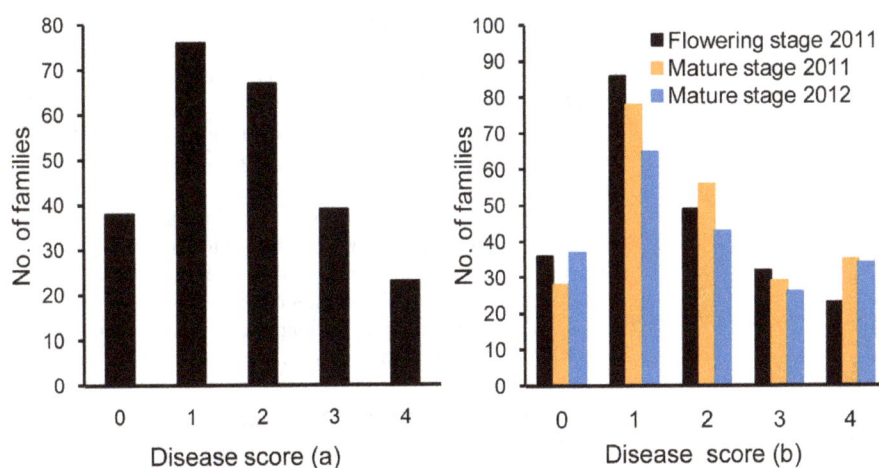

**Figure 1. The frequency distribution of disease scores.** (a) LHB22×JM11 F₂:₃ population was screened at seedling stage in greenhouse in 2010. (b) LHB22×NNG F₂:₃ population was evaluated in Verticillium wilt disease nursery at two stages in 2011 and 2012.

**Table 2.** QTL conferring resistance to Verticillium wilt detected in two $F_{2:3}$ populations of *G. hirsutum*.

| Population | Growing stage[a] | QTL | Nearest marker | Position (cM)[b] | LOD value | Additive effect (a) | Dominant effect (d) | d/a[c] | Variation explained (%)[d] | Donor parent[e] |
|---|---|---|---|---|---|---|---|---|---|---|
| LHB22×JM11 | S (2010) | qVW-c21-1 | DPL0050 | 20.6 | 3.4 | 0.31 | −0.79 | 2.54 | 14.44% | JM11 |
|  | S (2010) | qVW-c6-1 | MGHES18 | 28.8 | 3.7 | −0.66 | −0.65 | 0.98 | 22.90% | LHB22 |
| LHB22×NNG | F (2011) | qVW-c6-1 | MGHES18 | 34.5 | 4.2 | −0.72 | −0.78 | 1.08 | 25.70% | LHB22 |
|  | M (2011) | qVW-c6-1 | MGHES18 | 34.5 | 4.4 | −0.88 | −0.75 | 0.85 | 27.10% | LHB22 |
|  | M (2012) | qVW-c6-1 | MGHES18 | 34.5 | 3.9 | −0.70 | −0.75 | 1.07 | 23.10% | LHB22 |

a Verticillium wilt (VW) resistance of LHB22×JM11and LHB22×NNG population evaluated in a greenhouse and disease nursery, respectively. S = seedling stage, F = flowering stage, M = mature stage.
b Position of peak LOD score.
c Ratio of dominant effect and additive effect.
d Percentage of variance explained at peak LOD score.
e Parent that provided the positive allele of the QTL.

with LOD value 3.9–4.4, explaining 23.1–27.1% of phenotypic variation (Table 2 and Figure 2). As these three QTL shared the same nearest marker MGHES18 and all the positive alleles of these QTL originated from LHB22, it was concluded that they were the same QTL. Although there were only two common markers (GH433 and MGHES18) between the linkage groups of Chr. 6 from the two mapping populations, comparison of QTL positions could be roughly made through a bride consensus map [37] (Figure 2), which suggested that they could be the same QTL. This result also indicated that *qVW-c6-1* could be repeatedly detected in these two genetic backgrounds.

## VW resistance in the bulked $F_4$

The VW disease scores in bulked $F_4$ families were summarized in Table 3. There were statistically significant differences in disease scores at flowering stage and mature stage between genotypes as determined by one-way ANOVA ($F = 131.625$–$251.082$, $p = 0.008$-$0.000$). Yu2067, LHB22 and the two resistant bulks exhibited significantly lower disease scores than the two susceptible bulks and NNG at early flowering and mature stages. No significant difference in disease scores was detected between Bulk S1 and Bulk S2, or between Bulk R1 and Bulk R2. Although the disease scores of the two resistant bulks and LHB22 were lower than those of Yu2067, the differences among them were not statistically significant except for disease score at mature stage between Bulk R1 and Yu2067. The disease scores at flowering stage was positively correlated with that at mature stage ($r = 0.94$, $p = 0.000$). These results indicated that the VW resistance of LHB22 was stable from flowering stage to mature stage, and marker-assisted selection of the positive allele of QTL *qVW-c6-1* from LHB22 could increase VW resistance in $F_4$ generation effectively.

## Discussion

### VW resistance to isolate V991

The identification of gene/QTL conferring resistance/tolerance to VW is essential for the development of resistant/tolerant varieties in Upland cotton. The V991 is a prevailing defoliating *V. dahliae* isolate in the Yellow and the Yangtze River cotton growing regions in China [19]. As for the resistance to V991, six resistance QTL have been previously mapped, including three QTL on Chr. 2 and Chr. 16 in an $F_{2:3}$ population from Lumianyan22 × Luyuan343 [20], and three QTL on Chr. 9, Chr.17 and Chr.23 in an recombinant inbred lines (RILs) population from Prema×86-1 [19]. These results implied that different germplasms might possess different resistance QTLs for the same VW isolate V991, which was also investigated by VW resistance genes expression analyses [22–25]. In the present study, two QTL (*qVW-c6-1, qVW-c21-1*) conferring resistance to V991 were identified on Chr. 6 and Chr. 21 explaining 14.44–27.1% of the phenotypic variance and have seldom been reported as controlling VW thus far, which suggests that these two QTL are new QTL resistant to V991. The results of this work offered necessary complement to further understanding of the genetic base of VW resistance to V991 in cotton. These results together provide the possibility of developing new resistant cotton lines by QTL pyramiding through MAS.

## Possible reasons for the differences in QTL results between this research and a previous study [20]

It is necessary to discuss the possible reasons for the differences in QTL results (mentioned in the above paragraph) between this work and a previous research [20] because the resistant parent Lumianyan22 (L22) in that research is the recurrent parent of the

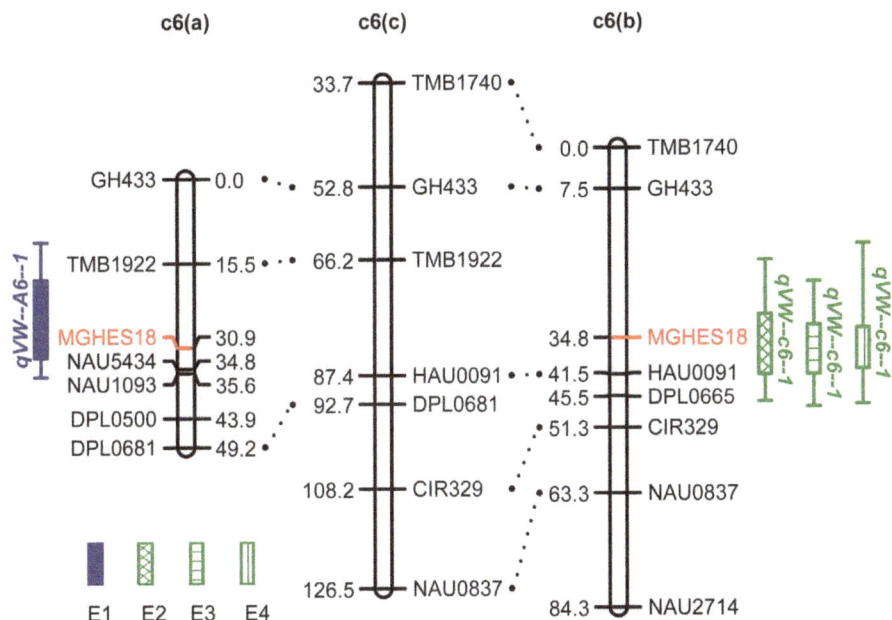

**Figure 2. A major QTL conferring resistance to Verticillium wilt detected on chromosome 6.** Linkage group c6 (a) was constructed from LHB22×JM11 $F_{2:3}$ population and linkage group c6 (b) was constructed from LHB22×NNG $F_{2:3}$ population. The c6 (c) map was drawn according to a consensus map of cotton [31] and only the related markers were selected. The QTL *qVW-c6-1* was detected in LHB22×JM11 population in 2010 at seedling stage in greenhouse (E1), and was further detected at flowering stage in 2011 (E2) and mature stage in 2011 (E3) and 2012 (E4) in LHB2×NNG $F_{2:3}$ population.

LHB22 used as resistant parent in the present study [27]. In that work, based on a genetic map of ten markers on three Chr., three QTL including two (*qVWR-c16-1*, *qVWR-c2-1b*) from L22 were detected at flowering stage in an $F_{2:3}$ population in artificially infected disease nursery with mixed VW isolates. In the present research, QTL mapping was performed in the LHB22×JM11 population at seedling stage under greenhouse conditions based on a genetic map of 141 markers distributed on 26 chromosomes. And after detecting the two QTL (*qVW-c6-1*, *qVW-c21-1*), we focused on validation of the *qVW-c6-1* in the LHB22×NNG population at flowering and mature stages only in Chr. 6. In other words, QTL detection was not performed on Chr. 2 and 16 and other chromosomes at flowering stage in the present research. We performed QTL mapping at genome level only at seeding stage under greenhouse conditions in LHB22×JM11 population. In addition to the introgression of genetic components from *G. bickii* in LHB22, the differences in disease infection methods, VW isolates used and marker coverage in these two researches might be possible reasons for not detecting those two QTLs (*qVWR-c16-1*, *qVWR-c2-1b*) at seedling stage in the LHB22×JM11 population in the present study, while these factors have been reported to seriously impact VW resistance QTL mapping results in previous researches [7,16,26,40]. For example, some Pima (*G. barbadense*) lines with high VW resistance under naturally infected field conditions were detected as susceptible in VW array under greenhouse conditions [7].

## The stability of QTL and its further use in breeding

The aim of QTL mapping is for MAS, so the stability and facility of QTL were really important. In this study, the major QTL *qVW-c6-1* could be detected in the JM11 background at seedling stage under greenhouse conditions and in the NNG background at flowering and mature stages under artificially infected disease nursery conditions. And the bulked $F_4$ experiment indicated the marker-assisted selection of the positive allele of this QTL with two flanking markers was also effective in the NNG background. It should be noted that our present results just indicated this QTL to be effective in these two susceptible variety/line backgrounds, and whether it is effective in other genetic backgrounds need to be tested in further researches. In view of the relative low QTL mapping solution in the present study, MAS using flanking markers on both sides combined with phenotypic screening in a large population was suggested in order to improve selection efficiency once the present results were used to transfer this QTL into other genetic backgrounds. This selection strategy

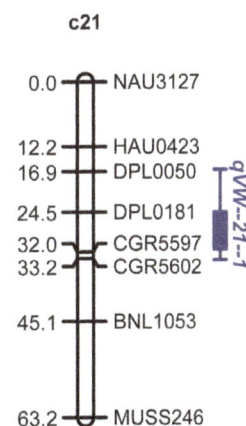

**Figure 3. A Verticillium wilt minor resistance QTL detected on chromosome 21 in LHB22×JM11 $F_{2:3}$ population.** The susceptible parent JM11 donates the positive allele of this QTL.

**Table 3.** The disease scores in bulked $F_4$ families of LHB22×NNG population.

| Genotype | Disease scores | |
| --- | --- | --- |
| | **Flowering stage** | **Mature stage** |
| Bulk S1 | 2.85 [Ab‡] | 3.16 [Ab] |
| Bulk S2 | 2.98 [Ab] | 3.26 [Ab] |
| Bulk R1 | 1.19 [Bc] | 1.27 [Bd] |
| Bulk R2 | 1.17 [Bc] | 1.34 [Bcd] |
| NNG | 3.44 [Aa] | 3.50 [Aa] |
| LHB22 | 1.20 [Bc] | 1.39 [Bcd] |
| Yu2067 | 1.22 [Bc] | 1.47 [Bc] |

‡Different lowercase letter and capital letter in the same column show difference significant at 0.05 and 0.01 level respectively.

has been proved effective in many researches [41,42]. In practice, all MAS schemes will be used in the context of the overall breeding program, and this will involve phenotypic selection at various stages [43]. In addition, the development of a RILs population from the LHB22×NNG cross is in progress and it will be applied to evaluate the QTL detected in more environments, to narrow the QTL confidence interval by combining SSR markers and other markers, such as single nucleotide polymorphisms (SNPs), and to detect putative QTLs located on other chromosomes.

## Supporting Information

**Figure S1 The linkage map constructed from the LHB22×JM11 F2:3 population.** This map contains 141 SSR marker loci on 26 linkage groups assigned to 23 chromosomes,

covering 1143.1 cM with an average distance of 8.11 cM between adjacent markers. No markers were mapped on Chr.2, 20 and 25.

## Acknowledgments

We thank Drs. Heqin Zhu of the Institute of Cotton Research and Guiliang Jian of Institute of Plant Protection of Chinese Academy of Agricultural Sciences (CAAS) for providing *V. dahliae* isolates and seeds of Jimian11 and Yu2067.

## Author Contributions

Conceived and designed the experiments: XLS XZS. Performed the experiments: XZ YY ZW XG YG SZ GZ. Analyzed the data: XZ XLS XZS. Contributed reagents/materials/analysis tools: JZ. Wrote the paper: XZ XLS XZS. XZ and YY contributed equally to this research.

## References

1. Song XL and Zhang TZ (2007) Identification of quantitative trait loci controlling seed physical and nutrient traits in cotton. Seed Sci Res 17:243–251.
2. Carpenter CW (1914) The Verticillium wilt problem. Phytopathology 4:393.
3. Bolek Y, El-Zik KM, Pepper AE, Bell AA, Magill CW, et al. (2005) Mapping of verticillium wilt resistance genes in cotton. Plant Sci 168: 1581–1590.
4. Wang Y, Chen DJ, Wang DM, Huang QS, Yao ZP, et al. (2004) Over expression of gastrodia anti-fungal protein enhances verticillium wilt resistance in colored cotton. Plant Breed 123: 454–459.
5. Zhou T and Dai X (2006) Research on physiological and biochemical mechanism of cotton against Verticillium wilt. Mol Plant Breed 14:593–600.
6. Fradin EF and Thomma BPHJ (2006) Physiology and molecular aspects of Verticillium wilt disease caused by *V. dahliae* and *V. albo-artrum*. Mol Plant Pathol 7:71–86.
7. Zhang JF, Sanogo S, Flynn R, Baral JB, Bajaj S, et al. (2012) Germplasm evaluation and transfer of Verticillium wilt resistance from Pima (*Gossypium barbadense*) to upland cotton (*G. hirsutum*). Euphytica 187:147–160.
8. Berlanger I and Powelson ML (2000) Verticillium wilt. The Plant Health Instructor. Avaliable: http://www.apsnet.org/edcenter/intropp/lessons/fungi/ascomycetes/Pages/VerticilliumWilt.aspx. doi: 10.1094/PHI-I-PHI-1-2000-0801-01. Updated 2005.
9. Tjamos EC, Tsitsiyannis DI, Tjamos SE (2000) Selection and evaluation of rhizosphere bacteria as biocontrol agents against *Verticillium dahliae*. In: Tjamos EC, Rowe RC, Heale JB, Fravel DR, editors. Advances in Verticillium Research and Disease Management. St. Paul: American Phytopathological Society (APS) Press. pp. 244–248.
10. Cai YF, He XH, Mo JC, Sun Q, Yang JP, et al. (2009) Molecular research and genetic engineering of resistance to Verticillium wilt in cotton: A review. Afr J Biotechnol 25: 7363–7372.
11. Du WS, Du XM, Ma ZY (2004) Studies on SSR markers of resistance gene of Verticillium wilt in cotton. J Northwest Sci Tech Univ Agri and For 3: 20–24.
12. Gao YQ, Nie YC, Zhang XL (2003) QTL mapping of genes resistant to Verticillium wilt in cotton. Cotton Sci 2:73–78.
13. Fang H, Zhou HP, Sanogo S, Flynn R, Percy RG, et al. (2013) Quantitative trait locus mapping for Verticillium wilt resistance in a backcross inbred line population of cotton (*Gossypium hirsutum*×*Gossypium barbadense*) based on RGA-AFLP analysis. Euphytica 194:79–91.

14. Fang H, Zhou HP, Sanogo S, Lipka AE, Fang DD, et al. (2013) Quantitative trait locus analysis of Verticillium wilt resistance in an introgressed recombinant inbred population of upland cotton. Mol Breed 33:7-9-720.
15. Wang HM, Lin ZX, Zhang XL, Chen W, Guo XP, et al. (2008) Mapping and quantitative trait loci analysis of Verticillium wilt resistance genes in cotton. J Integr Plant Biol 2:174–182.
16. Yang C, Guo WZ, Li GY, Gao F, Lin SS, et al. (2008) QTLs mapping for Verticillium wilt resistance at seedling and maturity stages in *Gossypium barbadense* L. Plant Sci 174: 290–298.
17. Zhen R, Wang XF, Ma ZY, Zhang GY, Wang X (2006) A SSR marker linked with the gene of Verticillium wilt resistance in *Gossypium barbadense*. Cotton Sci 5:269–272.
18. Jiang F, Zhao J, Zhuo L, Guo WZ and Zhang TZ (2009) Molecular mapping of Verticillium wilt resistance QTL clustered on chromosomes D7 and D9 in upland cotton. Sci China Ser C (life Sci) 9: 872–884.
19. Ning ZY, Zhao R, Chen H, Ai NJ, Zhang X, et al. (2013) Molecular tagging of a major quantitative trait locus for broad-spectrum resistance to Verticillium wilt in upland cotton cultivar Prema. Crop Sci 53:2304–2312.
20. Wang FR, Liu RZ, Wang LM, Zhang CY, Liu GD, et al. (2007) Molecular marker of Verticillium resistance in upland cotton (*Gossypium hirsutum* L.) cultivar and their effects on assisted phenotypic selection. Cotton Sci 6:424–430.
21. Yang C, Guo WZ and Zhang TZ (2007) QTL mapping for resistance to Verticillium wilt, fiber quality and yield traits in upland cotton (*Gossypium hirsutum* L). Mol Plant Breed 5(6):797–805.
22. Zhang WW, Wang SZ, Liu K, Si N, Qi FJ, et al. (2012) Comparative expression analysis in susceptible and resistant *Gossypium hirsutum* responding to *Verticillium dahliae* infection by cDNA-AFLP. Physiol Mol Plant Pathol 80: 50–57.
23. Sun Q, Jiang H, Zhu X, Wang W, He X, et al. (2013) Analysis of sea-island cotton and upland cotton in response to *Verticillium dahliae* infection by RNA sequencing. BMC Genomics 14:852. Available: http://www.biomedcentral.com/147-2164/14/852. Accessed December 5 2013.
24. Xu L, Zhu L, Tu L, Liu L, Yuan D, et al. (2011) Lignin metabolism has a central role in the resistance of cotton to the wilt fungus *Verticillium hahliae* as revealed by RNA-Seq-dependent transcriptional analysis and histochemistry. J Exp Bot 62: 5607–5621.
25. Gao W, Long L, Zhu L, Xu L, Gao W, et al. (2013) Proteomic and virus-induced gene silencing (VIGS) analyses reveal that gossypol, brassinosteroids and

jasmonic acid contribute to the resistance of cotton to *Verticillium dahliae*. Mol Cell Proteomics 12(12): 3690–703. Available: http://www.ncbi.nlm.nih.gov/pubmed/24019146. Accessed September 9 2013.

26. Bejarano-Alcázar J, Blanco-López MA, Melero-Vara JM, Jiménez-Díaz RM (1997) The influence of Verticillium wilt epidemics on cotton yield in southern Spain. Plant Pathology 46:168–178.

27. Gao MW, Wang LM, Wang J, Yu YJ, Wang XL, et al. (2008) Comparison of Cotton HB-red flower near-isogenic lines in the resistance to seedling diseases, Fusarium wilt and Verticillium wilt. Shandong Agri Sci 1:67–70.

28. Zhang XJ, Zhang MW, Wu MJ, Ma J, Wang F, et al. (2011) Physiological and genetic studies on resistance to Verticillium wilt of new cotton germplasm LU HB22. Shandong Agr Sci 7:64–68.

29. Liang Z, Jiang R, Zhong W (1996) Development of Hirsutum-Bickii hybrid germplasms with red flowers. Sci China Ser C (life Sci) 26(4):369–376.

30. Paterson AH, Brubaker CL, Wendel JF (1993) A rapid method for extraction of cotton (*Gossypium* spp.) genomic DNA suitable for RFLP or PCR analysis. Plant Mol Biol Rep 11: 122–127.

31. Guo WZ, Cai CP, Wang C, Zhao L, Wang L, et al. (2008) A preliminary analysis of genome structure and composition in *Gossypium hirsutum*. BMC Genomics 9:314. Available: http://www.biomedcentral.com/1471-2164/9/314. Accessed July 1 2008.

32. Yu JW, Yu SX, Lu CR, Wang W, Fan SL, et al. (2007) High-density Linkage Map of Cultivated Allotetraploid Cotton Based on SSR, TRAP, SRAP and AFLP Markers. J Integr Plant Biol 49 (5):716–724.

33. Wang PZ, Su L, Qin L, Hu BM, Guo WZ, et al. (2009) Identification and molecular mapping of a Fusarium wilt resistant gene in upland cotton. Theor Appl Genet 119:733–739.

34. Zhang J, Guo WZ, Zhang TZ (2002) Molecular linkage map of allotetraploid cotton (*Gossypium hirsutum* L×*Gossypium barbadense* L) with a haploid population. Theor Appl Genet 105: 1166–1174.

35. Van OJ, Voorrips RE. JoinMap Version 3.0, Software for the calculation of genetic linkage maps. Plant Research International, Wangeningen, the Netherlands, 2001.

36. Kosambi DD (1944) The estimation of map distances from recombination values. Ann Eugenic 12:172–175.

37. Blenda A, Fang DD, Rami JF, Garsmeur O, Luo F, et al. (2012) A high density consensus genetic map of tetraploid cotton that integrates multiple component maps through molecular marker redundancy check. PLoS ONE 7(9): e45739. doi:10.1371/journal.pone.0045739. AccessedSeptember 24 2012.

38. Qin HD, Guo WZ, Zhang YM and Zhang TZ (2008) QTL mapping of yield and fiber traits based on a four-way cross population in *Gossypium hirsutum* L. Theor Appl Genet 117: 883–894.

39. Shen XL, Guo WZ, Lu QX, Zhu XF, Yuan YL, et al. (2007) Genetic mapping of quantitative trait loci for fiber quality and yield trait by RIL approach in Upland cotton. Euphytica 155:371–380.

40. Devey ME and Rosielle AA (1986) Relationship between field and greenhouse ratings for tolerance to Verticillium wilt on cotton. Crop Sci 1:1–4.

41. Lande R, Thompson R (1990) Efficiency of marker-assisted selection in the improvement of quantitative traits. Genetics 124:743–756.

42. Zhou RH, Zhu ZD, Kong XY, Huo NX, Tian QZ, et al. (2005) Development of wheat near-isogenic lines for powdery mildew resistance. Theor Appl Genet 110:640–648.

43. Collard BCY and Mackill DJ (2008) Marker-assisted selection: an approach for precision plant breeding in the twenty-first century. Phil Trans R Soc B 363, doi: 10.1098/rstb.2007.2170.

# Upland Cotton Gene *GhFPF1* Confers Promotion of Flowering Time and Shade-Avoidance Responses in *Arabidopsis thaliana*

Xiaoyan Wang[1,2], Shuli Fan[2], Meizhen Song[2], Chaoyou Pang[2], Hengling Wei[2], Jiwen Yu[2], Qifeng Ma[1,2], Shuxun Yu[2]*

1 College of Agronomy, Northwest A&F University, Yangling, Shaanxi, People's Republic of China, 2 State Key Laboratory of Cotton Biology, Institute of Cotton Research of CAAS, Anyang, Henan, People's Republic of China

## Abstract

Extensive studies on floral transition in model species have revealed a network of regulatory interactions between proteins that transduce and integrate developmental and environmental signals to promote or inhibit the transition to flowering. Previous studies indicated *FLOWERING PROMOTING FACTOR 1* (*FPF1*) gene was involved in the promotion of flowering, but the molecular mechanism was still unclear. Here, *FPF1* homologous sequences were screened from diploid *Gossypium raimondii* L. (D-genome, n = 13) and *Gossypium arboreum* L. genome (A-genome, n = 13) databases. Orthologous genes from the two species were compared, suggesting that distinctions at nucleic acid and amino acid levels were not equivalent because of codon degeneracy. Six *FPF1* homologous genes were identified from the cultivated allotetraploid *Gossypium hirsutum* L. (AD-genome, n = 26). Analysis of relative transcripts of the six genes in different tissues revealed that this gene family displayed strong tissue-specific expression. *GhFPF1*, encoding a 12.0-kDa protein (Accession No: KC832319) exerted more transcripts in floral apices of short-season cotton, hinting that it could be involved in floral regulation. Significantly activated *APETALA 1* and suppressed *FLOWERING LOCUS C* expression were induced by over-expression of *GhFPF1* in the *Arabidopsis* Columbia-0 ecotype. In addition, transgenic *Arabidopsis* displayed a constitutive shade-avoiding phenotype that is characterized by long hypocotyls and petioles, reduced chlorophyll content, and early flowering. We propose that *GhFPF1* may be involved in flowering time control and shade-avoidance responses.

Editor: David D. Fang, USDA-ARS-SRRC, United States of America

Funding: This research was supported by the Earmarked Fund for China Agriculture Research System (CARS-18) and funded by the National High-tech Research and Development Projects of China (2011AA10A102). The funders had no role in study design, data collection and analysis, decision to publish, or preparation of the manuscript.

Competing Interests: The authors have declared that no competing interests exist.

* E-mail: yu@cricaas.com.cn

## Introduction

Cotton (*Gossypium* spp.) is one of the most important natural fiber crops in the world. In addition to its economic importance, cotton has attracted considerable scientific interest from plant breeders, taxonomists, developmental geneticists, and evolutionary biologists because of its unique reproductive developmental aspects and speciation history [1–3]. The genus *Gossypium* L. contains more than fifty species, which are cytogenetically differentiated into eight genomic groups (A–G, and K). Most of the species are diploid (n = 13), but five are allopolyploids (n = 26), originating from an interspecific hybridization event between A- and D-genome diploid species. Not only two diploids of A-genome, *Gossypium herbaceum* L. and *G. arboreum* L., but also two allopolyploids of AD-genome, *G. hirsutum* L. and *G. barbadense* L. were domesticated by humans for their fiber demands [3]. Because of environmental pressures, such as land use and climatic change, the earliness of cotton has become a vital subject for plant breeders. Several traits co-operate to influence the early ripeness of upland cotton (*G. hirsutum* L.), with flowering time being especially important. In the seed crop, floral transition is a key developmental switch in the life cycle of cotton as it contributes to the production of dry matter. Furthermore, shifting of the seasonal timing of reproduction is a major goal of plant breeding research as it will produce novel varieties better adapted to local environments and effects of climate change [4].

Plant growth originates from a small number of undifferentiated cells called meristems. The apical meristems being indeterminate or determinate contribute to the fate of shoot architecture. Indeterminate apical meristems retain a population of vegetative stem cells indefinitely which perform tissue and organ differentiation below and on the flanks of the main-stem. Meanwhile the determinate apical meristems undergo terminal differentiation, commonly in a flower or inflorescence. In annual plants, the floral induction process occurs when vegetative shoot meristems develop into inflorescence meristems, and give rise to flowers [5]. The use of the model species *Arabidopsis thaliana* and *Antirrhinum majus* has led to significant progress in the understanding of the floral transition. Numerous genes involved in the control of flowering time have been identified, and the roles they played in molecular and genetic pathways were also characterized. Previous studies uncovered that four main floral pathways, namely vernalization, photoperiod, gibberellin, and autonomous pathways converged to regulate floral integrator genes such as *LEAFY* (*LFY*), *APETALA 1* (*AP1*), *FLOWERING LOCUS T* (*FT*) and *SUPPRESSOR OF OVEREXPRESSION OF CONSTANS1* (*SOC1*), which in turn could activate genes required for reproductive development [6]. Another transcription factor *FLOWERING LOCUS C* (*FLC*) containing MADS domain was an inhibitor of flowering, acting as an important convergence point for the autonomous and vernalization pathways in *Arabidopsis thaliana* [7–9].

*FLOWERING PROMOTING FACTOR 1* (*FPF1*) gene was originally understood on account of its role in flowering. Over-expression of *FPF1* (Y11988) led to early flowering in *Arabidopsis* [10]. *FPF1* is expressed in apical meristems immediately after the photoperiodic induction of flowering in long-day plants that could flower in response to long days [10]. Previous studies indicated that *FPF1* might play an important role in modulating the competence of apical meristems to rapidly respond to the floral meristem identity genes *AP1* and *LFY*. During the transition to flowering in *Arabidopsis*, *FPF1* is normally activated at a similar time as *LFY*, and earlier than *AP1*. Over-expression of *AtAP1* and *AtFPF1* shows a synergistic effect in the shortening of the time to flowering both under long-day and short-day conditions [11]. Two closely related genes, *FLP1* and *FLP2* (*FPF1-Like* genes), have been identified in *Arabidopsis*. Constitutive over-expression of each gene causes earlier flowering under both long and short day conditions [12]. Up till now, homologous genes of *FPF1* have also been characterized in rice (*Oryza sativa*) and tobacco (*Nicotiana tabacum*) [13,14]. *OsRAA1*, a homolog of *FPF1* in rice, shares similarity of 58% with *AtFPF1* at amino acids level. Evidence revealed that over-expression of *OsRAA1* causes pleiotropic phenotypes in transgenic rice plants, including altered leaf shape, heading time and root development [13].

In summarize, *AtFPF1/OsRAA1* gene family takes parts in several aspects of plant development. Currently, little is known about the underlying mechanism of these, or of any crossover between this gene family and other floral molecules. This paper described our work to dissect these mechanisms. In our previous study, a high-quality, normalized, full-length cDNA library with a total of 14,373 unique ESTs was generated to provide sequence information for gene discovery related to flower development in upland cotton [15]. The publication of *G. raimondii* L. genome sequence by the Cotton Genome Project in 2012 facilitates gene excavation vital for the genetic improvement of cotton quality and productivity, as well as serving as a reference for the assembly of the tetraploid *G. hirsutum* L. genome [16]. *G. raimondii* L. genome (http://cgp.genomics.org.cn/) and *G. arboreum* L. genome databases (unpublished) were screened to identify and compare homologs of *AtFPF1* from the two diploid cotton species. The genes were cloned and characterized from *G. hirsutum* L. as one of the major cultivated species. In addition, their expression pattern in cotton, and ectopic expression in *Arabidopsis* were analyzed.

## Materials and Methods

### Plant Material and Growth Conditions

A genetic standard line TM-1 (*G. hirsutum* L.) and a short-season cotton variety CCRI 36 (*G. hirsutum* L.) were grown in a greenhouse (Anyang, China) with an optimal temperature of 28°C. In order to minimize experimental error, two varieties were planted in the soil possessing equal fertility and managed using standard agricultural practices. Shoot apices from about three hundred plants of the two varieties were harvested when three true leaves came out. Roots, stems, young leaves, mature leaves,

flowers, and fibers from mature CCRI 36 and TM-1 plants were collected and mixed together.

Seeds of *Arabidopsis thaliana* (Columbia-0 ecotype: wild-type or *GhFPF1* over-expressing transgenic plants) were sterilized in 75% (v/v) absolute alcohol for 30 s and 10% $H_2O_2$ (v/v) for 10 min. After four washes in sterilized double-distilled water (dd$H_2O$), seeds were sown onto sterilized 1/2 MS solid medium (PH 5.8) containing 1.5% sucrose, 0.7% agar (and 50 mg/L kanamycin for transgenic lines). 10-cm petri plates containing medium and seeds were chilled for 48–72 h at 4°C in the dark, and then transferred into an illuminating incubator with 100 $\mu$mol·m$^{-2}$ s$^{-1}$ fluorescent light at 22°C. Two weeks later, normal seedlings were transplanted into the soil in a growth chamber which provided the plants with continuous 150 $\mu$mol·m$^{-2}$ s$^{-1}$ fluorescent light at room temperature 22°C. As for the length of hypocotyl and petiole, chlorophyll content and flowering time assay, seeds were directly sowed into the soil in the growth chamber. Ten days later, seedlings were transplanted into flowerpots grown under the same conditions. All *Arabidopsis* plants were grown under long-day conditions with high red to far red (R/FR ratio: 4.5) provided by fluorescent lamps.

### Gene Screening and Sequence Comparative Analysis

Candidate sequences were obtained by performing Blast searches of genome databases of *G. raimondii* L. (http://cgp.genomics.org.cn/) and *G. arboreum* L. (unpublished) using *Arabidopsis thaliana* FPF1 gene as a query sequence. Multiple sequence alignment of the deduced amino acids was performed using the software ClustalX2 (http://www.ebi.ac.uk). The phylogenetic tree was constructed by Molecular Evolutionary Genetics Analysis (MEGA) software 4.1 (http://www.megasoftware.net). Sequence of promoter was analyzed in the database of plant cis-acting regulatory DNA elements (PlantCARE) (http://bioinformatics.psb.ugent.be/webtools/plantcare/html/).

### Gene Isolation and RNA Analysis

Different organs from upland cotton and *Arabidopsis* were harvested, and immediately plunged into liquid nitrogen. Total RNA was isolated from cotton and *Arabidopsis* using the EASY spin plus RNA reagent kit RN38 (AIDLAB, Beijing, China) following manufacturer's instructions. Poly (dT) cDNA was prepared from the total RNA using the Superscript III First-Strand Synthesis System (Invitrogen, USA) according to the manufacturer's protocol. Genomic DNA was extracted as described [17].

Open reading frames (ORFs) of *GhFPF1* and homologous genes were amplified from *G. hirsutum* L. CCRI 36 using the primers described in Table S1. The running conditions were as follows: denaturation at 94°C for 2 min, followed by 33 cycles at 94°C for 30 s, annealing at 58°C for 1 min and extension at 72°C for 30 s. To obtain the 5′-terminal and 3′-terminal sequences of *GhFPF1*, 5′- and 3′-RACE were performed using the SMART RACE cDNA Amplification Kit (Clontech, USA) according to the Kit's protocol in conjunction with primers GPS1 and GPS2 (Table S1).

Quantitative polymerase chain reactions (QRT -PCRs) were performed to measure relative expression of homologous genes using specific primers (Table S2), with *Histone 3* (AF024716) to be an internal control [18]. For *Arabidopsis*, *AtUBQ5* (AT3G62250) was used as the control. Transcriptional changes of flowering related genes *AtFPF1*, *AtLFY*, *AtAP1*, *AtCO*, *AtFT*, *AtSOC1*, *AtFLC* and *AtPHYB* were examined in the wild as well as transgenic plants. Each qRT-PCR experimental condition was independently repeated three times and in each of these three biological repetitions, three technical replicates were made. All of the

**Table 1.** Pair-wise alignment between *G. raimondii* L. and *G. arboreum* L. orthologous sequences on amino acid and nucleic acid levels.

| *G. raimondii* L. | DF10021325 | DF10039615 | DF10007455 | DF10000980 | DF10029023 | DF10009151 |
|---|---|---|---|---|---|---|
| *G. arboreum* L. | Garb08271 | Garb17683 | Garb19766 | Garb07734 | Garb34130 | Garb22630 |
| **Similarities of nucleic acid/amino acids (Identity %)** | 97/100 | 99/100 | 95/98 | 99/99 | 97/99 | 84/86 |

amplification of interested genes was analyzed on ABI 7500 system (Applied Biosystems, USA) with SYBR Green I (with Rox) reagents to detect the target sequences. The running conditions were as follows: holding stage at 50°C for 2 min, 94°C for 10 min, followed by 40 cycles at 95°C for 15 s, 60°C for 1 min. Then a melting curve was performed from 65 to 95°C to verify the specificity of the amplified product. QRT-PCR data was processed to measure the relative expression of genes in accordance with the $2^{-\Delta\Delta CT}$ method [19].

## Jasmonic Acid and Salicylic Acid Treatment

Jasmonic acid (JA; Sigma, USA) was dissolved in a small quantity of ethanol, and further diluted with $ddH_2O$ to a concentration of 200 $\mu mol \cdot L^{-1}$. Salicylic acid (SA; Sigma, USA) was directly dissolved into heated $ddH_2O$ water to a final concentration of 400 $\mu mol \cdot L^{-1}$. Upland cotton cultivar CCRI 36 was planted in the greenhouse with an optimal temperature of 28°C. One-month-old plants were sprayed with JA and SA solutions when four true leaves had developed [20,21,36]. The top two leaves were removed at 0, 0.5, 6, 12, 24, and 48 h time-point. Harvested leaves were immediately plunged into liquid nitrogen, and stored at –80°C for later RNA extraction and relative transcript analysis. *Nonexpressor of Pathogenesis-Related Genes 1* (*GhNPR1*, DQ409173), a SA- and JA-inducible gene in upland cotton was used as a positive control gene here [21].

## Vector Construction and Genetic Transformation

The open reading frames of *GhFPF1* (GenBank Accession No. KC832319) was cloned into the binary vector pBI121 using an In-Fusions ™ Advantage PCR Cloning Kit (Clontech, USA) via *Bam*HI and *Sac*I sites (New England BioLabs, USA). The recombinant plasmid containing CaMV35S::*GhFPF1* was introduced into *Agrobacterium tumefaciens* (strain LBA4404) and then transformed into *Arabidopsis* (Columbia-0 ecotype) according to the method of floral dip [22]. Transformants were selected on 1/2 MS medium containing 50 mg $l^{-1}$ kanamycin, and further confirmed at both the genomic DNA and transcriptional mRNA level. To confirm that *GhFPF1*had been integrated into the *Arabidopsis* genome, RT-PCR was performed with the 35S promoter primer (forward) and the *GhFPF1*-specific primer (reverse). To detect the mRNA expression of *GhFPF1*in T3 transgenic lines COL-3, COL-4 and COL-7, 18-day-old intact plants were sampled for quantitative RT-PCR.

## Flowering Time Calculation and Chlorophyll Content Determination

At least twenty individual plants of the homozygous T3 transgenic lines and wild-type, were governed under the long-day conditions (16 h light/8 h dark). Flowering time was measured by counting the numbers of rosette and cauline leaves and days of the appearance of the first flower. Statistical significance analysis was conducted using the software Sigma Stat 3.5 (Systat Software, San Jose, CA, USA) by One-Way ANOVA

analysis. Leaves to determine the content of chlorophyll were harvested from 24-day-old *Arabidopsis*. Chlorophyll was extracted through 95% acetone-alcohol according to the method as described previously [23] and monitored in a 96-well microplate reader (BioTek, USA).

## Ethics Statement

We did not make use of human or vertebrate animal subjects and/or tissue in our research.

## Results

### Gene Screening and Sequence Comparative Analysis

To investigate the homologs of *FLOWERING PROMOTING FACTOR 1* (*FPF1*) in cotton, the genome databases of *G. raimondii* L. and *G. arboreum* L. were screened with the coding region of *Arabidopsis FPF1* gene as the reference sequence. Six sequences, which covered complete ORFs, were selected from each database. The twelve ORFs were predicted to encode small proteins of 99 to 113 amino acids in length. Pair-wise alignment of these predicted proteins suggested that they shared 50–68% similarity to the *AtFPF1* protein. In addition, these proteins were rich in the three amino acids Ser, Val, and Leu.

Further, pair-wise alignment of the amino and nucleic acids was conducted to explore the differences between *G. raimondii* L. and *G. arboreum* L. orthologous genes. Nucleic acid alignment revealed that five of the six orthologous pairs exhibited 95–99% similarity to each other, and the other pair DF10009151- Garb22630, had 84% identity (Table 1). Two pairs among them, DF10021325-Garb08271 and DF10039615-Garb17683, owned the same deduced amino acids sequence because of codon degeneracy. One or two amino acids were different between the three pairs of orthologous DF10000980-Garb07734, DF10029023-Garb34130 and DF10007455-Garb19766. The remaining pair, DF10009151-Garb22630, showed greater divergences, with just 86% homology at amino acid level. Comparative analysis revealed that a short length of nucleic acids was missing from Garb22630 (Figure S1).

### Isolation of *FPF1* Homologous Genes in *G. hirsutum* L. and Sequence Analysis

Six genes were identified from CCRI36 and named as *GhFPF1*, *GhFLP-1*, *GhFLP-2*, *GhFLP-3*, *GhFLP-4*, and *GhFLP-5* corresponding to the sequences of DF10009151, DF10007455, DF10029023, DF10021325, DF10000980, and DF10039615 in *G. raimondii* L., respectively. The small *FPF1* gene family has been characterized in several species including *Arabidopsis thaliana*, white mustard (*Sinapis alba*), rice (*Oryza sativa*), tobacco (*Nicotiana tabacum*) and maize (*Zea mays*). Multiple alignment (Figure 1A) of amino acid sequences revealed that distinctions between the homologous proteins occurred in the N-terminus relative to C-terminus. Results suggested that there were three conserved domains present in the protein family. The first motif, -LGWERY- was located in the middle section of the protein, while the second and third conserved motifs, -D/HLISLP- and -MY/FDIVVKN-, were found closer to the C-terminus. Phylogenetic analysis of the *FPF1* gene family indicated that these proteins could be divided into three clades, represented by *NtFPF1*, *ZmFPF1/OsRAA1* and *AtFPF1*. *GhFPF1*, *GhFLP-1*, *GhFLP-2*, and *GhFLP-3* were placed into *NtFPF1* clade; *GhFLP-4* and *GhFLP-5* were in the same clade with *AtFPF1*. None of them were placed into the *ZmFPF1/OsRAA1* branch (Figure 1B).

### GhFPF1 had Higher Transcriptional Levels in the Floral Apices of CCRI 36

QRT-PCR was used to profile relative expression of above six genes in different tissues of upland cotton. Roots, stems, leaves, flowers and fibers are mixed samples from CCRI 36 and TM-1 while floral apices T and C represent floral apices from CCRI 36 and TM-1, respectively (Figure 2). *GhFPF1* gene family displayed tissue-specific expression because abundant transcripts of the six genes were found in roots, floral apices, flowers, and stems, but were barely detectable in leaves or fibers. *GhFLP-3*, *GhFLP-4* and *GhFLP-5*, in particular, had only some but not much expression in

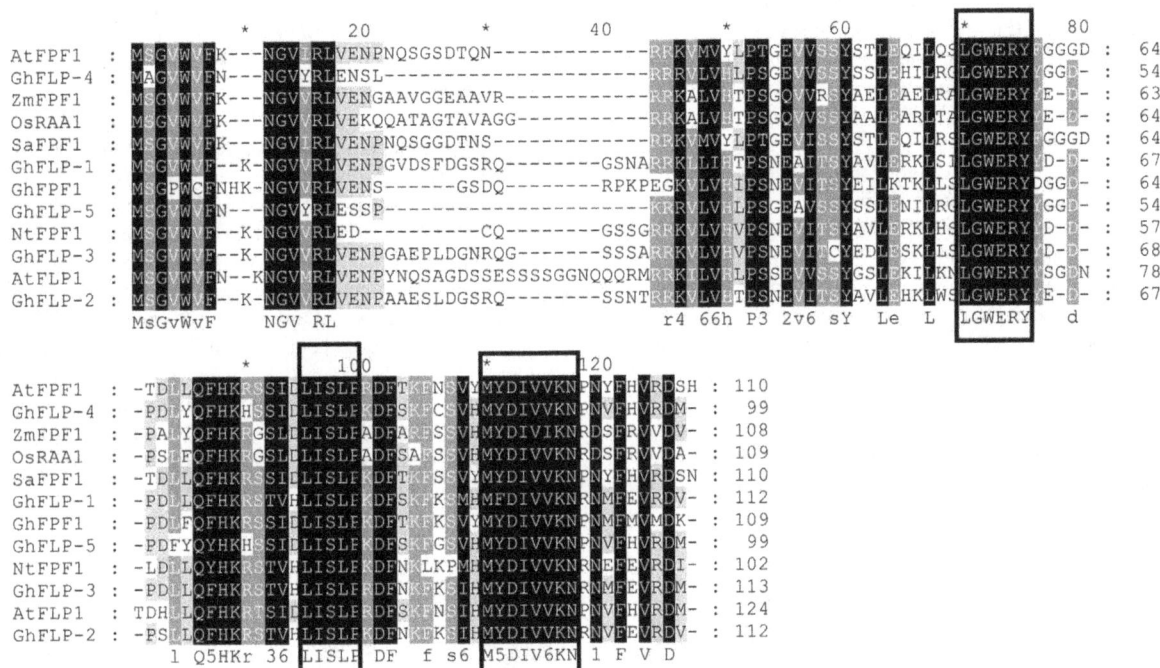

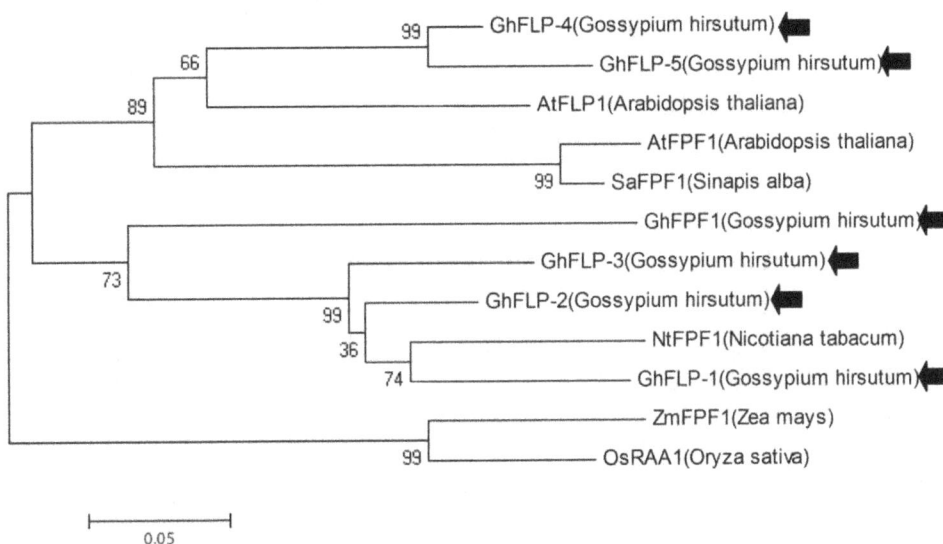

**Figure 1. Multiple sequence alignment of FPF1 protein family from *G. hirsutum* L., and other species.** A. Multiple alignment of FPF1 protein sequences in several species. AtFPF1 (Y11988) and AtFLP1 (AL353995) are Arabidopsis thaliana genes; SaFPF1 (Y11987), NtFPF1 (AY496934), ZmFPF1 (ACG44143) and OsRAA1 (AY659938) are from *Sinapis alba*, *Nicotiana tabacum*, *Zea mays* and *Oryza sativa*. B. Phylogenetic tree of the FPF1 proteins in the above plants as determined by the MEGA 4.1 software package.

floral apices relative to other tissues. More importantly, we focused on the contrastive analysis of gene expression in floral apices of CCRI 36 (a short-season cotton variety) and TM-1 (a genetic standard line). Results uncovered that *GhFPF1* had more than four-fold transcript levels in CCRI 36 than in TM-1. Higher expression of *GhFPF1* in the short-season cotton suggested that it was the most possible *FPF1* orthologous gene as *AtFPF1* involved in the promotion of flowering.

## Gene Structure Analysis of *GhFPF1* and its Response to JA and SA

A 5'- and 3'- RACE strategy was performed to gain transcription initiation and termination sites of *GhFPF1*. A full-length cDNA of 701 bp composed of 56 bp 5'-UTR, 315 bp 3'-UTR and 330 bp ORF was isolated from the cDNA pool of one-week-old seedlings (Figure 3A). Comparison of genomic and cDNA sequence revealed that there was no intron. Sequences of *GhFPF1* and *GhFPF1-like* genes were submitted to NCBI (Accession number: KC832319, KF830866-KF830870).

A 2076 bp promoter region of *GhFPF1* was identified from the genomic DNA of upland cotton, containing one and three response elements of methyl jasmonate and salicylic acid respectively. Treatment assay and the following qRT-PCR were performed to evaluate whether *GhFPF1* could be regulated by JA and SA plant hormones (Figure 3B). *NPR1* (*nonexpressor of pathogenesis-related genes 1*), a pathogen-related gene regulating SA-dependent defense response, systemic acquired resistance, and mediating crosstalk between SA and JA [24,25], was used as a SA- and JA-inducible reference gene. According to figure 3B, *GhFPF1* and *GhNPR1* exhibited the similar expression trend when plants were treated with SA and JA, respectively. After 12h treating with SA, the expression of *GhNPR1* reached its peak but peak

expression of *GhFPF1* occurred 12h later. However, *GhFPF1* was responsive to JA earlier than *GhNPR1*. *GhFPF1* transcripts were decreased 0.5h later and began to rise till 12h after the JA treatment, but the transition points of expression trend of *GhNPR1* was lagging. *GhFPF1* showed different expression fluctuations exposed to exogenous SA and JA but it was positively regulated by the two phytohormones over most time of 48h, suggesting that *GhFPF1* could be responsive to both JA and SA, the latter later.

## Over-expression of *GhFPF1* Promoted Flowering in *Arabidopsis*

To discuss whether *GhFPF1* could modulate flowering time in plants, the open reading frame of *GhFPF1*was transformed into the *Arabidopsis* Columbia-0 ecotype under the control of the cauliflower mosaic virus 35S promoter. Seven independent transgenic lines were obtained, and the *GhFPF1* transcript was analyzed in three transgenic lines COL-3, COL-4, and COL-7 (Figure 4D). Homozygous T3 lines and the wild-type were grown under long-day conditions of 16 h light/8 h dark. Twenty-four days after sowing, most plants of the transgenic lines had started bolting, while wild-type plants remained in vegetative growth with six rosette leaves developed (Figure 4A). Six days later, the transgenic plants flowered in succession, whereas the wild-type plants had just completed their basal rosette development at this point (Figure 4B and C). To determine flowering time, the number of rosette and cauline leaves and days of the first flower opening up were counted (Table 2). Results turned out that transgenic plant produced flower buds six days earlier than 36.8 days to flowering in the wild-type on average. This corresponds to a reduction in the number of leaves from 15.4 in wild-type to the fewest 11.5 in the transgenic line. Data demonstrated that over-expression of *GhFPF1* in *Arabidopsis* promoted flowering under inductive photoperiods.

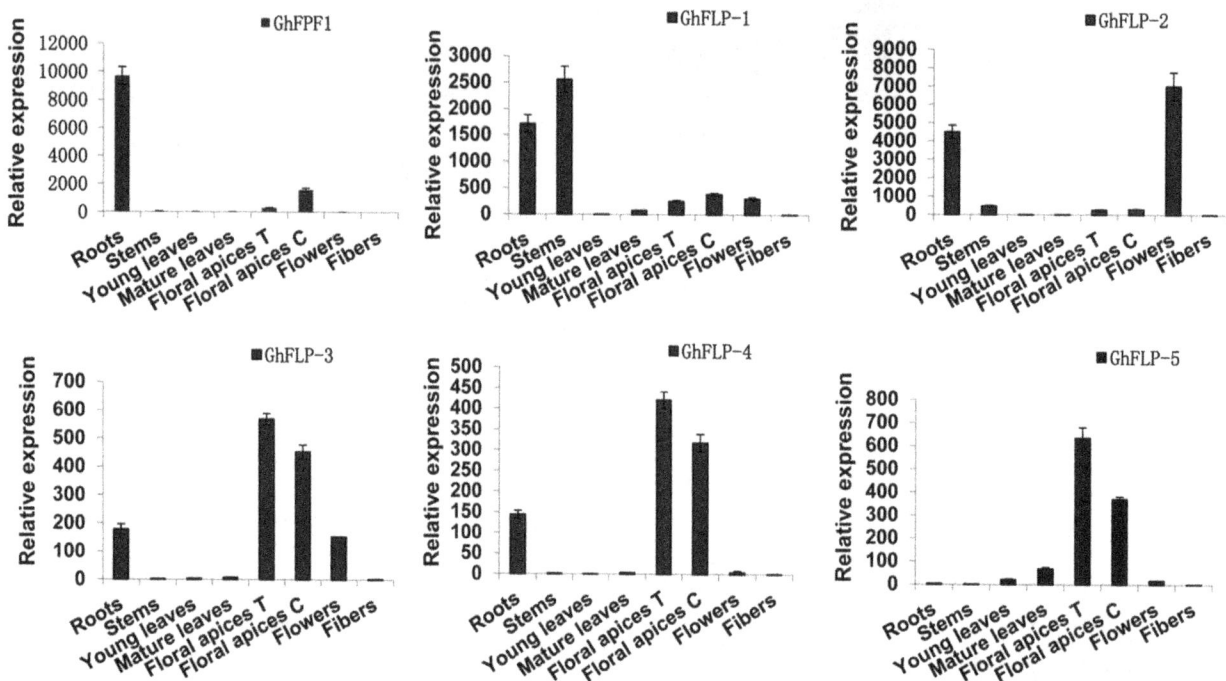

**Figure 2. Expression patterns of *GhFPF1*, *GhFLP-1*, *GhFLP-2*, *GhFLP-3*, *GhFLP-4*, and *GhFLP-5* in *G. hirsutum* L.** Relative expression of *GhFPF1* and its homologs were measured in different tissues of upland cotton CCRI 36 (a short-season cotton variety) and TM-1 (a genetic standard line) using qRT-PCR. Roots, stems, leaves, flowers and fibers stand for mixed samples from CCRI 36 and TM-1. Floral apices from CCRI 36 and TM-1 were harvested and named as floral apices T and C respectively. Error bars represent standard deviation (SD).

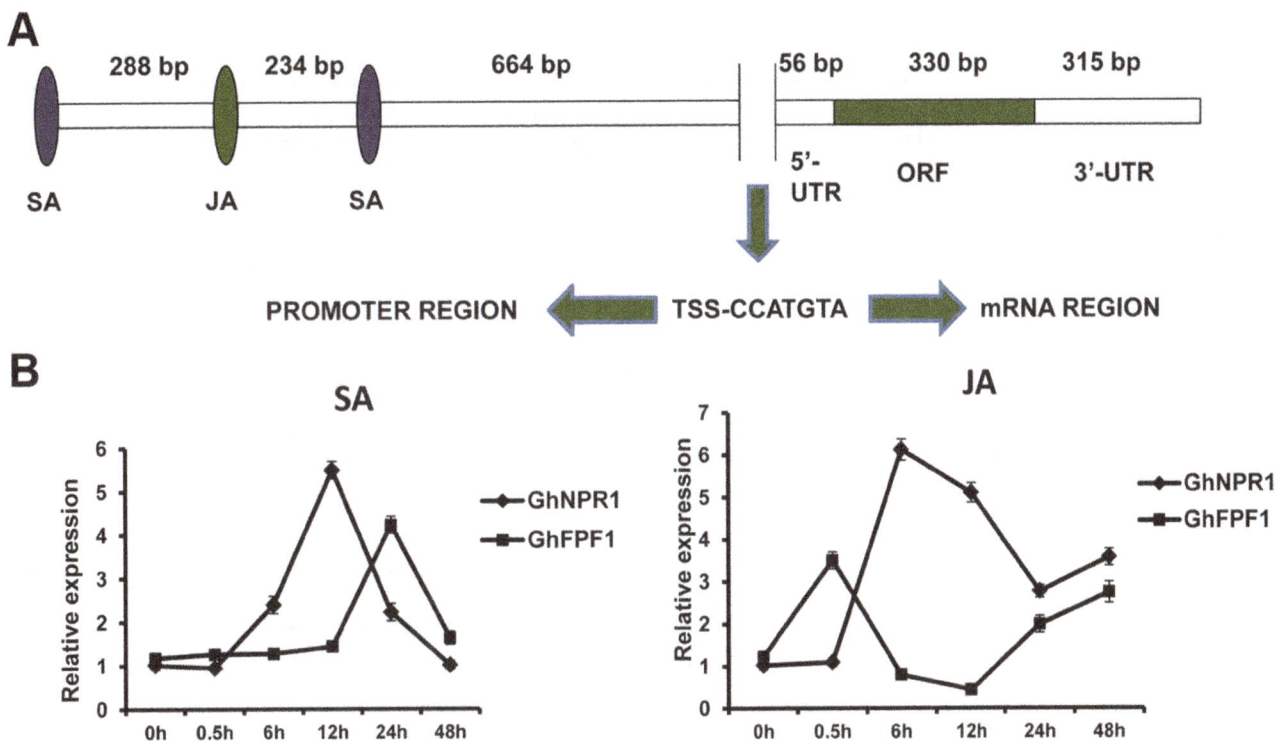

**Figure 3. Gene structure of *GhFPF1*, and its response to plant hormones.** A. The gene structure of *GhFPF1*. JA and SA represent response elements of JA and SA; TSS, transcriptional start site. B. *GhFPF1* and *GhNPR1* expression profiles in the first forty-eight hours after treatment with SA and JA. *GhNPR1* (DQ409173), a known JA- and SA-inducible gene, was used as a positive control here. Error bars represent SD.

To gain further insight into how *GhFPF1* regulates the floral transition, we examined the expression levels of six genes associated with the promotion of flowering of *Arabidopsis*: *FPF1*, *LFY*, *AP1*, *CONSTANS* (*CO*), *FT*, *SOC1*, and the flowering repressor, *FLC* both in wild-type and transgenic plants of vegetative (18-day-old) and bolting on (24-day-old). Relative

qRT-PCR indicated that the level of *AtFLC* transcripts in transgenic plants of 18-day-old was one-eighth of that in wild-type plants under long-day conditions. Meanwhile, *AtSOC1* and *AtAP1* transcripts were up-regulated by two to three-fold relative to the wild-type. There was no obvious change in the relative transcript amount of *AtFPF1*, *AtLFY*, *AtFT*, or *AtCO* (Figure 5A). In

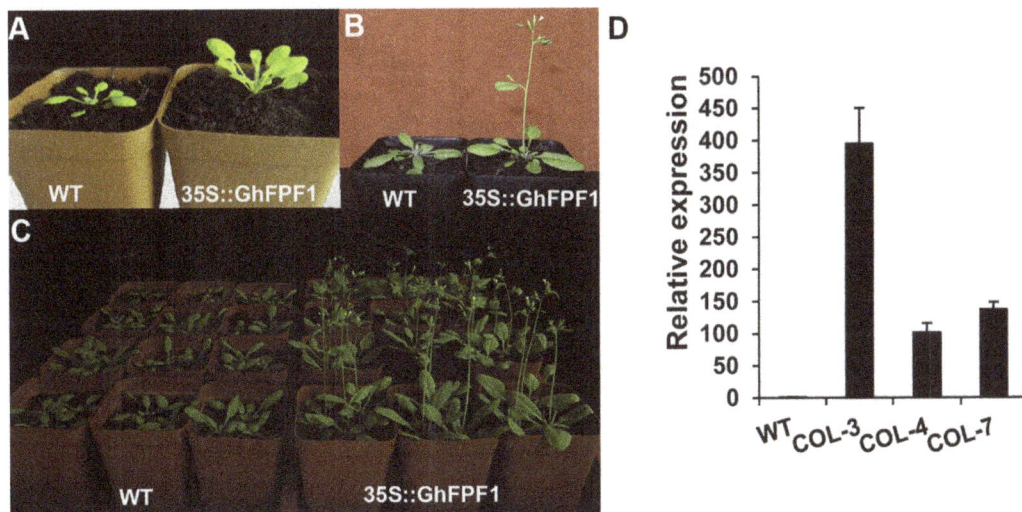

**Figure 4. over-expression of *GhFPF1* promoted flowering in *Arabidopsis*.** A. 24-day-old plants grown under long-day conditions. The wild-type control developed six small rosette leaves (left) by the time that the *GhFPF1* transgenic plant had started bolting on (right). B and C. One-month-old plants grown under long-day conditions. The wild-type control was still at the vegetative stage (left), whereas the transgenic plant(s) had flowered (right). D. Relative transcriptional analysis of *GhFPF1* in 18-day-old transgenic *Arabidopsis* lines COL-3, COL-4 and COL-7.

**Table 2.** Flowering time under long-day conditions, as measured by days to flowering and number of basal rosette and cauline leaves.

| Line NO | Flowering time (days after sowing) | P value | Rosette leaves + cauline leaves | P value |
|---------|-----------------------------------|---------|--------------------------------|---------|
| WT | 36.8±0.47 | | 15.4±0.55 | |
| COL1 | 30.0±0.44* | <0.05 | 11.5±0.57* | <0.05 |
| COL2 | 30.9±0.34* | <0.05 | 12.4±0.42* | <0.05 |
| COL3 | 30.3±0.53* | <0.05 | 11.9±0.42* | <0.05 |
| COL4 | 33.3±0.74* | <0.05 | 12.7±0.41* | <0.05 |
| COL5 | 32.2±0.72* | <0.05 | 13.3±0.57* | <0.05 |
| COL6 | 32.5±0.67* | <0.05 | 13.3±0.56* | <0.05 |
| COL7 | 30.7±0.66* | <0.05 | 12.2±0.32* | <0.05 |

Asterisks indicate significant variation differences between the wild-type and each 35S::GhFPF1 transgenic population line as determined by one-way ANOVA analysis. Means with SD from twenty plants of every line were shown.

24-day-old transgenic plants collected at the same time of 11 am (3 hours exposure to light), the transcripts of AtFPF1, AtLFY, AtFT, AtCO were up-regulated by four times more or less, but AtSOC1 transcripts remained at the same levels as in 18 days (Figure 5A and B). Remarkably, AtAP1 was activated highly with an increase of 18.5-fold compared to the wild-type; meanwhile, the expression of the flowering repressor AtFLC was almost completely suppressed due to the ectopic expression of GhFPF1. Combining these, it could be hypothesized that early flowering conferred by GhFPF1 over-expression in Arabidopsis might be mediated through AtAP1 and AtFLC.

## Over-expression of GhFPF1 Triggered Shade Avoidance Syndrome (SAS) in Arabidopsis

Except accelerated flowering time, compared with wild-type, transgenic Arabidopsis generated longer hypocotyls and petioles (Table 3), as well as the upward of movement of leaves. Also chlorophyll content was found to be reduced in transgenic plants (Figure 6). Taken together, transgenic plants were recognized as so-called shadow avoidance syndrome (SAS). The elongated appearance and early flowering response were similar to phenotype of phyB mutants. PHYB, acting as the major phytochrome in light-grown plants played a predominant role in shade avoidance syndrome [48]. Since the plants were provided sufficient fluorescent light with red to far-red ratio 4.5, transcripts of PHYB were measured using qPCR in transgenic and wild Arabidopsis further. The result revealed that expression of PHYB in transgenic plants was decreased by more than fifty percent (Figure 6E).

## Discussion

GhFPF1 gene belongs to a novel gene family that seems to be conserved in both higher and lower plants. Until now, as were characterized in several species, members of this gene family were short in length as well as lacking in intron of their genomic sequences [10–12]. Twelve FPF1 homologs were identified from the diploid cotton genomic databases of G. raimondii L. and G. arboretum L. Orthologous sequences from the two cotton species were compared with each other, suggesting that nucleic acid sequences of the six pairs of orthologs were distinct, though two pairs of orthologous genes possessed the same deduced protein sequence as a result of codon degeneracy. High constraints of

genetic divergence might occur during speciation for five genes had higher synonymous changes between the two species.

Previous studies have indicated that for each gene studied, allotetraploid species of cotton should have two homologs which represent descendants from the A-genome, and D-genome donors, evolving independently at the time of polyploidy formation [26]. Meanwhile, on allopolyploidy, genomic changes will take place, including chromosomal rearrangement and changes in gene expression. During the cloning and identification of the FPF1 homologous genes in G. hirsutum L., at least eight single clones were sequenced for every gene. It was discovered that five genes contained cDNA sequences identical to D-genome sequences from G. raimondii L., whereas only one gene GhFLP3 was found to have the same sequence as the A-genome sequence from G. arboreum L. The finding revealed that this gene family exhibited subgenome-specific expression bias to D-subgenome.

When plants undergo the transition to flowering, the vegetative shoot apical meristem will be transformed into an inflorescence meristem. Inflorescence meristems can respond to both environmental and endogenous flowering signals to give rise to floral meristems, which go on to produce the various floral organs in succession [27]. Axillary buds of upland cotton can be either vegetative or floral buds which develop into vegetative shoots or reproductive shoots, respectively. Two types of axillary buds are distinct from each other for apical meristems of the vegetative buds are small and either domed or tapered in shape, with one or two layers of tunica cells but the apical meristems of the floral buds are large and columnar, with two or three layers of tunica cells in floral buds with flat surfaces [28]. It was reported that the short-season cotton varieties of CCRI16 and CCRI36 initiated morphological differentiation of the floral bud at the developmental stage of two true leaves flattened, whereas the late-maturing CCRI12 activated this process when there were three true leaves expanded [28,29]. TM-1, a genetic standard line as well as a late-maturing variety was also found to begin morphological differentiation of the floral bud when three true leaves were expanded (Figure S2). To select specific genes that are involved in the floral development of short-season cotton, the relative expression of GhFPF1 and GhFPF1-like genes in the floral apices of CCRI 36 and TM-1 were analyzed and compared at the developmental stage of three true leaves expanded. Transcripts of GhFPF1 were discovered to be more abundant in the floral apices of CCRI 36, suggesting that GhFPF1 might be involved in the floral regulation of short-season cotton.

A

B

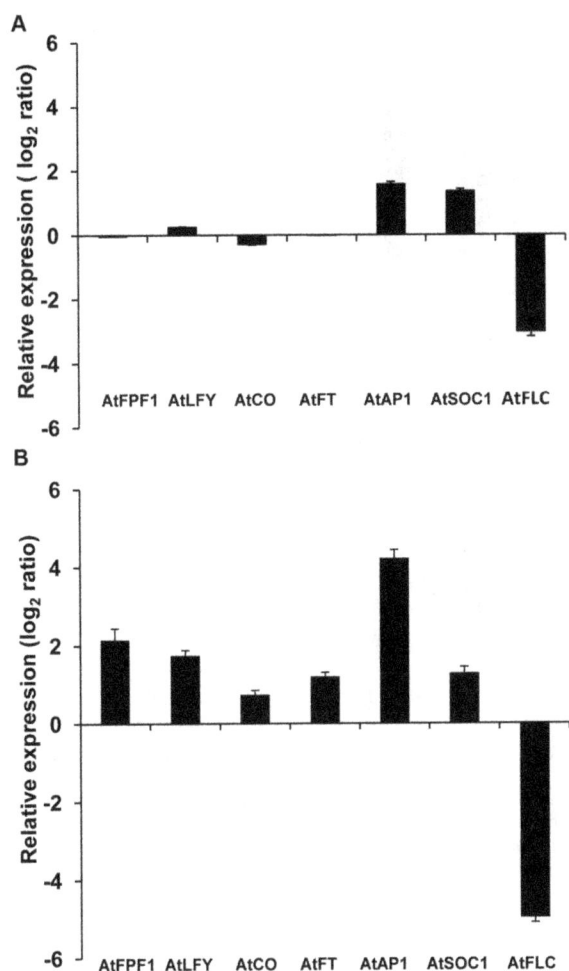

**Figure 5. Relative qRT-PCR analysis of genes involved in floral regulation in *Arabidopsis*.** The *AtUBQ5* gene was used for calibration and the histogram was drawn based on the $\log_2$ scale of the ratio of gene expression in transgenic plants relative to wild-type. Entire 18-day-old plants of vegetative (A) and the above-ground parts of 24-day-old plants of starting bolting on (B) were analyzed. Interested genes were *AtFPF1* (AT5G24860), *AtLFY* (AT5G61850), *AtCO* (AT5G15840), *AtFT* (AT1G65480), *AtAP1* (AT1G69120), *AtSOC1* (AT2G45660), and *AtFLC* (AT5G10140). All plants were grown under long-day conditions of 16 h light/8 h dark. Data in graph were mean values with standard deviation (error bar) from three replicates.

In the study we found that *GhFPF1* was predominantly expressed in roots and floral apices, the former more. In *Arabidopsis thaliana*, a number of genetic pathways controlling flowering time

have been identified. In fact, most of the genes involved were preferentially expressed in the shoot apical meristem (SAM) and the root tip, such as *PHYA*, *CRY2*, *FLC* and so on, but, surprisingly, only a few were expressed preferentially or exclusively in leaves for example *FT* [7,30,31]. Day length and light quality are essentially perceived by photosynthetic organs, expanded leaves and stem, whereas water and mineral availability are perceived by the roots. The root system is presumably capable of reacting to the critical environmental changes and, as a result, influences the flowering process to some extent. In *Sinapis*, analyses of changes in the contents of phloem and xylem saps during the floral transition have disclosed a complex shoot-to-root-to-shoot signalling loop involving both nutrients and hormones [32,33]. Higher transcripts of *GhFPF1* in roots implicates that it could be involved in multiple functions in *Gossypium hirsutum* L. Moreover, members of *GhFPF1* gene family displayed tissue specific expression, which is not rare. Most genes controlling flowering time were expressed across a wide range of organs and tissues, but a survey of available data on their spatial expression patterns revealed that many genes showed preferential expression in more limited areas such as SAM and RAM [34].

Jasmonic and salicylic acid response elements were found in the promoter region of *GhFPF1*, which suggested that *GhFPF1* may be regulated by the plant hormones JA and SA. Response of *GhFPF1* to JA (wound signal) and SA treatments could hint involvement of *GhFPF1* in plant defense responses. Previous studies implicated both positive and negative roles of SA in affecting flowering time via influencing the expression of flowering regulatory genes *FLC* and *FT* [35–38]. Also SA can links flowering time with some stress signalling coming from defense responses or poor-nutrition [36,38]. Recent studies provided new insights into the mechanisms of salicylic acid (SA) perception and *NPR1* was proposed to be with *NPR3*/*NPR4*, resembling the multi-receptor of SA in diverse immune responses such as basal defense, systemic acquired resistance establishment, and effector-triggered immunity (ETI) [39]. *GhFPF1* and *GhNPR1* shared the similar expression trend to exogenous SA but the association between *FPF1*, SA, plant defense responses and flowering time requires further investigation. Little has been reported about the relationship between JA and the regulation of flowering time. Thus, it seems that JA may not have a direct effect upon the transition to flowering. Because of domestication, upland cotton has become a compact day-neutral and an annual row-crop from a lanky photoperiodic and perennial plant. Moreover some growth characteristics of cotton are different from other species as vegetative growth will proceed after the initiation of reproductive growth. Also flowering and fruit set are not synchronized but continue through the growing season. These competing sinks in upland cotton may give rise to flowering mechanism different from *Arabidopsis*.

Ectopic expression of *GhFPF1* in *Arabidopsis* led to earlier flowering, and a decreased number of rosette and cauline leaves.

**Table 3.** Statistics of hypocotyl and petiole lengths of wild and transgenic plants grown under long-day conditions with fluorescent lamps.

| Line NO | Hypocotyl length (mm) | P value | Petiole length (mm) | P value |
|---|---|---|---|---|
| WT | 3.8±0.04 | | 7.0±0.18 | |
| 35S::*GhFPF1* | 6.5±0.07** | <0.01 | 11.2±0.21** | <0.01 |

Hypocotyl lengths of wild and transgenic lines were quantified from thirty ten-day-old plants. Petiole lengths of true leaves of the basal rosette were measured and the data were gathered from thirty wild type and transgenic 24-day-old plants, respectively. Mean values (±SD) were shown and statistical analysis was evaluated by the same way as table 2.

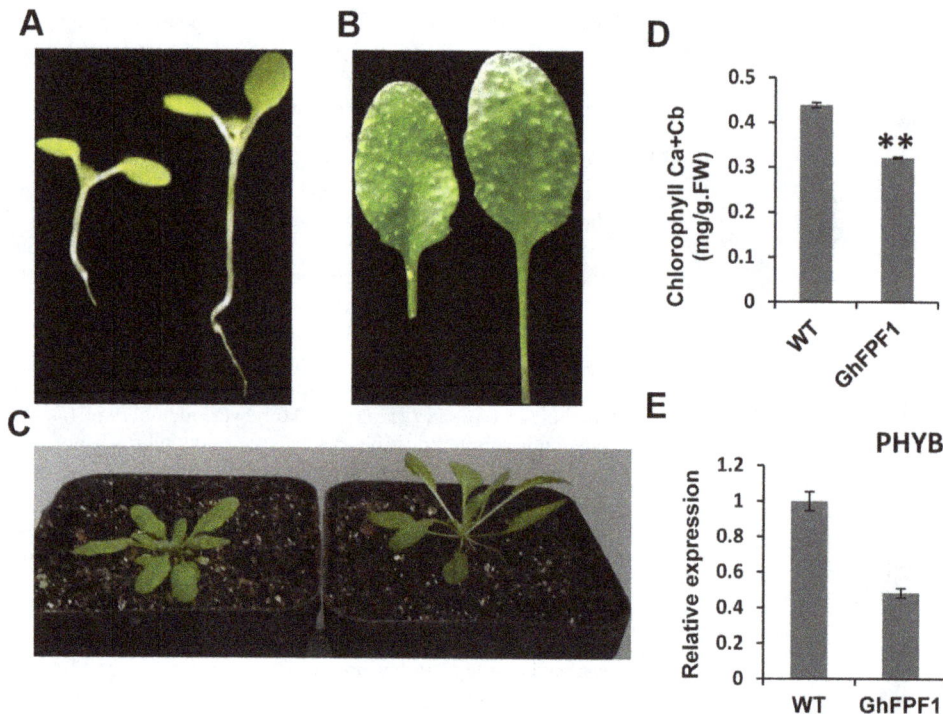

**Figure 6. Over-expression of *GhFPF1* led to shade-avoidance responses in transgenic plants.** A. Phenotype of hypocotyl of ten-day-old wild type (left) and transgenic plants (right) grown under long-day conditions. B. The seventh rosette leaves of wild type (left) and transgenic (right) 24-day-old plants grown under the same conditions as (A). C. Shade-avoidance responses in 24-day-old transgenic plants (right) which grew fast with upward leaves. All *Arabidopsis* plants were grown under long-day conditions with high red/far red (R/FR ratio: 4.5) light provided by fluorescent lamps. D. Chlorophyll content of leaves in transgenic plants was lower than that in wild-type and the difference was very significant (P<0.01) assessed by T-test. E. The transcript levels of *AtPHYB* (AT2G18790) in wild-type and *GhFPF1* over-expression transgenic plants. The *AtUBQ5* gene was used as calibrator.

When compared to the wild-type, transgenic plants had an increased expression of *AtAP1*, and suppressed expression of *AtFLC*. The MADS-domain transcription factor *AP1*, acts as a floral meristem identity gene that controls the onset of *Arabidopsis* flower development. *AP1* expression is first observed throughout the emerging floral primordia, and is later confined to the outer whorls of floral buds, where it is also involved in the specification of sepals and petals [40,41]. Another MADS-box transcription factor, *FLC* is a major repressor of flowering in *Arabidopsis*. It binds to the first intron of *FT*, and the promoter of *SOC1*, in each case inhibiting transcriptional activity. The FT protein interacts with FD to stimulate the activity of *AP1*. SOC1 can bind to the promoter of *LFY* to activate its transcription. The actions of *AP1* and *LFY* promote the development of the inflorescence meristem, which leads to the production of flowers [42–44]. Ectopic expression of *GhFPF1* in 18-day-old *Arabidopsis* caused slight increases in *AtAP1* and *AtSOC1*expression levels. *AtAP1* transcript was induced to a much higher level six days later, when the transgenic *Arabidopsis* had started bolting. *AtSOC1*expression remained at the same levels as before. *AtFLC* expression was suppressed at an extremely low level at both time points. This reveals that the promotion flowering of over-expression of *GhFPF1* in *Arabidopsis* is possibly dependent on *AtAP1* and *AtFLC*. Previous studies had confirmed that the *FPF1* gene family took a positive effect on flowering time regulation [10,13,14]. Until now, little was known about the molecular mechanism of *FPF1* in floral regulation pathways. It is expected that the role of *FPF1* in the transition to flowering will be the focus of future research.

To grow and develop optimally, all organisms need to perceive and process information from their environment. As sessile organisms, plants need to sense and respond to external stimuli more than most organisms. Therefore, plants have to adapt their developmental pattern to the environmental changes to ensure survival and reproduction. Light influences every developmental transition from seed germination and seedling emergence to flowering. For shade-intolerant plants, such as *Arabidopsis thaliana*, a reduction in the red to far-red (R: FR) ratio of incoming radiation, which is caused by absorption of red light and reflection of far-red radiation by canopy leaves, signals the proximity of neighboring plants and triggers the shade avoidance syndrome (SAS) [45]. A common phenotype of the SAS is re-allocation of energy resources from storage organs to stems and petioles so that the plant outgrows its competitors. Other responses induced by reduction in R: FR ratio include increased leaf angle, accelerated leaf senescence and reduced deposition of fixed carbon to storage organs [46]. Shade avoidance in higher plants is regulated by the action of multiple phytochrome (phy) species that detect changes in the red to far-red ratio (R: FR) of incident light to initiate a redirection of growth and an acceleration of flowering [47]. Phytochrome B (phyB) is clearly the most important photoreceptor in the vast majority of responses to shade, in some cases redundantly with other members of its clade. In *Arabidopsis*, *phyB* mutants display a constitutive shade-avoiding phenotype that is characterized by long hypocotyls and petioles, reduced chlorophyll content, early flowering [48]. In the previous study, transgenic *Arabidopsis* over-expressing *AtFPF1* was also deemed to share the similar phenotype to *phyB* mutant, and the authors deduced that the lack of phytochrome B leads to an enhanced responsiveness to GA [10]. Also they found that constitutively *FPF1*-expression plants contained slightly higher amounts of GA4 and GA20 than wild-type plants did. Here the transcripts of key genes in gibberellin biosynthesis *GA3$_{OXI}$* and *GA20$_{OXI}$* were analyzed and proved to show little difference between wild and transgenic plants. The expression of *PHYB* was checked to have only 44% amount of that in wild-type. In addition, the transgenic *Arabidopsis* presented typical shade avoidance responses such as early flowering, longer hypocotyls and petioles, reduced chlorophyll content under high R: FR ratio light conditions. Though functional complementation assay should be developed further, suppressed expression of *PHYB* in transgenic *Arabidopsis* of over-expressing of *GhFPF1* may be the reason why shade avoidance syndrome (SAS) was induced presumably.

In summary, identification of *FPF1* homologous genes in *G. raimondii* L. (D- genome) and *G. arboreum* L. (A- genome), also further comparing the sequences provided the groundwork necessary for understanding the distinctions and similarities between the sequences in the same genus of the different species. Strongly activated *AP1* and suppressed *FLC* expression in transgenic plants suggested some points for understanding the mechanism of *GhFPF1* to promote flowering. It seemed that transgenic *Arabidopsis* exhibited typical shade avoidance responses which might be caused by reduced expression of *PHYB*. Nevertheless, other proofs should be complemented in the future.

## Supporting Information

**Figure S1 Pair-wise alignment of *FPF1* homologs DF10009151 (*G. raimondii* L.) and Garb22630 (*G. arboreum* L.).**

Figure S2 Paraffin section analysis of flower bud differentiation in *G. hirsutum* L. TM-1. A, B, C and D corespond to four developmental stages in shoot apices when there were two cotyledons, one, two, and three true leaves flattened. Images of representative paraffin sections were shown here. Yellow arrows pointed out the shoot apical meristem (SAM) of every stage. White and red arrows indicated vegetative bud primordium (VP) and floral bud primordium (FP) respectively.

**Table S1 Cloning primers used for amplification of *GhFPF1* and homologous genes.**

**Table S2 Quantitative PCR primers of target genes used in the study.**

## Acknowledgments

We would like to thank doctoral candidate Xu DB for his help with the application of the laser confocal microscope. We wish to express our gratitude to teacher Suo TP for his guidance of the Scanning Electron Microscopy. Thanks for the seeds of CCRI 36 and TM-1 provided by teacher Wang L.

## Author Contributions

Conceived and designed the experiments: SY SF MS CP HW JY. Performed the experiments: XW. Analyzed the data: XW QM. Contributed reagents/materials/analysis tools: CP. Wrote the paper: XW. Edited the manuscript: MS CP.

## References

1. Adams KL, Cronn R, Percifield R, Wendel JF (2003) Genes duplicated by polyploidy show unequal contributions to the transcriptome and organ-specific reciprocal silencing. Proc Natl Acad Sci 100: 4649–4654.
2. Lightfoot DJ, Malone KM, Timmis JN, Orford SJ (2008) Evidence for alternative splicing of MADS-box transcripts in developing cotton fibre cells. Mol Genet Genomics 279: 75–85.
3. Wendel JF, Cronn RC (2003) Polyploidy and the evolutionary history of cotton. Adv Agron 78: 139–186.
4. Jung C, Muller AE (2009) Flowering time control and applications in plant breeding. Trends Plant Sci 14: 563–573.
5. Melzer S, Lens F, Gennen J, Vanneste S, Rohde A, et al. (2008) Flowering-time genes modulate meristem determinacy and growth form in Arabidopsis thaliana, Nat Genet 40: 1489–1492.
6. Simpson GG, Dean C (2002) Arabidopsis, the Rosetta stone of flowering time? Science 296: 285–289.
7. Michaels SD, Amasino RM (1999) FLOWERING LOCUS C encodes a novel MADS domain protein that acts as a repressor of flowering. Plant Cell 11: 949–956.
8. Michaels SD, Amasino RM (2001) Loss of FLOWERING LOCUS C activity eliminates the late-flowering phenotype of FRIGIDA and autonomous pathway mutations but not responsiveness to vernalization. Plant Cell 13: 935–941.
9. Sheldon CC, Rouse DT, Finnegan EJ, Peacock WJ, Dennis ES (2000) The molecular basis of vernalization: the central role of FLOWERING LOCUS C (FLC). Proc Natl Acad Sci 97: 3753–3758.
10. Kania T, Russenberger D, Peng S, Apel K, Melzer S (1997) FPF1 promotes flowering in Arabidopsis. Plant Cell 9: 1327–1338.
11. Melzer S, Kampmann G, Chandler J, Apel K (1999) FPF1 modulates the competence to flowering in Arabidopsis. Plant J 18: 395–405.
12. Roland Borner, Grit Kampmann, Klaus Apel, Siegbert Melzer (2000) The FPF1 gene family and flowering time control in Arabidopsis. 11th International Conference on Arabidopsis Research: June 24–28, Madison, Wisconsin, USA. Abstract can be downloaded from http://arabidopsis.org/servlets/TairObject?type=publication&id=1546989.
13. Ge L, Chen H, Jiang JF, Zhao Y, Xu ML, et al. (2004) Overexpression of OsRAA1 causes pleiotropic phenotypes in transgenic rice plants, including altered leaf, flower, and root development and root response to gravity. Plant Physiol 135: 1502–1513.
14. Smykal P, Gleissner R, Corbesier L, Apel K, Melzer S (2004) Modulation of flowering responses in different Nicotiana varieties. Plant Mol Biol 55: 253–262.
15. Lai D, Li H, Fan S, Song M, Pang C, et al. (2011) Generation of ESTs for flowering gene discovery and SSR marker development in upland cotton. PLoS ONE 6: e28676.
16. Wang K, Wang Z, Li F, Ye W, Wang J, et al. (2012) The draft genome of a diploid cotton Gossypium raimondii. Nat Genet 44: 1098–1103.
17. Attitalla IH (2011) Modified CTAB method for high quality genomic DNA extraction from medicinal plants. Pak J Biol Sci 14: 998–999.
18. Guan XY, Li QJ, Shan CM, Wang S, Mao YB, et al. (2008) The HD-Zip IV gene GaHOX1 from cotton is a functional homologue of the Arabidopsis GLABRA2. Physiol Plant 134: 174–182.
19. Livak KJ, Schmittgen TD (2001) Analysis of relative gene expression data using real-time quantitative PCR and the 2−ΔΔCT method. Methods 25: 402–408.
20. Chen LG, Zhang LP, Yu DQ (2010) Wounding-induced WRKY8 is involved in basal defense in Arabidopsis. Mol Plant Microbe In 5: 558–565.
21. Zhang Y, Wang X, Cheng C, Gao QQ, Liu JY, et al. (2008) Molecular cloning and characterization of GhNPR1, a gene implicated in pathogen responses from cotton (Gossypium hirsutum L.). Biosci Rep 28: 7–14.
22. Clough SJ, Bent AF (1998) Floral dip: a simplified method for Agrobacterium-mediated transformation of Arabidopsis thaliana. Plant J 16: 735–743.
23. Yang MW (2002) Study on Rapid Determination of Chlorophyll Content of Leaves. Chinese Journal of Spectroscopy Laboratory 19: 479–481.
24. Mukhtar MS, Nishimura MT, Dang J (2009) NPR1 in plant defense: it's not over 'til it's turned over. Cell 137: 804–806.
25. Spoel SH, Dong X (2008) Making sense of hormone crosstalk during plant immune responses. Cell Host Microbe 3: 348–351.
26. Cronn RC, Small RL, Wendel JF (1999) Duplicated genes evolve independently after polyploid formation in cotton. Proc Natl Acad Sci 96: 14406–14411.
27. Smyth DR, Bowman JL, Meyerowitz EM (1990) Early flower development in Arabidopsis. Plant Cell 2: 755–767.
28. Ren G, Chen Y, Dong H, Chen S (2000) Studies on flower bud differentiation and changes of endogenous hormones of Gossypium hirsutum. Acta Botanica Boreali-Occidentalia Sinica 20: 847–851.
29. Li J, Fan SL, Song MZ, Pang CY, Wei HL, et al. (2013) Cloning and characterization of a FLO/LFY ortholog in Gossypium hirsutum L. Plant Cell Rep 32: 1675–1686.
30. Tóth R, Kevei E, Hall A, Millar AJ, Nagy F, et al. (2001) Circadian clock-regulated expression of phytochrome and cryptochrome genes in Arabidopsis. Plant Physiol 127: 1607–1616.

31. Takada S, Goto K (2003) TERMINAL FLOWER 2, an Arabidopsis homolog of HETEROCHROMATIN PROTEIN1, counteracts the activation of FLOW-ERING LOCUS T by CONSTANS in the vascular tissues of leaves to regulate flowering time. Plant Cell 15: 2856–2865.

32. Havelange A, Lejeune P, Bernier G (2000) Sucrose/cytokinin interaction in Sinapis alba at floral induction: a shoot-to-root-to-shoot physiological loop. Physiol Plant 109: 343–350.

33. Bernier G, Corbesier L, Périlleux C (2002) The flowering process: on the track of controlling factors in Sinapis alba. Russ. J. Plant Physiol 49: 445–450.

34. Bernier G, Périlleux C (2005) A physiological overview of the genetics of flowering time control. Plant Biotechnol J 3: 3–16.

35. Korves TM, Bergelson J (2003) A developmental response to pathogen infection in Arabidopsis. Plant Physiol 133: 339–347.

36. Martinez C, Pons E, Prats G, Leon J (2004) Salicylic acid regulates flowering time and links defence responses and reproductive development. Plant J 37: 209–217.

37. Wada KC, Yamada M, Shiraya T, Takeno K (2010) Salicylic acid and the flowering gene FLOWERING LOCUS T homolog are involved in poor-nutrition stress-induced flowering of Pharbitis nil. J Plant Physiol 167: 447–452.

38. Wang GF, Seabolt S, Hamdoun S, Ng G, Park J, et al. (2011) Multiple roles of WIN3 in regulating disease resistance, cell death, and flowering time in Arabidopsis. Plant Physiol 156: 1508–1519.

39. Pajerowska-Mukhtar KM, Emerine DK, Mukhtar MS (2013) Tell me more: roles of NPRs in plant immunity. Trends in Plant Sci 18: 402–411.

40. Ferrandiz C, Gu Q, Martienssen R, Yanofsky MF (2000) Redundant regulation of meristem identity and plant architecture by FRUITFULL, APETALA1 and CAULIFLOWER. Development 127: 725–734.

41. Kaufmann K, Wellmer F, Muino JM, Ferrier T, Wuest SE, et al. (2010) Orchestration of floral initiation by APETALA1. Science 328: 85–89.

42. Corbesier L, Vincent C, Jang S, Fornara F, Fan Q, et al. (2007) FT protein movement contributes to long-distance signaling in floral induction of Arabidopsis. Science 316: 1030–1033.

43. Helliwell CA, Wood CC, Robertson M, James Peacock W, Dennis ES (2006) The Arabidopsis FLC protein interacts directly in vivo with SOC1 and FT chromatin and is part of a high-molecular-weight protein complex. Plant J 46: 183–192.

44. Searle I, He Y, Turck F, Vincent C, Fornara F, et al. (2006) The transcription factor FLC confers a flowering response to vernalization by repressing meristem competence and systemic signaling in Arabidopsis. Genes Dev 20: 898–912.

45. Ballar CL, Sanchez RA, Scopel AL, Casal JJ, Ghersa CM (1987) Early detection of neighbour plants by phytochrome perception of spectral changes in reflected sunlight. Plant Cell Environ 10: 551–57.

46. Ballare CL (1999) Keeping up with the neighbours: phytochrome sensing and other signalling mechanisms. Trends Plant Sci 4: 97–102.

47. Franklin KA (2005) Phytochromes and Shade-avoidance Responses in Plants. Annals of Botany 96: 169–175.

48. Reed JW, Nagpal P, Poole DS, Furuya M, Chory J (1993) Mutations in the gene for the red/far-red light receptor phytochrome B alter cell elongation and physiological responses throughout Arabidopsis development. Plant Cell 5: 147–157.

# Effects of Transgenic Cry1Ac + CpTI Cotton on Non-Target Mealybug Pest *Ferrisia virgata* and Its Predator *Cryptolaemus montrouzieri*

Hongsheng Wu[1,2], Yuhong Zhang[1], Ping Liu[1], Jiaqin Xie[1], Yunyu He[1], Congshuang Deng[1], Patrick De Clercq[2]*, Hong Pang[1]*

**1** State Key Laboratory of Biocontrol, School of Life Sciences, Sun Yat-sen University, Guangzhou, China, **2** Department of Crop Protection, Faculty of Bioscience Engineering, Ghent University, Ghent, Belgium

## Abstract

Recently, several invasive mealybugs (Hemiptera: Pseudococcidae) have rapidly spread to Asia and have become a serious threat to the production of cotton including transgenic cotton. Thus far, studies have mainly focused on the effects of mealybugs on non-transgenic cotton, without fully considering their effects on transgenic cotton and trophic interactions. Therefore, investigating the potential effects of mealybugs on transgenic cotton and their key natural enemies is vitally important. A first study on the effects of transgenic cotton on a non-target mealybug, *Ferrisia virgata* (Cockerell) (Hemiptera: Pseudococcidae) was performed by comparing its development, survival and body weight on transgenic cotton leaves expressing Cry1Ac (Bt toxin) + CpTI (Cowpea Trypsin Inhibitor) with those on its near-isogenic non-transgenic line. Furthermore, the development, survival, body weight, fecundity, adult longevity and feeding preference of the mealybug predator *Cryptolaemus montrouzieri* Mulsant (Coleoptera: Coccinellidae) was assessed when fed *F. virgata* maintained on transgenic cotton. In order to investigate potential transfer of Cry1Ac and CpTI proteins via the food chain, protein levels in cotton leaves, mealybugs and ladybirds were quantified. Experimental results showed that *F. virgata* could infest this bivalent transgenic cotton. No significant differences were observed in the physiological parameters of the predator *C. montrouzieri* offered *F. virgata* reared on transgenic cotton or its near-isogenic line. Cry1Ac and CpTI proteins were detected in transgenic cotton leaves, but no detectable levels of both proteins were present in the mealybug or its predator when reared on transgenic cotton leaves. Our bioassays indicated that transgenic cotton poses a negligible risk to the predatory coccinellid *C. montrouzieri* via its prey, the mealybug *F. virgata*.

**Editor:** Nicolas Desneux, French National Institute for Agricultural Research (INRA), France

**Funding:** This research was supported by grants from National Basic Research Program of China (973) (2013CB127605), National Natural Science Foundation of China (Grant No. 31171899) and the Youth Scientific Research Foundation of Guangdong Academy of Sciences (No. qnjj201206). The funders had no role in study design, data collection and analysis, decision to publish, or preparation of the manuscript.

**Competing Interests:** The authors have declared that no competing interests exist.

* E-mail: lsshpang@mail.sysu.edu.cn (HP); patrick.declercq@ugent.be (PDC)

## Introduction

Genetically modified (GM) crops hold great promise for pest control [1–4]. Most popular GM crops express one or more toxin genes from bacteria such as *Bacillus thuringiensis* (Bt), trypsin inhibitors such as cowpea trypsin inhibitor (CpTI), plant lectins, ribosome-inactivating proteins, secondary plant metabolites, vegetative insecticidal proteins and small RNA viruses [5–7]. So far Bt-cotton has been commercialized in the United States (1996), Mexico (1996), Australia (1996), China (1997), Argentina (1998), South Africa (1998), Colombia (2002), India (2002), Brazil (2005), and Burkina Faso (2008) and occupies 49% of the total global cotton area [8,9]. To delay the development of pesticide resistance in the major cotton pests [7], the bivalent transgenic cotton cultivar (CCRI41) expressing Cry1Ac and CpTI, has been commercially available since 2002 in China [10]. Currently, the cotton cultivar CCRI41 is planted at a large scale in the Yellow river cotton area in China [11]. However, with the rapid expansion in the commercial use of GM plants, there is an increasing need to understand their possible impact on non-target

organisms [12–14]. Non-target effects of several cultivars (Cry1Ac + CpTI cotton) on beneficial arthropods including pollinator insects have been recently studied [11,15–21].

Most studies on the potential ecological impacts of transgenic plants on phloem-feeding insects have focused on aphids or whiteflies [4,22–27]. Studies on the interactions between mealybugs and GM crops have not been previously reported. Like aphids and whiteflies, mealybugs are obligate phloem feeders. Several species of mealybugs have caused considerable economic damage to agricultural and horticultural plants in the tropics in the last few decades [28]. They also have the potential to become major cotton pests which is evident from the severe damage reported in different parts of Asia [29–31]. Particularly, *Phenacoccus solenopsis* Tinsley (Hemiptera: Pseudococcidae) has attracted much attention worldwide because of its harmful effects on cotton [30,32–35]. Indeed, this pest can successfully thrive on both Bt-cotton and non-Bt cultivars of cotton [36]. However, *P. solenopsis* is not the only mealybug species that infests cotton in Asia. Also *Maconellicoccus hirsutus* (Green) has increasingly been reported infesting cotton in India and Pakistan [37,38]. Mealybugs are attacked by a range of specialist predators and parasitoids. These non-target species can thus be exposed to GM toxins by feeding on or parasitizing their prey or host [39–41] and there may be side effects on the behavior of these natural enemies [12,42]. Therefore, there is a need to evaluate the potential effects of transgenic cotton on mealybugs and their key natural enemies.

The striped mealybug, *Ferrisia virgata* (Cockerell) (Hemiptera: Pseudococcidae), is also a cosmopolitan and polyphagous species that attacks a wide variety of crops including cotton [34,43,44]. The adult female is wingless, and has an elongated body covered by a powdery white wax, with a pair of dark longitudinal stripes on the dorsum and white wax threads extending from the posterior end resembling tails [34]. In cotton, *F. virgata* occurs in patches and feeds on all parts of a plant, particularly on growing tips or on leaves [33]. The species has been found infesting colored fiber cotton and has emerged as a serious pest in the Northeast of Brazil [34]. Given that mealybugs like *P. solenopsis*, *M. hirsutus* and *F. virgata* are aggressive invasive pests that seriously threaten cotton production, significant concern over their potential effects on transgenic cotton should be raised. At present, only the cotton mealybug *P. solenopsis* has been reported to damage Bt cotton. However, whether other mealybug species can infest transgenic cotton is yet to be determined.

The mealybug destroyer, *Cryptolaemus montrouzieri* Mulsant (Coleoptera: Coccinellidae), is a ladybird native to Australia and has been used in many biological control programs as one of the most efficient natural enemies to suppress mealybug outbreaks around the world [45–47]. Both the adults and larvae of the ladybird prey on a variety of mealybugs [47]. *C. montrouzieri* has also been used as a biological control agent in areas where outbreaks of *F. virgata* and *P. solenopsis* occur [38,48–50]. These predators can encounter transgene products expressed by plants (Bt toxins) when feeding on plant material such as pollen, nectar, or leaf exudates and when preying on organisms that have consumed transgenic plant tissue or toxin-loaded prey [51–53].In the present study, bioassays were performed to assess the development, reproduction and feeding choices of *C. montrouzieri* presented with mealybugs reared on the cotton cultivar CCRI41 versus its near-isogenic non-transgenic line. To study whether Cry1Ac and CpTI proteins can pass through the trophic chain up to a natural enemy, quantification of Cry1Ac and CpTI proteins in leaves, mealybugs and ladybirds was also done.

This study is the first report on tritrophic relationships involving a non-target pest mealybug (*F. virgata*), its predator (*C. montrouzieri*) and a transgenic cotton cultivar expressing Cry1Ac (Bt toxin) and CpTI (Cowpea Trypsin Inhibitor).

## Materials and Methods

### Plants

Bivalent transgenic cotton cultivar CCRI 41 (Bt+CpTI cotton) and non-transgenic cotton cultivar CCRI 23 (control) were used as the host plants in all experiments. CCRI 41 was bred by introducing the synthetic Cry1Ac gene and modified CpTI (cowpea trypsin inhibitor) gene into the elite cotton cultivar CCRI 23 by way of the pollen tube pathway technique [54]. Seeds of transgenic Cry1Ac and CpTI cotton cultivar CCRI 41 and its near-isogenic CCRI 23 were obtained from the Institute of Cotton Research, Chinese Academy of Agricultural Sciences. Both cultivars were planted singly in plastic pots (16×13 cm) with the same soil. All plants were individually grown from seeds in climate chambers (25±1°C, 75±5% RH, 16: 8 h (L: D)) and they were

five weeks old (about five to eight true leaves) at the start of experiments.

### Insects

Stock cultures of *C. montrouzieri* and *F. virgata* were originally obtained from the State Key Laboratory of Biocontrol, Sun Yat-sen University, Guangzhou, China. Cultures of *C. montrouzieri* were reared on *Planococcus citri* Risso (Hemiptera: Pseudococcidae) and *F. virgata*, which were both produced on pumpkin fruits (*Cucurbita moschata* (Duch.ex Lam.) Duch. ex Poiretand) in metal frame cages (45×36×33 cm) covered with fine-mesh nylon gauze. The colony of *F. virgata* was maintained on plastic trays (40×30 cm) containing pumpkins as food. Environmental conditions at the insectarium were 26±2°C, 50±10% RH and a photoperiod of 16: 8 h (L: D). Both *C. montrouzieri* and *F. virgata* cultures used in these experiments had been maintained at our facilities for at least six years.

### Bioassay with *F. Virgata*

**Effects of transgenic Cry1Ac and CpTI cotton on development and survival of *F. virgate*.** Development and survival of *F. virgata* on the leaves of transgenic and non-transgenic cotton plants was studied in climate chambers (25±1°C, 75±5% RH, 16: 8 h (L: D)). The experiment was subdivided into two stages: crawlers (first instars) of the mealybug were reared for the first 5 days in 6-cm diameter plastic containers to preclude escape, whereas in a second stage larger plastic bags were used to accommodate the later instars. In the first stage of the experiment 20 newly emerged first-instar nymphs (<24 h) springing from the same female were placed in a plastic container (6.0×1.5 cm) covered with a fine-mesh nylon gauze using a soft paintbrush. Each plastic container had a small hole in it allowing a leaf to be inserted. A piece of cotton wool was wrapped around the petiole to prevent *F. virgata* from escaping through the hole in the container. To encourage crawlers to settle, the environmental chamber was maintained in complete darkness for 24 h [55,56]. All plastic containers were fixed on live cotton plants by small brackets. Mealybugs on each cotton plant represented a cohort or a replicate. A total of 15 cohorts (replicates) were prepared for both the treatments with transgenic and control cotton plants.

In the second stage of the experiment the mealybugs were kept in transparent plastic bags (15×10 cm) with several small holes for ventilation. The transparent plastic bag together with cotton wool wrapped around the petiole could also prevent mealybugs from escaping or dropping off. The mealybug cohorts on each leaf (still attached to the plant) were examined every 12 h, and the development and survival of each nymphal instar were recorded. Successful development from one instar to the next was determined by the presence of exuviae. Survival rate of each stage was calculated as the percentage of individuals that successfully developed to the next stage in a cohort [56]. The sex of individual mealybugs could not be determined at the crawler stage. Therefore, sex was determined during the latter part of the second instar when males change their color from yellow to dark. At this point, the developmental times of males and females were recorded separately [55].

**Effects of transgenic Cry1Ac and CpTI cotton on body weight of *F. virgate*.** To assess the body weights of *F. virgata*, 200 second-instar nymphs were collected at the same time from stock cultures reared on pumpkin. Ten mealybugs per cotton plant were placed as a cohort on the leaves of 10 non-transgenic or transgenic cotton plants using a soft paintbrush. Thus, a total of 10 cohorts (replicates) were prepared for both the treatments with transgenic and control cotton plants. To prevent mealybugs from escaping or dropping off, each leaf infested with *F. virgata* was placed in a transparent plastic bag (15×10 cm) with several small holes for ventilation. To encourage the nymphs to settle, the environmental chamber was maintained in complete darkness for 24 h. Thereafter, the plants were kept in an environmental chamber as described above. Surviving mealybugs from the initial 10 individuals on each plant were weighed individually after 10 and 20 days using an electronic balance (Sartorius BSA124S, Germany) with a precision of 0.1 mg.

## Tritrophic Bioassay with *C. Montrouzieri*

**Effects of transgenic Cry1Ac and CpTI cotton on the development and survival of immature *C.montrouzieri*.** Two plastic boxes (12.0×5.0×4.0 cm, covered with fine-mesh nylon gauze for ventilation) each containing 50 *C. montrouzieri* eggs (<12 h old) collected from the stock colony were placed in a climate chamber (25±1°C, 75±5% RH, 14:10 h (L:D) photoperiod). The eggs were observed carefully every 12 h and numbers of larvae that hatched were recorded. Newly hatched first-instar larvae from 50 *C. montrouzieri* eggs (<12 h old) were individually transferred to the leaves of non-transgenic (45 larvae) or transgenic cotton (46 larvae), which were previously infested with *F. virgata* (~60–100 mealybugs per leaf). Each cotton plant received two or three *C. montrouzieri* larvae which were distributed on different leaves. Pieces of cotton wool were wrapped around the stem or petiole to prevent the larvae from leaving the cotton leaves. Predator larvae were randomly moved to newly infested plants when mealybug prey was depleted. In total, about 60 non-transgenic or transgenic cotton plants were used for the experiment. Larvae of *C. montrouzieri* were checked every 12 h for molting, which was determined by the presence of exuviae. The developmental time and survival of each immature stage of *C. montrouzieri* were also recorded up to adulthood.

**Effects of transgenic Cry1Ac and CpTI cotton on reproduction and adult longevity.** After adult emergence, *C. montrouzieri* females and males were single paired and each pair was transferred to a transparent plastic bag (15×10 cm) with several small holes for ventilation. A total of 12 and 16 pairs (replicates) were set up for non-transgenic and transgenic cotton plants, respectively. A piece of cotton was placed in the bag for oviposition. A leaf of non-transgenic or transgenic cotton infected with *F. virgata* (~60–100 mealybugs per leaf) was also placed in this bag. The bag containing *C. montrouzieri* adults was transferred to a new freshly infested leaf on the same plant every 3 days. The pre-oviposition period, number of eggs and survival of the mating pairs of *C. montrouzieri* were checked every day until the death of all adults.

**Effects of transgenic Cry1Ac and CpTI cotton on body weight of *C. montrouzieri*.** In order to determine fresh body weight during each developmental stage, 50 newly hatched first instar *C. montrouzieri* (<12 h old) were individually transferred to the leaves of non-transgenic or transgenic cotton using a soft hairbrush and placed in close vicinity to the prey. The leaf with mealybugs (~60–100 mealybugs per leaf) was replaced every 3 days and *C. montrouzieri* larvae were checked every 12 h for molting and development. Newly emerged 1st, 2nd, 3rd, and 4th instar larvae, pupae and adults of *C. montrouzieri* were weighed individually after 24 h using an electronic balance (Sartorius BSA124S, Germany) with a precision of 0.1 mg to record their body mass.

**Feeding performance of *C. montrouzieri* on mealybugs reared on non-transgenic versus transgenic cotton leaves.** Metal frame cages (45×36×33 cm) covered with fine-mesh nylon gauze were used in these experiments with five cages or replicates each. In each cage, 20 *C. montrouzieri* adults (10 males

and 10 females, <1 month old) were taken from the laboratory stock and starved for 24 h. Three pots each of non-transgenic and transgenic cotton (with one cotton plant per pot) were placed in a cage. Each non-transgenic or transgenic cotton plant was previously infested with 20 similar-sized female adult mealybugs. Every day, the plants infected with 20 mealybugs were replaced with newly infested plants. The experiment continued for 9 days and the numbers of consumed mealybugs were recorded every day.

## Quantification of Toxins in Leaves, *F. Virgata* and *C. Montrouzieri*

To confirm Cry1Ac and CpTI expression of the transgenic cotton plants (8-leaf stage) used in both bioassays, five leaf samples were collected from five different cotton plants. Each sample was obtained from a middle-upper leaf of a transgenic or control plant [57]. Approximately 100 mg fresh weight (f.w.) of the transgenic or control cotton leaves was collected.

To quantify the level of Cry1Ac and CpTI in *F. virgata*, a group of approximately sixty gravid females from the laboratory culture were allowed to settle on cotton leaves and reproduce. After 24 h, about 100 newborn nymphs were brushed carefully onto each transgenic or control cotton leaf and the leaf was covered with a transparent plastic bag (15×10 cm) with several small holes for ventilation. A piece of cotton wool was wrapped around the petiole to prevent *F. virgata* from escaping from the leaf. To encourage crawlers to settle, the environmental chamber was maintained in complete darkness for 24 h. Three weeks later, five samples of *F. virgata* larvae (with a total fresh weight of 60–100 mg) were collected from plants of either variety.

To assess the potential transfer of Cry1Ac and CpTI proteins via the food chain, a transgenic or control cotton leaf (still attached to the plant) which was previously infested with *F. virgata* as described above and a newly molted 2nd instar larva or an adult (< 1 month old) of *C. montrouzieri* were kept in a ventilated plastic bag (15×10 cm). Ten transgenic or control cotton plants were used. After 3 days, five samples of individual *C. montrouzieri* larvae or adults were collected for analysis.

All experiments described above were conducted in a growth chamber at 25±1°C, 75±5% RH and a photoperiod of 16:8 h (L:D). All samples were weighed and transferred to 1.5-ml centrifuge tubes. Samples were kept at −20°C until quantification of Cry1Ac and CpTI proteins.

The amount of Cry1Ac protein in the leaf and insect material was measured using an enzyme linked immuno-sorbent assay (ELISA). Envirologix Qualiplate Kits (EnviroLogix Quantiplate Kit, Portland, ME, USA) were used to estimate Cry1Ac quantities. The quantitative detection limit of the Cry1Ac kit was 0.1 ng ml$^{-1}$. The ELISA polyclonal kits used to detect CpTI protein were obtained from the Center for Crop Chemical Control, China Agricultural University (Beijing, China). The method has been validated [58] and the limit of detection and working range of the assay were 0.21 and 1–100 ng ml$^{-1}$, respectively [59]. Prior to analysis, all insects were washed in phosphate buffered saline with Tween-20 (PBST) buffer to remove any Cry1Ac and CpTI toxin from their outer surface. After adding PBST to the samples at a ratio of about 1:10 (mg sample: μl buffer) in 1.5 ml centrifuge tubes, the samples were fully ground by hand using a plastic pestle. To detect Cry1Ac protein, samples were centrifuged for 5 min at 13,000×g and leaf samples were diluted to 1:10 with PBST (insect samples were not diluted). For analysis of CpTI protein, the tubes were centrifuged at 10,000×g for 15 min. The supernatants were used to detect targeted proteins. ELISA was performed based on the manufacturer's instructions.

ODs were calibrated by a range of concentrations of Cry1Ac or CpTI made from purified toxin solution.

## Data Analysis

For the studied parameters in the bioassay with *F. virgata*, the average values of each cohort were used as replicates for the data analyses. The duration of the immature stages, survival and weight on transgenic and non-transgenic cotton were compared using independent t-tests. For the tritrophic bioassay with *C. montrouzieri*, a Mann–Whitney U test was performed for the duration of the immature stages and preoviposition period. Weights, fecundity, oviposition period, and adult longevity were analyzed using independent t-tests. The percentages of total survival and egg hatch were compared by logistic regression, which is a generalized linear model using a probit (log odds) link and a binomial error function [60]. Each test consists of a regression coefficient that is calculated and tested for being significantly different from zero, for which P-values are presented [61]. Consumption rates in the feeding performance test were compared using a general linear model for repeated measures analysis of variance (ANOVA) followed by a LSD test. All datasets were first tested for normality and homogeneity of variances using a Kolmogorov-Smirnov test and Levene test, respectively, and transformed if necessary. SPSS software (IBM SPSS Statistics, Ver. 20) was used for all statistical analyses. For all tests, the significance level was set at P≤0.05.

## Results

### Bioassay with *F. Virgata*

**Effects of transgenic Cry1Ac and CpTI cotton on the developmental duration of *F. virgate*.** *F. virgata* nymphs completed their development when reared on non-transgenic cotton CCRI 23 and its near-isogenic transgenic cotton CCRI 41 (Table 1). However, there was no significant difference in the developmental duration of female or male *F. virgata* larvae reared on transgenic or non-transgenic cotton except during the first and fourth instars. The duration of first instar development was longer on transgenic cotton. In contrast, fourth instar males reared on transgenic cotton had shorter development compared to those reared on non-transgenic cotton. No significant differences were observed in the developmental durations of the second instar, third instar and in cumulative developmental time.

**Effects of transgenic Cry1Ac and CpTI cotton on nymphal survival of *F. virgate*.** No significant difference was observed in the survival rate of female or male *F. virgata* nymphs reared on transgenic or non-transgenic cotton except in the first instar (Table 2). The survival rate of the first instars was lower when reared on transgenic cotton. No significant differences in the survival rates of the second instar, third instar, fourth instar of male and in cumulative survival rate were observed.

**Effects of transgenic Cry1Ac and CpTI cotton on body weight of *F. virgate*.** The weight of all *F. virgata* nymphs increased when reared on transgenic or non-transgenic cotton leaves for 10 or 20 days. However, nymphal weights were not significantly influenced by cotton variety (P>0.05, independent t-tests). Mean weights (± SE) of adult *F. virgata* reared on non-transgenic cotton (77 and 52, respectively) and transgenic cotton (87 and 48, repectively) leaves for 10 days were 1.29±0.15 mg and 1.30±0.10 mg (t = −0.091; df = 18; P = 0.928), and for 20 days were 2.30±0.22 mg and 2.06±0.21 mg (t = 0.799; df = 18; P = 0.434), respectively.

**Table 1.** Mean number of days (±SE) for each developmental stage of *F. virgata* reared non-transgenic or transgenic cotton leaves.

| Developmental time per stage (days) | | | | | | | | Cumulative (days) | |
|---|---|---|---|---|---|---|---|---|---|
| | First[†] | Second | | Third | | Fourth* | | | |
| Cotton cultivar | | Female | Male | Female | Male | Male | | Female | Male |
| Non-transgenic cotton | 8.24±0.16a | 5.31±0.19a | 6.34±0.28a | 6.43±0.07a | 2.48±0.09a | 5.68±0.08a | | 19.98±0.23a | 22.55±0.24a |
| Transgenic cotton | 8.72±0.14b | 5.10±0.16a | 6.62±0.23a | 6.59±0.14a | 2.39±0.07a | 5.26±0.14b | | 20.41±0.36a | 22.87±0.22a |
| t | −2.218 | 0.846 | −0.776 | −0.982 | −0.063 | 2.654 | | −1.007 | −1.008 |
| df | 28 | 28 | 28 | 28 | 28 | 28 | | 28 | 28 |
| P | 0.035 | 0.405 | 0.444 | 0.334 | 0.405 | 0.013 | | 0.323 | 0.322 |

Means ± SE within a column followed by the same letter are not significantly different (P>0.05; independent t-test). The experiment was started with 15 cohorts (replicates) per treatment.
[†]Sex could not be determined before the second instar.
*Female mealybugs have only three nymphal instars while males have four nymphal instars.

**Table 2.** Mean (±SE) survival rate (%) of each developmental stage of *F. virgata* reared on non-transgenic or transgenic cotton leaves.

| Cotton cultivar | First[†] | Second | Third | | Fourth* | Total survival |
| --- | --- | --- | --- | --- | --- | --- |
| | | | Female | Male | Male | |
| Non-transgenic cotton | 75.67±3.68a | 83.56±3.71a | 94.93±3.06a | 87.98±4.53a | 97.38±1.86a | 57.00±4.05a |
| Transgenic cotton | 61.33±4.15b | 86.96±4.20a | 89.79±2.93a | 97.22±1.94a | 97.00±2.06a | 49.00±4.37a |
| t | 2.583 | −0.592 | 1.213 | 1.878 | 0.006 | 1.343 |
| df | 28 | 28 | 28 | 28 | 28 | 28 |
| P | 0.015 | 0.501 | 0.235 | 0.071 | 0.892 | 0.190 |

Means ± SE within a column followed by the same letter are not significantly different (P>0.05; independent t-test). The experiment was started with 15 cohorts (replicates) per treatment.
[†]Sex could not be determined before the second instar.
*Female mealybugs have only three nymphal instars while males have four nymphal instars.

## Tritrophic Bioassay with *C. Montrouzieri*

**Effects of transgenic Cry1Ac and CpTI cotton on development and survival of immature *C. montrouzieri*.** The developmental time of all immature stages and total survival did not differ when reared on transgenic or its near-isogenic non-transgenic cotton (Table 3). There was no significant difference in immature stages and survival.

**Effects of transgenic Cry1Ac and CpTI cotton on body weight of *C. montrouzieri*.** When reared on transgenic cotton, first instar (t = −1.579; df = 8; P = 0.153), second instar (t = 1.941; df = 98; P = 0.055), third instar (t = −0.343; df = 97; P = 0.733) and fourth instar larvae (t = 0.782; df = 95; P = 0.436), pupae (t = 0.659; df = 90; P = 0.512), and male (t = −1.795; df = 39; P = 0.080) and female (t = −0.421; df = 34; P = 0.677) adults showed no significant difference in their body weight upon emergence compared with their counterparts reared on non-transgenic cotton (Figure 1).

**Reproduction and longevity of *C. montrouzieri* reared on non-transgenic or transgenic cotton leaves.** Preoviposition period (U = 68; df = 1; P = 0.906), fecundity (t = 0.390; df = 21; P = 0.700), number of eggs laid per female per day (t = 1.581; df = 21; P = 0.129), egg hatch ($\chi^2$ = 1.753; df = 1; P = 0.185), male longevity (t = 0.148; df = 26; P = 0.883) and female longevity (t = −1.183; df = 26; P = 0.247) were not significantly affected by treatment (Table 4).

**Feeding performance of *C. montrouzieri* on mealybugs reared on non-transgenic versus transgenic cotton leaves.** Daily consumption of mealybugs by *C. montrouzieri* adults on non-transgenic cotton was not different from that on transgenic cotton during the entire 9-day test period (F = 0.111; df = 1; P = 0. 748) (Figure 2). The interaction between the factors cotton type and time was also not significant, meaning that differential consumption of mealybugs between transgenic cotton and non-transgenic cotton was not a function of time (F = 0.692; df = 8; P = 0.697). However, *C. montrouzieri* consumed a decreasing number of mealybugs on both cotton varieties over the course of the experiment (F = 5.098; df = 8; P<0.001).

## Quantification of Toxins in Leaves, *F. Virgata* and *C. Montrouzieri*

Expressed levels of the Cry1Ac and CpTI proteins in CCRI41 cotton leaves averaged 5.76±0.33 μg Cry1Ac/g f.w. and 14.28±1.70 ng CpTI/g f.w. (means ± SE), respectively. ELISA revealed that *F. virgata* maintained on transgenic cotton did not contain detectable amounts of the Cry1Ac and CpTI proteins. Similarly, no Cry1Ac or CpTI protein was detected in *C.*

*montrouzieri* larvae and adults. None of the non-transgenic cotton leaves, or of the mealybug and ladybird samples reared on control plants were found to contain any Cry1Ac or CpTI protein.

## Discussion

*F. virgata* is a widely spread mealybug and is reported in more than 100 countries around the world, including the USA, Argentina, Canada, India, China, Brazil, and Pakistan [44], where transgenic cotton is being cultivated. Our results demonstrate that *F. virgata* nymphs completed their development when reared on leaves of both non-transgenic and transgenic cotton. Overall, no significant differences were detected in the total survival, cumulative developmental duration and body weight of the immature stages of *F. virgata* reared on transgenic and non-transgenic cotton. Higher mortality was observed during the first instar on transgenic cotton but the difference was small and total mortality from first instar to adult did not differ between treatments. These results indicate that the transgenic Bt+CpTI cotton had negligible adverse effects on the development of *F. virgata*, which is consistent with previous reports by Dutt [36] and Zhao et al. [18] stating that the mealybug *P. solenopsis* was able to infest Bt and Bt+CpTI transgenic cotton without negative effects on its fitness.

Further, ELISA analyses revealed that none of the mealybug samples from the Bt+CpTI cotton contained detectable Bt protein despite high expression levels in leaves. Like aphids and whiteflies, mealybugs are obligate phloem sap feeding insects. We postulate that *F. virgata* was not exposed to the Bt endotoxins expressed in the cotton plants given its phloem feeding habit. In previous studies on transgenic maize, Bt toxins were not detected or only in negligible amounts in the phloem sap, or in aphids that had fed on the maize [62,63]. In transgenic Bt cotton fields the density of sap-feeding insects, such as whiteflies, aphids and leafhoppers, has been reported to be higher than in non-transgenic cotton fields [64,65]. Lawo et al. [26] noted that Indian Bt cotton varieties had no effect on aphids, leading them to conclude that Bt cotton poses a negligible risk for aphid antagonists and that the aphids should remain under natural control in Bt cotton fields.

On the other hand, it was expected that any impact of transgenic Bt+CpTI cotton on mealybugs may be largely attributed to the CpTI gene encoding the cowpea trypsin inhibitor, which acts on insect gut digestive enzymes and inhibits protease activity [66]. The cysteine protease inhibitor, oryzacystatin I (OC-I), was detected in both leaves and phloem sap of

**Table 3.** Developmental time (days) and total survival rate (%) of the immature stages of *C. montrouzieri* reared non-transgenic or transgenic cotton leaves.

| Cotton cultivar | Developmental time per stage (days)* | | | | | | Total survival (%)† |
|---|---|---|---|---|---|---|---|
| | 1st instar | 2nd instar | 3rd instar | 4th instar | Pupa | Total immature | |
| Non-transgenic cotton | 3.12±0.04 | 2.74±0.05 | 3.22±0.06 | 5.50±0.07 | 8.71±0.07 | 23.40±0.13 | 80.00±0.06 |
| Transgenic cotton | 3.11±0.03 | 2.81±0.05 | 3.27±0.04 | 5.49±0.07 | 8.66±0.09 | 23.31±0.09 | 86.96±0.05 |
| $U/\chi^2$ | 942.5 | 776.0 | 820.0 | 878.0 | 640.5 | 672.5 | 0.798 |
| df | 1 | 1 | 1 | 1 | 1 | 1 | 1 |
| P | 0.793 | 0.293 | 0.538 | 0.973 | 0.610 | 0.610 | 0.372 |

No significant difference was observed between the control and treated groups within the same column (means ± SE) (P>0.05; *Mann-Whitney U test or †Wald $\chi^2$ test); 45 and 46 larvae were initially tested for non-transgenic and transgenic cotton plants, respectively.

transgenic oilseed rape, which significantly inhibited growth of *Aphis gossypii* Glover, *Acyrthosiphon pisum* (Harris), and *Myzus persicae* (Sulzer) in vitro, despite low levels of proteolysis in the guts of these homopterans [67]. Although in the present study no CpTI protein could be detected by ELISA in *F. virgata* samples, the effects of the CpTI protein on the mealybug cannot be fully excluded. Low amounts of the cowpea trypsin inhibitors (CpTI) ingested by *F. virgata*, could act as an anti-feedant to the mealybugs, which may explain lower survival rates in the first instar. In fact, Han et al. [11] demonstrated an antifeedant effect of CCRI41 cotton pollen (Bt+CpTI) on the honey bee *Apis mellifera* L. Feeding behaviour of the bees was disturbed and they consumed significantly less CCRI41 cotton pollen than in the control group given conventional cotton pollen. The antifeedant affect may have led to insufficient food uptake and malnutrition for the larvae and newly emerged bees [11,68,69]. Further, according to an EPG (Electric Penetration Graph) signal, Liu et al.[23] found that the frequencies of moving and searching for feeding sites, and probing activity of the aphid *A. gossypii* reared on CCRI 41 cotton were significantly higher than those on control cotton. Given their high mobility 1st instar mealybugs are responsible for plant colonization in the field [70]. When 1st instars of *F. virgata* select their feeding site on transgenic cotton a succession of walks and stops is observed. Consequently, the 1st instar mealybugs in our study may have spent more energy in finding and probing for food on transgenic cotton leaves than on non-transgenic leaves, which might have negatively affected the survival rates in the first instar. However, if present, this antifeedant effect to *F. virgata* appears limited because no significant difference was found in total survival and developmental duration. Besides, the *F. virgata* clones used in the present study were not resistant to the transgenic plants, as they had been maintained exclusively on pumpkin for at least 6 years without any contact with cotton. Due to inadvertent adaptations to laboratory conditions, host finding and acceptance behaviors of mass produced insects may be changed over the generations [71–73]. Colonization effects may therefore have influenced the responses of the mealybug to cotton as a host plant and it may be warranted to investigate the interactions between transgenic cotton and wild or recently colonized mealybugs.

The mealybug destroyer, *C. montrouzieri*, might ingest toxins expressed by transgenic plants that accumulate in the mealybugs feeding on these plants. In this context, we conducted tritrophic bioassays to investigate the potential effects of CCRI 41 cotton on *C. montrouzieri* by using *F. virgata* as prey. These experiments did not reveal any adverse effects on the fitness of *C. montrouzieri* after ingestion of *F. virgata* that fed on Bt+CpTI cotton leaves compared with those that fed on the corresponding non-transgenic cotton leaves. Besides a longer oviposition period on transgenic cotton than on non-transgenic cotton, there were no differences in reproductive parameters. This finding is consistent with other studies which reported no or little adverse effects on various predators or parasitoids after feeding on different Bt + CpTI cottons, including a ladybird [74] and two hymenopteran parasitoids [15,75].

Several possible mechanisms can explain the observed results. Firstly, *C. montrouzieri* may not be sensitive to Cry1Ac proteins. Porcar et al. [76] reported no statistical differences in mortality of *C. montrouzieri* adults and *Adalia bipunctata* L. larvae fed on artificial diets with or without Cry1Ab and Cry3Aa toxins. Duan et al. [77] and Lundgren and Wiedenmann [78] found no significant adverse effects when Bt maize pollen were fed to larvae of the ladybird *Coleomegilla maculata* DeGeer. The same ladybird species was also found to be unaffected by Bt cotton or higher amounts of Cry2Ab and Cry1Ac proteins indicating that Bt cotton poses a negligible

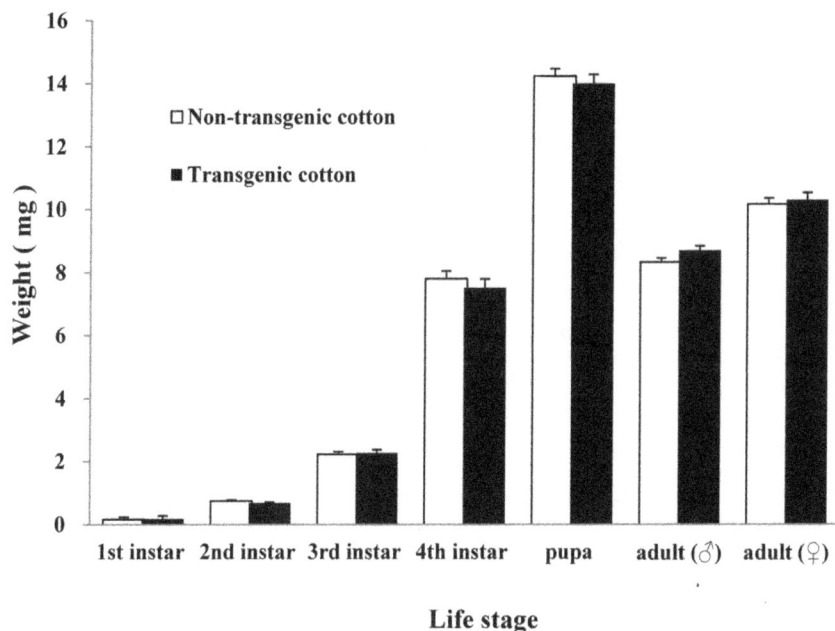

**Figure 1. Weight upon molting (means ± SE) of different life stages of** *C. montrouzieri* **reared on non-transgenic or transgenic cotton leaves.** No significant difference was observed between the control and treated groups in each life stage (P>0.05; independent t-test). The experiment was started with 50 larvae per treatment.

risk to *C. maculata* [57]. In addition, no negative effects of Bt-transgenic plants were observed on the development, survival, and reproduction of the ladybirds *Hippodamia convergens* (Guérin-Méneville) and *Propylea japonica* (Thunberg) through their aphid prey that fed on the Bt plants [25,79].

In the field, no significant differences were observed in the abundance of coccinellid beetles on Bt-transgenic and non-transgenic cottons [80]. Pollen from Cry1Ac+CpTI transgenic cotton (CCRI41) did not affect the pollinating beetle *Haptoncus luteolus* (Erichson) in the field and in the laboratory [19]. Xu et al. [81] found that CCRI41 cotton did not affect the population dynamics of non-target pests and predators including ladybirds and spiders in Xinjiang, China. Zhang et al. [82] observed negative effects on the ladybird *P. japonica* when offered young *Spodoptera litura* (F.) larvae reared on Bt-transgenic cotton expressing Cry1Ac toxin; however, adverse effects on the ladybird were attributed to poor prey quality. Lumbierres et al. [83] investigated the effects of Bt maize on aphid parasitism and the aphid–parasitoid complex in field conditions on three transgenic varieties and found that Bt maize did not alter the aphid–parasitoid associations and had no effect on aphid parasitism and hyperparasitism rates.

Ramirez-Romero et al. [84] concluded that Bt-maize did not affect the development of the non-target aphid *Sitobion avenae* (F.) and Cry1Ab toxin quantities detected in these aphids were nil, indicating that none or negligible amounts of Cry1Ac are passed on from the aphids to higher trophic levels. Probably, the amount of Cry1Ac/CpTI proteins ingested by the mealybugs in our study was too low to be effective. Indeed, ELISA measurements indicated that Bt+CpTI cotton-reared *F. virgata* and its predator did not contain detectable amounts of the Cry1Ac and CpTI protein. Because the commercial ELISA kit for determining CpTI expression was not available [19] or the amount of CpTI proteins was lower than the lowest limit of quantification [11,21] there are few earlier reports on tritrophic interactions involving CpTI protein. On the contrary, many studies related to the transfer of Bt toxic proteins to higher trophic levels have been carried out. For example, ELISA analyses revealed no or only trace amounts of Bt protein in sap-sucking insects of the order Hemiptera after feeding on different Bt plants, including maize [63,84–86] and cotton [26,87]. Trace amounts of Bt toxins were detected in *A. gossypii* feeding on Bt cotton cultivars and ladybirds preying on Bt-fed aphids [41]. Another possible reason for the weak effect of Cry1Ac/CpTI proteins is that ladybirds may digest or excrete the

**Table 4.** Reproduction and longevity of *C. montrouzieri* females reared on non-transgenic or transgenic cotton leaves.

| Cotton cultivar | Preoviposition period (days)‡ | Fecundity (eggs/♀)* | Oviposition rate (eggs/ ♀/day)* | Egg hatch (%)† | Longevity (days)* | |
|---|---|---|---|---|---|---|
| | | | | | ♂ | ♀ |
| Non-transgenic cotton | 7.00±0.93 | 823.80±84.25 | 7.21±0.83 | 90.60±0.01 | 160.96±17.36 | 131.42±9.82 |
| Transgenic cotton | 7.18±0.58 | 766.46±110.99 | 5.44±0.74 | 90.80±0.11 | 157.88±11.62 | 152.91±13.85 |

No significant difference was observed between the control and treated groups within the same column (Means ± SE) (P>0.05; *independent t-test, ‡Mann-Whitney U test or †Wald χ² test); 12 and 16 pairs of *C. montrouzieri* were used for non-transgenic and transgenic cotton plants, respectively.

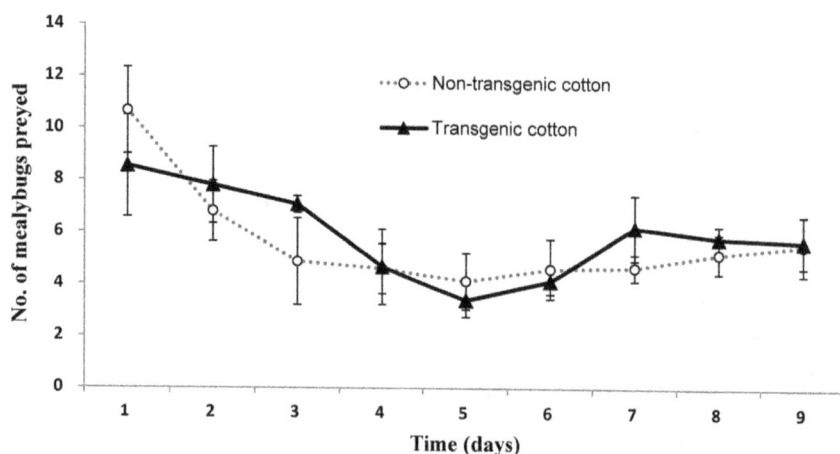

**Figure 2. Feeding performance of *C. montrouzieri* on *F. virgata* mealybugs reared on non-transgenic or transgenic cotton.** The data represent numbers (±SE) of mealybugs consumed per individual predator over a 9 day period (P>0.05; repeated measures analysis of variance (ANOVA) followed by a LSD test).

toxins taken up via their prey. For example, Li and Romeis [88] fed the ladybird *Stethorus punctillum* (Weise) with spider mites, *Tetranychus urticae* (Koch), reared on Cry3Bb1-expressing Bt maize. Subsequent bioassays revealed that the Cry protein concentrations in the ladybird beetle larvae and adults were 6- and 20-fold lower, respectively, than the levels in the spider mite prey. Cry1 proteins were also detected in *C. maculata* when offered *Trichoplusia ni* (Hübner) larvae reared on Bt-cotton, but the Bt protein levels were 21-fold lower for Cry2Ab and 6-fold lower for Cry1Ac compared to the concentrations in the prey [57].

In summary, our study indicates that *F. virgata* can successfully develop on bivalent transgenic cotton CCRI41expressing Cry1Ac+CpTI and thus can pose a risk for this crop. The finding that not only *P. solenopsis* but also other mealybugs like *F. virgata* can easily infest transgenic cotton plants has important implications for pest management in this cropping system. Further, our study demonstrates that transgenic cotton poses a negligible risk to the predatory coccinellid *C. montrouzieri* via its mealybug prey.

However, further field studies assessing the impact of transgenic cotton on the mealybug pest and its key natural enemies are needed.

## Acknowledgments

We wish to thank our colleagues of Sun Yat-sen University for support and input during all stages of the work. Dr. Fenglong Jia, Dandan Zhang and Binglan Zhang provided useful suggestions on an earlier design of the experiment. Ruixin Jiang helped with the statistical analysis and Lijun Ma is thanked for assistance with insect rearing and cotton planting.

## Author Contributions

Conceived and designed the experiments: HW HP PDC. Performed the experiments: HW YZ PL JX YH CD. Analyzed the data: HW HP PDC. Contributed reagents/materials/analysis tools: HW YZ PL JX YH. Wrote the paper: HW HP PDC.

## References

1. Kos M, van Loon JJ, Dicke M, Vet LE (2009) Transgenic plants as vital components of integrated pest management. Trends in biotechnology 27: 621–627.
2. Ferry N, Edwards M, Gatehouse J, Capell T, Christou P, et al. (2006) Transgenic plants for insect pest control: a forward looking scientific perspective. Transgenic Research 15: 13–19.
3. Poppy GM, Sutherland JP (2004) Can biological control benefit from genetically-modified crops? Tritrophic interactions on insect-resistant transgenic plants. Physiological Entomology 29: 257–268.
4. Lu Y, Wu K, Jiang Y, Guo Y, Desneux N (2012) Widespread adoption of Bt cotton and insecticide decrease promotes biocontrol services. Nature 487: 362–365.
5. Hilder VA, Boulter D (1999) Genetic engineering of crop plants for insect resistance - a critical review. Crop Protection 18: 177–191.
6. Sharma H, Ortiz R (2000) Transgenics, pest management, and the environment. Current Science 79: 421–437.
7. Lundgren JG, Gassmann AJ, Bernal J, Duan JJ, Ruberson J (2009) Ecological compatibility of GM crops and biological control. Crop Protection 28: 1017–1030.
8. James C (2009) Global status of commercialized biotech/GM crops: 2009. ISAAA Brief No, 39 (International Service for the Acquisition of Agri-Biotech Applications, Ithaca, NY, USA).
9. Dhillon M, Sharma H (2013) Comparative studies on the effects of Bt-transgenic and non-transgenic cotton on arthropod diversity, seedcotton yield and bollworms control. Journal of Environmental Biology 34: 67–73.
10. Cui JJ (2003) Effects and mechanisms of the transgenic Cry1Ac plus CpTI (cowpea trypsin inhibitor) cotton on insect communities. Dissertation, Chinese Academy of Agricultural Sciences.

11. Han P, Niu CY, Lei CL, Cui JJ, Desneux N (2010) Quantification of toxins in a Cry1Ac + CpTI cotton cultivar and its potential effects on the honey bee *Apis mellifera* L. Ecotoxicology 19: 1452–1459.
12. Faria CA, Wackers FL, Pritchard J, Barrett DA, Turlings TC (2007) High susceptibility of Bt maize to aphids enhances the performance of parasitoids of lepidopteran pests. PLoS One 2: e600.
13. Romeis J, Bartsch D, Bigler F, Candolfi MP, Gielkens MM, et al. (2008) Assessment of risk of insect-resistant transgenic crops to nontarget arthropods. Nature biotechnology 26: 203–208.
14. Desneux N, Bernal JS (2010) Genetically modified crops deserve greater ecotoxicological scrutiny. Ecotoxicology 19: 1642–1644.
15. Liu XX, Sun CG, Zhang QW (2005) Effects of transgenic Cry1A+ CpTI cotton and Cry1Ac toxin on the parasitoid, *Campoketis chlorideae* (Hymenoptera: Ichneumonidae). Insect Science 12: 101–107.
16. Liu B, Shu C, Xue K, Zhou K, Li X, et al. (2009) The oral toxicity of the transgenic Bt+ CpTI cotton pollen to honeybees (*Apis mellifera*). Ecotoxicology and Environmental Safety 72: 1163–1169.
17. Xu Y, Wu KM, Li HB, Liu J, Ding RF, et al. (2012) Effects of Transgenic Bt + CpTI Cotton on Field Abundance of Non-Target Pests and Predators in Xinjiang, China. Journal of Integrative Agriculture 11: 1493–1499.
18. Zhao XN, Cui XH, Chen L, Hung WF, Zheng D, et al. (2012) Ontogenesis and Adaptability of Mealybug *Phenacoccus solenopsis* Tinsley (Hemiptera:Pseudococcidae) on Different Varieties of Cotton. Cotton Science 24: 496–502.
19. Chen L, Cui J, Ma W, Niu C, Lei C (2011) Pollen from Cry1Ac/CpTI-transgenic cotton does not affect the pollinating beetle Haptoncus luteolus. Journal of Pest Science 84: 9–14.
20. Han P, Niu CY, Lei CL, Cui JJ, Desneux N (2010) Use of an innovative T-tube maze assay and the proboscis extension response assay to assess sublethal effects

of GM products and pesticides on learning capacity of the honey bee *Apis mellifera* L. Ecotoxicology 19: 1612–1619.

21. Han P, Niu CY, Biondi A, Desneux N (2012) Does transgenic Cry1Ac + CpTI cotton pollen affect hypopharyngeal gland development and midgut proteolytic enzyme activity in the honey bee *Apis mellifera* L.(Hymenoptera, Apidae)? Ecotoxicology 21: 2214–2221.

22. Ashouri A, Michaud D, Cloutier C (2001) Unexpected effects of different potato resistance factors to the Colorado potato beetle (Coleoptera: Chrysomelidae) on the potato aphid (Homoptera: Aphididae). Environmental entomology 30: 524–532.

23. Liu XD, Zhai BP, Zhang XX, Zong JM (2005) Impact of transgenic cotton plants on a non-target pest, *Aphis gossypii* Glover. Ecological Entomology 30: 307–315.

24. Zhou FC, Du YZ, Ren SX (2005) Effects of transgenic cotton on population of the piercing-sucking mouthparts insects. Entomological Journal of East China 14: 132–135.

25. Zhu S, Su J, Liu X, Du L, Yardim EN, et al. (2006) Development and reproduction of *Propylaea japonica* (Coleoptera: Coccinellidae) raised on *Aphis gossypii* (Homoptera: Aphididae) fed transgenic cotton. Zoological Studies 45: 98–103.

26. Lawo NC, Wäckers FL, Romeis J (2009) Indian Bt cotton varieties do not affect the performance of cotton aphids. PLoS One 4: e4804.

27. Porcar M, Grenier A-M, Federici B, Rahbé Y (2009) Effects of *Bacillus thuringiensis* δ-Endotoxins on the Pea Aphid (*Acyrthosiphon pisum*). Applied and environmental microbiology 75: 4897–4900.

28. Beltrà A, Soto A, Germain J, Matile-Ferrero D, Mazzeo G, et al. (2010) The Bougainvillea mealybug *Phenacoccus peruvianus*, a rapid invader from South America to Europe. Entomol Hell 19: 137–143.

29. Solangi GS, Mahar GM, Oad FC (2008) Presence and abundance of different insect predators against sucking insect pest of cotton. Journal of Entomology 5: 31–37.

30. Wang YP, Watson GW, Zhang RZ (2010) The potential distribution of an invasive mealybug *Phenacoccus solenopsis* and its threat to cotton in Asia. Agricultural and Forest Entomology 12: 403–416.

31. Khuhro S, Lohar M, Abro G, Talpur M, Khuhro R (2012) Feeding potential of lady bird beetle, *Brumus suturalis* Fabricius (Coleopteran: Coccinellidae) on cotton mealy bug *Phenococcus solenopsis* (Tinsley) in laboratory and field. Sarhad J Agric 28: 259–265.

32. Hodgson C, Abbas G, Arif MJ, Saeed S, Karar H (2008) *Phenacoccus solenopsis* Tinsley (Sternorrhyncha: Coccoidea: Pseudococcidae), an invasive mealybug damaging cotton in Pakistan and India, with a discussion on seasonal morphological variation. Zootaxa 1: 1913.

33. Nagrare V, Kranthi S, Kumar R, Dhara Jothi B, Amutha M, et al. (2011) Compendium of cotton mealybugs. Shankar Nagar, Nagpur, India: CICR.

34. Silva-Torres C, Oliveira M, Torres J (2013) Host selection and establishment of striped mealybug, *Ferrisia virgata*, on cotton cultivars. Phytoparasitica 41: 31–40.

35. Hanchinal S, Patil B, Basavanagoud K, Nagangoud A, Biradar D, et al. (2011) Incidence of invasive mealybug (*Phenacoccus solenopsis* Tinsley) on cotton. Karnataka Journal of Agricultural Sciences 24.

36. Dutt U (2007) Mealy Bug Infestation in Punjab: Bt. Cotton Falls Flat. Countercurrents org Available: http://www.countercurrents.org/dutt210807.htm. Accessed 2013 Jun 20.

37. Hanchinal S, Patil B, Bheemanna M, Hosamani A (2010) Population dynamics of mealybug, *Phenacoccus solenopsis* Tinsley and it's natural enemies on Bt cotton. Karnataka Journal of Agricultural Sciences 23: 137–139.

38. Khan HAA, Sayyed AH, Akram W, Raza S, Ali M (2012) Predatory potential of *Chrysoperla carnea* and *Cryptolaemus montrouzieri* larvae on different stages of the mealybug, *Phenacoccus solenopsis*: A threat to cotton in South Asia. Journal of Insect Science 12: 1–12.

39. Azzouz H, Cherqui A, Campan EDM, Rahbé Y, Duport G, et al. (2005) Effects of plant protease inhibitors, oryzacystatin I and soybean Bowman-Birk inhibitor, on the aphid *Macrosiphum euphorbiae* (Homoptera, Aphididae) and its parasitoid *Aphelinus abdominalis* (Hymenoptera, Aphelinidae). Journal of insect physiology 51: 75–86.

40. Ramirez-Romero R, Bernal J, Chaufaux J, Kaiser L (2007) Impact assessment of Bt-maize on a moth parasitoid, *Cotesia marginiventris* (Hymenoptera: Braconidae), via host exposure to purified Cry1Ab protein or Bt-plants. Crop Protection 26: 953–962.

41. Zhang GF, Wan FH, Lövei GL, Liu WX, Guo JY (2006) Transmission of Bt toxin to the predator *Propylaea japonica* (Coleoptera: Coccinellidae) through its aphid prey feeding on transgenic Bt cotton. Environmental entomology 35: 143–150.

42. Desneux N, Ramírez-Romero R, Bokonon-Ganta AH, Bernal JS (2010) Attraction of the parasitoid *Cotesia marginiventris* to host (*Spodoptera frugiperda*) frass is affected by transgenic maize. Ecotoxicology 19: 1183–1192.

43. Schreiner I (2000) Striped mealybug [*Ferrisia virgata* (Cockrell)]. Available: http://wwwadaphawaiiedu/adap/Publications/ADAP_pubs/2000-18pdf. Accessed 2013 Jun 20.

44. Ben-Dov Y, Miller DR, Gibson GAP (2005) ScaleNet: A Searchable Information System on Scale Insects. Available: http://wwwselbarcusdagov/scalenet/scalenethtm. Accessed 2013 Jun 20.

45. Bartlett BR (1974) Introduction into California of cold-tolerant biotypes of the mealybug predator *Cryptolaemus montrouzieri*, and laboratory procedures for testing natural enemies for cold-hardiness. Environmental entomology 3: 553–556.

46. Li LY (1993) The research and application prospects of *Cryptolaemus montrouzieri* in China. Nature Enemies Insects 15: 142–152.

47. Jiang RX, Li S, Guo ZP, Pang H (2009) Research status of *Cryptolaemus montrouzieri* Mulsant and establishing its description criteria. Journal of Environmental Entomology 31: 238–247.

48. Mani M, Krishnamoorthy A, Singh S (1990) The impact of the predator, *Cryptolaemus montrouzieri* Mulsant, on pesticide-resistant populations of the striped mealybug, *Ferrisia virgata*(Ckll.) on guava in India. Insect Science and its Application 11: 167–170.

49. Mani M, Krishnamoorthy A (2008) Biological suppression of the mealybugs *Planococcus citri* (Risso), *Ferrisia virgata* (Cockerell) and *Nipaecoccus viridis* (Newstead) on pummelo with *Cryptolaemus montrouzieri* Mulsant in India. Journal of Biological Control 22: 169–172.

50. Kaur H, Virk J (2012) Feeding potential of *Cryptolaemus montrouzieri* against the mealybug *Phenacoccus solenopsis*. Phytoparasitica 40: 131–136.

51. Harwood JD, Wallin WG, Obrycki JJ (2005) Uptake of Bt endotoxins by nontarget herbivores and higher order arthropod predators: molecular evidence from a transgenic corn agroecosystem. Molecular Ecology 14: 2815–2823.

52. Zwahlen C, Andow DA (2005) Field evidence for the exposure of ground beetles to Cry1Ab from transgenic corn. Environmental Biosafety Research 4: 113–117.

53. Schmidt JE, Braun CU, Whitehouse LP, Hilbeck A (2009) Effects of activated Bt transgene products (Cry1Ab, Cry3Bb) on immature stages of the ladybird *Adalia bipunctata* in laboratory ecotoxicity testing. Archives of environmental contamination and toxicology 56: 221–228.

54. Li FG, Cui JJ, Liu CL, Wu ZX, Li FL, et al. (2000) The studies on Bt+CpTI cotton and its resistance. Scientia Agricultura Sinica 33: 46–52.

55. Amarasekare KG, Mannion CM, Osborne LS, Epsky ND (2008) Life history of *Paracoccus marginatus* (Hemiptera: Pseudococcidae) on four host plant species under laboratory conditions. Environmental entomology 37: 630–635.

56. Chong JH, Roda AL, Mannion CM (2008) Life history of the mealybug, *Maconellicoccus hirsutus* (Hemiptera: Pseudococcidae), at Constant temperatures. Environmental entomology 37: 323–332.

57. Li Y, Romeis J, Wang P, Peng Y, Shelton AM (2011) A comprehensive assessment of the effects of Bt cotton on *Coleomegilla maculata* demonstrates no detrimental effects by Cry1Ac and Cry2Ab. PLoS One 6: e22185.

58. Rui YK, Wang BM, Li ZH, Duan LS, Tian XL, et al. (2004) Development of an enzyme immunoassay for the determination of the cowpea trypsin inhibitor (CpTI) in transgenic crop. Scientia Agricultura Sinica 37: 1575–1579.

59. Tan GY, Nan TG, Gao W, Li QX, Cui JJ, et al. (2013) Development of Monoclonal Antibody-Based Sensitive Sandwich ELISA for the Detection of Antinutritional Factor Cowpea Trypsin Inhibitor. Food Analytical Methods 6: 614–620.

60. Quinn GP, Michael JK (2002) Experimental design and data analysis for biologists. Cambridge, UK: Cambridge University Press.

61. McCullagh P, Nelder JA (1989) Generalized linear models. London, UK: Chapman & Hall.

62. Raps A, Kehr J, Gugerli P, Moar W, Bigler F, et al. (2001) Immunological analysis of phloem sap of *Bacillus thuringiensis* corn and of the nontarget herbivore *Rhopalosiphum padi* (Homoptera: Aphididae) for the presence of Cry1Ab. Molecular Ecology 10: 525–533.

63. Dutton A, Klein H, Romeis J, Bigler F (2002) Uptake of Bt-toxin by herbivores feeding on transgenic maize and consequences for the predator *Chrysoperla carnea*. Ecological Entomology 27: 441–447.

64. Cui JJ, Xia JY (2000) Effects of Bt (*Bacillus thuringiensis*) transgenic cotton on the dynamics of pest population and their enemies. Acta Phytophylacica Sinica 27: 141–145.

65. Lumbierres B, Albajes R, Pons X (2004) Transgenic Bt maize and *Rhopalosiphum padi* (Hom., Aphididae) performance. Ecological Entomology 29: 309–317.

66. Lawrence PK, Koundal KR (2002) Plant protease inhibitors in control of phytophagous insects. Electronic Journal of Biotechnology 5: 5–6.

67. Rahbe Y, Deraison C, Bonade-Bottino M, Girard C, Nardon C, et al. (2003) Effects of the cysteine protease inhibitor oryzacystatin (OC-I) on different aphids and reduced performance of *Myzus persicae* on OC-I expressing transgenic oilseed rape. Plant science 164: 441–450.

68. Desneux N, Decourtye A, Delpuech J-M (2007) The sublethal effects of pesticides on beneficial arthropods. Annu Rev Entomol 52: 81–106.

69. Decourtye A, Mader E, Desneux N (2010) Landscape enhancement of floral resources for honey bees in agro-ecosystems. Apidologie 41: 264–277.

70. Renard S, Calatayud PA, Pierre JS, Rü BL (1998) Recognition Behavior of the Cassava Mealybug *Phenacoccus manihoti* Matile-Ferrero (Homoptera: Pseudococcidae) at the Leaf Surface of Different Host Plants. Journal of Insect Behavior 11: 429–450.

71. Kölliker-Ott UM, Bigler F, Hoffmann AA (2003) Does mass rearing of field collected *Trichogramma brassicae* wasps influence acceptance of European corn borer eggs? Entomologia experimentalis et applicata 109: 197–203.

72. Geden C, Smith L, Long S, Rutz D (1992) Rapid deterioration of searching behavior, host destruction, and fecundity of the parasitoid Muscidifurax raptor (Hymenoptera: Pteromalidae) in culture. Annals of the Entomological Society of America 85: 179–187.

73. Joyce AL, Aluja M, Sivinski J, Vinson SB, Ramirez-Romero R, et al. (2010) Effect of continuous rearing on courtship acoustics of five braconid parasitoids, candidates for augmentative biological control of *Anastrepha species*. BioControl 55: 573–582.

74. Lu Y, Xue L, Zhou ZT, Dong JJ, Gao XW, et al. (2011) Effects of Transgenic Bt Plus CpTI cotton on Predating Function Response of *Coccinella septempunctata* to *Aphis gossypii* Glover. Acta Agriculturae Boreali-Sinica 26: 163–167.

75. Geng JH, Shen ZR, Song K, Zheng L (2006) Effect of pollen of regular cotton and transgenic Bt+ CpTI cotton on the survival and reproduction of the parasitoid wasp *Trichogramma chilonis* (Hymenoptera: Trichogrammatidae) in the laboratory. Environmental entomology 35: 1661–1668.

76. Porcar M, García-Robles I, Domínguez-Escribà L, Latorre A (2010) Effects of *Bacillus thuringiensis* Cry1Ab and Cry3Aa endotoxins on predatory Coleoptera tested through artificial diet-incorporation bioassays. Bulletin of entomological research 100: 297.

77. Duan JJ, Head G, McKee MJ, Nickson TE, Martin JW, et al. (2002) Evaluation of dietary effects of transgenic corn pollen expressing Cry3Bb1 protein on a non-target ladybird beetle, *Coleomegilla maculata*. Entomologia experimentalis et applicata 104: 271–280.

78. Lundgren JG, Wiedenmann RN (2002) Coleopteran-specific Cry3Bb Toxin from Transgenic Corn Pollen Does Not Affect The Fitness of a Nontarget Species, *Coleomegilla maculata* DeGeer (Coleoptera: Coccinellidae). Environmental entomology 31: 1213–1218.

79. Dogan E, Berry R, Reed G, Rossignol P (1996) Biological parameters of convergent lady beetle (Coleoptera: Coccinellidae) feeding on aphids (Homoptera: Aphididae) on transgenic potato. Journal of Economic Entomology 89: 1105–1108.

80. Sharma HC, Arora R, Pampapathy G (2007) Influence of transgenic cottons with *Bacillus thuringiensis* cry1Ac gene on the natural enemies of *Helicoverpa armigera*. BioControl 52: 469–489.

81. Xu Y, Wu KM, Li HB, Liu J, Ding RF, et al. (2012) Effects of Transgenic Bt+ CpTI Cotton on Field Abundance of Non-Target Pests and Predators in Xinjiang, China. Journal of Integrative Agriculture 11: 1493–1499.

82. Zhang GF, Wan FH, Wan XL, Guo JY (2006) Early Instar Response to Plant Derived Bt-Toxin in a Herbivore (*Spodoptera litura*) and a Predator (*Propylaea japonica*). Crop Protection 25: 527–533.

83. Lumbierres B, Starý P, Pons X (2011) Effect of Bt maize on the plant-aphid-parasitoid tritrophic relationships. BioControl 56: 133–143.

84. Ramirez-Romero R, Desneux N, Chaufaux J, Kaiser L (2008) Bt-maize effects on biological parameters of the non-target aphid *Sitobion avenae* (Homoptera: Aphididae) and Cry1Ab toxin detection. Pesticide Biochemistry and Physiology 91: 110–115.

85. Head G, Brown CR, Groth ME, Duan JJ (2001) Cry1Ab protein levels in phytophagous insects feeding on transgenic corn: implications for secondary exposure risk assessment. Entomologia experimentalis et applicata 99: 37–45.

86. Obrist L, Dutton A, Albajes R, Bigler F (2006) Exposure of arthropod predators to Cry1Ab toxin in Bt maize fields. Ecological Entomology 31: 143–154.

87. Torres JB, Ruberson JR, Adang MJ (2006) Expression of *Bacillus thuringiensis* Cry1Ac protein in cotton plants, acquisition by pests and predators: a tritrophic analysis. Agricultural and Forest Entomology 8: 191–202.

88. Li Y, Romeis J (2010) Bt maize expressing Cry3Bb1 does not harm the spider mite, *Tetranychus urticae*, or its ladybird beetle predator, *Stethorus punctillum*. Biological Control 53: 337–344.

# Association Mapping for Epistasis and Environmental Interaction of Yield Traits in 323 Cotton Cultivars under 9 Different Environments

Yinhua Jia[1,9], Xiwei Sun[2,9], Junling Sun[1,9], Zhaoe Pan[1], Xiwen Wang[1], Shoupu He[1], Songhua Xiao[3], Weijun Shi[4], Zhongli Zhou[1], Baoyin Pang[1], Liru Wang[1], Jianguang Liu[3], Jun Ma[4], Xiongming Du[1]*, Jun Zhu[2]*

1 Institute of Cotton Research of Chinese Academy of Agricultural Sciences (ICR, CAAS), State Key Laboratory of Cotton Biology, Key Laboratory of Cotton Genetic Improvement, Ministry of Agriculture, Anyang, China, 2 Key Laboratory of Crop Germplasm Resource of Zhejiang Province, Zhejiang University, Hangzhou, China, 3 Institute of industrial Crops, Jiangsu Academy of Agricultural Sciences, Nanjing, China, 4 Economic Crop Research Institute, Xinjiang Academy of Agricultural Science, Urumqi1, China

## Abstract

Improving yield is a major objective for cotton breeding schemes, and lint yield and its three component traits (boll number, boll weight and lint percentage) are complex traits controlled by multiple genes and various environments. Association mapping was performed to detect markers associated with these four traits using 651 simple sequence repeats (SSRs). A mixed linear model including epistasis and environmental interaction was used to screen the loci associated with these four yield traits by 323 accessions of *Gossypium hirsutum* L. evaluated in nine different environments. 251 significant loci were detected to be associated with lint yield and its three components, including 69 loci with individual effects and all involved in epistasis interactions. These significant loci explain ~ 62.05% of the phenotypic variance (ranging from 49.06% ~ 72.29% for these four traits). It was indicated by high contribution of environmental interaction to the phenotypic variance for lint yield and boll numbers, that genetic effects of SSR loci were susceptible to environment factors. Shared loci were also observed among these four traits, which may be used for simultaneous improvement in cotton breeding for yield traits. Furthermore, consistent and elite loci were screened with $-\text{Log}_{10}$ (*P*-value) >8.0 based on predicted effects of loci detected in different environments. There was one locus and 6 pairs of epistasis for lint yield, 4 loci and 10 epistasis for boll number, 15 loci and 2 epistasis for boll weight, and 2 loci and 5 epistasis for lint percentage, respectively. These results provided insights into the genetic basis of lint yield and its components and may be useful for marker-assisted breeding to improve cotton production.

**Editor:** Baohong Zhang, East Carolina University, United States of America

**Funding:** This work has been funded by the National Science and Technology Support Program of China (2006BAD13B04), and National Basic Research Program of China (2010CB12600, 2011CB109306). The funders had no rule in study design, data collection and analysis, decision to publish, or preparation of the manuscript.

**Competing Interests:** The authors have declared that no competing interests exist.

* E-mail: jzhu@zju.edu.cn (JZ); duxm@cricaas.com.cn (XD)

⑨ These authors contributed equally to this work.

## Introduction

Cotton is one of the most important economic crops in the world and provides large amounts of raw materials for textile industry. High yield is always remained the primary focus of cotton breeding programs. Boll number, boll weight, and lint percentage are three yield component traits, which are complex traits controlled by multiple genes with gene × gene interaction and gene × environment interaction [1–2]. Understanding genetic architecture of lint yield and its component traits at the molecular level is very important for efficiently improving cotton yield traits. Identifying and charactering the key QTLs and genes affecting yield traits in cotton has been a research frontier over the past few decades. On the basis of linkage analysis in the QTL mapping populations, several QTLs controlling lint yield and its component traits have been obtained [1–8]. However, these detected QTLs

involved in large regions of the chromosome, where candidate genes or genetic variants are difficult to distinguish. In addition, the mapping results have only limit utilization because of lower genetic diversity in the special population used for QTL mapping and can not fully elucidate the genetic basis of lint yield and its component traits.

Genome-wide association studies (GWAS), which are based on the linkage disequilibrium (LD), provide the opportunity to systematically identify the genetic components of complex traits of crops in recent years [9–14]. It has the potential to detect single nucleotide polymorphisms (SNPs) or other types of molecular markers such as single sequence repeats (SSRs) within or nearby a gene. It can extensively exploit historical recombination and natural genetic diversity in natural population, and overcome the limitations of conventional linkage mapping for crop breeding. Recently, Zhang *et al* performed association mapping using 81 accessions of Upland cotton based on 121 SSRs to map 12 agronomical and fiber quality traits, and identified 180 loci only with additive effects significantly associated with the traits [14]. However, current GWAS predominately focus on examination of SNPs at a single environment and fails to detect the gene × gene interaction as well as gene × environment interaction across multiple environments. The genetic variation of complex traits is contributed in part by epistasis and environmental interaction effects. Therefore, searching for only major effects may miss other key genetic variants specific to environment factors and cannot provide reliable estimates for genetic effects [3].

Due to a heavy computational burden, current methods of GWAS have not been used for detecting gene × gene interaction and gene × environment interaction for breeding populations in multiple environments. In this study, we used mixed linear model approach and a newly developed mapping software of **QTXNetwork** based on GPU parallel computation to identify individual and epistasis loci and their environmental interactions for lint yield and its component traits. We also identified candidate genes related to markers of these target traits. These results and genetic architecture identified in this study could provide more precise and reliable information for marker-assisted selection in crop breeding.

## Materials and Methods

### Plant Materials

Mapping populations representing Upland cotton cultivars and genetic resources were mainly obtained from China and some from foreign countries including the United States of America, Russia, Australia etc (Table S1), which are available from the National Mid-term Genebank of the Institute of Cotton Research, Chinese Academy of Agricultural Sciences (ICR-CAAS) after signing the Material Transfer Agreement (MTA). These cultivars are suitable for association mapping since they have abundant genetic resources. 323 sampled cultivars of Upland cotton were planted and evaluated for lint yield and yield components during 2007 to 2009 in three locations: 1) Anyang of Henan province (annual frost free period = 180–230 days, average active accumulated temperature = 3800–4900°C, annual rainfall = 500–1000 mm) in the Yellow River valley cotton growing region; 2) Kuche of Xinjiang province (annual frost free period = 170–230 days, average active accumulated temperature = 3000–5400°C, annual rainfall = 15–380 mm) in the northwestward cotton growing region; 3) Nanjing of Jiangsu province (annual frost free period = 227–278 days, average active accumulated temperature = 3500–5500°C, annual rainfall = 1000–1600 mm) in the Yangtze River valley cotton growing region. This research is regular cotton regional trial test conducted in farm field of cotton research institute, and not related with protected area of land and protection of wildlife. It does not need specific official permission. We regarded combination of each location and each year as an individual environment, setting a total of nine environments. A randomized complete block design with three replications was employed in the filed trials, and each block was settled with two rows and every row was kept in a plot with 5 m long and 0.7 m wide. The field management utilized conventional field production management techniques adjusted to local practice. Lint yield and three yield components were evaluated: number of bolls per plant, boll weight, and lint percentage.

### DNA Extraction and Marker Analysis

The genomic DNA was extracted from fresh, young leaves of the 323 cultivars using CTAB method. SSR primers sequences including BNL, JESPR, TMB, CIR, HAU, DPL, GH, NAU,

MGHES etc. can be available from Cotton Microsatellite Database (CMD, http://www.cottonssr.org). We selected 20 cultivars of large morphological variations from the research natural population to screen for polymorphisms primers. In total, 5,600 SSR primers were chosen to survey the polymorphisms of the 323 cultivars. Only 651 polymorphisms SSR alleles were detected, which were used for association mapping (Table S2).

### Genetic Models and Statistical Methods

Association mapping were performed using the mixed linear model, including environment ($e$) as fixed effects, SSR loci effects ($a$, $aa$) and loci by environment interaction ($ae$, $aae$) as random effects. The genetic model for phenotypic value of the $k$-th genotypes in the $h$-th environment ($y_{hk}$) can be expressed by the following mixed linear model,

$$y_{hk} = \mu + e_h + \sum_i a_i u_{ik} + \sum_{i<j} aa_{ij} u_{ik} u_{jk} + \sum_i ae_{hi} u_{hik}$$
$$+ \sum_{i<j} aae_{hij} u_{hik} u_{hjk} + \varepsilon_{hk} \tag{1}$$

where $\mu$ is the population mean; $e_h$ is the fixed effect of the $h$-th environment; $a_i$ is the $i$-th additive effect with coefficient $u_{ik}$; $aa_{ij}$ is the $i$-th additive by $j$-th additive epistasis effect with coefficient $u_{ik} u_{jk}$; $ae_{hi}$ is the $i$-th additive by the $h$-th environment interaction effect with coefficient $u_{hik}$; $aae_{hij}$ is the $aa_{ij}$ by the $h$-th environment interaction effect with coefficient $u_{hik} u_{hjk}$; and $\varepsilon_{hk}$ is the random residual effect of the $k$-th breeding line in the $h$-th environment.

The mixed linear model can be presented in matrix notation,

$$\mathbf{y} = \mathbf{Xb} + \mathbf{U}_A \mathbf{e}_A + \mathbf{U}_{AA} \mathbf{e}_{AA} + \mathbf{U}_{AE} \mathbf{e}_{AE} + \mathbf{U}_{AAE} \mathbf{e}_{AAE} + \mathbf{e}_\varepsilon$$
$$= \mathbf{Xb} + \sum_{v=1}^{4} \mathbf{U}_v \mathbf{e}_v + \mathbf{e}_\varepsilon \tag{2}$$
$$\sim MVN(\mathbf{Xb}, \sum_{v=1}^{4} \sigma_v^2 \mathbf{U}_v \mathbf{U}_v^T + \mathbf{I}\sigma_\varepsilon^2)$$

where $\mathbf{y}$ is an $n \times 1$ column vector of phenotypic values and $n$ is the sample size of observations; $\mathbf{b}$ is a column vector of $u$ and environments in the experiment; $\mathbf{X}$ is the known incidence matrix relating to the fixed effects; $\mathbf{U}_v$ is the known coefficient matrix relating to the $v$-th random vector $\mathbf{e}_v$; $\mathbf{e}_\varepsilon \tilde{M} VN(0, \mathbf{I}\sigma_\varepsilon^2)$ is an $n \times 1$ column vector of residual effects.

The phenotypic variance $V_p$ is considered as the sum of additive variance $V_A$, epistasis variance $V_{AA}$, additive by environment interaction variance $V_{AE}$, epistasis by environment interaction variance $V_{AAE}$, and residual variance $V_\varepsilon$.

$$V_p = V_A + V_{AA} + V_{AE} + V_{AAE} + V_\varepsilon \tag{3}$$

Heritability is defined as the relative contribution of genetic variance to phenotypic variance with the following estimation model:

$$h_{G+GE}^2 = h_A^2 + h_{AA}^2 + h_{AE}^2 + h_{AAE}^2$$
$$= \sum h_a^2 + \sum h_{aa}^2 + \sum h_{ae}^2 + \sum h_{aae}^2 \tag{4}$$

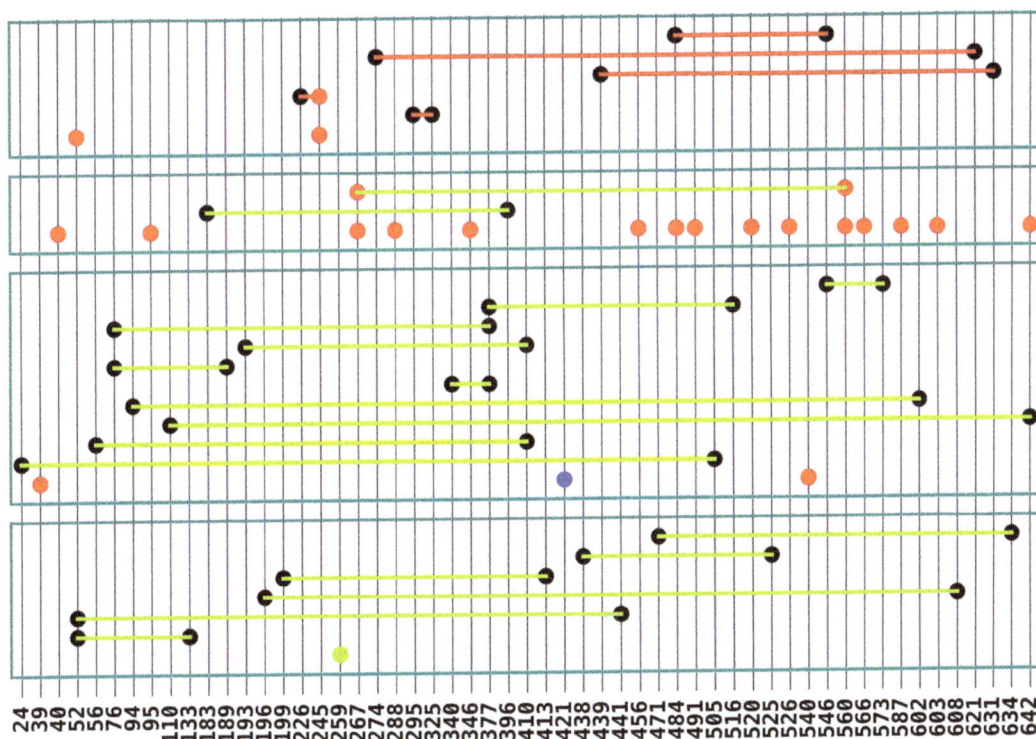

**Figure 1. Network plot of highly significant loci detected for yield and yield components.** Notes: The bottom axis is the SNP ID for QTS. Red dot = loci with additive effect, green dot = loci with additive × environment interaction effect, blue dot = loci with both additive and environment-specific effects, black dot with a line = loci with epistasis but no individual effect. Red line between two dot = *aa* epistasis, Green line between two dot = *aae* environment-specific epistasis.

where $h^2_{G+GE}$ = total heritability; $h^2_A$ = heritability contributed by sum of individual locus, $h^2_{AA}$ = heritability contributed by sum of pair-wise epistasis, $h^2_{AE}$ = additive by environment interaction heritability contributed by sum of individual additive by environment interaction effects, $h^2_{AAE}$ = epistasis by environment interaction heritability contributed by sum of pair-wise epistasis by environment interaction effects; $h^2_a$ = additive heritability of individual locus, $h^2_{aa}$ = epistasis heritability of pair-wise loci, $h^2_{ae}$ = heritability of additive by environment interaction effect, $h^2_{aae}$ = heritability of epistasis by environment interaction effect.

## Association Mapping

A GPU parallel computing software **QTXNetwork** (http://ibi.zju.edu.cn/software/QTXNetwork/) was used to dissect the genetic architecture of lint yield and its component traits. There were 651 SSR markers used for association mapping for yield traits of 323 representative Upland cotton cultivars in nine different environments. Significant SSR markers were screened and the estimation of fixed effects (*e*) and prediction of random effects (*a*, *aa*, *ae*, and *aaè*) of loci were obtained.

## Gene Locations Based on the Associated Markers by Reference D Genome

Based on the experiment-wise type I error ($\alpha_{EW} < 0.05$) setting by **QTXNetwork**, the SSR loci associated with the fiber yield traits were selected. The primer information was acquired from cotton marker database (http://www.cottonmarker.org/cmd_downloads/ssr_project_data/CMD_PRIMER_ALL.xls). Only one SSR location for each marker on the genome D (ftp://ftp.

ncbi.nlm.nih.gov/genbank/genomes/Eukaryotes/plants/Gossypium_raimondii/) was acquired by bowtie method which 2 bases not matched was permitted (bowtie -a -v 2–fr cottonPD -f -1 primerForward.fasta -2 primerReverse.fasta –best –strata bowtieResult_PD.txt). The SSR motifs of the searched SSR sequences on the *D* genome were scanned by MISA. The location between SSR locus and the related gene was decided based the star location of the motif on the genome.

## Results

In total, there were 94 significant SSRs (14 loci with individual effects and 93 loci involved in epistasis interactions) associated with lint yield, 95 SSRs (28 individual loci) associated with boll number, 91 SSRs (21 individual loci) associated with boll weight, and 88 SSRs (19 individual loci) associated with lint percentage. All identified loci associated with lint yield components were involved in epistasis interactions.

As to the lint yield, the total heritability of genotype × environment interaction effects ($h^2_{GE} \triangleq 38.26\%$) was mainly contributed due to epistasis × environment interaction ($h^2_{AAE} \triangleq 34.05\%$), and was much larger than the total genotype heritability ($h^2_G \triangleq 10.80\%$) (Table 1). We listed highly significant genetic effects of loci ($-\text{Log}_{10}P > 8.0$ and $h^2_{GE} > 1.0\%$, $h^2_G > 0.5\%$) and consistent loci with environment interaction which were stable in a particular ecological area across at least two years for lint yield and three yield traits (Figure 1, Table 2–5). Among the loci associated with lint yield (Table 2), 1 locus (NAU3011-2) and 6 pairs of epistasis loci were highly significant with environmental interaction at Nanjing in 2007 and 2009. It was suggested that improving lint yield could be expected by selecting major-allele genotype $QQ$ (NAU3011−197) and $QQ \times QQ$ (MUCS101−330×MGHES18−228, NAU3325−238×TMB1989−255, NAU3608−245×HAU773−170), as well as minor-allele genotype $qq \times qq$ (HAU1794−320×TMB10−375, NAU1362−225×NAU3774−248, and NAU3325−238×HAU423−175) at Nanjing (Table 2).

For boll number trait, the heritability of epistasis × environment interaction effects ($h^2_{AAE} \triangleq 41.93\%$) was larger than other trait heritability (Table 1), which indicated that boll number was mainly controlled by environment-specific epistasis effects. 3 loci (HAU1385-3, NAU1102-3, MGHES41-6) with additive heritability were found to be more than 0.5%, and 10 pair epistasis loci with environment-specific epistasis heritability showed more than 1.0% genetic effects (Figure 1, Table 3), indicating that the improvement of boll number in all 9 environments could be predictable ($h^2_A \triangleq 1.83\%$) by selecting genotype $QQ$ for HAU1385−150, and $qq$ for other 2 loci (MGHES41−333, NAU1102−241). Further improving boll number could be expected ($h^2_{AE_3} \triangleq 11.31\%$) through selecting HAU1029−197 × NAU1125−243 in both Anyang and Nanjing areas and other 9 pair epistasis loci only in Nanjing area, among these 7 epistasis loci with genotype of $QQ \times QQ$ showed positive effects, and 3 pair epistasis loci with genotype of $qq \times qq$ showed negative effects. The total heritability ($h^2_{G+GE} \triangleq 62.85\%$) of boll weight was mostly due to additive effects ($h^2_A \triangleq 21.48\%$), epistasis effects ($h^2_{AA} \triangleq 22.01\%$) and environment-specific additive effects ($h^2_{AAE} \triangleq 18.73\%$). It was suggested that most genetic effects were fairly stable across environments ($h^2_G \triangleq 43.49\%$) for boll weight. There were 15 loci with heritability of additive effect larger than 0.5% and 2 pair epistasis loci with heritability of environment-specific epistasis effects larger than 1.0% for boll weight (Figure 1, Table 4). With regard to little impact of environmental factors on boll weight, relevant robust loci could be selected for improvement of lint yield across different ecological areas.

For lint percentage, the total heritability ($h^2_{G+GE} \triangleq 72.29\%$) was mostly attributed to epistasis effects ($h^2_{AA} \triangleq 57.78\%$). Thus, it was concluded that lint percentage was very stable across various environments and mainly controlled by epistasis effects. Compared with lint yield and other yield components, most genetic effects of individual loci and epistasis loci explained a small proportion of total phenotypic variance for lint percentage. Only 2 loci with additive effects and 5 pair epistasis loci with epistasis effects had heritability larger than 0.5% (Figure 1 and Table 5). It was apparent that lint percentage was mostly controlled by many loci with smaller effects as compared with lint yield and other component traits. In this population, further genetic gain could be projected ($h^2_{A+AA} \triangleq 6.3\%$) based on selecting 2 individual loci with additive effects (NAU3325−238, NAU3519−200) and 5 pair loci

with epistasis effects for increasing lint percentage across different environments.

It has been observed that some loci were associated with more than one trait. 4 loci (DPL910-120, HAU639-175, NAU1155-205, NAU4042−175) were found to be associated with all of the four traits, 22 loci were detected to be associated with three of the four traits, and 61 loci were associated with two of the four traits. Generally, these common loci made it possible for us to simultaneously improve multiple traits via breeding programs. However, 41 loci controlling lint yield were not detected in other component traits, which suggested that lint yield, as the result of complex biosynthesis pathways, may comprise some unknown components affected by polygenes.

It had been observed that epistasis is the genetic base of lint yield and yield components. In the results, most of the loci with individual effects were involved in epistasis interactions for lint yield and yield components. Altogether 293 digenic loci with epistasis effect ($aa$) and/or epistasis × environment interaction effect ($aae$) were identified to be associated with lint yield. There were 193 digenic epistasis interactions with their total heritability ($h^2_{AA+AAE} \triangleq 30.71\%$) of epistasis effects and environment-specific epistasis effects involved in the loci without individual effects. We also observed this phenomenon in yield components. It was shown that many loci, even without controlling yield and yield components on their own, could affect the traits in combination with other loci.

It was observed that many loci could interact with several other loci for controlling phenotypic variation of lint yield and its component traits. For example, NAU2126-195 could affect lint yield in combination with other 4 loci (HAU773−170, NAU3377−180, TMB10−375, GH501-265), respectively. Magnitude of epistasis effects of them differed from each other. NAU2126-195 may be regarded as core locus of epistasis interaction jointly with other casual genes for controlling target traits. This phenomenon might indicate that higher order interaction could also exist for association with lint yield and its components.

## Discussion

Association mapping is a powerful method to detect loci underlying complex traits and more efficient compared to linkage analysis, because it exploited abundant genetic variation in diverse genetic backgrounds [15]. However, in most of the current GWAS, the epistasis and environmental interaction effects were not detectable due to absence of appropriate statistical methods and a heavy computational burden [16–18]. The models without epistasis and environmental interaction for GWAS may cause some problems: 1) It may result in biased estimation of effects on loci and decrease the precision and power of loci detection [18–

**Table 1.** Estimated heritability for lint yield and yield components.

| Trait | $h^2_A$ (%) | $h^2_{AA}$ (%) | $h^2_{AE}$ (%) | $h^2_{AAE}$ (%) | $h^2_{G+GE}$ (%) |
|---|---|---|---|---|---|
| Lint Yield | 0.20 | 10.6 | 4.21 | 34.05 | 49.06 |
| Boll Number | 3.87 | 9.28 | 8.89 | 41.93 | 63.98 |
| Boll Weight | 21.48 | 22.01 | 0.63 | 18.73 | 62.85 |
| Lint Percentage | 6.58 | 57.78 | 0.29 | 7.64 | 72.29 |

**Note:** $h^2_A$ = heritability for additive effects, $h^2_{AA}$ = heritability for additive epistasis effects, $h^2_{AE}$ = heritability for additive by environment interaction effects, $h^2_{AAE}$ = heritability for additive epistasis by environment effects, $h^2_{G+GE}$ = total heritability.

**Table 2.** Predicted genetic effects with significance and heritability of lint yield for cotton regional trials in three locations and three years.

| Locus | Effect | Predict | $-\text{Log}_{10}P$ | $h^2$ (%) | Candidate Gene |
|---|---|---|---|---|---|
| NAU3011−197 | $ae_3y_1$ | 1.01 | 11.32 | 1.04 | IPR000679 |
| | $ae_3y_3$ | 1.56 | 25.93 | | |
| HAU1794−320×TMB10−375 | $aae_3y_1$ | −1.58 | 26.17 | 0.96 | – × Gorai.004G126800 |
| | $aae_3y_3$ | −0.99 | 10.82 | | |
| MUCS101−330×MGHES18−228 | $aae_3y_1$ | 0.98 | 10.97 | 0.65 | Cotton_A_29394×IPR001128 |
| | $aae_3y_3$ | 1.06 | 12.52 | | |
| NAU1362−225×NAU3774−248 | $aae_3y_1$ | −0.84 | 8.05 | 0.79 | IPR001764×IPR002913 |
| | $aae_3y_2$ | −1.15 | 14.61 | | |
| | $aae_3y_3$ | −1.04 | 12.03 | | |
| NAU3325−238×HAU423−175 | $aae_3y_1$ | −1.03 | 11.93 | 0.72 | IPR000778×IPR001461 |
| | $aae_3y_3$ | −1.25 | 17.12 | | |
| NAU3325−238×TMB1989−255 | $aae_3y_1$ | 1.44 | 21.74 | 1.00 | IPR000778×Cotton_A_24169 |
| | $aae_3y_3$ | 1.15 | 14.23 | | |
| NAU3608−245×HAU773−170 | $aae_3y_1$ | 1.10 | 13.26 | 0.73 | IPR000719×IPR007087 |
| | $aae_3y_3$ | 0.89 | 8.99 | | |

**Note:** $ae_3y_1$ = additive by environment interaction effect at Nanjing in 2007; $ae_3y_3$ = additive by environment interaction effect at Nanjing in 2009; $aae_3y_1$ = epistasis by environment interaction effect at Nanjing in 2007; $aae_3y_2$ = epistasis by environment interaction effect at Nanjing in 2008; $aae_3y_3$ = epistasis by environment interaction effect at Nanjing in 2009; $-\text{Log}_{10}P$ = minus $\log_{10}$ ($P$-value), $h^2$ (%) = heritability (%).

19]; 2) The heritability of complex traits would be missed in the condition of reduced models [20–21].

Our proposed methodology is based on mixed linear model including epistasis and environmental interaction, but it can also be used for reduced models ignoring epistasis and environmental interaction. We also performed GWAS of various reduced models for lint yield and yield components. As compared with the full model (Table 1), the total heritability of lint yield and yield components decreased in the additive model ($h_G^2 \triangleq 18.35\%$ for lint yield, 6.53% for boll number, 28.65% for boll weight, and 22.95% for lint percentage), and also in the additive with environmental interaction model ($h_{G+GE}^2 \triangleq 17.31\%$ for lint yield, 23.76% for boll number, 32.25% for boll weight, and 25.08% for lint percentage). The results also showed steep decrease in numbers and effects of loci associated lint yield and yield components between reduced models and full model. This may explain the reasons for missing heritability reported recently [21].

Because of low LD associated with higher frequencies of cross-pollination, cotton species are amenable for association mapping of agronomic traits with a relatively large numbers of markers. Previous studies involving nearly 1000 polymorphic markers would ensure the coverage of whole cotton genome [14]. A total of 651 SSR markers and 323 Upland cotton cultivars under 9 environments in this study can significantly improve the power of association mapping. However, with the rapid development of next generation sequencing, a substantial number of single nucleotide polymorphisms (SNPs) will facilitate the accurate and fine association analysis.

It is shown by the results of GWAS that lint yield, boll number and lint percentage are inherited in a mainly epistasis manner, whereas boll weight is predominately controlled by both additive and epistasis effects. In addition, lint yield and boll number are particularly susceptible to a broad range of environments, as compared to boll weight and lint percentage, which has also been observed in previous studies [5,22–23]. Both lint yield and

component traits were complex traits controlled by many genes and gene-network cascades interacting with each other and also with the environment factors (Figure 1). Many loci (ten or more in the present study) associated with lint yield, boll number, and boll weight showed much larger effects and heritability (additive or epistasis effects), while few loci associated with lint percentage could have large effects. This was indicated that improving cotton lint yield could be difficult through increasing lint percentage. All these information obtained from GWAS could provide useful reference for MAS in cotton breeding programs.

Although the effects of many loci appeared to be consistent across various environments, the magnitude of effects and even direction of loci may vary depending on environmental factors because of interactions between loci and environment [3,18,24]. We can classify loci identified in the present study into three types. The first category of loci is called constituted loci with large major effects ($a$, $aa$) and can be consistently detected across different environments. This type of loci is very useful for improving cotton traits across different ecological locations. Another type of loci is called environment-specific loci with only environment-special effects identified in a special ecological region but stable in different years. For example, TMB312−212 × GH132-180 was stably found in the Northwestward region and NAU3013−245 × NAU5163−216 was stably scanned in the Yangtze River valley region. This type of loci could be used for maker-assisted breeding in a particular ecological area. The third type of loci is called environment-sensitive loci that could not be consistently detected in various grown regions. This kind of loci could only be used in specific conditions.

SSRs widely used in major crop were highly reliable, polymorphic, simple and cheap. The SSRs associated with nearby genes were extremely useful for MAS. In the present study, we successfully detected 251 significant loci associated with lint yield and its three components, including 69 loci with individual effects. Most of these significant loci generally had small impacts on cotton

**Table 3.** Predicted genetic effects with significance and heritability of boll number for cotton regional trials in three locations and three years.

| Locus | Effect | Predict | $-\text{Log}_{10}P$ | $h^2$ (%) | Candidate Gene |
|---|---|---|---|---|---|
| HAU1385−150 | $a$ | 0.3 | 42.29 | 0.68 | IPR003245 |
| MGHES41−330 | $a$ | −0.27 | 34.72 | 0.55 | IPR000297 |
| NAU1102−241 | $a$ | −0.28 | 37.61 | 0.6 | IPR003329 |
| | $ae_3y_1$ | −0.38 | 8.35 | 0.63 | |
| | $ae_3y_2$ | −0.46 | 11.99 | | |
| | $ae_3y_3$ | −0.52 | 14.8 | | |
| GH111−245×NAU874−215 | $aae_3y_1$ | 0.37 | 8.25 | 0.37 | − ×IPR001128 |
| | $aae_3y_3$ | 0.39 | 8.93 | | |
| GH111−245×NAU5433−330 | $aae_3y_1$ | 0.55 | 16.75 | 0.96 | − ×IPR001865 |
| | $aae_3y_3$ | 0.84 | 37.43 | | |
| HAU1029−197×NAU1125−243 | $aae_1y_2$ | 0.47 | 12.16 | 1.59 | IPR001012×IPR016196 |
| | $aae_1y_3$ | 0.55 | 16.85 | | |
| | $aae_3y_1$ | 0.93 | 46.13 | | |
| | $aae_3y_3$ | 0.63 | 21.32 | | |
| HAU1969−375×JESPR42−128 | $aae_3y_1$ | 1.08 | 60.86 | 2.43 | IPR000181×Gorai.009G322400 |
| | $aae_3y_3$ | 1.27 | 83.72 | | |
| HAU1639−320×NAU5099−233 | $aae_3y_1$ | 0.38 | 8.55 | 0.53 | IPR000217×IPR001214 |
| | $aae_3y_3$ | 0.5 | 14.18 | | |
| NAU2715−184×JESPR42−128 | $aae_3y_2$ | −0.63 | 22.12 | 0.65 | IPR004022×Gorai.009G322400 |
| | $aae_3y_3$ | −0.38 | 8.43 | | |
| NAU2873−352×NAU5433−330 | $aae_3y_1$ | −0.50 | 13.93 | 0.77 | IPR003311×IPR001865 |
| | $aae_3y_3$ | −0.77 | 32.56 | | |
| NAU4042−175×GH111−245 | $aae_3y_1$ | 0.47 | 12.53 | 0.64 | Cotton_A_04013×− |
| | $aae_3y_3$ | 0.68 | 25.62 | | |
| STV61-131×NAU3305−155 | $aae_3y_1$ | −0.91 | 43.35 | 1.49 | DUF1685×IPR002913 |
| | $aae_3y_2$ | −0.65 | 22.48 | | |
| TMB1181−220×NAU3563−149 | $aae_1y_1$ | 0.6 | 19.26 | 1.25 | Gorai.002G088500×IPR003311 |
| | $aae_1y_2$ | 0.42 | 9.91 | | |

**Note:** $a$ = additive effect; $ae_3y_1$, $ae_3y_3$, $aae_3y_1$, $aae_3y_2$, $aae_3y_3$ as defined in **Table 2**; $ae_3y_2$ = additive by environment interaction effect at Nanjing in 2008, $aae_1y_1$ = epistasis by environment interaction effect at Anyang in 2007, $aae_1y_2$ = epistasis by environment interaction effect at Anyang in 2008, $aae_1y_3$ = epistasis by environment interaction effect at Anyang in 2009; $-\text{Log}_{10}P$ = minus $\log_{10}$ ($P$-value), $h^2$ (%) = heritability (%).

breeding due to small effects and easily affected by environment [23]. In order to develop effective MAS schemes, reliable and elite loci associated with lint yield and yield components have been screened. These loci, which consist of constituted loci and environment-specific loci, are reliable to predict corresponding phenotype. Constituted loci (such as NAU3588−315 and NAU3325−238) can be applied for MAS in various cotton-growing region, whereas application of environment-specific loci (examples are NAU1102−241 and NAU1362−225 × NAU3774−248) are limited to specific cotton-growing region. Both of them have substantial potential to improve the efficiency and precision of conventional cotton breeding.

With the information from Cotton Microsatellite Database (CMD) and published report, several markers identified in our present study had been located within or nearby the reported QTL in previous linkage analysis and association mapping [1−8,14]. For example, Zhang et al [14] independently detected QTLs called qLP-A1-1 (between JESPR101 and BNL3590) using linkage method and a locus called JESPR101 using association method, which affected lint percentage. We also identified this marker ($-\text{Log}_{10}P$-value = 15.09) associated with the same (Table 5). In addition, most of these consistent loci were involved in interaction in our present results (examples are NAU5433−330 for numbers of boll in Table 3). It indicated that these loci were stably transferred and can be further used in MAS.

The traditional QTL was mainly located between two markers with the scale of cM (centi-Mogan). Based on the physical mapping of reference genome, the fine mapping of the markers and the genes related to the fiber yield traits become possible. In this study, candidate gene locations of yield fiber traits were identified based on the associated SSR markers analyzed by the **QTXNetwork**. For the lint yield, one major-allele genotype $QQ$ (NAU3011−197) related to zinc finger (GATA-type) gene (IPR000679) was identified. 3 epistasis loci with positive effects ($QQ×QQ$ genotype) could be related to the interactions with the genes of cytochrome P450 (IPR001128), Cytochrome b245 (IPR000778), zinc finger with $C2H2$-type (IPR007087), and protein kinase (IPR000719) etc. 3 minor-allele with negative effects ($qq×qq$ genotype) was linked to the interactions with the genes of glycoside hydrolase (IPR001764), lipid-binding $START$

**Table 4.** Predicted genetic effects with significance and heritability of boll weight for cotton regional trials in three locations and three years.

| Locus | Effect | Predict | $-\text{Log}_{10}P$ | $h^2$(%) | Candidate Gene |
|---|---|---|---|---|---|
| NAU3744−200 | $a$ | −0.07 | 66.02 | 1.29 | IPR020828 |
| TMB312−212 | $a$ | −0.06 | 54.75 | 1.07 | IPR011335 |
| TMB1296−230 | $a$ | 0.07 | 60.15 | 1.18 | IPR002591 |
| BNL1313-180 | $a$ | 0.08 | 95.32 | 1.89 | – |
| NAU3588−315 | $a$ | −0.15 | 299.91 | 6.14 | – |
| GH132−180 | $a$ | −0.06 | 47.17 | 0.92 | IPR001611 |
| JESPR274−129 | $a$ | −0.06 | 43.33 | 0.84 | – |
| JESPR101−122 | $a$ | 0.05 | 42.82 | 0.83 | IPR004813 |
| NAU3305−155 | $a$ | −0.05 | 42.63 | 0.83 | IPR002913 |
| NAU5099−280 | $a$ | 0.05 | 42.79 | 0.83 | IPR001214 |
| TMB1963−243 | $a$ | −0.05 | 37.99 | 0.73 | – |
| HAU1385−155 | $a$ | 0.05 | 36.04 | 0.7 | IPR003245 |
| BNL3033−175 | $a$ | −0.05 | 33.68 | 0.65 | IPR008540 |
| NAU2931−250 | $a$ | −0.04 | 27.91 | 0.53 | IPR001841 |
| NAU1163-160 | $a$ | −0.04 | 26.63 | 0.51 | – |
| TMB312−212×GH132−180 | $aae_2y_1$ | −0.08 | 11.95 | 0.35 | IPR011335×IPR001611 |
|  | $aae_2y_2$ | −0.07 | 8.79 |  |  |
| NAU3013−245×NAU5163−216 | $aae_3y_1$ | −0.09 | 12.53 | 0.66 | Cotton_A_34510×IPR002917 |
|  | $aae_3y_3$ | −0.08 | 10.01 |  |  |

**Note:** $a$ = additive effect, $aa$ = additive by additive epistasis effect; $aae_2y_1$ = epistasis by environment interaction effect at Kuche in 2007; $aae_2y_2$ = epistasis by environment interaction effect at Kuche in 2008; $aae_3y_1$ = epistasis by environment interaction effect at Nanjing in 2007. $aae_3y_3$ = epistasis by environment interaction effect at Nanjing in 2009; $-\text{Log}_{10}P$ = minus $\log_{10}$ (P-value), $h^2$ (%) = heritability (%).

(IPR002913), and peptidase aspartic (IPR021109) at Nanjing areas (Table 2). It seems that the genes catalyzed the oxidation of organic substances will have positive effects on cotton lint yield, but the decompose genes of glycoside, lipid and protein maybe have negative effects for lint yield.

For boll number, there were 3 additive SSR loci (HAU1385-150, NAU1102−241, MGHES41−333) related to genes of peptidyl-prolyl cis-trans isomerases (PPIase, IPR000297), plasto-cyanin-like (IPR003245), and acylneuraminate cytidylyltransferase (IPR003329). 7 epistasis loci with positive effects for boll number ($QQ \times QQ$ genotype) might be related to gene interactions of cytochrome $P450$ (IPR001128), $UBX$ domain-containing protein (IPR001012), tubulin (IPR000217), ribosomal protein (IPR001865), major facilitator superfamily (IPR016196), formyl-methionine deformylase (IPR000181), and SET domain of SUVH4 (IPR001214) etc. There were 3 epistasis loci with negative effects ($qq \times qq$ genotype) related to gene interactions of $DDT$ domain superfamily (IPR004022), $AUX/IAA$ protein (IPR003311), ribosomal protein (IPR001865), protein of unknown function (DUF1685), lipid-binding $START$ (IPR002913) etc. It seems that the gene interactions for the metabolism and transport of cysteine, prolyl and formylmethionine could increase the boll number. But the regulation of the gene interactions of the lipid-binding DNA binding and auxin-binding protein could also affect the boll number.

**Table 5.** Predicted genetic effects with significance and heritability of lint percentage for cotton regional trials in three locations and three years.

| Loci | Effect | Predict | $-\text{Log}_{10}P$ | $h^2$ (%) | Candidate Gene |
|---|---|---|---|---|---|
| NAU3325−238 | $a$ | −0.44 | 164.11 | 2.47 | IPR000778 |
| NAU3519−200 | $a$ | −0.23 | 46 | 0.67 | IPR000719 |
| NAU3519−220× TMB1268−157 | $aa$ | 0.29 | 71.31 | 1.04 | IPR000719× IPR002048 |
| HAU1185−174×TMB1791−217 | $aa$ | −0.21 | 38.63 | 0.55 | IPR000194 ×IPR000209 |
| TMB1638−189×DPL513−320 | $aa$ | 0.21 | 37.87 | 0.54 | IPR002109 ×− |
| NAU3110−292×NAU2862−248 | $aa$ | −0.20 | 36.04 | 0.52 | IPR003329 ×IPR001841 |
| JESPR101−122×BNL1231−228 | $aa$ | −0.20 | 35.61 | 0.51 | IPR004813 × − |

**Note: Genetic effect:** $a$ = additive effect, $aa$ = additive by additive epistasis effect; $-\text{Log}_{10}P$ = minus $\log_{10}$ (P-value), $h^2$ (%) = heritability (%).

For boll weight, there were 15 additive SSR loci with positive effects relate to following genes: Type I phosphodiesterase (IPR002591), *SET* domain (IPR001214) of *SUVH4*, plastocyanin-like (IPR003245), oligopeptide transporter (OPT) superfamily (IPR004813). While the genes could have negative effects, such as glyceraldehyde 3-phosphate dehydrogenase (IPR020828), restriction endonuclease (IPR011335), leucine-rich repeat (IPR001611), STAR-related lipid-transfer (*START*) domain (IPR002913), brassinosteroid signalling positive regulator (*BZR1*) family protein(IPR008540), and RING-type zinc finger (IPR001841), respectively. Except the main additive effects, there are also 2 epistasis loci with negative effects (*qq*×*qq* genotype) which maybe resulted from the interactions of leucine-rich repeat (IPR001611), GTP-binding protein (IPR002917) etc. It was inferred that boll weight is mainly controlled by the genes with the phosphorus, lipid and signalling transport, which may be further regulated by the zinc finger, leucine-rich repeat, and the methylation protein.

For lint percentage, 2 SSR loci with negative additive effects could be related to protein kinase (IPR000719) and cytochrome b245 (IPR000778), respectively. There were 2 epistasis loci with positive effects (*QQ*×*QQ* genotype) related to the interactions of the genes of protein kinase (IPR000719), calcium-binding EF-hand (IPR002048), glutaredoxin (IPR002109). Three epistasis loci with

negative effects (*qq*×*qq* genotype) relate to the interactions of the genes of peptidase S8/S53 (IPR000209), zinc finger with RING-type (IPR001841), and oligopeptide transporter OPT superfamily (IPR004813).

## Supporting Information

**Table S1   Variety germplasm of Upland cotton used in association mapping for fiber traits.** Note: New lines were new breeding lines and genetic stocks bred by Institute of Cotton Research of Chinese Academy of Agricultural Sciences (ICR, CAAS) in recent years.

**Table S2   651 primer alleles that were polymorphic in the tested population.**

## Author Contributions

Conceived and designed the experiments: XMD. Performed the experiments: YHJ XWS JLS ZEP XWW SPH SHX WJS ZLZ BYP LRW JGL JM. Analyzed the data: JZ. Contributed reagents/materials/analysis tools: XMD JZ. Wrote the paper: XWS JZ XMD.

## References

1. He DH, Lin ZX, Zhang XL, Nie YC, Guo XP, et al. (2005) Mapping QTLs of traits contributing to yield and analysis of genetic effects in tetraploid cotton. Euphytica 144: 141–149.
2. Shen XL, Zhang TZ, Guo WZ, Zhu XF, Zhang XY (2006) Mapping Fiber and Yield QTLs with Main, Epistatic, and QTL × Environment Interaction Effects in Recombinant Inbred Lines of Upland Cotton. Crop Science 46: 61.
3. Murcray CE, Lewinger JP, Gauderman WJ (2008) Gene-Environment Interaction in Genome-Wide Association Studies. American Journal of Epidemiology 169: 219–226.
4. Chen H, Qian N, Guo WZ, Song QP, Li BC, et al. (2010) Using three selected overlapping RILs to fine-map the yield component QTL on Chro.D8 in Upland cotton. Euphytica 176: 321–329.
5. Liu RZ, Wang BH, Guo WZ, Qin YS, Wang LG, et al. (2011a) Quantitative trait loci mapping for yield and its components by using two immortalized populations of a heterotic hybrid in Gossypium hirsutum L. Mol Breeding 29: 297–311.
6. Yadava SK, Arumugam N, Mukhopadhyay A, Sodhi YS, Gupta V, et al. (2012) QTL mapping of yield-associated traits in Brassica juncea: meta-analysis and epistatic interactions using two different crosses between east European and Indian gene pool lines. Theor Appl Genet 125: 1553–1564.
7. Yu JW, Zhang K, Li SY, Yu SX, Zhai HH, et al. (2012) Mapping quantitative trait loci for lint yield and fiber quality across environments in a Gossypium hirsutum × Gossypium barbadense backcross inbred line population. Theor Appl Genet 126: 275–287.
8. Lacape JM, Gawrysiak G, Cao TV, Viot C, Llewellyn D, et al. (2013) Mapping QTLs for traits related to phenology, morphology and yield components in an inter-specific Gossypium hirsutum×G. barbadense cotton RIL population. Field Crops Research 144: 256–267.
9. Huang XH, Wei XH, Sang T, Zhao Q, Feng Q, et al. (2010) Genome-wide association studies of 14 agronomic traits in rice landraces. Nature Genetics 42: 961–967.
10. Poland JA, Bradbury PJ, Buckler ES, Nelson RJ (2011) Genome-wide nested association mapping of quantitative resistance to northern leaf blight in maize. Proc Natl Acad Sci USA 108: 6893–6898.
11. Kump KL, Bradbury PJ, Wisser RJ, Buckler ES, Belcher AR, et al. (2011) Genome-wide association study of quantitative resistance to southern leaf blight

in the maize nested association mapping population. Nature Genetics 43: 163–168.
12. Tian F, Bradbury PJ, Brown PJ, Hung H, Sun Q, et al. (2011) Genome-wide association study of leaf architecture in the maize nested association mapping population. Nature Genetics 43: 159–162.
13. Li H, Peng ZY, Yang XH, Wang WD, Fu JJ, et al. (2012) Genome-wide association study dissects the genetic architecture of oil biosynthesis in maize kernels. Nature Genetics: 1–10.
14. Zhang TZ, Qian N, Zhu XF, Chen H, Wang S, et al. (2013) Variations and Transmission of QTL Alleles for Yield and Fiber Qualities in Upland Cotton Cultivars Developed in China. PLoS ONE 8: e57220.
15. Ziegler A, König IR, Thompson JR (2008) Biostatistical Aspects of Genome-Wide Association Studies. Biom J 50: 8–28.
16. Carlborg O, Haley CS (2004) Epistasis: too often neglected in complex trait studies? Nat Rev Genet 5: 618–625.
17. Phillips PC (2008) Epistasis – the essential role of gene interactions in the structure and evolution of genetic systems. Nature Publishing Group 9: 855–867.
18. van Os J, Rutten BPF (2009) Gene-environment-wide interaction studies in psychiatry. Am J Psychiatry 166: 964–966.
19. Culverhouse R, Suarez BK, Lin J, Reich T (2002) A perspective on epistasis: limits of models displaying no main effect. Am J Hum Genet 70: 461–471.
20. Brachi B, Morris GP, Borevitz JO (2011) Genome-wide association studies in plants: the missing heritability is in the field. Genome Biol 12: 232.
21. Zuk O, Hechter E, Sunyaev SR, Lander ES (2012) The mystery of missing heritability: genetic interactions create phantom heritability. Proc Natl Acad Sci USA 109: 1193–1198.
22. Godoy S, Palomo G (1999) Genetic analysis of earliness in upland cotton. II. Yield and fiber properties. Euphytica 105: 161–166.
23. Campbell BT, Jones MA (2005) Assessment of genotype × environment interactions for yield and fiber quality in cotton performance trials. Euphytica 144: 69–78.
24. Collard BCY, Mackill DJ (2008) Marker-assisted selection: an approach for precision plant breeding in the twenty-first century. Philosophical Transactions of the Royal Society B: Biological Sciences 363: 557–572.

# 20

# Transcriptome Sequencing and *De Novo* Analysis of Cytoplasmic Male Sterility and Maintenance in JA-CMS Cotton

**Peng Yang[1,2], Jinfeng Han[1], Jinling Huang[3]***

**1** Department of Agronomy, Henan Agricultural University, Zhengzhou, Henan, China, **2** Department of Rural Development, Shanxi Agricultural University, Taigu, Shanxi, China, **3** Department of Agronomy, Shanxi Agricultural University, Taigu, Shanxi, China

## Abstract

Cytoplasmic male sterility (CMS) is the failure to produce functional pollen, which is inherited maternally. And it is known that anther development is modulated through complicated interactions between nuclear and mitochondrial genes in sporophytic and gametophytic tissues. However, an unbiased transcriptome sequencing analysis of CMS in cotton is currently lacking in the literature. This study compared differentially expressed (DE) genes of floral buds at the sporogenous cells stage (SS) and microsporocyte stage (MS) (the two most important stages for pollen abortion in JA-CMS) between JA-CMS and its fertile maintainer line JB cotton plants, using the Illumina HiSeq 2000 sequencing platform. A total of 709 (1.8%) DE genes including 293 up-regulated and 416 down-regulated genes were identified in JA-CMS line comparing with its maintainer line at the SS stage, and 644 (1.6%) DE genes with 263 up-regulated and 381 down-regulated genes were detected at the MS stage. By comparing the two stages in the same material, there were 8 up-regulated and 9 down-regulated DE genes in JA-CMS line and 29 up-regulated and 9 down-regulated DE genes in JB maintainer line at the MS stage. Quantitative RT-PCR was used to validate 7 randomly selected DE genes. Bioinformatics analysis revealed that genes involved in reduction-oxidation reactions and alpha-linolenic acid metabolism were down-regulated, while genes pertaining to photosynthesis and flavonoid biosynthesis were up-regulated in JA-CMS floral buds compared with their JB counterparts at the SS and/or MS stages. All these four biological processes play important roles in reactive oxygen species (ROS) homeostasis, which may be an important factor contributing to the sterile trait of JA-CMS. Further experiments are warranted to elucidate molecular mechanisms of these genes that lead to CMS.

**Editor:** Jinfa Zhang, New Mexico State University, United States of America

**Funding:** This project was supported by Shanxi Scientific and Technological Project Grant 20130311004–3 to JH and Shanxi Agricultural University Ph.D. fund. The funders had no role in study design, data collection and analysis, decision to publish, or preparation of the manuscript.

**Competing Interests:** The authors have declared that no competing interests exist.

\* Email: huangjl@sxau.edu.cn

## Introduction

Cytoplasmic male sterility (CMS) is a maternally inherited trait in higher plants incapable of producing functional pollen [1]. However, CMS plants exhibit normal vegetative growth and female fertility, and this trait can be restored by nuclear genes known as *restorer-of-fertility* (*Rf*) genes [2]. On one hand, it has been widely accepted that CMS is closely related to mitochondrial genome rearrangement, which creates chimeric genes that disturb normal pollen development [3]. More than 50 mitochondrial genes have been identified as CMS-related in various plants [4–7], which are valuable in producing F1 hybrid cultivars with heterosis. On the other hand, changes in mitochondrial function trigger altered nuclear gene expression by a mysterious process called mitochondrial retrograde regulation (MRR) [8]. Much is still unknown about signalling pathways involved in this process, MRR targets, and how cells switch to programmed cell death (PCD) mode in the case of CMS.

Transcriptomic analysis using DNA microarray or RNA-seq technology has been implemented to investigate molecular markers or differentially expressed (DE) genes implicated in a variety of traits. Suzuki et al. [9] presented the first genome-wide transcriptome analysis of CMS and restoration in cotton using Affymetrix GeneChips Cotton Genome Array. They compared DE genes of floral buds at the meiosis stage between CMS-D8 and its restorer line, and found 458 (1.9%) DE genes including 127 up-regulated and 331 down-regulated ones. The most frequent DE gene group was involved in cell wall expansion [9].

In recent years, the emergence and advancement of next generation sequencing technology has made a breakthrough in life sciences, by offering unprecedented speed and cost efficiency to study genomic and transcriptomic data [10,11]. Comparing with DNA microarray, RNA-seq has very low, if any, background signal, and has also been shown to be highly accurate for quantifying expression levels [12]. Researchers have used RNA-seq to study CMS in sweet orange (*Citrus sinensis*) [13], rape

(*Brassica napus*) [14], and chili pepper (*Capsicum annuum* L.) [15], to give a few examples.

Zheng et al. [13] compared nuclear gene expression profiles of male sterile cybrid and fertile pummelo floral buds by RNA-seq analysis. *APETALA3* and *PISTILLATA* transcripts, which encode key transcription factors for stamen identification and are restricted to normal floral whorls, were repressed in the sterile line. As the authors stated, these citrus class-B MADS-box genes are likely to be targets for CMS retrograde signalling [13]. In *Brassica napus*, 3,231 genes of *Brassica rapa* and 3,371 genes of *Brassica oleracea* were detected with significantly different expression levels from young floral buds of sterile and fertile plants, which were derived from self-pollinated offspring of the $F_1$ hybrid from novel restorer line NR1 and *Nsa* CMS line [14]. Altered expressions of genes involved in carbon metabolism, tricarboxylic acid cycle (TCA cycle), oxidative phosphorylation, oxidoreductase activity and pentatricopeptide repeat (PPR) proteins were observed by Yan et al. [14]. Liu et al. [15]'s study profiled anther transcriptomes of a chili pepper CMS line 121A and its restorer line 121C, and found the top three pathways covering the most DE unigenes were starch and sucrose metabolism, oxidative phosphorylation and plant-pathogen interaction [15].

However, unbiased transcriptome sequencing analysis of CMS in cotton is lacking. To narrow this gap, we isolated, sequenced and quantified transcriptomes of floral buds at the sporogenous cells stage (SS) and microsporocyte stage (MS), which were demonstrated to be the two most important stages for pollen sterility [16] in JA-CMS, a CMS line cultivated by Cotton Breeding Lab of Shanxi Agricultural University (Please see materials and methods for details about JA-CMS). The obtained transcriptomic sequences were then assembled, annotated, and analysed to discover DE genes between the two lines at the two specific stages, to find pathways that were enriched with these DE genes, and to further determine potential key genes functional in CMS.

## Materials and Methods

### Plant materials and RNA extraction

JA-CMS and its maintainer line JB cotton plants were grown in the fields of Shanxi Agricultural University, Taigu, Shanxi, China. The two lines were isolated by repeated backcrossing of (*Gossypium hirsutum*×*G. thurberi*)×(*G. arboreum*×*G. hirsutum*) for more than 10 generations using MB177, one kind of *G. hirsutum*, as the recurrent parent. With 100% sterile plants, JA-CMS is unique in the following three ways: it has the genetic background of both *G. thurberi* and *G. arboreum*; it matures around 20 days earlier than 104-7A [17], another CMS line cultivated by Chinese scientists through backcrossing of *G. hirsutum* and *G. barbadense*; it is conditioned by *G. hirsutum* cytoplasm, which is different from CMS lines of *G. harknessii*, 104-7A and *G. trilobum*, CMS-D8, based on origins and phenotypes. Please refer to the supplementary note for a detailed comparison of JA-CMS, 104-7A, CMS-D2 and CMS-D8. Our group has studied morphological and cellular characteristics of JA-CMS, determined major development stages for pollen abortion, and conducted biochemical and molecular biological research on CMS. In collaboration with the Institute of Cotton Study, Shanxi Agricultural Academy, a restorer line of JA-CMS named JB was also cultivated [16,18].

Because of the stable correlation between development stages of pollen mother cells and floral bud morphology [19], following the method described by Zhao and Huang [18], three floral buds, with diameters less than 2 mm or between 2 and 3 mm, were collected from five sterile and fertile plants at the same time, respectively, to confirm the correlation in our samples using cytological methods. As the results suggested, for JA-CMS and its maintainer line, a floral bud is at the SS stage when its diameter is between 1.5–2.2 mm, and at the MS stage when its diameter reaches 2.2–2.6 mm.

Floral buds were harvested from at least 50 plants of each line at apporoximately the same time of the day; 2 g of floral buds were mixed as a pool based on different stages and cotton lines. Pooled materials were snap-frozen in liquid nitrogen and stored at −80°C for RNA preparation. Total RNA was isolated using an RN37-EASYspin RNA extraction kit (Aidlab Biotechnologies, China) according to the manufacturer's protocol. The integrity of total RNA was checked by 1% agarose gel electrophoresis; the concentration and purity were determined using NanoDrop (Thermo Scientific, USA) and Agilent 2100 Bioanalyzer (Agilent, USA). A total amount of 3 μg RNA per sample was used as input material for the RNA sample preparations. All four samples had RIN (RNA Integrity Number) values above 8.

### Library preparation and sequencing

Sequencing libraries were generated using Illumina TruSeq RNA Sample Preparation Kit (Illumina, USA) following manufacturer's recommendations and four index codes were added to attribute sequences to each sample. Briefly, after integrity, concentration and purity check, poly-(T) oligo-attached magnetic beads were utilized for mRNA enrichment. Then fragmentation buffer was added to break mRNA into short fragments, which served as templates for the first strand cDNA synthesis with random hexamer-primer. Buffer, dNTPs, RNase H and DNA polymerase I were added afterwards to synthesize the second strand. Remaining overhangs were converted into blunt ends via exonuclease/polymerase activities and enzymes were removed. The double-stranded cDNA was purified with AMPure XP beads, followed by end repair, poly-(A) addition and sequencing adaptor ligation. Fragments preferentially 200 bp in length were enriched using Illumina PCR Primer Cocktail in a 10 cycle PCR amplification to obtain the cDNA library. To ensure quality of the library, Qubit 2.0 (Life Technologies, USA), Agilent 2100 (Agilent, USA) and quantitative PCR were used for initial quantification, insert size check and effective concentration determination (library effective concentration>2 nM), respectively.

The clustering of the index-coded samples was performed on a cBot Cluster Generation System using TruSeq PE Cluster Kit v3-cBot-HS (Illumina, USA) according to the manufacturer's instructions. After cluster generation, the library was sequenced on Illumina HiSeq 2000 platform; 100 bp paired-end reads were generated. Low-quality reads containing ambiguous nucleotides or adaptor sequences were removed from the raw reads to gain clean reads, which were then assembled using Trinity to construct unique consensus sequences without a reference genome [20].

### Bioinformatics analysis

Seven databases were used for unigene annotation, including the NCBI protein database (http://www.ncbi.nlm.nih.gov/protein), the NCBI nucleotide database (http://www.ncbi.nlm.nih.gov/nucleotide), the Pfam database (http://pfam.xfam.org/), the euKaryotic Orthologous Groups (KOG; http://www.ncbi.nlm.nih.gov/COG/), the UniProt Knowledgebase (UniProtKB)/Swiss-Prot (http://www.ebi.ac.uk/uniprot), the KEGG Orthology database (KO; http://www.genome.jp/kegg/ko.html), and the Gene Ontology (GO; http://www.geneontology.org/). We performed these annotations by using a combination of BLAST (http://blast.ncbi.nlm.nih.gov/Blast.cgi), HMMER (http://hmmer.janelia.org/), Blast2GO (http://www.blast2go.com/b2ghome), and KEGG Automatic Annotation Server (KAAS; http://www.genome.jp/kegg/kaas/).

Using the transcriptome assembled by Trinity [20] as reference, clean reads of each sample were mapped to it by RSEM (RNA-Seq by Expectation-Maximization) [21]. For biologically non-duplicating samples, mapped read counts were normalized through applying trimmed mean of M values (TMM) method [22]. Then normalized counts were analysed and unigenes showing significant expression differences were determined using DEGseq [23], with FDR q-value threshold set to 0.005 and absolute value of expression fold change set to 2.

After selection of unigenes with significant transcript abundance difference (DE genes), we conducted enrichment analyses, i.e. GO and KEGG analysis, to elucidate distributions of these unigenes based on their functions and attached biological pathways. GO analysis was performed using Goseq [24], which is based on Wallenius non-central hypergeometric distribution. KEGG analysis utilizes hypergeometric test to search DE genes that are significantly enriched in certain KEGG pathways compared with other genes in the genome. KOBAS 2.0 [25] was used for KEGG analysis. GO categories and KEGG pathways with FDR q-value ≤ 0.05 were selected as significantly enriched.

KEGG analysis results are presented graphically using scatter plots, with three parameters for enrichment level assessment, which are rich factor, the ratio between counts of DE genes and all annotated genes enriched in a certain pathway (higher the ratio, more significant the pathway); qvalue, P value after multiple hypothesis testing correction with a range between 0 and 1 (closer to 0, more significant the pathway); and number of genes enriched. Twenty most significant pathways were plotted, when more than twenty pathways were identified for each of the four comparisons.

## Quantitative RT-PCR

Quantitative RT-PCR (qRT-PCR) was performed to validate transcript abundance of unigenes observed by RNA-seq. Primer-BLAST (http://www.ncbi.nlm.nih.gov/tools/primer-blast/) was used to design gene-specific primers for seven randomly selected DE genes plus elongation factor-1α gene as the internal control (Table S1). Reactions were carried out using the SYBR *Premix Ex Taq* II (Tli RnaseH Plus) Kit (Takara, USA) in the iQ5 Multicolour Real-Time PCR Detection System (Bio-Rad, USA), according to the manufacturer's instructions.

The first-strand cDNA was synthesized using the PrimeScript RT Master Mix Kit (Takara, USA) under the following conditions: 37°C for 15 min (reverse transcription) and 85°C for 5 s (denaturing reverse transcriptase). Then amplification reactions of 20 μL volume consisting of 10 μL SYBR *Premix Ex Taq* II, 0.8 μL of 10 μM forward and reverse primers each, 1 μL cDNA template and sterile water, were conducted, with cycling parameters as follows: 95°C for 30 s, 40 cycles of 95°C for 5 s and 60°C for 30 s. Relative expression levels of unigenes from different samples were calculated using $2^{-\Delta\Delta CT}$ method [26]. The qRT-PCR was conducted with 3 replicates for each sample, and data were indicated as means ± standard errors (n = 3) in Figure S1.

## Data access

The transcriptome sequencing data from this study have been deposited in the NCBI SRA database and are accessible through accession numbers SRX547770, SRX547777, SRX547779, and SRX547781 (http://www.ncbi.nlm.nih.gov/sra).

## Results

### RNA sequencing and gene annotations

A total of $2.8 \times 10^8$ raw reads were obtained from the Illumina sequencing platform. Each stage of either sterile or fertile line had around 6.8 ± 0.6 giga base pairs (Gb) of sequencing data. After data quality control as described in materials and methods, 97.54% ± 0.26% of the raw reads were clean reads for each of the four samples (Table 1). Trinity [20] reconstructed all the clean reads into a *de novo* reference transcriptome; the longest transcript for each gene was recognized as the unigene. Following these steps, 206,496 transcripts and 86,093 unigenes were identified, length distributions of which are presented in Figure S2 and S3. The transcripts and unigenes are between 201 bp and 17,233 bp in length; the mean lengths of the transcripts and unigenes are 1,184 bp and 755 bp, respectively. Around 36.3% of the transcripts and 59.8% of the unigenes are within the range of 200–500 bp.

The percentage of successfully annotated unigenes differs among the seven databases specified in the Materials and Methods part, with the NCBI protein database having the highest, 42.05% and KO providing the lowest, 11.76%. Only 5.15% of the unigenes are annotated in all the databases, while 49.47% are annotated in at least one database. Annotations are provided whenever available.

### Genome-wide analysis of DE genes

Gene expression levels at the SS or MS stage for JA-CMS and JB floral buds were quantified and compared using RSEM [21], TMM [22], and DEGseq [23]. At the SS stage, there were 11,111 genes uniquely expressed in JA-CMS plants and 7,811 in JB plants; 40,318 genes were shared between the two lines at this stage. At the MS stage, 8,502 and 8,636 genes were specifically expressed in JA-CMS and JB lines, and the number of commonly expressed genes was 39,967. Regarding gene expression level comparison between the two stages of the same plant material, JA-CMS plants had 9,334 genes uniquely expressed at the SS stage and 6,374 at the MS stage; 42,905 genes were expressed at both stages; and there were 7,366 and 7,840 genes expressed at the SS and MS stages of JB plants, respectively, while 40,763 genes were commonly found at both stages (Figure 1).

Due to the large number of genes, we used FDR q value<0.005 combined with |log₂(fold change)|>1 to select DE genes between the two materials at the same stage, or between different stages of the same material. By comparing with the JB line, JA-CMS plants had 709 genes with significantly different expression levels at the SS stage, among which 293 were up-regulated and 416 were down-regulated (Table S2). At the MS stage, 644 genes passed the significance threshold, with 263 up-regulated and 381 down-regulated in JA-CMS plants (Table S3). Fewer genes showed significant expression difference between the two stages of the same material. Seventeen genes, with 8 up-regulated and 9 down-regulated, had significantly different expression levels at the MS stage of JA-CMS comparing with the SS stage (Table S4). There were 38 DE genes showing significance in JB line between the two stages, among which 29 were up-regulated and 9 were down-regulated at the MS stage (Table S5).

All the 854 DE genes that showed significance in at least one of the four comparisons have 80.76 ± 158.10 and 78.07 ± 150.42 RPKM (Reads per kilobase per million) values at the SS and MS stages of JA-CMS, respectively. The RPKM values for JB plants have larger means and standard deviations than JA-CMS at both stages: 91.40 ± 301.55 (SS stage) and 96.25 ± 304.16 (MS stage). These values indicate reasonable coverage for the DE genes.

**Table 1.** Transcriptome sequencing data quality.

| Sample | Raw reads | Clean reads | Clean bases | Error (%) | Q20 (%) | Q30 (%) | GC (%) |
|---|---|---|---|---|---|---|---|
| JA-CMS at the SS stage | 66695222 | 64995444 | 6.50 Gb | 0.03 | 98.05 | 93.60 | 43.79 |
| JA-CMS at the MS stage | 77998708 | 75844362 | 7.58 Gb | 0.03 | 98.02 | 93.57 | 43.65 |
| JB at the SS stage | 71264156 | 69566718 | 6.96 Gb | 0.03 | 98.01 | 93.45 | 43.74 |
| JB at the MS stage | 62779022 | 61431768 | 6.14 Gb | 0.03 | 98.11 | 93.71 | 43.75 |

Notes: Gb = Giga base pair; Q20 represents an error rate of 1 in 100, with a corresponding call accuracy of 99%; Q30 represents an error rate of 1 in 1000, with a corresponding call accuracy of 99.9%.

Please refer to Figure S4 for the density distributions of log10(RPKM) for the four samples.

## Enrichment analysis of DE genes

Enrichment analysis was performed for the purpose of identifying major functions or pathways associated with CMS when genes function co-ordinately under certain biological mechanisms instead of individually to cause specific phenotype, which is the case for CMS. This type of analysis prioritizes DE genes for further investigation, if many DE genes are identified, and significantly increases statistical power to find predisposing genes, if few are available [27].

**GO analysis.** GO covers three domains: biological process (BP), cellular component (CC) and molecular function (MF). DE genes between the two materials at the SS stage were significantly enriched in 27 GO categories. Processes, components and functions related to reduction-oxidation (redox) reactions, photosynthesis, and transcription factors showed up in two out of the three GO domains. Oxidation-reduction process (GO:0055114) encompassed 100 DE genes with the smallest q value of all the 27 categories, which equals to $7.9 \times 10^{-4}$ (Table 2). We also conducted GO analysis on up-regulated and down-regulated DE genes at the SS stage, separately (Figure S5 and S6). For the up-regulated genes, their product properties were mostly enriched in the biological process of photosynthesis and cellular components of photosynthetic membranes with both q values equal to $6.31 \times 10^{-6}$; besides these two categories, other processes, components and functions related to photosynthesis were also prominent, such as photosystem ($q = 7.48 \times 10^{-6}$), thylakoid part and thylakoid ($q = 1.50 \times 10^{-5}$; Figure S5). On the other hand, down-regulated gene product properties were over-represented in only four categories, which are all pertinent to redox reactions (Figure S6).

For the MS stage, product properties of DE genes between JA-CMS and JB lines were separated into 12 categories (Table 2). The most significant categories ($q = 2.43 \times 10^{-4}$) were oxidation-reduction process, oxidoreductase activity (acting on paired donors, with incorporation or reduction of molecular oxygen), cation binding and metal ion binding. Same as the SS stage, redox reactions and photosynthesis covered most of the GO categories. Additional GO analysis on up-regulated DE genes showed that, besides photosynthesis related categories, 7 out of the 43 significantly enriched classes were pertinent to transcription (Figure S7). Consistent with the SS stage analysis results, every category identified in GO analysis on down-regulated DE genes at the MS stage was involved in redox reactions (Figure S8). By comparing JA-CMS and JB lines at the two stages, it became obvious that genes implicated in redox reactions were persistently down-regulated, while genes related to photosynthesis were continuously up-regulated.

GO analysis did not find any significantly enriched category concerning up-regulated DE genes at the MS stage comparing with the SS stage of sterile plants; however, 24 categories were over-represented by down-regulated DE gene products, all of which were related to ATP biosynthesis (Figure S9). Combined analysis of up-regulated and down-regulated DE genes revealed the same category distribution as the down-regulated analysis. No GO enrichment was detected for DE genes between the two development stages of fertile plants.

**KEGG analysis.** There were 134 pathways determined for DE genes between JA-CMS and JB plants at the SS stage, among which the most significant ones were "photosynthesis" (gene number [N] = 15), "flavonoid biosynthesis" (N = 12), "metabolic pathways" (N = 112), "DNA replication" (N = 9), and "alpha-linolenic acid metabolism" (N = 9; Figure 2A). Consistent with

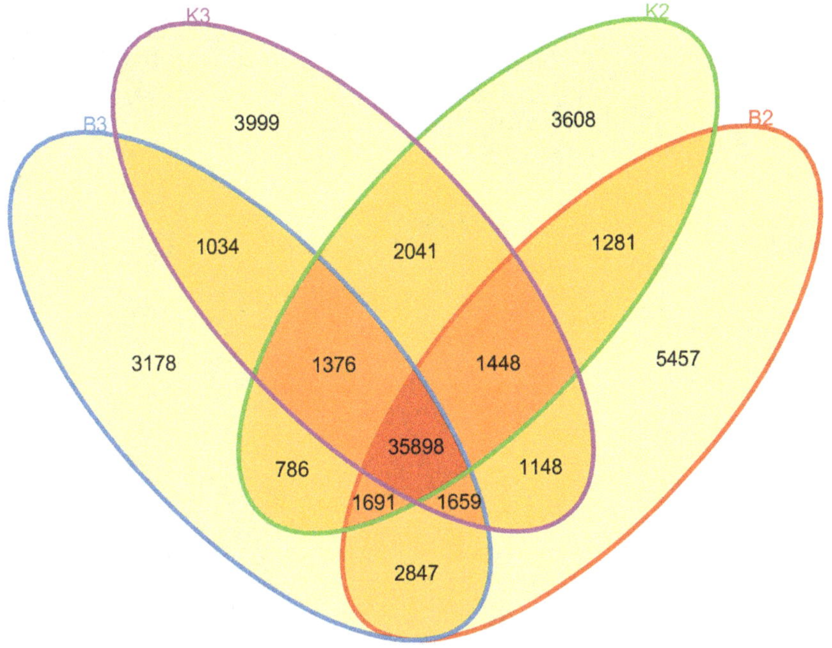

**Figure 1. Venn diagram of DE gene counts at the SS and MS stages for JA-CMS and JB floral buds.** B2 = SS stage of JA-CMS, K2 = SS stage of JB, K3 = MS stage of JB, B3 = MS stage of JA-CMS (from right to left).

GO analysis, "photosynthesis" was significantly enriched with up-regulated DE genes, as well as "flavonoid biosynthesis" and "DNA replication" (Figure S10). Down-regulated DE genes were over-represented in "alpha-linolenic acid metabolism" (Figure S11). Four out of the five pathways identified by KEGG for the MS stage were the same as the SS stage, which are "photosynthesis" (N = 14), "flavonoid biosynthesis" (N = 10), "DNA replication" (N = 9), and "metabolic pathways" (N = 99). Another pathway showing significance was "drug metabolism" (N = 7) among the total 140 pathways identified at the MS stage (Figure 2B). The three pathways significantly enriched with up-regulated genes at this stage were the same as the SS stage (Figure S12); "Metabolism of xenobiotics by cytochrome P450" and "drug metabolism" were enriched with down-regulated genes (Figure S13).

Because fewer DE genes were determined between the two stages of the same material, no significantly enriched pathway was detected.

### qRT-PCR verification on gene expression patterns

Although quantitative differences existed between qRT-PCR analysis results and sequencing data for the selected seven DE genes (Table S1 and Figure S1), overall expression tendencies were the same (Figure 3). For example, comp54402_c0 is 1-Aminocyclopropane-1-carboxylic acid synthase (ACS) gene included in amino acid metabolism pathway, expression of which was down-regulated in sterile plants. Sequencing data showed 16-fold difference of transcript abundance between JA-CMS and JB lines at the SS stage, while qRT-PCR analysis yielded around 14-fold difference. Chlorophyll a/b-binding protein gene, comp70789_c0, was up-regulated in sterile plants compared with fertile counterparts at the MS stage based on sequencing data (15-fold), which was confirmed by the 11-fold increase shown in the corresponding qRT-PCR analysis result.

### Discussion

As a result of our unbiased transcriptome sequencing analysis of JA-CMS and its maintainer line, four candidate functional categories or pathways related to CMS have been identified, which include two down-regulated gene groups in JA-CMS: genes related to redox reactions and alpha-linolenic acid metabolism, and two up-regulated gene groups: genes pertaining to photosynthesis and flavonoid biosynthesis. Discussion of these genes would not be possible without mentioning the preceding study results of JA-CMS.

According to previous cytological studies, premature microspore abortion and tapetum degradation at the SS stage is one prominent characteristic of JA-CMS, and enzymatic activities of peroxidase, superoxide dismutase, cytochrome oxidase and succinate dehydrogenase were found significantly lower in JA-CMS comparing with its fertile counterpart [16]. Tapetum PCD plays an important role in pollen development [28–30], and abnormal reactive oxygen species (ROS) level is believed to be a signal promoting PCD in plants [31]. Disrupted ROS homeostasis was found in several CMS systems, e.g., excessive accumulation of $O_2^-$, $H_2O_2$ and malondialdehyde (MDA) existed in CMS cotton at the abortion peak [32], and mitochondria of a rice CMS line suffered from serious oxidative stress, which was induced by abnormal increased ROS at the meiosis stage [33].

In plants, chloroplasts are a major site of ROS generation, where photosynthetic electron transfer chains produce $O_2^-$ [34]. Because ROS can cause damage to proteins, lipids, and DNA, its production and removal must be strictly controlled [35]. Plants possess complex antioxidative defense system comprising of nonenzymatic and enzymatic components to scavenge ROS. Nonenzymatic components include the major cellular redox buffers ascorbate (AsA) and glutathione (γ-glutamyl-cysteinyl-glycine, GSH), which act as cofactors for enzymes of AsA-GSH cycle ascorbate peroxidase (APX), monodehydroascorbate reductase (MDHAR), dehydroascorbate reductase (DHAR), glutathione

**Table 2.** GO analysis results of DE genes at the SS and MS stages between JA-CMS and JB.

| GO accession ID | Description | Domain | SS stage | | | MS stage | | |
|---|---|---|---|---|---|---|---|---|
| | | | P value | Q value | DE gene # | P value | Q value | DE gene # |
| GO:0055114 | Oxidation-reduction process | BP | 1.53E-07 | 7.90E-04 | 100 | 6.79E-08 | 2.43E-04 | 92 |
| GO:0016705 | Oxidoreductase activity, acting on paired donors, with incorporation or reduction of molecular oxygen | MF | 6.46E-07 | 1.66E-03 | 35 | 1.02E-07 | 2.43E-04 | 34 |
| GO:0001071 | Nucleic acid binding transcription factor activity | MF | 1.58E-06 | 2.04E-03 | 48 | Not significant | | |
| GO:0003700 | Sequence-specific DNA binding transcription factor activity | MF | 1.58E-06 | 2.04E-03 | 48 | Not significant | | |
| GO:0016491 | Oxidoreductase activity | MF | 2.73E-06 | 2.44E-03 | 96 | 4.37E-07 | 4.51E-04 | 89 |
| GO:0005667 | Transcription factor complex | CC | 2.83E-06 | 2.44E-03 | 55 | Not significant | | |
| GO:0009521 | Photosystem | CC | 4.40E-06 | 3.24E-03 | 17 | 6.62E-05 | 3.10E-02 | 14 |
| GO:0034357 | Photosynthetic membrane | CC | 5.60E-06 | 3.61E-03 | 18 | 6.58E-05 | 3.10E-02 | 15 |
| GO:0046872 | Metal ion binding | MF | 8.76E-06 | 4.86E-03 | 116 | 1.89E-07 | 2.43E-04 | 110 |
| GO:0043169 | Cation binding | MF | 9.43E-06 | 4.86E-03 | 118 | 1.83E-07 | 2.43E-04 | 112 |
| GO:0005654 | Nucleoplasm | CC | 1.20E-05 | 5.13E-03 | 58 | Not significant | | |
| GO:0044451 | Nucleoplasm part | CC | 1.20E-05 | 5.13E-03 | 58 | Not significant | | |
| GO:0015979 | Photosynthesis | BP | 1.40E-05 | 5.54E-03 | 26 | 1.07E-04 | 4.59E-02 | 22 |
| GO:0008152 | Metabolic process | BP | 1.57E-05 | 5.77E-03 | 394 | Not significant | | |
| GO:0044436 | Thylakoid part | CC | 2.34E-05 | 8.05E-03 | 18 | Not significant | | |
| GO:0009579 | Thylakoid | CC | 2.81E-05 | 8.58E-03 | 18 | Not significant | | |
| GO:0016706 | Oxidoreductase activity, acting on paired donors, with incorporation or reduction of molecular oxygen, 2-oxoglutarate as one donor, and incorporation of one atom each of oxygen into both donors | MF | 2.83E-05 | 8.58E-03 | 14 | 7.40E-06 | 5.45E-03 | 14 |
| GO:0031981 | Nuclear lumen | CC | 7.38E-05 | 2.11E-02 | 63 | Not significant | | |
| GO:0032787 | Monocarboxylic acid metabolic process | BP | 8.31E-05 | 2.25E-02 | 24 | Not significant | | |
| GO:0009538 | Photosystem I reaction center | CC | 9.21E-05 | 2.37E-02 | 5 | 5.10E-05 | 2.92E-02 | 5 |
| GO:0044428 | Nuclear part | CC | 1.07E-04 | 2.62E-02 | 75 | Not significant | | |
| GO:0043233 | Organelle lumen | CC | 1.21E-04 | 2.71E-02 | 63 | Not significant | | |

**Table 2.** Cont.

| GO accession ID | Description | Domain | SS stage | | | MS stage | | |
|---|---|---|---|---|---|---|---|---|
| | | | P value | Q value | DE gene # | P value | Q value | DE gene # |
| GO:0070013 | Intracellular organelle lumen | CC | 1.21E-04 | 2.71E-02 | 63 | Not significant | | |
| GO:0044710 | Single-organism metabolic process | BP | 1.47E-04 | 3.15E-02 | 184 | Not significant | | |
| GO:0031974 | Membrane-enclosed lumen | CC | 1.55E-04 | 3.19E-02 | 63 | Not significant | | |
| GO:0051213 | Dioxygenase activity | MF | 1.90E-04 | 3.77E-02 | 15 | 3.09E-06 | 2.65E-03 | 17 |
| GO:0048037 | Cofactor binding | MF | 2.17E-04 | 4.13E-02 | 38 | Not significant | | |
| GO:0046914 | Transition metal ion binding | MF | Not significant | | | 2.03E-05 | 1.31E-02 | 83 |

Notes: 1) BP = Biological process, CC = Cellular component, MF = Molecular function; Q value = FDR corrected p value. 2) Only GO categories with FDR q value ≤ 0.05 are presented here. 3) GO categories are arranged based on q values from smallest to largest at the SS stage.

reductase (GR), glutathione peroxidase (GPX), and glutathione S transferase (GST) [36].

Based on both GO and KEGG analysis results, up-regulated genes in JA-CMS were significantly enriched in photosynthetic categories and pathways, while oxidation-reduction process and oxidoreductase activity categories were significantly over-represented with down-regulated DE genes at the SS and MS stages in sterile plants. There are 14 DE genes up-regulated in JA-CMS belonged to the KEGG pathway of "photosynthesis", among which Photosystem I subunit PsaO encoding gene (comp65579_c0) showed the maximum fold change of expression at both SS (14-fold) and MS (12-fold) stages (Table 3).

Down-regulated oxidoreductase genes in this study comprise 8 iron/ascorbate family oxidoreductase genes, 1 MDHAR gene, 1 GPX gene and 5 GST genes, which explains the significant enrichment of "drug metabolism" in KEGG analysis of down-regulated genes at the MS stage (Table 4). Within these genes, MDHAR gene had the maximum 17-fold decrease of its transcript abundance level in JA-CMS, which may be a major player of glutathione-ascorbate cycle that causes ROS damages. Thioredoxin 1 and thioredoxin reductase (NADPH) were down-regulated in JA-CMS as well (Table 4). Although not directly involved in glutathione-ascorbate cycle, thioredoxins are involved in oxidative damage avoidance by supplying reducing power to reductases detoxifying lipid hydroperoxides or repairing oxidized proteins [37]. Overall, increased production of ROS through photosynthesis plus decreased enzymatic activity of ROS-scavenging enzymes may dramatically perturb ROS homeostasis in JA-CMS.

The "alpha-linolenic acid metabolism" pathway was another one that was significantly over-represented in KEGG analysis of down-regulated DE genes at the SS stage in JA-CMS. Although this pathway was only detected for the SS stage, genes encoding key enzymes in this pathway were significantly down-regulated in JA-CMS at both SS and/or MS stages, which include allene oxide synthase gene (AOS), 2 allene oxide cyclase genes (AOC), OPDA (12-oxo-phytodienoic acid) reductase gene, OPC-8:0 (3-oxo-2((2Z)-pentenyl)-cyclopentane-1-octanoic acid) CoA ligase gene, acyl-CoA oxidase gene (AOX), acetyl-CoA acyltransferase gene (ACAA), and α-dioxygenase gene (Table 4). Among these genes, α-dioxygenase gene is worth noting, because its transcript abundance decreased around 50-fold in JA-CMS comparing with its counterpart at the SS stage; and it was reported to protect against oxidative stress and cell death in *Arabidopsis* [38], lack of which may exacerbate ROS imbalance as aforementioned.

The "alpha-linolenic acid metabolism" pathway belongs to jasmonic acid biosynthesis module according to KEGG pathway database. Multiple studies have presented the association between jasmonic acid biosynthesis and CMS. McConn and Browse [39] found that *Arabidopsis* triple mutants that contain negligible levels of trienoic fatty acids were male sterile, and exogenous applications of α-linoleate or jasmonate restored fertility. An *Arabidopsis* knock-out mutant defective in the AOS gene *CYP74A* also showed severe male sterility due to defects in anther and pollen development; this phenotype was completely rescued by exogenous application of methyl jasmonate and by complementation with constitutive expression of the AOS gene [40]. In addition, aberrant jasmonic acid pathway was detected in CMS rice during the development of microspores [41].

Besides photosynthesis related genes, the "flavonoid biosynthesis" pathway were over-represented with up-regulated DE genes in JA-CMS at the SS and MS stages as well. This pathway comprises 10 up-regulated genes in this study, which are 4 chalcone synthase genes, one of the key enzymes for flavonoid biosynthesis, 1 anthocyanidin reductase gene, 1 leucoanthocyanidin dioxygenase

A

B

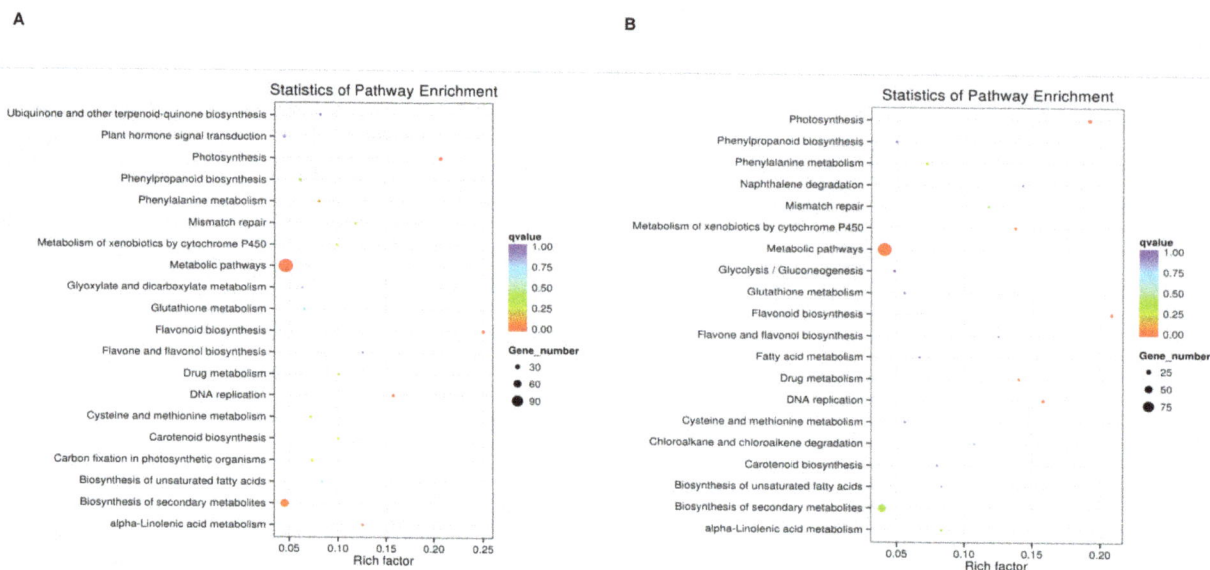

**Figure 2. KEGG analysis results of DE genes at the SS and MS stages comparing JA-CMS and JB lines.** Rich factor is the ratio between counts of DE genes and all annotated genes enriched in a certain pathway; qvalue is P value after multiple hypothesis testing correction with a range between 0 and 1. Twenty most significant pathways were plotted, when more than 20 pathways were identified. A) Results for the SS stage. B) Results for the MS stage.

gene, 1 bifunctional dihydrofavonol 4-reductase/flavanone 4-reductase gene, 1 trans-cinnamate 4-monooxygenase gene, 1 flavonoid 3′, 5′-hydroxylase gene and 1 naringenin 3-dioxygenase gene (Table 3). However, two flavonoid biosynthesis related genes were found down-regulated in JA-CMS simultaneously; they are 1 chalcone synthase gene with a 4-fold expression decrease $(q = 2.21 \times 10^{-3})$ at the SS stage, and 1 caffeoyl-CoA

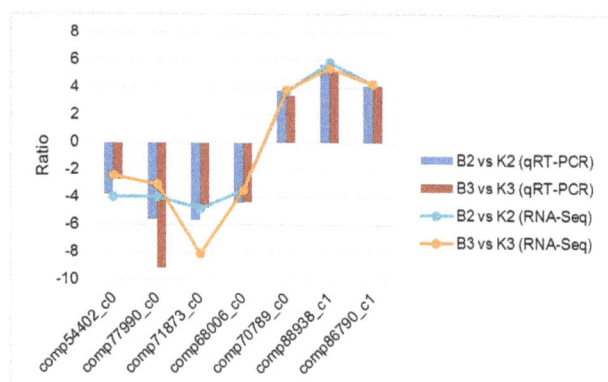

**Figure 3. Quantitative RT-PCR validations for expression profiles of seven selected DE genes.** Comp54402_c0 is 1-aminocyclopropane-1-carboxylate synthase gene, comp77990_c0 is alpha-expansin 11 precursor gene, comp71873_c0 is a predicted protein gene, comp68006_c0 is a conserved hypothetical protein gene, comp70789_c0 is chloroplast chlorophyll A–B binding protein gene, comp88938_c1 is transcription factor LHY gene, comp86790_c1 is *G. hirsutum* cab gene for chlorophyll A–B binding protein. Ratio stands for log2 ratio of relative expression levels for qRT-PCR and RPKM (Reads per kilobase per million) values for RNA sequencing, respectively. B2 = SS stage of JA-CMS, K2 = SS stage of JB, B3 = MS stage of JA-CMS, K3 = MS stage of JB.

O-methyltransferase gene down-regulated at both SS (2-fold, $q = 1.05 \times 10^{-5}$) and MS stages (2-fold, $q = 3.44 \times 10^{-7}$).

In literature, several nuclear genes pertaining to flavonoid biosynthesis were reported to be inhibited in Ogura CMS of *Raphanus sativus*. The expression of chalcone synthase gene was strongly inhibited in the later stages of anther development in sterile cytoplasm [42]. Wei et al. [43] implemented digital gene expression analysis to compare gene expressions of wild type and genetic male sterility (GMS) mutant cotton at the meiosis, tetrad, and uninucleate microspore stages, respectively. Genes involved in flavonoid metabolism, such as CHS, flavonoid 3′, 5′-hydroxylase, anthocyanidin reductase, and leucoanthocyanidin reductase were down-regulated at the meiosis and uninucleate microspore stages, but were up-regulated at the tetrad stage, in GMS mutant anthers. Differences between the current study and these findings may lie in diverse plant materials, different sterility mechanisms and distinct development stages. Nevertheless contradictory results, flavonoids have been suggested to constitute a secondary ROS-scavenging system in plants exposed to severe stress conditions [44] and genes in flavonoid biosynthesis are important in anther and pollen development. Further studies are warranted to disentangle specific functions of these genes in balancing ROS homeostasis at different development stages of sterile plants.

In addition, in GO analysis of DE genes between the two stages of the same material, ATP biosynthesis related categories were significantly enriched in sterile plants. Three out of the 9 down-regulated genes identified in this comparison were involved in ATP biosynthesis. The three DE genes, only detected in JA-CMS, are proton-transporting ATP synthase complex, coupling factor $F_o$ (comp73972_c0; 3-fold expression decrease at the MS stage, $q = 5.88 \times 10^{-85}$), ATPase subunit 1 (comp84388_c5; 3-fold, $q = 1.58 \times 10^{-5}$), and ATP synthase CF1 alpha subunit (comp85965_c1; 2-fold, $q = 4.19 \times 10^{-15}$). Same expression patterns of ATP synthase related genes were observed in studies of Dong et al. [45] and Liu et al. [15].

**Table 3.** Up-regulated photosynthesis and flavonoid biosynthesis related genes in JA-CMS at the SS and/or MS stages.

| Category | Function | Gene ID | SS stage | | | MS stage | | |
|---|---|---|---|---|---|---|---|---|
| | | | Fold | P value | Q value | Fold | P value | Q value |
| Photosynthesis | Photosystem I subunit X | comp53760_c0 | 2.53 | 2.23E-08 | 2.32E-06 | 2.56 | 1.23E-07 | 1.23E-05 |
| | Plastocyanin | comp65386_c0 | 3.02 | 1.48E-30 | 7.37E-28 | 2.61 | 6.19E-24 | 2.34E-21 |
| | Photosystem I subunit PsaO | comp65579_c0 | 3.77 | 8.58E-17 | 1.99E-14 | 3.61 | 4.29E-14 | 8.67E-12 |
| | Photosystem II PsbW protein | comp65639_c0 | 2.34 | 5.34E-07 | 4.77E-05 | 2.35 | 1.28E-06 | 1.14E-04 |
| | Photosystem II 10kDa protein | comp68020_c0 | 1.17 | 3.48E-10 | 4.54E-08 | Not significant | | |
| | Photosystem II PsbY protein | comp68793_c0 | 1.72 | 4.38E-07 | 3.96E-05 | Not significant | | |
| | Photosystem I subunit XI | comp70643_c0 | 1.83 | 3.25E-15 | 6.79E-13 | 1.60 | 9.63E-11 | 1.44E-08 |
| | Ferredoxin | comp71277_c0 | 3.11 | 4.21E-18 | 1.07E-15 | 2.68 | 9.12E-15 | 1.94E-12 |
| | Photosystem I subunit IV | comp72189_c0 | 1.36 | 1.59E-08 | 1.70E-06 | 1.07 | 1.35E-05 | 1.01E-03 |
| | Photosystem II oxygen-evolving enhancer protein 1 | comp73151_c0 | 2.96 | 7.52E-57 | 9.24E-54 | 2.74 | 1.66E-45 | 1.70E-42 |
| | Photosystem II 22kDa protein | comp73229_c0 | 2.28 | 2.04E-14 | 4.05E-12 | 2.14 | 4.45E-12 | 7.67E-10 |
| | Photosystem II oxygen-evolving enhancer protein 2 | comp73376_c0 | 1.18 | 2.53E-10 | 3.38E-08 | Not significant | | |
| | Photosystem I subunit III | comp73879_c0 | 1.83 | 2.89E-20 | 8.55E-18 | 1.65 | 1.73E-14 | 3.61E-12 |
| | Photosystem II oxygen-evolving enhancer protein 3 | comp84774_c2 | 1.83 | 4.26E-10 | 5.51E-08 | 1.55 | 2.13E-07 | 2.10E-05 |
| Flavonoid biosynthesis | Chalcone synthase | comp72843_c0 | 2.23 | 2.61E-41 | 2.07E-38 | 1.63 | 1.96E-29 | 1.03E-26 |
| | | comp72843_c1 | 2.11 | 6.78E-13 | 1.18E-10 | 1.48 | 4.46E-09 | 5.49E-07 |
| | | comp84628_c1 | 2.06 | 5.90E-62 | 8.22E-59 | 1.56 | 2.09E-43 | 2.00E-40 |
| | | comp84628_c0 | 1.88 | 2.55E-78 | 5.51E-75 | 1.56 | 2.23E-68 | 4.13E-65 |
| | Anthocyanidin reductase | comp73018_c0 | 1.27 | 1.48E-39 | 1.09E-36 | Not significant | | |
| | Leucoanthocyanidin dioxygenase | comp75804_c0 | 2.17 | 1.28E-94 | 4.02E-91 | 1.89 | 3.33E-81 | 8.49E-78 |
| | Bifunctional dihydroflavonol 4-reductase/flavanone 4-reductase | comp75884_c0 | 1.75 | 2.37E-69 | 3.81E-66 | 1.54 | 8.80E-60 | 1.35E-56 |
| | Trans-cinnamate 4-monooxygenase | comp80855_c0 | 1.47 | 1.42E-71 | 2.54E-68 | 1.09 | 7.25E-44 | 7.05E-41 |
| | Flavonoid 3',5'-hydroxylase | comp81439_c0 | 2.50 | 1.74E-93 | 4.97E-90 | 1.86 | 3.08E-57 | 4.39E-54 |
| | Naringenin 3-dioxygenase | comp82651_c1 | 1.67 | 5.00E-94 | 1.49E-90 | 1.25 | 6.40E-67 | 1.12E-63 |

Note: Fold = $\log_2$(fold change). Significant DE genes were determined based on |$\log_2$(fold change)|>1 and FDR q value<0.005.

**Table 4.** Down-regulated oxidoreductase and alpha-linolenic acid metabolism genes in JA-CMS at the SS and/or MS stages.

| Category | Function | Gene ID | SS stage | | | MS stage | | |
|---|---|---|---|---|---|---|---|---|
| | | | Fold | P value | Q value | Fold | P value | Q value |
| Oxidoreductase | Iron/ascorbate family oxidoreductase | comp74308_c0 | −2.17 | 1.52E-19 | 4.38E-17 | −2.46 | 1.15E-30 | 6.70E-28 |
| | | comp75973_c0 | −1.29 | 1.58E-14 | 3.19E-12 | −1.60 | 6.00E-28 | 2.92E-25 |
| | | comp76619_c0 | −1.97 | 2.58E-69 | 4.05E-66 | −1.58 | 3.31E-40 | 2.66E-37 |
| | | comp76821_c0 | −1.55 | 6.12E-11 | 8.77E-09 | −1.12 | 1.53E-06 | 1.34E-04 |
| | | comp77552_c0 | −3.93 | 8.85E-91 | 2.22E-87 | −2.97 | 1.16E-42 | 1.06E-39 |
| | | comp78972_c0 | −4.08 | 8.28E-77 | 1.67E-73 | −2.89 | 8.24E-43 | 7.64E-40 |
| | | comp79641_c0 | −1.96 | 5.14E-05 | 3.22E-03 | −2.35 | 6.26E-07 | 5.81E-05 |
| | | comp79893_c0 | −1.16 | 1.52E-13 | 2.81E-11 | −1.02 | 9.31E-12 | 1.54E-09 |
| | Monodehydroascorbate reductase (MDHAR) | comp73498_c0 | −4.21 | 1.43E-07 | 1.37E-05 | Not significant | | |
| | Glutathione peroxidase (GPX) | comp73626_c0 | −1.45 | 2.69E-80 | 6.02E-77 | −1.55 | 5.17E-97 | 1.58E-93 |
| | Glutathione S-transferase (GST) | comp40493_c0 | −2.71 | 1.54E-11 | 2.36E-09 | −2.46 | 3.63E-13 | 6.88E-11 |
| | | comp74280_c0 | −1.66 | 3.00E-14 | 5.83E-12 | −1.32 | 1.37E-10 | 2.00E-08 |
| | | comp76601_c0 | −1.85 | 3.65E-15 | 7.60E-13 | −1.77 | 2.14E-16 | 5.29E-14 |
| | | comp81581_c0 | −1.54 | 1.40E-05 | 1.01E-03 | −1.36 | 4.44E-05 | 2.95E-03 |
| | | comp92611_c0 | −2.96 | 1.61E-06 | 1.35E-04 | −2.32 | 3.14E-06 | 2.60E-04 |
| | Thioredoxin 1 | comp63917_c0 | Not significant | | | −1.57 | 4.86E-07 | 4.58E-05 |
| | Thioredoxin reductase (NADPH) | comp87684_c1 | −1.17 | 1.99E-12 | 3.35E-10 | −1.20 | 1.21E-13 | 2.37E-11 |
| Alpha-linolenic acid metabolism | Allene oxide synthase (AOS) | comp75493_c0 | −1.45 | 1.04E-127 | 5.93E-124 | −1.02 | 1.57E-58 | 2.34E-55 |
| | Allene oxide cyclase (AOC) | comp70218_c0 | −2.68 | 1.51E-10 | 2.09E-08 | −2.03 | 1.63E-08 | 1.83E-06 |
| | | comp80154_c0 | −1.75 | 4.08E-59 | 5.11E-56 | −1.61 | 7.73E-53 | 9.09E-50 |
| | OPDA (12-oxophytodienoic acid) reductase | comp79354_c1 | −2.83 | 8.17E-85 | 1.97E-81 | −2.84 | 1.73E-71 | 3.53E-68 |
| | OPC-8:0 CoA ligase 1 | comp82201_c0 | −1.01 | 8.43E-29 | 3.95E-26 | Not significant | | |
| | Acyl-CoA oxidase (AOX) | comp86147_c0 | −1.80 | 4.97E-105 | 1.64E-101 | −1.63 | 4.80E-79 | 1.18E-75 |
| | Acetyl-CoA acyltransferase 1 (ACAA) | comp80628_c2 | −1.09 | 7.33E-17 | 1.73E-14 | Not significant | | |
| | Alpha-dioxygenase | comp66758_c0 | −5.66 | 2.09E-11 | 3.15E-09 | | | |

Note: Fold = $\log_2$(fold change). Significant DE genes were determined based on |$\log_2$(fold change)|>1 and FDR q value<0.005.

ATP synthase, located within the plant mitochondria, consists of $F_o$ and $F_1$ regions, which plays an important role in providing energy. The $F_o$ portion is within the membrane and functions as a proton pore; it consists of orf25, orfB, subunit 6 (*atp6*) and subunit 9 (*atp9*) [46], among which loci within or up/downstream *atp6* and *atp9* have been associated with CMS in a variety of plant species [47–50]. Comp73972_c0 in this study may be one of these genes. There have been at least 16 CMS-related ORFs containing structural genes, especially the ones encoding ATP-$F_o$ and ATP-$F_1$ [2,51–54]. Cotton CMS studies have also demonstrated step by step that the *atp1* and *atp6* genes in CMS-D8 could be the candidates of CMS-associated genes in the mitochondrial genome [55], the abnormal sequence and expression of *atpA* gene is associated with CMS expression in Upland cotton [56], and PCR-based SNP markers in genes encoding ATP synthase subunits are useful in discriminating CMS-D8, CMS-D2 and Upland cotton cytoplasms [57]. Because of the increased energy demand during pollen development, one or a few gene product(s) that interferes with mitochondrial $F_oF_1$-ATP synthase function is likely to induce pollen sterility [58]. But based on the results presented here, dysfunctional ATP synthase may not be the major cause of CMS in this study.

In summary, as the first study using unbiased transcriptome sequencing technology to investigate CMS-associated genes in cotton, a substantial number of biologically meaningful DE genes, especially the ones related with photosynthesis, redox reactions, alpha-linolenic acid metabolism, and flavonoid biosynthesis, were identified. Profiling expression patterns of these genes across all stages of pollen development may provide additional evidence for their importance. Further experiments are warranted to elucidate molecular mechanisms of these genes that lead to CMS.

## Supplementary Note

Several CMS systems have been reported in cotton which include *Gossypiumar boreum*, *G. anomalum*, *G. harknessii*, *G. trilobum*, *G. barbadense* and *G. hirsutum*. CMS-D2 and CMS-D8 are the two most commonly studied systems among them.

The male sterile characteristic of JA-CMS is maintained by *G. hirsutum*, and its *restorer-of-fertility* (*Rf*) genes come from *G. barbadense*. The primitive type and natural outcrossing hybrids of JA-CMS resemble its female parent (*G. hirsutum*) [59]. However, the original *Rf* genes of CMS-D2 and 104-7A come from *G. harknessii* [17,60]. And CMS-D8 is known to have its *Rf* genes come from its sister line [61].

Meanwhile, the mtDNA RFLP and PCR analysis results revealed that JA-CMS possesses a novel mitotype compared with the mitochondrial genome of either 104-7A or *G. harknessii* CMS systems [62,63].

These evidence at genetic, morphological and molecular levels demonstrated that factors and mechanisms associated with CMS induction in JA-CMS may be different from CMS-D2, CMS-D8 and 104-7A.

## Supporting Information

**Figure S1** Results of qRT-PCR. Columns and bars represented the means and standard errors (n = 3), respectively. B2 = SS stage of JA-CMS, K2 = SS stage of JB, B3 = MS stage of JA-CMS, K3 = MS stage of JB.

**Figure S2** Transcript length distribution. Minimum, N90, mean, N50, and maximum lengths of transcripts are marked on the figure. The N90/N50 value is a weighted median and defined as the length of the smallest transcript S in the sorted list of all transcripts where the cumulative length from the largest transcript to S is at least 90%/50% of the total length. There are 206,496 transcripts in total; most of the transcripts (74,998 [36.3%]) have lengths between 200–500 base pairs.

**Figure S3** Unigene length distribution. Minimum, N90, mean, N50, and maximum lengths of unigenes are marked on the figure. The N90/N50 value is a weighted median and defined as the length of the smallest unigene S in the sorted list of all unigenes where the cumulative length from the largest unigene to S is at least 90%/50% of the total length. A total of 86,093 unigenes were identified; 59.8% (51,491) of the unigenes have lengths between 200–500 base pairs.

**Figure S4** RPKM density distributions of 854 DE genes. B2 = SS stage of JA-CMS, K2 = SS stage of JB, B3 = MS stage of JA-CMS, K3 = MS stage of JB.

**Figure S5** GO analysis results of up-regulated DE genes in JA-CMS at the SS stage comparing with JB. BP = Biological process, CC = Cellular component, MF = Molecular function; B2 = SS stage of JA-CMS, K2 = SS stage of JB.

**Figure S6** GO analysis results of down-regulated DE genes in JA-CMS at the SS stage comparing with JB. BP = Biological process, CC = Cellular component, MF = Molecular function; B2 = SS stage of JA-CMS, K2 = SS stage of JB.

**Figure S7** GO analysis results of up-regulated DE genes in JA-CMS at the MS stage comparing with JB. BP = Biological process, CC = Cellular component, MF = Molecular function; B3 = MS stage of JA-CMS, K3 = MS stage of JB.

**Figure S8** GO analysis results of down-regulated DE genes in JA-CMS at the MS stage comparing with JB. BP = Biological process, CC = Cellular component, MF = Molecular function; B3 = MS stage of JA-CMS, K3 = MS stage of JB.

**Figure S9** GO analysis results of down-regulated DE genes at the MS stage comparing with the SS stage in JA-CMS. BP = Biological process, CC = Cellular component, MF = Molecular function; B3 = MS stage of JA-CMS, B2 = SS stage of JA-CMS.

**Figure S10** KEGG analysis results of up-regulated DE genes in JA-CMS at the SS stage comparing with JB. Rich factor is the ratio between counts of DE genes and all annotated genes enriched in a certain pathway; qvalue is P value after multiple hypothesis testing correction with a range between 0 and 1. Twenty most significant pathways were plotted, when more than 20 pathways were identified.

**Figure S11** KEGG analysis results of down-regulated DE genes in JA-CMS at the SS stage comparing with JB. Rich factor is the ratio between counts of DE genes and all annotated genes enriched in a certain pathway; qvalue is P value after multiple hypothesis testing correction with a range between 0 and 1. Twenty most significant pathways were plotted, when more than 20 pathways were identified.

**Figure S12**   KEGG analysis results of up-regulated DE genes in JA-CMS at the MS stage comparing with JB. Rich factor is the ratio between counts of DE genes and all annotated genes enriched in a certain pathway; qvalue is P value after multiple hypothesis testing correction with a range between 0 and 1. Twenty most significant pathways were plotted, when more than 20 pathways were identified.

**Figure S13**   KEGG analysis results of down-regulated DE genes in JA-CMS at the MS stage comparing with JB. Rich factor is the ratio between counts of DE genes and all annotated genes enriched in a certain pathway; qvalue is P value after multiple hypothesis testing correction with a range between 0 and 1. Twenty most significant pathways were plotted, when more than 20 pathways were identified.

**Table S1**   Gene-specific primers for qRT-PCR.

**Table S2**   DE genes between JA-CMS and JB plants at the SS stage.

**Table S3**   DE genes between JA-CMS and JB plants at the MS stage.

**Table S4**   DE genes between the SS and MS stages in JA-CMS.

**Table S5**   DE genes between the SS and MS stages in JB.

## Author Contributions

Conceived and designed the experiments: PY J. Han J. Huang. Performed the experiments: PY J. Huang. Analyzed the data: PY J. Huang. Contributed reagents/materials/analysis tools: J. Han J. Huang. Contributed to the writing of the manuscript: PY J. Han J. Huang.

## References

1. Young EG, Hanson MR (1987) A fused mitochondrial gene associated with cytoplasmic male sterility is developmentally regulated. Cell 50: 41–49.
2. Chase CD (2007) Cytoplasmic male sterility: a window to the world of plant mitochondrial-nuclear interactions. Trends Genet 23: 81–90.
3. Linke B, Borner T (2005) Mitochondrial effects on flower and pollen development. Mitochondrion 5: 389–402.
4. Unseld M, Marienfeld JR, Brandt P, Brennicke A (1997) The mitochondrial genome of Arabidopsis thaliana contains 57 genes in 366,924 nucleotides. Nat Genet 15: 57–61.
5. Kubo T, Nishizawa S, Sugawara A, Itchoda N, Estiati A, et al. (2000) The complete nucleotide sequence of the mitochondrial genome of sugar beet (Beta vulgaris L.) reveals a novel gene for tRNA(Cys)(GCA). Nucleic Acids Res 28: 2571–2576.
6. Notsu Y, Masood S, Nishikawa T, Kubo N, Akiduki G, et al. (2002) The complete sequence of the rice (Oryza sativa L.) mitochondrial genome: frequent DNA sequence acquisition and loss during the evolution of flowering plants. Mol Genet Genomics 268: 434–445.
7. Sugiyama Y, Watase Y, Nagase M, Makita N, Yagura S, et al. (2005) The complete nucleotide sequence and multipartite organization of the tobacco mitochondrial genome: comparative analysis of mitochondrial genomes in higher plants. Mol Genet Genomics 272: 603–615.
8. Rhoads DM, Subbaiah CC (2007) Mitochondrial retrograde regulation in plants. Mitochondrion 7: 177–194.
9. Suzuki H, Rodriguez-Uribe L, Xu J, Zhang J (2013) Transcriptome analysis of cytoplasmic male sterility and restoration in CMS-D8 cotton. Plant Cell Rep 32: 1531–1542.
10. Schuster SC (2008) Next-generation sequencing transforms today's biology. Nat Methods 5: 16–18.
11. Ansorge WJ (2009) Next-generation DNA sequencing techniques. N Biotechnol 25: 195–203.
12. Wang Z, Gerstein M, Snyder M (2009) RNA-Seq: a revolutionary tool for transcriptomics. Nat Rev Genet 10: 57–63.
13. Zheng BB, Wu XM, Ge XX, Deng XX, Grosser JW, et al. (2012) Comparative transcript profiling of a male sterile cybrid pummelo and its fertile type revealed altered gene expression related to flower development. PLoS One 7: e43758.
14. Yan X, Dong C, Yu J, Liu W, Jiang C, et al. (2013) Transcriptome profile analysis of young floral buds of fertile and sterile plants from the self-pollinated offspring of the hybrid between novel restorer line NR1 and Nsa CMS line in Brassica napus. BMC Genomics 14: 26.
15. Liu C, Ma N, Wang PY, Fu N, Shen HL (2013) Transcriptome sequencing and De Novo analysis of a cytoplasmic male sterile line and its near-isogenic restorer line in chili pepper (Capsicum annuum L.). PLoS One 8: e65209.
16. Huang J, Yang P, Li B (2001) The Microstructural and Ultrastructural Study on Microsporogenesis of Cytoplasmic Male Sterile Cotton Line Jin A. Cotton Science 13: 259–263.
17. Jia ZC (1990) Selection and cultivation of CMS 104-7A cotton and its maintainer and restorer line. China Cotton 17: 11.
18. Zhao H, Huang J (2012) Study on Microspore Abortion of Male Sterile Cotton Yamian A and Yamian B. Scientia Agricultura Sinica 45: 4130–4140.
19. Hou L (2001) Detection and isolation of genes differentially expressed during anther development of cotton by cDNA-AFLP.Chongqing: Southwestern Agricultural University.
20. Grabherr MG, Haas BJ, Yassour M, Levin JZ, Thompson DA, et al. (2011) Full-length transcriptome assembly from RNA-Seq data without a reference genome. Nat Biotechnol 29: 644–652.
21. Li B, Dewey CN (2011) RSEM: accurate transcript quantification from RNA-Seq data with or without a reference genome. BMC Bioinformatics 12: 323.
22. Robinson MD, Oshlack A (2010) A scaling normalization method for differential expression analysis of RNA-seq data. Genome Biol 11: R25.
23. Wang L, Feng Z, Wang X, Zhang X (2010) DEGseq: an R package for identifying differentially expressed genes from RNA-seq data. Bioinformatics 26: 136–138.
24. Young MD, Wakefield MJ, Smyth GK, Oshlack A (2010) Gene ontology analysis for RNA-seq: accounting for selection bias. Genome Biol 11: R14.
25. Xie C, Mao X, Huang J, Ding Y, Wu J, et al. (2011) KOBAS 2.0: a web server for annotation and identification of enriched pathways and diseases. Nucleic Acids Res 39: W316–322.
26. Livak KJ, Schmittgen TD (2001) Analysis of relative gene expression data using real-time quantitative PCR and the 2(-Delta Delta C(T)) Method. Methods 25: 402–408.
27. Subramanian A, Tamayo P, Mootha VK, Mukherjee S, Ebert BL, et al. (2005) Gene set enrichment analysis: a knowledge-based approach for interpreting genome-wide expression profiles. Proc Natl Acad Sci U S A 102: 15545–15550.
28. Balk J, Leaver CJ (2001) The PET1-CMS mitochondrial mutation in sunflower is associated with premature programmed cell death and cytochrome c release. Plant Cell 13: 1803–1818.
29. Diamond M, McCabe P (2011) Mitochondrial Regulation of Plant Programmed Cell Death. In: Kempken F, editor. Plant Mitochondria: Springer New York. 439–465.
30. Papini A, Mosti S, Brighigna L (1999) Programmed-cell death events during tapetum development of angiosperms. Protoplasma 207: 213–221.
31. Doyle SM, Diamond M, McCabe PF (2010) Chloroplast and reactive oxygen species involvement in apoptotic-like programmed cell death in Arabidopsis suspension cultures. Journal of Experimental Botany 61: 473–482.
32. Jiang PD, Zhang XQ, Zhu YG, Zhu W, Xie HY, et al. (2007) Metabolism of reactive oxygen species in cotton cytoplasmic male sterility and its restoration. Plant Cell Rep 26: 1627–1634.
33. Wan CX, Li SQ, Wen L, Kong J, Wang K, et al. (2007) Damage of oxidative stress on mitochondria during microspores development in Honglian CMS line of rice. Plant Cell Rep 26: 373–382.
34. Gechev TS, Van Breusegem F, Stone JM, Denev I, Laloi C (2006) Reactive oxygen species as signals that modulate plant stress responses and programmed cell death. Bioessays 28: 1091–1101.
35. Moller IM (2001) Plant mitochondria and oxidative stress: Electron transport, NADPH turnover, and metabolism of reactive oxygen species. Annual Review of Plant Physiology and Plant Molecular Biology 52: 561–591.
36. Sharma P, Jha AB, Dubey RS, Pessarakli M (2012) Reactive Oxygen Species, Oxidative Damage, and Antioxidative Defense Mechanism in Plants under Stressful Conditions. Journal of Botany 2012: 26.
37. Dos Santos CV, Rey P (2006) Plant thioredoxins are key actors in the oxidative stress response. Trends in Plant Science 11: 329–334.
38. De Leon IP, Sanz A, Hamberg M, Castresana C (2002) Involvement of the Arabidopsis alpha-DOX1 fatty acid dioxygenase in protection against oxidative stress and cell death. Plant J 29: 61–62.
39. McConn M, Browse J (1996) The critical requirement for linolenic acid is pollen development, not photosynthesis, in an arabidopsis mutant. Plant Cell 8: 403–416.
40. Park JH, Halitschke R, Kim HB, Baldwin IT, Feldmann KA, et al. (2002) A knock-out mutation in allene oxide synthase results in male sterility and defective wound signal transduction in Arabidopsis due to a block in jasmonic acid biosynthesis. Plant Journal 31: 1–12.

41. Liu G, Tian H, Huang YQ, Hu J, Ji YX, et al. (2012) Alterations of Mitochondrial Protein Assembly and Jasmonic Acid Biosynthesis Pathway in Honglian (HL)-type Cytoplasmic Male Sterility Rice. Journal of Biological Chemistry 287.

42. Yang S, Terachi T, Yamagishi H (2008) Inhibition of chalcone synthase expression in anthers of Raphanus sativus with Ogura male sterile cytoplasm. Annals of Botany 102: 483–489.

43. Wei M, Song M, Fan S, Yu S (2013) Transcriptomic analysis of differentially expressed genes during anther development in genetic male sterile and wild type cotton by digital gene-expression profiling. BMC Genomics 14: 97.

44. Fini A, Brunetti C, Di Ferdinando M, Ferrini F, Tattini M (2011) Stress-induced flavonoid biosynthesis and the antioxidant machinery of plants. Plant Signal Behav 6: 709–711.

45. Dong X, Feng H, Xu M, Lee J, Kim YK, et al. (2013) Comprehensive analysis of genic male sterility-related genes in Brassica rapa using a newly developed Br300K oligomeric chip. PLoS One 8: e72178.

46. Heazlewood JL, Whelan J, Millar AH (2003) The products of the mitochondrial orf25 and orfB genes are FO components in the plant F1FO ATP synthase. FEBS Lett 540: 201–205.

47. Kempken F, Howard W, Pring DR (1998) Mutations at specific atp6 codons which cause human mitochondrial diseases also lead to male sterility in a plant. FEBS Lett 441: 159–160.

48. de la Canal L, Crouzillat D, Quetier F, Ledoigt G (2001) A transcriptional alteration on the atp9 gene is associated with a sunflower male-sterile cytoplasm. Theoretical and Applied Genetics 102: 1185–1189.

49. Kurek I, Ezra D, Begu D, Erel N, Litvak S, et al. (1997) Studies on the effects of nuclear background and tissue specificity on RNA editing of the mitochondrial ATP synthase subunits alpha, 6 and 9 in fertile and cytoplasmic male-sterile (CMS) wheat. Theoretical and Applied Genetics 95: 1305–1311.

50. Szklarczyk M, Oczkowski M, Augustyniak H, Borner T, Linke B, et al. (2000) Organisation and expression of mitochondrial atp9 genes from CMS and fertile carrots. Theoretical and Applied Genetics 100: 263–270.

51. Hanson MR, Bentolila S (2004) Interactions of mitochondrial and nuclear genes that affect male gametophyte development. Plant Cell 16 Suppl: S154–169.

52. Kim DH, Kang JG, Kim BD (2007) Isolation and characterization of the cytoplasmic male sterility-associated orf456 gene of chili pepper (Capsicum annuum L.). Plant Mol Biol 63: 519–532.

53. Yamamoto MP, Kubo T, Mikami T (2005) The 5′-leader sequence of sugar beet mitochondrial atp6 encodes a novel polypeptide that is characteristic of Owen cytoplasmic male sterility. Mol Genet Genomics 273: 342–349.

54. Zhang H, Li S, Yi P, Wan C, Chen Z, et al. (2007) A Honglian CMS line of rice displays aberrant F0 of F0F1-ATPase. Plant Cell Rep 26: 1065–1071.

55. Wang F, Feng CD, O'Connell MA, Stewart JM, Zhang JF (2010) RFLP analysis of mitochondrial DNA in two cytoplasmic male sterility systems (CMS-D2 and CMS-D8) of cotton. Euphytica 172: 93–99.

56. Wu JY, Gong YC, Cui MH, Qi TX, Guo LP, et al. (2011) Molecular characterization of cytoplasmic male sterility conditioned by Gossypium harknessii cytoplasm (CMS-D2) in upland cotton. Euphytica 181: 17–29.

57. Suzuki H, Yu JW, Wang F, Zhang JF (2013) Identification of mitochondrial DNA sequence variation and development of single nucleotide polymorphic markers for CMS-D8 in cotton. Theoretical and Applied Genetics 126: 1521–1529.

58. Li J, Pandeya D, Jo YD, Liu WY, Kang BC (2013) Reduced activity of ATP synthase in mitochondria causes cytoplasmic male sterility in chili pepper. Planta 237: 1097–1109.

59. Yuan J, Zhang Y, Liu X, Hao R, Wang H (1996) Discovery and study of JA cytoplasmic male sterility. China Cotton 23: 6–7.

60. Wei Z, Luo C (1987) Preliminary studies of the cytoplasm influence on economic traits of upland cotton Cotton Breeding Basic Research Papers: 103–107.

61. Stewart JM (1992) A new cytoplasmic male sterile and restorer. Prod Beltw ide Cotton Prod Res Conf.

62. Huang J (2003) Genetic study on JinA cytoplasmic male sterile line in cotton. Taigu: Shanxi Agricultural University.

63. Ma W, Ren A, Cui M, Han Q, Zhang Y, et al. (2012) The difference of mitochondrial DNA of three cotton cytoplasmic male sterility lines. Cotton Science 34: 10–15.

# 21

# Effects of Gibberellic Acid and N, N-Dimethyl Piperidinium Chloride on the Dose of and Physiological Responses to Prometryn in Black Nightshade (*Solanum nigrum* L.)

**Hailan Jiang**[1,2◯], **Xiaoxia Deng**[1,2◯], **Jungang Wang**[1]*, **Jing Wang**[1], **Jun Peng**[1], **Tingting Zhou**[1]

1 College of Agriculture, Shihezi University, Shihezi, Xinjiang, People's Republic of China, 2 Kuerle Agricultural Technology Extension Center, Kuerle, Xinjiang, People's Republic of China

## Abstract

The use of gibberellic acid ($GA_3$) and N, N-dimethyl piperidinium chloride (DPC) in combination with prometryn would likely increase the control of black nightshade in cotton fields. Experiments were designed to investigate the physiological and biochemical responses of black nightshade at the three- to four-leaf stage to prometryn applied at different rates, either alone or in combination with $GA_3$ or DPC, in a greenhouse environment. These studies demonstrated that prometryn applied in combination with DPC at low rates (7.2 g ai ha$^{-1}$) led to increased fresh weight and visible injury of black nightshade compared with prometryn applied alone or in combination with $GA_3$; however, at rates of 36, 180, and 900 g ai ha$^{-1}$, prometryn in combination with DPC caused the least visible injury among all treatments and prometryn in combination with $GA_3$ caused the greatest visible injury. These results suggest that black nightshade suffered more severe damage when prometryn was applied in combination with $GA_3$, which is supported by the reduced soluble protein content, lower antioxidant enzyme activities, and higher malondialdehyde (MDA) content in the plants treated with prometryn plus $GA_3$. These results indicate that the application of $GA_3$ in combination with prometryn to black nightshade may have the potential to lower the levels of prometryn tolerance in these plants.

**Editor:** Manuel Reigosa, University of Vigo, Spain

**Funding:** This work was supported by the Special Fund for Agro-scientific Research in the Public Interest (201303022), corps doctoral Fund (2013BB004) and National Natural Science Foundation Project of China (31260435). The funders had no role in study design, data collection and analysis, decision to publish, or preparation of the manuscript.

**Competing Interests:** The authors have declared that no competing interests exist.

* E-mail: jungangwang98@163.com

◯ These authors contributed equally to this work.

## Introduction

Black nightshade (*Solanum nigrum* L.) is widely distributed throughout the world and is an important weed in many countries [1]. In recent decades, black nightshade has become increasingly problematic as a field crop weed [2,3]. It is also one of the dominant weeds in cotton fields [4,5] and seriously affects the growth of cotton.

Prometryn is a triazine herbicide in the same family as simetryn. Cotton is tolerant to certain triazine herbicides, such as prometryn, when they are applied at pre-emergence or post-emergence; however, severe injury can occur following post-emergence application [6,7]. The triazine herbicides control of a wide spectrum of weeds in crop and non-crop areas under many environmental and edaphic conditions. Prometryn has been widely used as a residual soil-applied and post-emergence-directed herbicide in cotton [8,9] and provides season-long control of annual grasses [10,11].

Plant growth regulators, such as N, N-dimethyl piperidinium chloride (DPC) and gibberellic acid ($GA_3$), are widely used in cotton production and have become indispensable tools in cotton cultivation [12,13]. Although variations in the application of these growth regulators can occur among regions, years, and environmental conditions, they are applied throughout the growing season up to several weeks prior to harvest. DPC has been known to reduce the synthesis of $GA_3$, resulting in the suppression of cell enlargement [14]. Thus, DPC-treated plants tend to be short and narrow with thick, small leaves [15], while $GA_3$-treated plants present an opposite phenotype. Recently, there has been increased interest in the co-application of herbicides with plant growth regulators [16]. Applying these products in combination is also preferred by growers due to the convenience, reduced time investment, and reduced application and labour costs.

Due to the availability of prometryn-tolerant cotton, it is likely that this herbicide will be applied simultaneously with plant growth regulators and will effectively control black nightshade when it is applied to cotton at early post-emergence or late post-emergence [17,18,19]. Despite these benefits, incompatibility can occur with certain pesticides. Therefore, defining the interactions of agrochemicals is important when considering the simultaneous application of multiple agrochemicals [20,21]. Because previous studies have seldom focused on the physiological responses of plants to the combined application of herbicides and plant growth regulators, it is desirable to devise an index of resistance that reflects physiological responses under different treatments. We hypothesised that the foliar application of plant growth regulators to black nightshade would strengthen the effect of prometryn on weed control. Therefore, the current study was designed to evaluate the efficacy of prometryn when applied alone or in combination with $GA_3$ or DPC in controlling black nightshade and to study the physiological and biochemical response of the weed to co-application in a greenhouse.

## Materials and Methods

### Experimental Conditions and Treatments

The experiments were conducted in a greenhouse at Shihezi University in Xinjiang in 2012. The greenhouse temperature was maintained between 22°C at night and 29°C during the day. Separate experiments were conducted to investigate the control of black nightshade and to assess the physiological and biochemical changes in this plant. The plants were watered daily, grown under natural sunlight, and fertilised every 7 d to maintain active growth. Treatments for the black nightshade included 5 levels of the non-selective herbicide prometryn (7.2, 36, 180, 900, and 4500 g ai $ha^{-1}$) applied alone or in combination with a plant growth regulator, either DPC (22.05 kg ai $ha^{-1}$) or $GA_3$ (14.7 kg ai $ha^{-1}$). For the physiological and biochemical response study, the treatments consisted of prometryn (900 g ai $ha^{-1}$) applied alone or in combination with a plant growth regulator, either DPC (22.05 kg ai $ha^{-1}$) or $GA_3$ (14.7 kg ai $ha^{-1}$). All chemicals were applied at the manufacturer's recommended rate. A non-treated control was included in these experiments. The experimental design for the control of black nightshade experiment and for the physiological and biochemical response experiment was a randomised complete block with four treatment replications.

Black nightshade seed samples (provided by Shihezi University) were stored in a cold room at 4°C prior to the initiation of the study. The black nightshade seeds were planted in a 20×15×5 cm plastic crisper in the greenhouse to accelerate germination. Once the black nightshade seedlings had germinated, 5 seedlings were transplanted into plastic pots (16 cm diameter, 11 cm height) filled with potting soil (fine-loamy, mixed, Typic Argiudoll) at pH 6.4 containing 1.9% organic matter. Three seedlings were removed prior to the treatments, and the treatments were applied when the black nightshade seedlings reached the three- to four-leaf stage.

### Dose Response Study

The black nightshade plants were excised at the soil level 6 d after the treatments to determine the fresh weight and calculate the percent reduction. Visual estimates of percent injury were recorded 6 d after treatment using a scale of 0 to 100, where 0 = no foliar injury and 100 = plant death. Foliar chlorosis, necrosis, tissue distortion, swelling, malformation, and plant stunting were considered when visually estimating the percent injury.

### Physiological and Biochemical Response Study

All the measurements were recorded close to midday on the uppermost fully expanded main stem leaf located three nodes below the terminus of the plant. The third main stem leaf was then collected 3 d and 6 d after application to observe changes in black nightshade physiological parameters, especially chlorophyll content, malondialdehyde content, and soluble protein content, as well as changes in biochemical parameters and antioxidant enzyme activity (superoxide dismutase (SOD), catalase (CAT), and peroxidise (POD)). Prior to the physiological and biochemical

measurements, a single leaf of each plant was collected and immediately stored in an ultra freezer (−80°C) for antioxidant enzyme and protein extraction. A UV spectrophotometer (UV-1800; Shimadzu Co., Ltd., Japan) was used for measuring the changes in the physiological and biochemical parameters.

The chlorophyll (CHL) content of samples was measured using the method specified by Arnon [22]. Lipid peroxidation was defined as the content of all 2-thiobarbituric acid-reactive substances and was expressed as equivalents of MDA, as reported by Heath and Packer [23].

Frozen leaves (0.5 g) were homogenised in 50 mM potassium phosphate buffer (pH 7.8) containing 1 mM EDTA, 3 mM 2-mercaptoethanol, and 5% (w/v) polyvinylpyrrolidone (PVP) in a chilled mortar and pestle. The homogenate was centrifuged at 12,000×g for 20 min at 4°C, and the supernatant was used for enzyme assays.

Soluble proteins were measured according to the technique described by Bradford [24].

The SOD activity was assayed by the nitroblue tetrazolium (NBT) method [25]. The 3 ml reaction mixture contained 50 mM Na-phosphate buffer (pH 7.3), 13 mM methionine, 75 mM NBT, 0.1 mM EDTA, 4 mM riboflavin, and 0.2 ml of enzyme extract. This reaction was initiated by the addition of riboflavin, and the glass test tubes were shaken and placed under fluorescent lamps (160 μmol $m^{-2} s^{-1}$). The reaction was allowed to proceed for 5 min and was then stopped by switching off the light. The absorbance was measured at 560 nm. Blanks and controls were treated in the same manner but lacked illumination and enzyme, respectively.

The activities of POD and CAT were assayed using the method of Chance and Maehly [26] with some modifications. The POD reaction solution (3 ml) contained 50 mM phosphate buffer (pH 7.8), 25 mM guaiacol, 200 mM $H_2O_2$, and 0.5 ml of enzyme extract. Changes in the absorbance of the reaction solution at 470 nm were determined every 30 s. One unit of POD activity was defined as an absorbance change of 0.01 unit $min^{-1}$. The CAT reaction solution (3 ml) contained 50 mM phosphate buffer (pH 7.0), 200 mM $H_2O_2$, and 0.5 ml of enzyme extract. The reaction was initiated by the addition of the extract. Changes in the absorbance of the reaction solution at 240 nm were measured every 30 s.

### Data Analysis

Differences between the plant growth regulator treatments were detected using analysis of variance (ANOVA). All data analyses were conducted using SPSS Statistics 17.0 with estimates of means and standard errors, and Duncan's multiple range test was used for determining significant differences.

## Results

### Dose-response Study

In general, the fresh weight observed at 6 d after application decreased as the herbicide rate increased, and the fresh weight was decreased to a similar extent across rates for all three treatments (Fig. 1). All herbicide treatments at rates of 7.2 and 36 g ai $ha^{-1}$ caused reductions in fresh weight of less than 50%. The fresh weight reductions caused by prometryn and prometryn plus DPC were not drastically different. Prometryn applied in combination with $GA_3$ at 7.2, 36, and 180 g ai $ha^{-1}$ provided greater control than prometryn applied alone or combined with DPC; however, minimal differences were found when the rates of all treatments were higher than 180 g ai $ha^{-1}$.

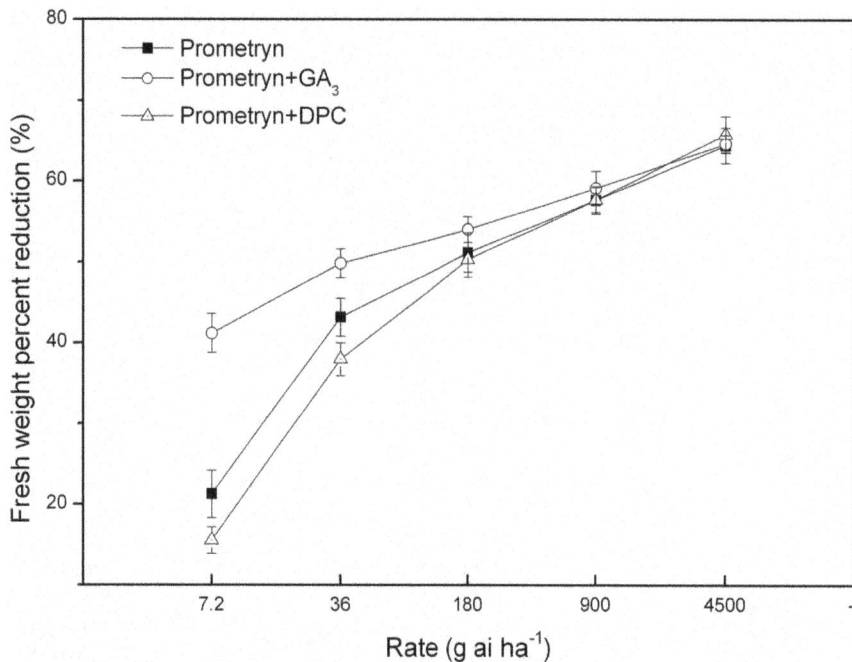

**Figure 1. Percent reduction in fresh weight caused by prometryn applied alone or in combination with GA₃ or DPC.** Measurements were performed 6 d after application. The data are the means of 4 replicates ± standard error.

Visible injury increased as the herbicide rates increased, with prometryn plus $GA_3$ causing the greatest injury at 6 d after treatment (Fig. 2). All treatments caused less than 20% visible injury at 7.2 g ai ha$^{-1}$. Prometryn combined with DPC at 7.2 g ai ha$^{-1}$ provided the best control of black nightshade; however, it caused the least injury at 36 and 180 g ai ha$^{-1}$. Although the visible injury caused by prometryn plus $GA_3$ was more serious than the injury caused by prometryn alone, the visible injuries were similar at all rates.

### Physiological and Biochemical Response Study

A similar trend was observed in the changes in chlorophyll a and chlorophyll b contents after all treatments. After the prometryn treatment, the chlorophyll levels decreased continuously and drastically by 31.09% (CHL a) and 36.61% (CHL b) 3 d after application and by 84.01% (CHL a) and 42.61% (CHL b) 6 d after application (Fig. 3). When prometryn was applied in combination with $GA_3$, the change in chlorophyll content was more rapid, decreasing by 14.61% (CHL a) and 10.13% CHL b) 3 d after application and by 21.83% (CHL a) and 14.58% (CHL b) 6 d after application, compared to the treatment with prometryn alone (Fig. 3); however, 6 d after application, the effects of these two treatments were not significant. Although the chlorophyll content of the plants treated with prometryn plus DPC also decreased, the change was not significant compared to the control 3 d after application. The chlorophyll content of the plants treated with prometryn plus DPC 6 d after application was significantly lower than that of the control; however, it was still significantly higher than the chlorophyll content of plants treated with prometryn or prometryn plus $GA_3$.

The soluble protein content was significantly different between the prometryn and the control treatments 3 and 6 d after application (Fig. 4), and the soluble protein content of all treated plants was significantly lower than that of the untreated plants. The soluble protein content of plants treated with prometryn plus

$GA_3$ was significantly lower than that of plants treated with prometryn plus DPC 3 d after application; however, the soluble protein content of plants exposed to these treatments was not significantly different from that of plants treated with prometryn alone. No significant difference was observed in the soluble protein content of all prometryn-treated plants 6 d after application. The percent reductions in soluble protein content were 66.68%, 71.57%, and 69.49% for the prometryn, prometryn plus $GA_3$, and prometryn plus DPC treatments, respectively, compared with the control.

After prometryn treatment, the MDA content of the plants increased drastically by 34.07% and 42.61% 3 d and 6 d after application, respectively (Fig. 5). When prometryn was applied in combination with $GA_3$, the increase in MDA content was quicker than prometryn treatment alone. The highest MDA content was detected in plants treated with prometryn plus $GA_3$ 3 d and 6 d after application; however, no significant differences were found between the prometryn and prometryn plus $GA_3$ treatments. Compared with the control, the MDA content in plants treated with prometryn, prometryn plus $GA_3$, and prometryn plus DPC was increased by 78.71%, 92.03%, and 48.82%, respectively.

### Changes in Antioxidant Enzyme Activities

Prometryn applied alone or in combination with either $GA_3$ or DPC had a significant effect on SOD activity; i.e., herbicide-treated plants exhibited significantly lower SOD activities compared with the control plants 3 d and 6 d after application (Fig. 6). The SOD activity of plants treated with prometryn plus DPC decreased 17.07% (3 d) and 18.37% (6 d) compared to the control. The lowest SOD activity was found in the plants treated with prometryn plus $GA_3$; however, no significant difference was found between plants treated with prometryn alone and those treated with prometryn plus $GA_3$.

Prometryn applied alone or in combination with either $GA_3$ or DPC had a significant effect on POD activity; i.e., herbicide-

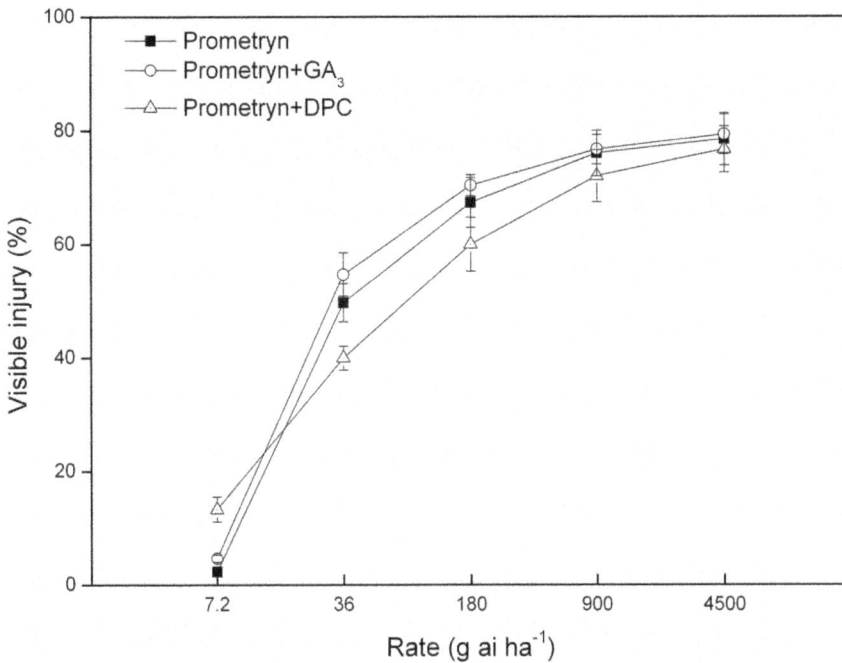

**Figure 2. Visible injury caused by prometryn applied alone or in combination with GA₃ or DPC.** Measurements were performed 6 d after application. The data are the means of 4 replicates ± standard error.

treated plants exhibited significantly lower POD activities compared with the control plants 3 d and 6 d after application. However, no significant differences were observed between any of the herbicide treatments 6 d after application (Fig. 7). The POD activity of plants treated with prometryn plus DPC decreased 14.53% (3 d) and 59.34% (6 d) compared to the control. The lowest POD activities were noted in the plants treated with prometryn (3 d) and prometryn plus GA₃ (6 d); in these plants, POD activity decreased by 54.68% and 68.35%, respectively, compared with the control.

Prometryn applied alone or in combination with either GA₃ or DPC also had a significant effect on CAT activity; i.e., the herbicide-treated plants exhibited significantly lower CAT

activities compared with the control plants 3 d and 6 d after application. However, no significant difference was observed between any of the herbicide treatments 3 d and 6 d after application (Fig. 8). Compared to the control, the CAT activities in plants subjected to all treatments were decreased by more than 95%.

## Discussion

The proper use of growth-regulating chemicals requires a detailed understanding of the effects of their interaction with herbicides on plants. The results of this study suggest that the application of prometryn in combination with GA₃ led to a greater

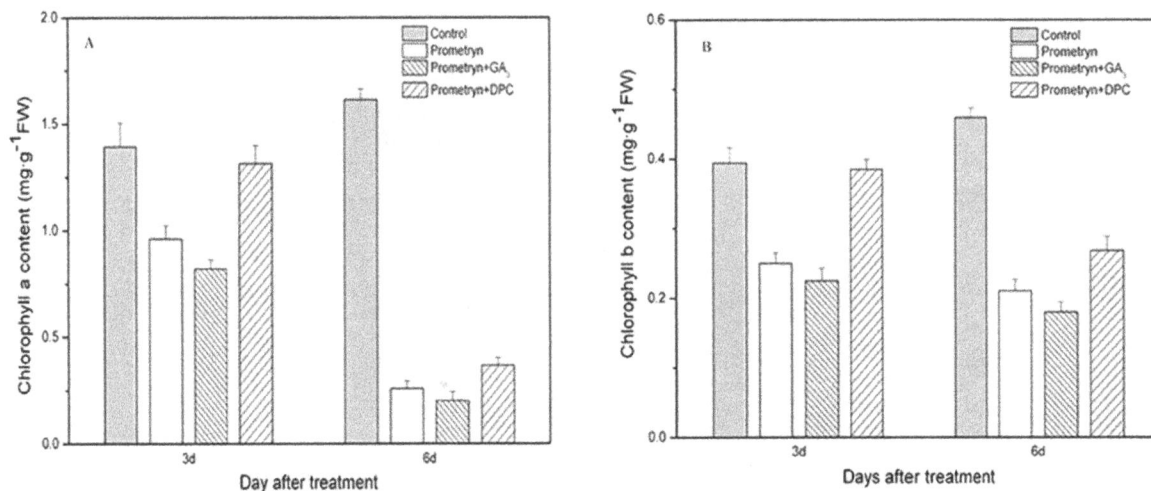

**Figure 3. Effect of prometryn applied alone or in combination with GA₃ or DPC on chlorophyll content.** A: chlorophyll a content. B: chlorophyll b content. Measurements were performed 3 and 6 d after application. The data are the means of 4 replicates + standard error.

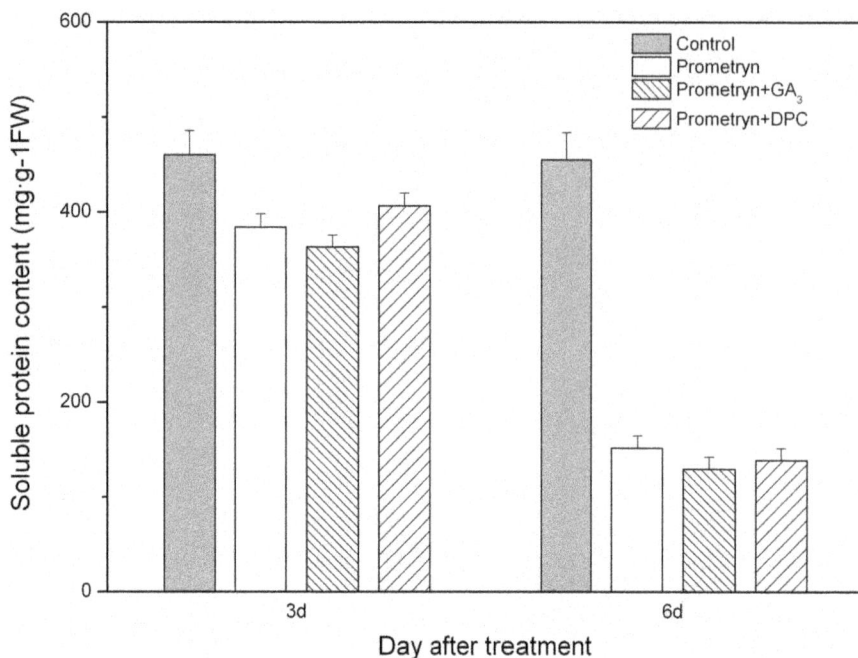

**Figure 4. Effect of prometryn applied alone or in combination with GA₃ or DPC on soluble protein content.** Measurements were performed 3 and 6 d after application. The data are the means of 4 replicates + standard error.

reduction of plant fresh weight compared with prometryn alone and caused the greatest visual injury, indicating that prometryn had a stronger effect when it was combined with GA₃. Changes in physiological and biochemical parameters after the foliar application of prometryn plus plant growth regulators help to explain the greater visible injury caused by prometryn in combination with GA₃ in black nightshade.

CHL is vital for photosynthesis, which allows plants to obtain energy from light. The destruction and degradation of CHL directly results in a reduced photosynthetic rate and leads to leaf senescence. It has been reported that the CHL content was decreased in plants subjected to herbicide stress [27,28]. The current study also showed that the CHL a and b contents were significantly decreased in black nightshade treated with prometryn

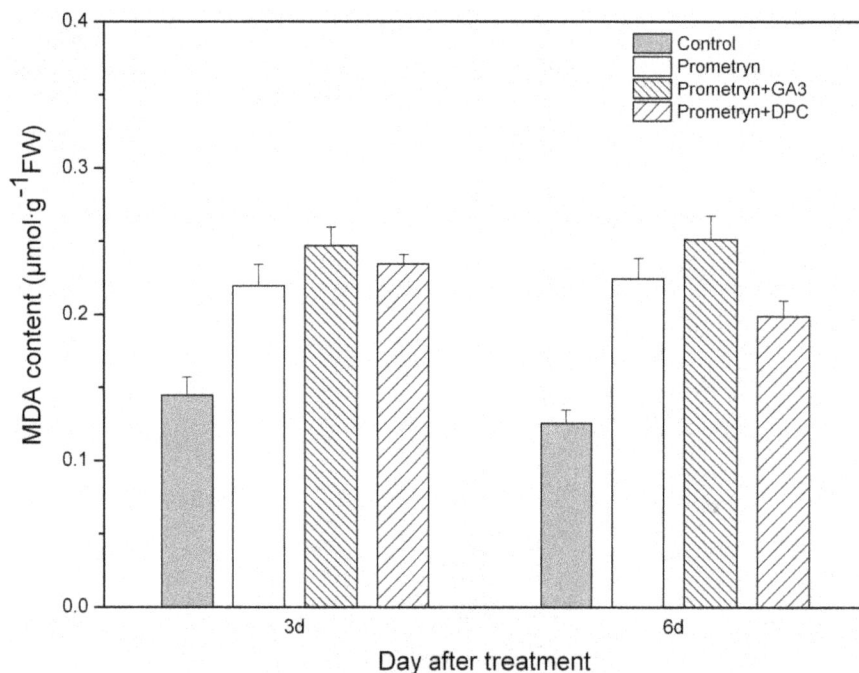

**Figure 5. Effect of prometryn applied alone or in combination with GA₃ or DPC on MDA content.** Measurements were performed 3 and 6 d after application. The data are the means of 4 replicates + standard error.

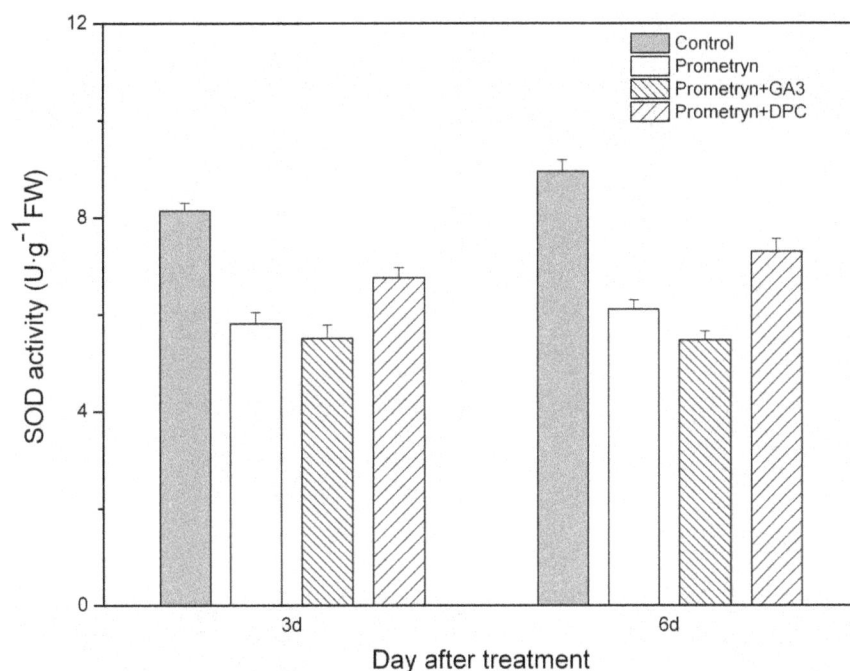

**Figure 6. Effect of prometryn applied alone or in combination with GA$_3$ or DPC on SOD activity.** Measurements were performed 3 and 6 d after application. The data are the means of 4 replicates + standard error.

3 d and 6 d after application. Application of prometryn in combination with GA$_3$ caused a larger decrease in the CHL content compared with application of prometryn alone.

The soluble proteins in plants can enhance resistance to cellular stress conditions, and the amount of soluble protein directly affects metabolism [29]. Guy [30] concluded that an increase in the soluble protein content during stress appears to be an indicator of plant tolerance. We found that the soluble protein content was decreased in plants treated with prometryn applied alone or in combination with either GA$_3$ or DPC. Compared with all other treatments, the application of prometryn in combination with GA$_3$ resulted in the largest decrease in soluble protein content.

Under stress conditions, the formation of reactive oxygen species (ROS) is enhanced, thus inducing protective responses and

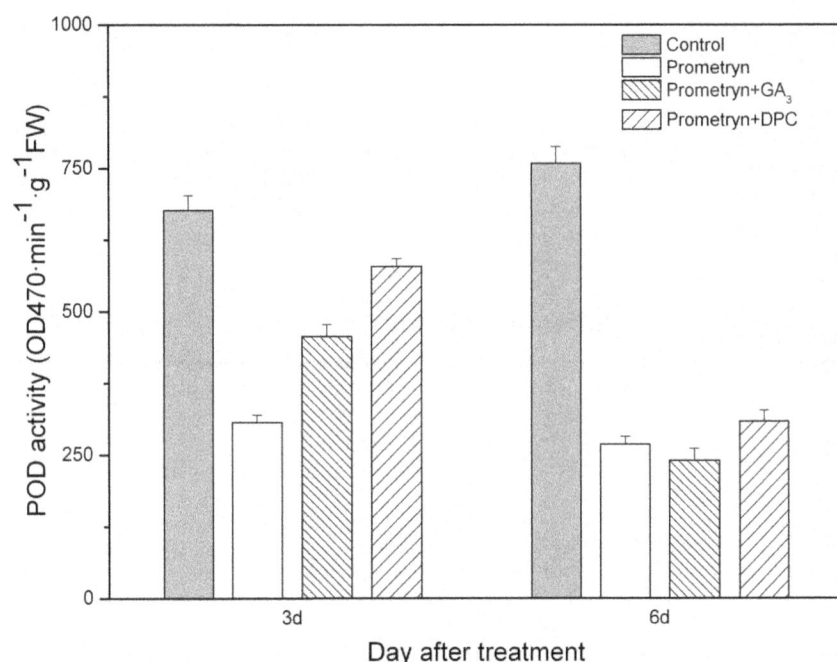

**Figure 7. Effect of prometryn applied alone or in combination with GA$_3$ or DPC on POD activity.** Measurements were performed 3 and 6 d after application. The data are the means of 4 replicates + standard error.

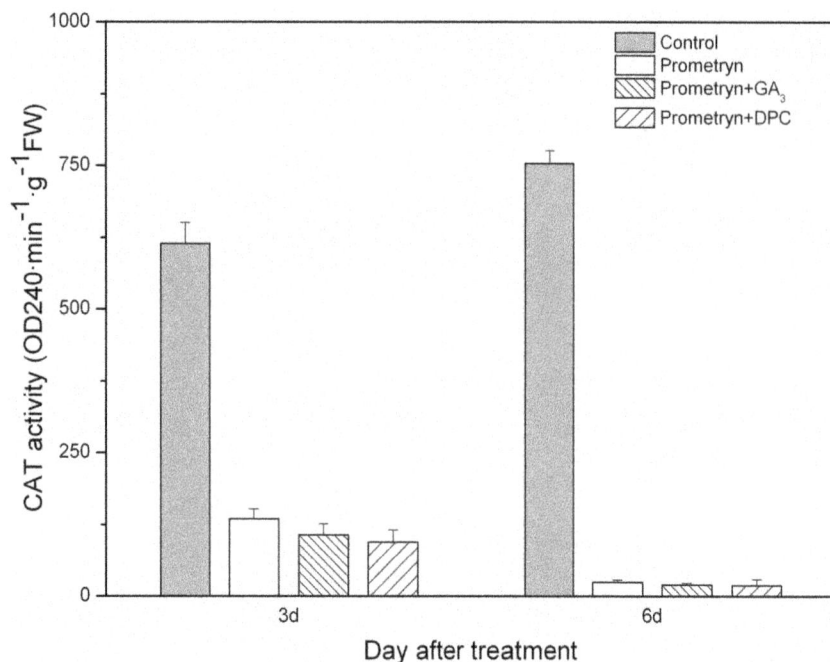

**Figure 8. Effect of prometryn applied alone or in combination with GA$_3$ or DPC on CAT activity.** Measurements were performed 3 and 6 d after application. The data are the means of 4 replicates + standard error.

cellular damage [31]. Membrane damage is caused by lipid peroxidation, which is in turn caused by increased concentrations of H$_2$O$_2$ [32,33]. MDA is a product of lipid peroxidation in cell membranes [34,35], and its content in vivo can indicate the extent of oxidative stress in plants and membrane homeostasis. We have demonstrated that herbicide treatments cause a dramatic increase in the leaf MDA content in black nightshade, indicating the presence of oxidation. Moreover, the exogenous application of prometryn plus GA$_3$ was effective in increasing the levels of MDA in the leaves of black nightshade in a time-dependent manner. In contrast, we found that the MDA content in plants treated with prometryn plus DPC was lower than that in plants treated with prometryn alone 6 d after application. This result suggests that the prometryn treatment causes damage to the leaves and implies that GA$_3$ could increase this damage when applied with prometryn.

The improvement of stress tolerance in plants is often related to antioxidant activities. An increased ROS level always triggers the upregulation of antioxidant enzyme activity, which in turn protects plants from oxidative stress [36,37]. SOD, POD, and CAT are the most important detoxifying enzymes that promote ROS scavenging [38,39]. SOD is responsible for the scavenging of toxic O$^{2-}$ in different cellular organelles [40]. Furthermore, H$_2$O$_2$ might be detoxified by CAT, POD, and the ascorbate–glutathione cycle [41]. Researchers have found that the antioxidant enzyme activity in plants treated with herbicides decreases with time [42,43].

Here, we also found that the activities of SOD, POD, and CAT in prometryn-treated leaves decreased over time. The antioxidant system in plants is a protective mechanism that confers tolerance to environmental stresses [44]. We demonstrated that the exogenous application of GA$_3$ decreased the enzyme activities in black nightshade leaves, thus strengthening prometryn-induced oxidative damage. However, compared to plants treated with prometryn alone, higher enzyme activities were observed in plants treated with prometryn plus DPC.

In conclusion, GA$_3$ effectively controls black nightshade. The combination of prometryn and GA$_3$ positively reduced the herbicide tolerance of black nightshade at least in part by decreasing the content of leaf CHL and the soluble protein content, enhancing membrane lipid peroxidation and weakening the activities of SOD, POD, and CAT. The results presented in this study provide a basic understanding of the role of GA$_3$ when applied in combination with prometryn and indicate its potential use in weed management.

## Author Contributions

Conceived and designed the experiments: Jungang Wang HLJ XXD. Performed the experiments: HLJ XXD Jing Wang JP TTZ. Analyzed the data: HLJ XXD. Contributed reagents/materials/analysis tools: Jungang Wang. Wrote the paper: HLJ XXD.

## References

1. Holm LG, Plucknett DL, Pancho JV, Herberger JP (1991) The World's Worst Weeds. Distribution and Biology. Honolulu, HI: University of Hawaii Press. 430–435.
2. Wuolo A, and Jönsson B (2002) Nattskatta är ett problemogäs. Potatis Grönsaker Nr4.
3. Shrestha A, Fidelibus M (2005) Grapevine row orientation affects light environment, growth, and development of black nightshade (Solanum nigrum). Weed Sci 53(6): 802–812.
4. Economou G, Bilalis D, Avgoulas C (2005) Weed Flora Distribution in Greek Cotton Fields and Its Possible Influence by Herbicides. Weed Sci 33(4): 406–419.
5. Kaloumenos NS, Veletza VG, Papantoniou AN, Kadis SG, Eleftherohorinos IG (2005) Influence of Pyrithiobac Application Rate and Timing on Weed Control and Cotton Yield in Greece. Weed Technol 19(1): 207–216.
6. Keeling JW, Verett KS, Reed JD, Dotray PA (2011) Cotton (Gossypium hirsutum) Tolerance to Propazine Applied Preand Postemergence. Weed Technol. 25: 178–182.

7. Dunk P, John WW, Shawn DA (2002) Weed management with CGA-362622, fluometuron, and prometryn in cotton. Weed Sci, 50(5): 642–647.

8. Byrd JD (2000) Report of the 1999 Cotton Weed Loss Committee. 2000 Proc. Beltwide Cotton Conf. 2: 1455–1458.

9. Walter ET, Britton TT, Clewis SB, Askew SD, Wilcut JW (2006) Glyphosate-Resistant Cotton (Gossypium hirsutum) Response and Weed Management with Trifloxysulfuron, Glyphosate, Prometryn, and MSMA. Weed Technol 20(1): 6–13.

10. Porterfield D, Wilcut JW, Askew SD (2002) Weed management with CGA-362622, fluometuron, and prometryn in cotton. Weed Sci 50: 642–647.

11. Thomas WE, Britton TT, Clewis SB, Askew SD, Wilcut JW (2006) Glyphosate-resistant cotton (Gossypium hirsutum L.) response and weed management with trifloxysulfuron, glyphosate, prometryn, and MSMA. Weed Technol 20: 6–13.

12. Reddy AR, Reddy KR, Hodges HF (1996) Mepiquat chloride (PIX)-induced changes in photosynthesis and growth of cotton. Plant Growth Regulation 20: 179–183.

13. Feng GY, Yao YD, Du MW, Tian JS, Luo HH, et al. (2012) Dimethyl Piperidinium Chloride (DPC) Regulation of Canopy Architecture and Photosynthesis in a Cotton Hybrid in an Arid Region. Cotton Sci 24(1): 44–51.

14. Hake K, Kerby T, McCarty W, O'Neal D, Supak J (1991) Physiology of PIX. In: Physiology Today, Vol. 2, No. 6. Memphis, USA: National Cotton Council of America.

15. Gausman HW, Walter H, Stein E, Rittig FR, Learner RW, et al. (1979) Leaf CO2 uptake and chlorophyll ratios of PIX-treated cotton. In: Proc 6th Ann Meeting of Plant Growth Reg Working Group, Las Vegas, 117-12.5. USA: PGRWG, Longmount, CO.

16. Gurinderbir SC, Johnson WG (2012) Influence of Glyphosate or Glufosinate Combinations with Growth Regulator Herbicides and Other Agrochemicals in Controlling Glyphosate-Resistant Weeds. Weed Technol 26(4): 638–643.

17. Jerry LC, Shawn DA, Dunk P, John WW (2002) Bromoxynil, Prometryn, Pyrithiobac, and MSMA Weed Management Systems for Bromoxynil-Resistant Cotton (Gossypium hirsutum). Weed Technol 16(4): 712–718.

18. Scott BC, Wilcut JW (2007) Economic Assessment of Weed Management in Strip- and Conventional-Tillage Nontransgenic and Transgenic Cotton. Weed Technol 21(1): 45–52.

19. Scott BC, Miller DK, Koger CH, Baughman TA, Price AJ, et al. (2008) Weed Management and Crop Response with Glyphosate, S-Metolachlor, Trifloxysulfuron, Prometryn, and Msma in Glyphosate-Resistant Cotton. Weed Technol 22(1): 160–167.

20. Green JM (1989) Herbicide antagonism at the whole plant level. Weed Technol 3: 217–226.

21. Hatzios KK, Penner D (1985) Interaction of herbicides with other agricultural chemicals in higher plants. Rev. Weed Sci 1: 1–64.

22. Arnon DI (1949) Copper enzymes in isolated chloroplasts. polyphenol oxidase in Beta vulgaris. Plant Physiol 24: 1–15.

23. Heath RL, Packer L (1968) Photoperoxidation in isolated chloroplasts. I. Kinetics and stoichiometry of fatty acid peroxidation. Arch Biochem Biophys 125: 189–198.

24. Bradford MM (1976) A rapid and sensitive method for the quantification of microgram quantities of protein utilizing the principle of protein-dye binding. Anal Biochem 72: 248–254.

25. Dhindsa RS, Plumb-Dhindsa P, Throne TA (1981) Leaf senescence: correlated with increased levels of membrane permeability and lipid peroxidation and decreased levels of superoxide dismutase and catalase. J Exp Bot 32: 93–101.

26. Chance M, Maehly AC (1955) Assay of catalases and peroxidases. Methods Enzymol 2: 764–775.

27. Wang ZG, Feng CN, Guo WS, Xia YR, Zhu XK, et al. (2010) Effects of Herbicides on Physiology and Biochemistry of Weak-Gluten Wheat. J Agro-Environ Sci 29(6): 1027–1032.

28. Zhang DY, Yang WD, Dang JY, Miao GY (2007) Effects of Herbicides on Grain Yield and Physiological Characteristics of Strong Gluten Wheat. J Applied & Environ Biol 13(3): 294–300.

29. Zhang MS, Du JC, Xie B, Tan F,Yang YH (2004) Relationship between osmoregulation substance in sweet potato under water stress and variety drought resistance. J Nanjing Agric University 27(4): 123–125.

30. Guy CL (1990) Cold acclimation and freezing stress tolerance: role of protein metabolism. Annu Rev Plant Physiol Plant Mol Biol 41: 187–223.

31. Blokhina O, Virolainen E, Fagerstedt KV (2003) Antioxidants, oxidative damage and oxygen deprivation stress: a review. Ann Bot 91: 179–194.

32. Sairam RK, Srivastava GC (2002) Changes in antioxidant activity in sub-cellular fractions of tolerant and susceptible wheat genotypes in response to long term salt stress. Plant Sci 162: 897–904.

33. Ma X, Ma F, Mi Y, Ma Y, Shu H (2008a) Morphological and physiological responses of two contrasting Malus species to exogenous abscisic acid application. Plant Growth Regul 56: 77–87.

34. Bailly C, Benamar A, Corbineau F, DÔme D (1996) Changes in malondialdehyde content and in superoxide dismutase, catalase and glutathione reductase activities in sunflower seed as related to deterioration during accelerated aging. Plant Physiol 97: 104–110.

35. Shah K, Kumar RG, Verma S, Dubey RS (2001) Effect of cadmium on lipid peroxidation, superoxide anion generation and activities of antioxidant enzymes in growing rice seedlings. Plant Sci 161: 1135–1144.

36. Davey MW, Montagu MV, Inzé D, Sanmatin M, Kanellis A, et al. (2000) Plant L-ascorbic acid: chemistry, function, metabolism, bioavailability and effects of processing. J Sci Food Agric 80: 825–860.

37. Ma YH, Ma FW, Zhang JK, Li MY, Wang YH, et al. (2008b) Effects of high temperature on activities and gene expression of enzymes involved in ascorbate–glutathione cycle in apple leaves. Plant Sci 175: 761–766.

38. Hernandez JA, Ferrer MA, Jiménez A, Barceló AR, Sevilla F (2001) Antioxidant systems and O²⁻/H₂O₂ production in the apoplast of pea leaves. Its relation with salt-induced necrotic lesions in minor veins. Plant Physiol 127: 817–831.

39. Parida AK, Das AB, Mohanty P (2004) Defense potentials to NaCl in a mangrove, Bruguiera parviflora: Differential changes of isoforms of some antioxidative enzymes. J Plant Physiol 161: 531–554.

40. Fridovich I (1986) Biological effects of superoxide radical. Arch Biochem Biophys 247(1): 1–11.

41. Molassiotis A, Sotiropoulos T, Tanou G, Diamantidis G, Therios L (2006) Boron-induced oxidative damage and antioxidant and nucleolytic responses in shoot tips culture of the apple rootstock EM 9 (Malus domestica Borkh.). Environ Exp Bot 56: 54–62.

42. Kenyon W, Duke S (1985) Effects of Acifluorfen on Endogenous Antioxidants andProtective Enzymes in Cucumber (Cucumis sativus L.)CotyledonsPlant. Physiol 79, 862–866.

43. Nemat MM, Hassan NM, El-Bastawisy M (2008) Changes in antioxidants and kinetics of glutathione-S-transferase of maize in response to isoproturon treatment, Plant Biosystems 142: 5–16.

44. Noctor G, Foyer CH (1998) Ascorbate and glutathione: keeping active oxygen under control. Annu Rev Plant Physiol Plant Mol Biol 49: 249–279.

# mRNA-seq Analysis of the *Gossypium* arboreum transcriptome Reveals Tissue Selective Signaling in Response to Water Stress during Seedling Stage

**Xueyan Zhang**[1⊙], **Dongxia Yao**[2⊙], **Qianhua Wang**[1⊙], **Wenying Xu**[2⊙], **Qiang Wei**[2], **Chunchao Wang**[2], **Chuanliang Liu**[1], **Chaojun Zhang**[1], **Hong Yan**[2], **Yi Ling**[2], **Zhen Su**[2*], **Fuguang Li**[1*]

**1** State Key Laboratory of Cotton Biology, Institute of Cotton Research, Chinese Academy of Agriculture Sciences (CAAS), Anyang, Henan, China, **2** State Key Laboratory of Plant Physiology and Biochemistry, College of Biological Sciences, China Agricultural University, Beijing, China

## Abstract

The cotton diploid species, *Gossypium arboreum*, shows important properties of stress tolerance and good genetic stability. In this study, through mRNA-seq, we *de novo* assembled the unigenes of multiple samples with 3h $H_2O$, NaCl, or PEG treatments in leaf, stem and root tissues and successfully obtained 123,579 transcripts of *G. arboreum*, 89,128 of which were with hits through BLAST against known cotton ESTs and draft genome of *G. raimondii*. About 36,961 transcripts (including 1,958 possible transcription factor members) were identified with differential expression under water stresses. Principal component analysis of differential expression levels in multiple samples suggested tissue selective signalling responding to water stresses. Venn diagram analysis showed the specificity and intersection of transcripts' response to NaCl and PEG treatments in different tissues. Self-organized mapping and hierarchical cluster analysis of the data also revealed strong tissue selectivity of transcripts under salt and osmotic stresses. In addition, the enriched gene ontology (GO) terms for the selected tissue groups were differed, including some unique enriched GO terms such as photosynthesis and tetrapyrrole binding only in leaf tissues, while the stem-specific genes showed unique GO terms related to plant-type cell wall biogenesis, and root-specific genes showed unique GO terms such as monooxygenase activity. Furthermore, there were multiple hormone cross-talks in response to osmotic and salt stress. In summary, our multidimensional mRNA sequencing revealed tissue selective signalling and hormone crosstalk in response to salt and osmotic stresses in *G. arboreum*. To our knowledge, this is the first such report of spatial resolution of transcriptome analysis in *G. arboreum*. Our study will potentially advance understanding of possible transcriptional networks associated with water stress in cotton and other crop species.

**Editor:** Jinfa Zhang, New Mexico State University, United States of America

**Funding:** This work was supported by grants from the National Science Fund for Distinguished Young Scholars (31125020), the National Natural Science Foundation of China (31171276 and 90817006), and the Innovation Scientists and Technicians Troop Construction Projects of Henan Province. The funders had no role in study design, data collection and analysis, decision to publish, or preparation of the manuscript.

**Competing Interests:** The authors have declared that no competing interests exist.

* E-mail: zhensu@cau.edu.cn (ZS); aylifug@hotmail.com (FL)

⊙ These authors contributed equally to this work.

## Introduction

Cotton is an essential crop for producing fiber used in textiles and is also a major oil source. Cotton yield is dramatically reduced under drought and high salinity conditions [1,2,3,4,5,6]. Water stress (mainly including both salt and drought stresses) is a major environmental stress that many plants have to cope with during their whole life cycle [7,8,9,10,11,12]. The water stress signals stimulate leaf abscission [13], and enhance root extension into deeper and moist soil, adjusting the root system architecture (RSA) [14]. There is a functional balance between root-based water uptake and shoot-based photosynthesis [15,16,17]. Generally, high salinity disturbs cytoplasmic $K^+/Na^+$ homeostasis and can result in ion toxicity and osmotic stress, as well as altering growth regulation, etc [8,9,10,18,19].

Compared to stress-susceptible species such as the model plant Arabidopsis (*Arabidopsis thaliana*), cotton is moderately to fairly salt tolerant [20]. In agriculture, plant breeders normally use two tetraploid species (*Gossypium hirsutum* L. and *G. barbadense* L.) and two diploid species (*G. arboreum* L. and *G. herbaceum* L.). The diploid species, especially Asiatic desi cotton (*G. arboreum*), commonly called tree cotton, can be cultivated in severely dry and hot climates, and shows great potential against abiotic and biotic stresses, with good genetic stability and important property in stress tolerance [4,21,22]. *Gossypium arboreum* is an essential source of stress resistance genes [23], e.g. one heat-shock protein GHSP26 from *G. arboreum* was introduced into *G. hirsutum* and transgenic cotton plants showed an enhanced drought tolerance phenotype [24]. Recently, two research groups constructed *G. arboreum* cDNA libraries: one related to drought stress [25], and the other concerning biotic and abiotic stress up-regulated ESTs [23].

Some previous studies have reported possible mechanisms related to cotton water-stress response [15,16,22,26,27,28,29,30,31,32,33,34]. However, the possible regulatory pathways involved in water stress are not well understood in cotton. When exposed to water stress, many plant genes are induced to directly protect against stress or regulate expression of other target genes. Plant transcriptome mapping studies have become a powerful way to reveal the possible mechanism involved in water stress and to dissect the water stress signal transduction pathways and predict genes with biological functions. [32,35,36,37,38,39,40,41,42,43,44,45,46,47,48,49,50].

The key aim of transcriptomics is to catalog all species of transcripts and quantify the changing expression levels of each transcript during development and under different environmental conditions. Microarrays have already become a main platform for profiling gene expression. During the past decade, $\geq 100$ publications have used microarrays to study transcriptomic responses to water stress in about 28 plant species [37]. Our previous work showed an overview of the transcription map of cotton (*G. hirsutum*) roots under salt stress [51]. In addition, increasing number of groups have studied the spatiotemporal dynamic regulation of transcriptional responses to environmental stimuli. For example, Kreps et al. used microarrays to study the transcriptome changes for Arabidopsis in response to salt, osmotic and cold stresses in leaves and roots after 3- and 27-h stress treatments [40]. Nevertheless, the development of high-through-put technology advanced transcriptome analysis for environmental stress together with cell and developmental-stage-specific profiling, leading to identification of high-confidence transcriptional modules. For example, Dinneny et al. developed a comprehensive view of cell-type-specific abiotic stress responses [36]. Their results indicated that the cell identity mediates the abiotic stress response in Arabidopsis roots by studying the transcriptional response to high salinity and iron deprivation in different Arabidopsis root cell layers and developmental stages. Thus, during transcriptome analysis, the spatial and temporal dynamic changes should be considered.

Microarray data involves thousands of plant samples and this platform is anticipated to have a wide range of applications in future transcriptome studies. However there are some limitations during microarray-based transcriptome analysis, e.g. relatively lower intensity, lower dynamic range, higher background, some non-specific hybridization, and biases of labeling. In the mean-while, the defined probe sets of microarrays should use existing genome sequences as reference. The recent application of massively parallel cDNA sequencing (RNA-seq) has complement-ed microarray-based methods for characterization and quantifi-cation of the transcriptome, providing more complete descriptions of transcriptomes and more efficient ways to measure transcrip-tome data with deep coverage and base-level resolution in different organisms [37,52,53,54,55,56,57,58,59,60,61,62,63,64,65]. RNA-seq can also be used on a much wider range of species in studies of water stress, especially for some plant species whose whole genome sequences are not finished yet.

In this study, the diploid cotton species, *G. arboreum*, was selected for transcriptome analysis due to its important properties of stress tolerance. To elucidate possible mechanisms regulating the water stress response of *G. arboreum*, we applied Illumina sequencing technology based mRNA deep sequencing (mRNA-seq) to *de novo* construct transcriptome profiling and to gain a more comprehen-sive understanding of transcriptional processes during water stress in cotton seedlings.

## Results

### De Novo mRNA-seq Assembly Across Different Expression Levels of Leaf, Stem and Root Tissues Under Normal and Water Stress Conditions

The cotton genotype, *G. arboreum* cv. Shixiya, was chosen for this study because of its great potential against abiotic and biotic stresses. The seedling plants were treated by 17% polyethylene glycol (PEG) and 150 mM NaCl (water as mock, CK) for 3 hours, and three tissues including root, stem (including hypocotyl), and leaf, were respectively harvested for mRNA-seq analysis. The experimental design and mRNA-seq procedures are shown in Fig. S1. The total RNA of each sample was isolated individually, and the transcriptome profiles generated through the standard Illumina protocol (detailed description in Materials and Methods). To maximize transcript coverage, we pooled the Illumina read sequences from nine biological conditions during the *G. arboretum* seedling stage: leaf, stem and root tissues treated by all of CK, PEG, and NaCl, respectively. We obtained approximately total 271.6 million clean reads (or 135.8 million paired-end reads) and total roughly 23 Gb nucleotides which passed the Illumina quality filtering (the number of clean reads for each sample is shown in Table 1).

The cotton whole-genome sequencing results are not publically available, thus the *de novo* assembly was carried out using SOAPdenovo, a short reads assembling program [66]. SOAPde-novo firstly combined clean reads from each sample with 29-mer overlap to form contigs; secondly connected the contigs to make scaffolds with the insertion information of the paired-end reads; then sequence clustering software (TGICL: http://sourceforge. net/projects/tgicl/) was used to connect scaffolds to unigenes which could not be extended on either end; finally, unigenes from each sample's assembly were taken into TGICL again to acquire non-redundant All-Unigenes (here termed 'transcripts'). The number of contigs, scaffolds, and unigenes for each sample are shown in Table 1. There were 56–76 k unigenes for each sample and the total length of assembly unigenes were 27–38 M, and the sequence depth of each sample was from 62× to 90×. In total, we got 123,579 transcripts with lengths $\geq 200$ bp. The total length of all transcripts was approximately 76.6 Mb (we obtained sequence depth of about 300×), the N50 was 1,065 bp, and there were 21,253 transcripts of $\geq 1$ kb in length. The length distribution of these 123,579 transcripts is shown in Fig. 1A and gap distribution in Fig. 1B. There were $\geq 60\%$ transcripts without gaps (Ns), and $\leq 5\%$ of transcripts with $\geq 20\%$ gaps.

### Functional Annotation of the Assembled Cotton Transcripts

Following the mRNA-seq *de novo* assembly, the functional annotation process for these transcripts was mainly based on homolog search. To obtain the possible annotation and predict the sequence direction, BLAST (blastx alignment, e-value cutoff as $10^{-6}$) was used to search the best aligning results for transcripts against protein databases like nr (in NCBI), Swiss-Prot (in UniProt), KEGG (www.genome.jp/kegg/) and COG (http://www.ncbi.nlm.nih.gov/COG/). For transcripts with no homolog hit, ESTScan [67] was applied to predict the coding regions as well as to determine sequence direction. In the total of 123,579 transcripts, there were 81,369 sequences with determinable direction. In addition, we compared the transcripts with known cotton ESTs from NCBI and DFCI (http://compbio.dfci.harvard. edu/tgi/plant.html); there were 75,855 transcripts matching the known cotton ESTs. There were also 74,573 transcripts matching the protein-coding genes in the recently published draft genome of

**Table 1.** The data quality of mRNA-seq and assembly.

| Samples | Total reads | Total nucleotides (nt) | All contig | All scafford | All unigene | Length of all unigene (nt) |
|---|---|---|---|---|---|---|
| Leaf-Mock | 2.671E+07 | 2.404E+09 | 334,935 | 98,372 | 60,608 | 38,613,473 |
| Leaf-PEG | 2.756E+07 | 2.480E+09 | 484,539 | 111,592 | 66,371 | 35,425,037 |
| Leaf-NaCl | 2.673E+07 | 2.406E+09 | 481,884 | 109,592 | 65,608 | 37,796,363 |
| Stem-Mock | 2.676E+07 | 2.408E+09 | 378,039 | 113,832 | 71,548 | 35,658,125 |
| Stem-PEG | 3.852E+07 | 3.467E+09 | 415,789 | 126,267 | 76,860 | 38,206,259 |
| Stem-NaCl | 2.960E+07 | 2.664E+09 | 368,169 | 113,447 | 70,597 | 34,352,402 |
| Root-Mock | 3.229E+07 | 2.422E+09 | 180,867 | 96,615 | 57,723 | 29,210,215 |
| Root-PEG | 3.107E+07 | 2.330E+09 | 163,186 | 91,888 | 56,054 | 29,180,862 |
| Root-NaCl | 3.237E+07 | 2.428E+09 | 175,985 | 93,498 | 56,780 | 27,481,197 |

*G. raimondii* [68]. Take this into account, there were 34,451 remaining transcripts which may be considered as newly discovered transcripts.

KEGG annotation provides information of transcripts related to metabolic process and functions in cellular processes. We summarized the KEGG pathway distribution of the transcripts (Fig. 2C), which showed that 20,071 transcripts (several transcripts hit multiple pathways) mapped to 117 pathways belong to all five categories of KEGG, including metabolism, genetic information processing, environmental information processing, cellular processes, and organismal systems.

Gene ontology (GO) is an international standardized gene functional classification which can provide a biological foundation on global characterization of *de novo* assembly transcripts. With nr annotation, we used Blast2GO program [69] to obtain GO annotation of the transcripts. There were 20,008 transcripts with GO annotation (total 96,911 matches). The major categories of GO category distribution were Biological Process (BP), Molecular Function (MF), and Cellular Component (CC) (Fig. 2D).

To compare the expression level of individual transcripts in different samples, we used the *de novo* assembled sequences as a reference for short-read mapping. SOAPaligner/soap2, a short-read alignment program [66], was used to map the uniquely aligned reads on to the 123,579 transcript sequences. To eliminate the influence of different transcript length and sequencing level on the calculation, the RPKM method (Reads Per kb per Million reads) [70] was used for normalization and the result directly used for comparing the difference of gene expression between samples. Of individual samples, about 60–70% of transcripts could be detected, in which there at least one read was uniquely aligned to the transcript sequence.

## Principal Component Analysis (PCA) of Differential Expression Levels in Multiple Samples and Real-time RT-PCR Validation

During mRNA-seq transcriptome analysis, the spatial resolution of cotton response to water stress was investigated due to the tissues exhibiting different levels of gene expression. The transcripts fell along the diagonal region of pair-wise scatter plots between the nine samples (Fig. S2) indicating no major variation between the pairs; whereas some transcripts fell above or below diagonal lines, indicating their differential expression level during different tissue sample and treatment conditions. The largest difference among the nine samples was observed between tissues.

The principal component analysis (PCA) for the samples, based on the raw expression level (Fig. 1E), showed the first three principal components accounted for 63.8% of the variation; all three treatments (mock, PEG, and NaCl) in each cotton tissue (leaf, stem, and root) were clustered together in different vertical planes, the distances between treatments in root samples were much greater than for leaf and stem tissues.

To validate the mRNA-seq results, we selected 12 transcripts with differential expression patterns for real-time RT-PCR analysis and made one-by-one comparisons of each transcript between real-time RT-PCR and mRNA-seq results (Fig. 2). We calculated Pearson correlation coefficient (r) between the real-time RT-PCR values and RPKMs of mRNA-seq across nine samples for each transcript. Among these 12 transcripts, the correlation coefficient range was 0.78–0.995 ($P \leq 0.05$), and seven transcripts had the correlation coefficient >0.9. The majority of real-time RT-PCR results matched the mRNA-seq expression patterns.

## Differential Expression Analysis of Assembled Cotton Transcripts Under NaCl and PEG Treatments in Different Tissues

We further conducted differential expression analysis for the transcripts responding to PEG and salt stresses in cotton leaf, stem, and root samples, respectively. Referring to Audic's algorithm [71], we calculated the FDR (False Discovery Rate) based on the *P*-value which corresponds to differential expression tests of transcripts. Using "FDR $\leq 0.001$ and the absolute value of $\log_2$Ratio $\geq 1$" as the threshold, we identified total 36,961 transcripts either up- or down-regulated under 150 mM NaCl or 17% PEG treatment conditions in at least one tissue sample of cotton leaf, stem, and root. The numbers for six comparisons and the detailed information for each transcript are separately shown in Fig. S2 and Table S2. In cotton leaf samples, 8,981 transcripts were up-regulated and 5,109 transcripts down-regulated under PEG treatment; and 4,884 up-regulated and 3,692 down-regulated under salt treatment. In cotton stem samples, 1,430 transcripts were up-regulated and 1,717 down-regulated under PEG treatment; and 2,462 up-regulated and 3,859 down-regulated under salt treatment. In cotton root samples, 4,137 transcripts were up-regulated and 8,568 down-regulated under PEG treatment; and 6,303 up-regulated and 11,068 down-regulated under salt treatment. Among the three tissues, there were much greater numbers of differentially expressed transcripts under water stress in root samples, and the lowest in stem samples. In cotton root

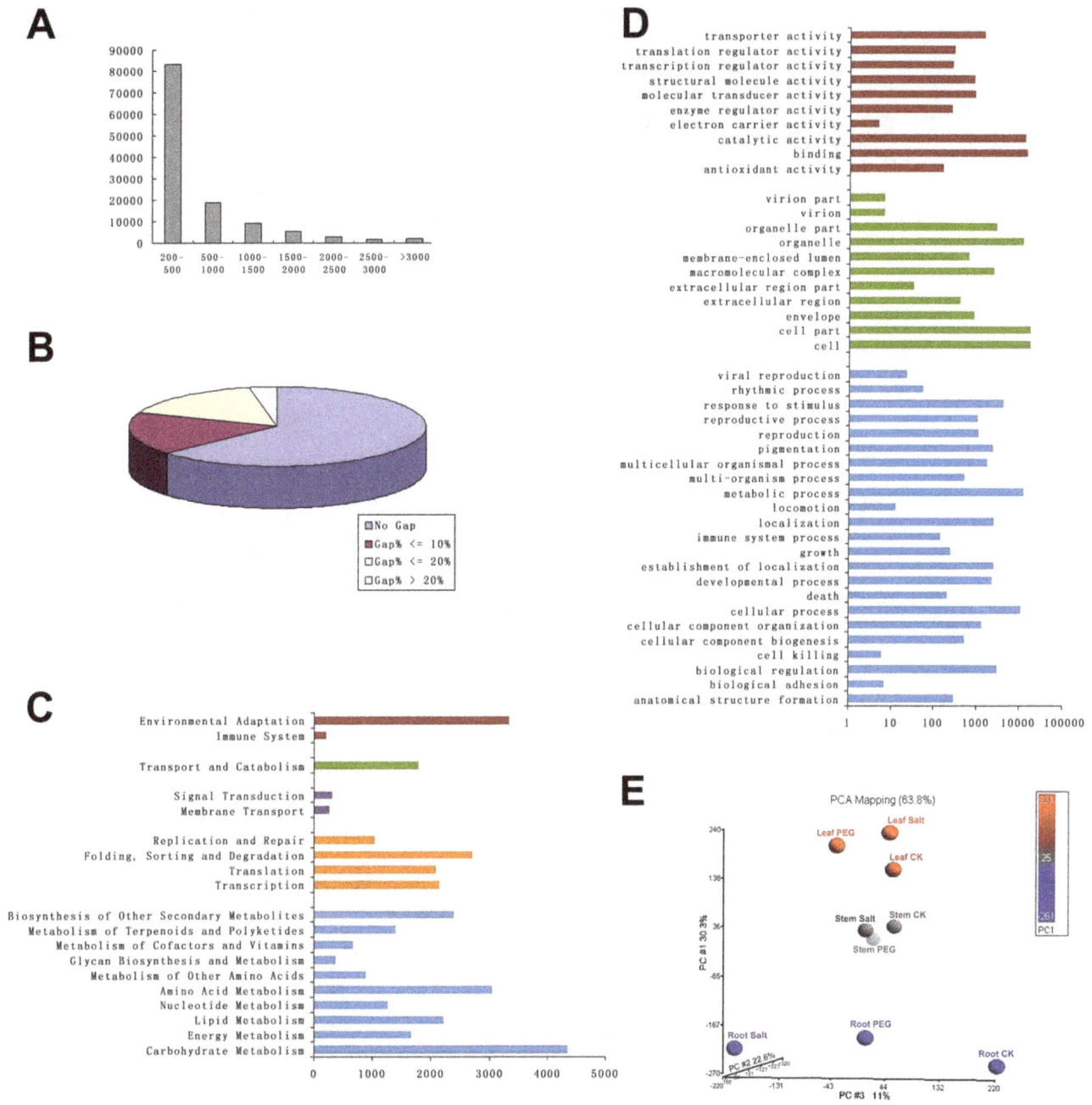

**Figure 1. Data quality and annotation of transcripts assembled with mRNA-seq. A:** Length distribution of transcripts. **B:** Gap distribution of transcripts. **C:** Function annotation of the transcripts based on KEGG classification. The numbers of transcripts mapped to each pathway group are shown in the bar chart. The color indicates different KEGG categories: blue for metabolism, orange for genetic information processing, purple for environmental information processing, green for cellular processes, and red for organismal systems. **D:** Gene ontology (GO) classification of the transcripts. Each bar represents the number of transcripts mapped to each GO category. The color indicates different GO categories: blue for biological process (BP), green for cellular component (CC), and red for molecular function (MF). **E:** Principle components analysis (PCA) for the samples based on the raw reads of the transcripts. The red balls represent leaf samples, the gray balls represent stem samples, and the blue balls represent root samples. The color indicates the number range in PC #1 (Principal Component 1) shown in the color bar.

samples, more transcripts were down-regulated by PEG and salt stress than transcripts up-regulated; whereas in leaf samples, more transcripts were up-regulated by PEG and salt stress than those down-regulated.

Venn diagram analysis showed the specificity and intersection of transcripts' response to NaCl and PEG treatments among the leaf, stem, and root tissues (Fig. 3A). For the different tissues, large

numbers of the transcripts' response to NaCl and PEG treatments overlapped in the same direction. For example, in cotton root tissue, the overlap number of up-regulated transcripts was 1,829 (about 44% of the 4,137 PEG up-regulated transcripts); for the down-regulated transcripts, the overlap number was 5,166 transcripts (>60% of the 8,568 PEG down-regulated transcripts); very few numbers (227 plus 150) of transcripts responded to NaCl

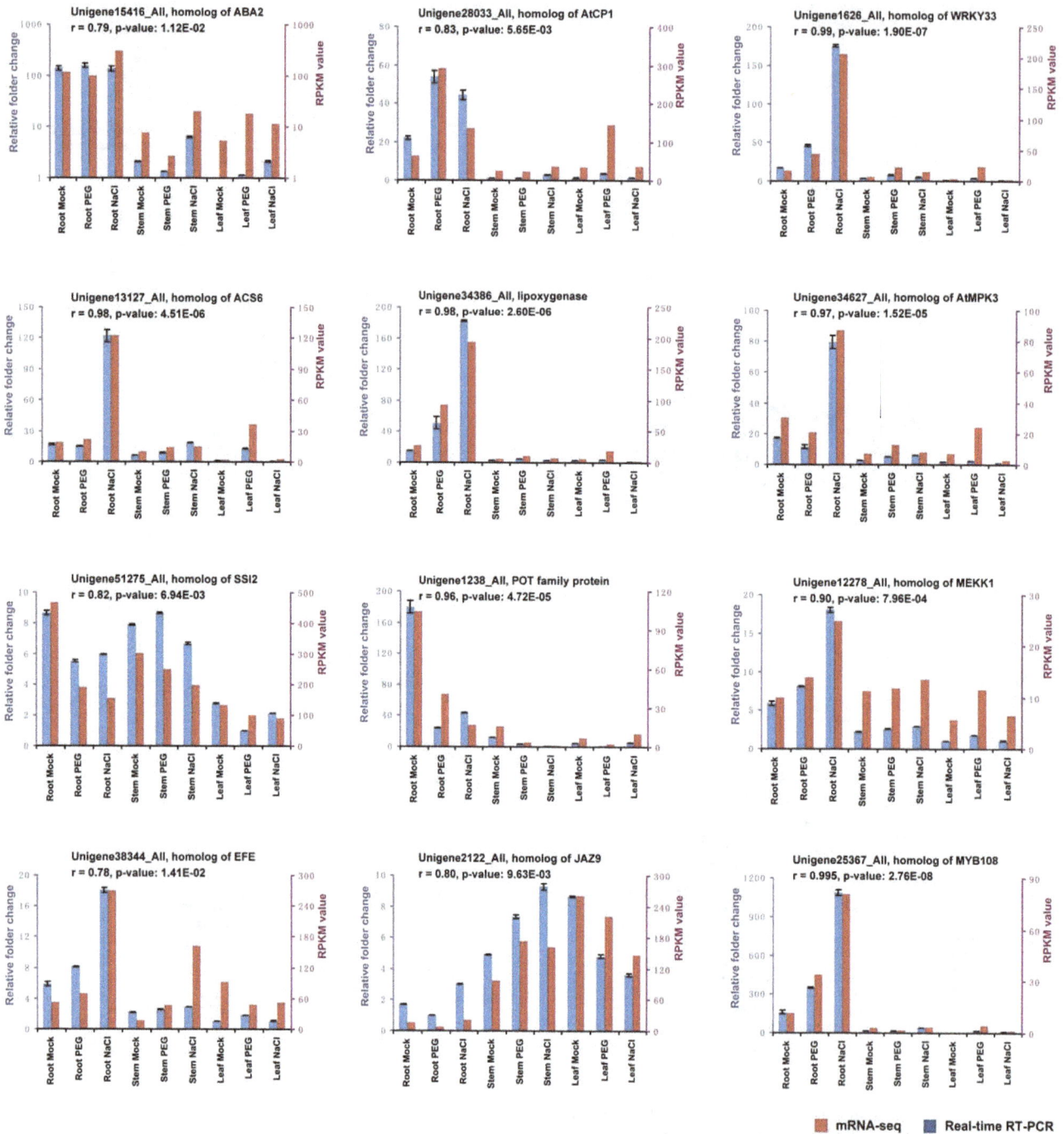

**Figure 2. Real-time RT-PCR validation for selected transcripts.** Twelve transcripts were selected for real-time RT-PCR to validate the expression patterns in different samples and treatments. The blue bars represent the relative intensity of real-time RT-PCR from independent biological replicates (using the left y-axis), the red bars represent the expression level (RPKM) of the transcript (using the right y-axis). The correlation coefficient (r) and its P-value between the RT-PCR values and RPKMs for each transcript are listed in each individual chart. The transcripts are: Unigene15416_All hits AT1G52340.1 (ABA2); Unigene28033_All hits AT5G49480.1 (ATCP1); Unigene1626_All hits AT2G38470.1 (WRKY33); Unigene13127_All hits AT4G11280.1 (ACS6); Unigene34386_All hits AT1G72520.1 (lipoxygenase); Unigene34627_All hits AT3G45640.1 (ATMPK3); Unigene51275_All hits AT2G43710.1 (SSI2); Unigene1238_All hits AT1G32450.1 (POT family protein); Unigene12278_All hits AT4G08500.1 (MEKK1); Unigene38344_All hits AT1G05010.1 (EFE); Unigene2122_All hits AT1G70700.3 (JAZ9); and Unigene25367_All hits AT3G06490.1 (MYB108). The real-time RT-PCR primers for each transcript are listed in Table S1.

and PEG treatments in opposite ways. The intersection trends were similar for stem and leaf to that in root tissue.

## Clustering and GO Analysis of the Differentially Expressed Transcripts

To further identify the co-expressed transcripts with similar response patterns to the NaCl and PEG treatments in different

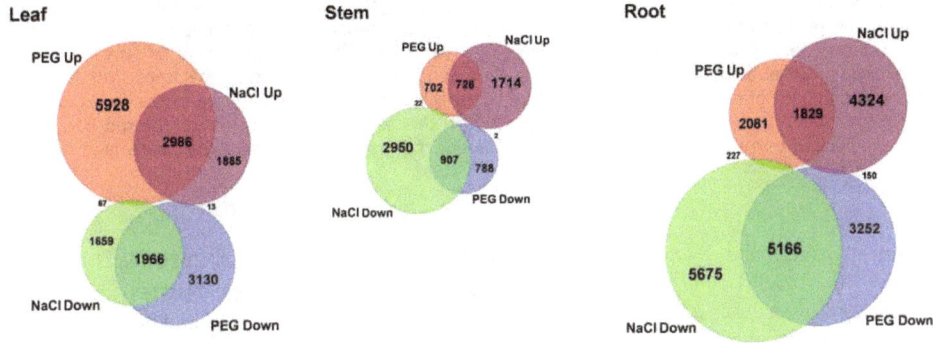

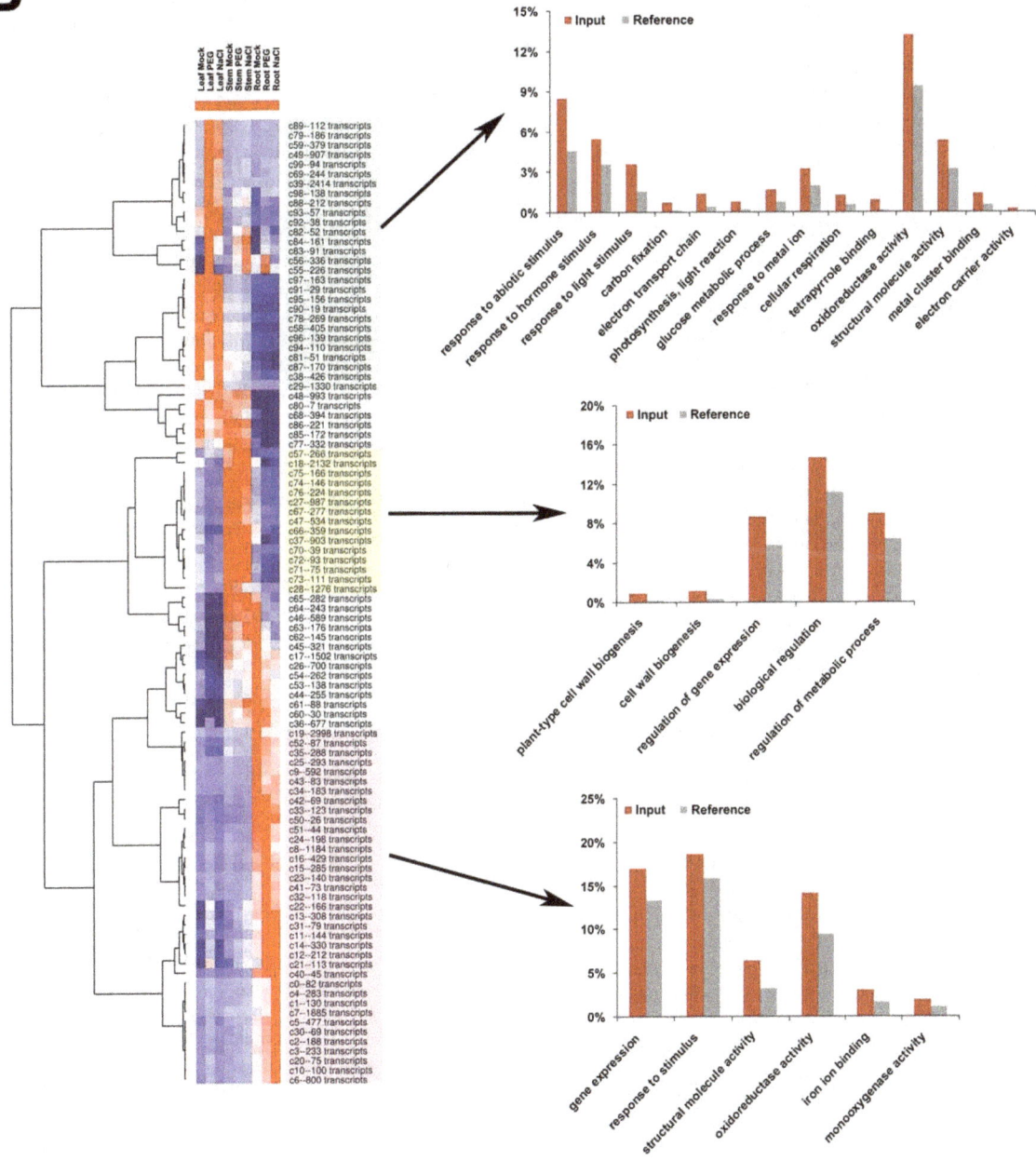

**Figure 3. Summary of the differential expression transcripts across cotton tissue samples and treatments. A:** Venn diagrams illustrate the differential expression transcripts under PEG and NaCl treatment in root, stem, and leaf samples. The red and blue colors represent the up-regulated and down-regulated transcripts under PEG treatment, respectively. The purple and green colors represent the up-regulated and down-regulated transcripts under NaCl treatment, respectively. **B:** Cluster and gene ontology (GO) analysis of the transcripts' response to PEG or NaCl treatments in different cotton tissues. The overview hierarchical cluster result of the centroids of SOM cluster (listed in Fig. S4), the red (high) and blue (low) colors represent the relative expression level across the samples. The marked centroid groups represent the transcripts in these clusters preferentially expressed in the leaf (marked in light green background), stem (light yellow background), or root (light purple background) tissue. In the comparison of the enriched GO terms in the transcripts preferentially expressed in the leaf, stem, or root tissue, the red bars represent the percentage of transcripts belonging to enriched terms in the query list (Input), whereas the gray bars represent the percentage in all transcripts (Reference).

tissues, both SOM (self-organized mapping) and hierarchical methods were used for clustering the 36,961 differentially expressed transcripts. A 10×10 SOM cluster was applied to divide the 36,961 transcripts into 100 clusters: the cluster name and the number of transcripts in each cluster were listed, the expression pattern of the centroid presented and the variances shown (Fig. S3). The SOM classified that the 36,961 transcripts into different groups with various centroids, some of which showed very similar patterns. Further classification of these centroids may provide a much clearer picture of tissue selective signalling in response to water stress in cotton. We applied hierarchical method to cluster the centroids (Fig. 3B). A heat map represented the relative expression level of the centroids in nine cotton samples. Intuitively, these centroids could be grouped into multiple clusters, for example, the top 28 centroids (marked in light green; Fig. 3B), represent the transcripts that mainly responded to NaCl or PEG treatment in leaf tissue; whereas the bottom 37 centroids (marked in light purple; Fig. 3B), representing the transcripts that mainly responded to NaCl or PEG treatment in root tissue.

In addition, we applied GO analysis (using agriGO; http://bioinfo.cau.edu.cn/agriGO/) to the selected centroids' groups (Fig. 3B). Interestingly, many of the enriched GO terms for the selected three groups differed. For the centroids' groups representing the transcripts that were preferentially expressed in stem tissue (marked in light yellow; Fig. 3B), the enriched GO terms were mainly related to plant cell wall (plant-type cell wall biogenesis, FDR $P$-value: 5.80E−04; and cell wall biogenesis, FDR $P$-value: 4.40E−03) and regulation of metabolic process (FDR $P$-value: 5.60E−03). Regarding the transcripts preferentially expressed in leaf tissue (the 28 clusters including 8914 transcripts), the enriched biology process type GO terms include "response to abiotic stimulus" (FDR $P$-value: 2.00E−10), "response to light stimulus" (FDR $P$-value: 3.90E−07), "response to hormone stimulus" (FDR $P$-value: 2.60E−03), "carbon fixation" (FDR $P$-value: 2.00E−04), "photosynthesis, light reaction" (FDR $P$-value: 1.40E−03), "electron transport chain" (FDR $P$-value: 2.70E−04), "glucose metabolic process" (FDR $P$-value: 1.20E−02), "response to metal ion" (FDR $P$-value: 2.20E−02), "cellular respiration" (FDR $P$-value: 2.40E−02). The enriched molecular function type GO terms include "tetrapyrrole binding" (FDR $P$-value: 1.20E−05), "oxidoreductase activity" (FDR $P$-value: 3.60E−05), "structural molecule activity" (FDR $P$-value: 2.60E−04), "metal cluster binding" (FDR $P$-value: 9.20E−03), "electron carrier activity" (FDR $P$-value: 4.20E−02). Of transcripts preferentially expressed in root tissue, the enriched GO terms mainly include "response to stimulus" (FDR $P$-value: 4.50E−02), "structural molecule activity" (FDR $P$-value: 2.30E−15), "oxidoreductase activity" (FDR $P$-value: 5.00E−14), "iron ion binding" (FDR $P$-value: 6.10E−05), and "monooxygenase activity" (FDR $P$-value: 2.20E−02), etc.

## Transcription Factors Responding to NaCl and PEG Treatments in Different Cotton Tissues

Transcription is a dynamic process and transcription factors are essential for regulation of gene expression. In this study, we also performed global transcription factor classification for differentially expressed transcripts and identified a total of 4,002 transcripts (56 transcription factor families; Table S2). About 49% of transcription factor members responded to NaCl or PEG treatment in cotton tissues. Several key regulatory gene families involved in responding to abiotic and biotic sources of stress such as AP2-EREBP (62.03%), WRKY (61.78%), ABI3VP1 (58.73%), Tify (55.88%), bHLH (55.45%), MYB (54.46%), NAC (52.48%), bZIP (52.50%), EIL (50.00%), HSF (50.00%), and C2C2-YABBY(50.00%), were largely up- or down-regulated under NaCl or PEG in at least one tissue (Table 2). Several transcription factor family genes; e.g. WRKY, NAC, and Tify (including the cotton JAZ genes) were up-regulated by salt and PEG treatments. For example, there were 64 WRKY members up-regulated versus 12 down-regulated under NaCl treatment in root tissue; there were 50 members up-regulated versus nine down-regulated under PEG treatment in leaf tissue. In some other transcription factor families, e.g. C2C2-GATA, more family members were down-regulated by NaCl and PEG treatments. For MYB in root tissue, the up- and down-regulated members were evenly balanced.

## Transcripts Related to Hormone Signaling Pathways, Responding to NaCl and PEG Treatment in Different Cotton Tissues

Phytohormones play essential roles in the ability of plants to adapt to water stress by mediating various adaptive responses. We mapped the assembled mRNA-seq transcripts to eight plant hormones' signalling transduction pathways in KEGG, including auxin, cytokinine (CK), gibberellin (GA), abscisic acid (ABA), ethylene, brassinosteroid (BR), jasmonate (JA), and salycylic acid (SA). The overview of gene expression patterns under NaCl and PEG treatment in three cotton tissues (Fig. 4) showed the proportion of the transcripts related to hormone signaling up-regulated and down-regulated under NaCl and PEG treatments in cotton tissues.

ABA signaling and ABA- responsive genes were among the most studied topics in the response of plants to water stress. In the ABA signal transduction pathway, a large number of assembled cotton transcripts showed significantly differential expression under NaCl and PEG treatments with no tissue selectivity (Fig. 4D). The majority of changed *PYR/PYL* homologs were down-regulated, while most *PP2C* and *ABF* homologs were up-regulated under NaCl and PEG treatments in all cotton seedling tissues. Many homologs of *SnRK2* were up-regulated by NaCl and PEG in leaf tissue, and mainly down-regulated by NaCl in root tissue.

As well as the well-known stress-responsive ABA, other plant hormones are also involved in salt and osmotic stresses; and the roles of other hormones during water stress are emerging. In the

**Table 2.** Transcription factor (TF) members that responded to PEG and NaCl treatments in cotton tissues.

| Transcription factor family | Total number | Changed members (%) | Change | Leaf | | Stem | | Root | |
|---|---|---|---|---|---|---|---|---|---|
| | | | | NaCl | PEG | NaCl | PEG | NaCl | PEG |
| AP2-EREBP | 316 | 196 (62.03%) | Up | 23 | 74 | 37 | 23 | 90 | 51 |
| | | | Down | 32 | 19 | 19 | 11 | 21 | 35 |
| WRKY | 191 | 118 (61.78%) | Up | 9 | 50 | 21 | 17 | 64 | 48 |
| | | | Down | 7 | 9 | 8 | 6 | 12 | 7 |
| GRF | 44 | 27 (61.36%) | Up | 2 | 7 | 0 | 4 | 2 | 1 |
| | | | Down | 11 | 7 | 6 | 2 | 5 | 12 |
| C2C2-GATA | 54 | 33 (61.11%) | Up | 1 | 6 | 1 | 1 | 0 | 2 |
| | | | Down | 10 | 5 | 7 | 1 | 15 | 18 |
| CSD | 18 | 11 (61.11%) | Up | 3 | 4 | 0 | 1 | 1 | 3 |
| | | | Down | 2 | 2 | 1 | 0 | 3 | 3 |
| ABI3VP1 | 63 | 37 (58.73%) | Up | 3 | 5 | 1 | 2 | 5 | 5 |
| | | | Down | 4 | 6 | 5 | 3 | 11 | 11 |
| DBP | 12 | 7 (58.33%) | Up | 0 | 4 | 2 | 1 | 7 | 3 |
| | | | Down | 0 | 0 | 0 | 0 | 0 | 0 |
| DBB | 19 | 11 (57.89%) | Up | 0 | 2 | 1 | 0 | 6 | 4 |
| | | | Down | 1 | 2 | 0 | 0 | 1 | 1 |
| zf-HD | 26 | 15 (57.69%) | Up | 0 | 2 | 0 | 3 | 4 | 2 |
| | | | Down | 2 | 1 | 5 | 1 | 0 | 0 |
| Tify | 34 | 19 (55.88%) | Up | 6 | 9 | 5 | 2 | 12 | 1 |
| | | | Down | 0 | 0 | 0 | 0 | 0 | 4 |
| bHLH | 330 | 183 (55.45%) | Up | 8 | 21 | 18 | 11 | 35 | 26 |
| | | | Down | 30 | 39 | 39 | 14 | 42 | 49 |
| C2C2-CO-like | 67 | 37 (55.22%) | Up | 4 | 8 | 5 | 6 | 10 | 8 |
| | | | Down | 2 | 7 | 10 | 4 | 2 | 5 |
| ARR-B | 40 | 22 (55.00%) | Up | 2 | 1 | 5 | 1 | 3 | 0 |
| | | | Down | 4 | 4 | 2 | 1 | 7 | 8 |
| MYB | 303 | 165 (54.46%) | Up | 25 | 41 | 25 | 17 | 43 | 38 |
| | | | Down | 14 | 21 | 29 | 10 | 45 | 37 |
| PLATZ | 32 | 17 (53.13%) | Up | 5 | 7 | 0 | 0 | 4 | 2 |
| | | | Down | 1 | 2 | 2 | 1 | 5 | 6 |
| bZIP | 160 | 84 (52.50%) | Up | 11 | 26 | 12 | 8 | 23 | 23 |
| | | | Down | 15 | 16 | 18 | 12 | 16 | 5 |
| NAC | 202 | 106 (52.48%) | Up | 15 | 37 | 18 | 8 | 38 | 45 |
| | | | Down | 9 | 9 | 15 | 11 | 20 | 18 |
| C2C2-YABBY | 12 | 6 (50.00%) | Up | 0 | 0 | 0 | 3 | 0 | 0 |
| | | | Down | 2 | 3 | 0 | 0 | 0 | 0 |
| EIL | 22 | 11 (50.00%) | Up | 1 | 9 | 0 | 0 | 1 | 2 |
| | | | Down | 0 | 0 | 0 | 0 | 0 | 0 |
| HSF | 66 | 33 (50.00%) | Up | 8 | 12 | 5 | 3 | 15 | 10 |
| | | | Down | 4 | 1 | 3 | 1 | 4 | 4 |
| C2C2-Dof | 87 | 43 (49.43%) | Up | 4 | 10 | 3 | 0 | 5 | 2 |
| | | | Down | 5 | 13 | 14 | 3 | 11 | 4 |
| TCP | 55 | 27 (49.09%) | Up | 1 | 3 | 0 | 0 | 1 | 0 |
| | | | Down | 5 | 9 | 11 | 0 | 8 | 4 |
| G2-like | 100 | 48 (48.00%) | Up | 3 | 6 | 1 | 0 | 15 | 7 |
| | | | Down | 4 | 10 | 8 | 5 | 12 | 9 |
| SBP | 55 | 26 (47.27%) | Up | 4 | 4 | 0 | 1 | 4 | 2 |
| | | | Down | 0 | 3 | 12 | 1 | 12 | 1 |

**Table 2.** Cont.

| Transcription factor family | Total number | Changed members (%) | Change | Leaf NaCl | Leaf PEG | Stem NaCl | Stem PEG | Root NaCl | Root PEG |
|---|---|---|---|---|---|---|---|---|---|
| GRAS | 140 | 66 (47.14%) | Up | 3 | 13 | 9 | 5 | 15 | 10 |
|  |  |  | Down | 11 | 14 | 11 | 3 | 24 | 15 |
| C2H2 | 245 | 114 (46.53%) | Up | 10 | 17 | 9 | 16 | 34 | 21 |
|  |  |  | Down | 15 | 24 | 18 | 12 | 36 | 23 |
| HB | 282 | 127 (45.04%) | Up | 11 | 36 | 10 | 8 | 29 | 24 |
|  |  |  | Down | 15 | 16 | 29 | 5 | 30 | 21 |
| LIM | 23 | 10 (43.48%) | Up | 2 | 3 | 0 | 0 | 0 | 0 |
|  |  |  | Down | 1 | 2 | 3 | 0 | 5 | 3 |
| MADS | 67 | 29 (43.28%) | Up | 10 | 9 | 10 | 8 | 5 | 4 |
|  |  |  | Down | 1 | 2 | 3 | 3 | 4 | 3 |
| MYB-related | 137 | 57 (41.61%) | Up | 8 | 11 | 5 | 2 | 18 | 9 |
|  |  |  | Down | 7 | 11 | 8 | 7 | 13 | 12 |
| CCAAT | 72 | 29 (40.28%) | Up | 5 | 11 | 2 | 1 | 4 | 8 |
|  |  |  | Down | 2 | 2 | 3 | 1 | 9 | 2 |
| TAZ | 18 | 6 (33.33%) | Up | 1 | 0 | 1 | 0 | 5 | 0 |
|  |  |  | Down | 0 | 0 | 0 | 0 | 0 | 1 |
| CAMTA | 22 | 7 (31.82%) | Up | 0 | 2 | 0 | 0 | 4 | 2 |
|  |  |  | Down | 0 | 1 | 0 | 0 | 1 | 0 |
| E2F-DP | 22 | 7 (31.82%) | Up | 0 | 0 | 0 | 0 | 0 | 0 |
|  |  |  | Down | 5 | 3 | 2 | 0 | 3 | 3 |
| C3H | 202 | 64 (31.68%) | Up | 9 | 24 | 9 | 7 | 9 | 3 |
|  |  |  | Down | 3 | 9 | 1 | 2 | 13 | 15 |
| ARF | 94 | 28 (29.79%) | Up | 1 | 3 | 2 | 1 | 6 | 0 |
|  |  |  | Down | 3 | 5 | 5 | 1 | 9 | 7 |
| BES1 | 14 | 4 (28.57%) | Up | 0 | 1 | 0 | 0 | 0 | 0 |
|  |  |  | Down | 0 | 0 | 3 | 0 | 1 | 1 |
| Trihelix | 62 | 17 (27.42%) | Up | 0 | 2 | 1 | 1 | 6 | 4 |
|  |  |  | Down | 3 | 3 | 1 | 1 | 6 | 5 |
| Other | 294 | 111 (37.76%) | Up | 7 | 21 | 6 | 4 | 25 | 14 |
|  |  |  | Down | 13 | 14 | 21 | 2 | 28 | 22 |
| **Total** | **4002** | **1958 (48.93%)** | **Up** | **205** | **501** | **224** | **165** | **548** | **384** |
|  |  |  | **Down** | **243** | **294** | **322** | **124** | **434** | **374** |

auxin signal transduction pathway (Fig. 4A), we found that most homologs of *ARF* and *AUX1* were down-regulated under both NaCl and PEG treatments; transcripts for *TIR1* were up-regulated by NaCl and down-regulated by PEG in root tissue. Recently, an interactive feedback loop between auxin and CK signaling was discovered, possibly balancing CK and auxin concentration in developing root and shoot tissues. The *AHP* transcripts showed tissue selectivity during NaCl and PEG treatments, with up-regulation in leaf tissue and most down-regulation in root tissue. The majority of changed *ARR* transcripts were down-regulated (Fig. 4B). For both GA and BR signals, very few genes were differentially expressed under NaCl and PEG treatments in stem tissue. Although there were relatively lower numbers of differentially expressed transcripts in our data sets for the GA signal transduction pathway (Fig. 4C), the transcripts of *GID1* were significantly up-regulated under both NaCl and PEG treatments in leaf tissue. For the BR signal transduction pathway (Fig. 4F), the changed transcripts showed tissue selectivity and were mostly down-regulated under NaCl and PEG treatments, except that *BKI1* was up-regulated.

For the ethylene signal transduction pathway (Fig. 4E), the majority of changed transcripts were up-regulated, e.g. homologs of *ETR*, *CTR*, *MPK*, and *EIN3*. There was also tissue selectivity: e.g., the most differentially expressed *ETR* homologs were up-regulated under NaCl treatment in stem and root, but not leaf tissue; while in response to PEG treatment, the changed *ETR* homologs were only up-regulated in leaf and root tissues. Many *EIN3* homologs were up-regulated by PEG in leaf tissue, whereas they rarely responded to stresses in other tissues. In the JA signal transduction pathway (Fig. 4G), a large proportion of JA-related transcripts were up-regulated under NaCl in three tissues. However, for PEG treatment, there was tissue selectivity, especially for *JAZ*s, with most changed transcripts up-regulated in leaf and down-regulated in root tissue. Some transcripts for the

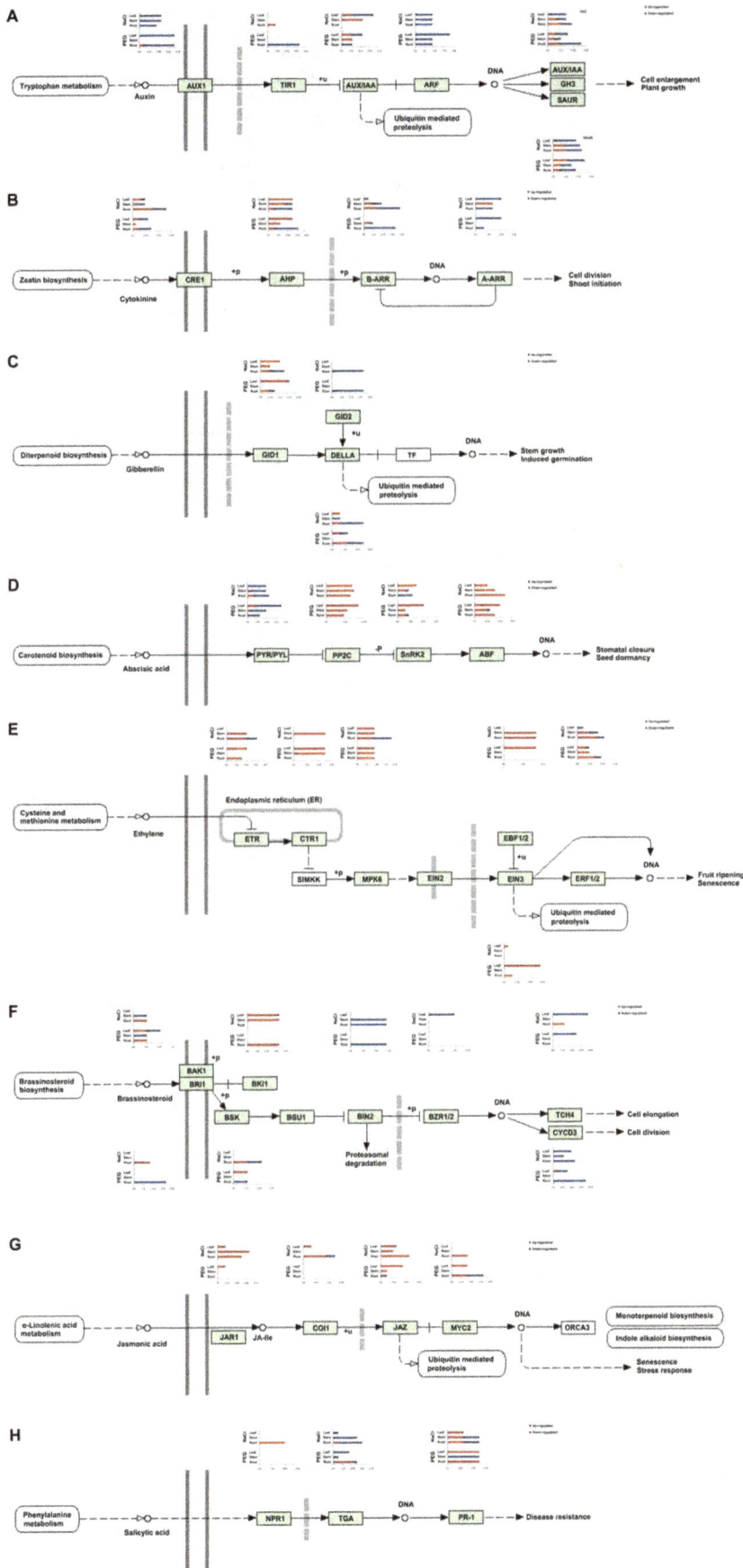

**Figure 4. Summary of transcripts related to plant hormone signal transduction pathways and their response to PEG and NaCl treatments across cotton tissue samples.** The colored bars represent the percentage of the transcripts in each bin (re-annotated to MapMan classification) whether up-regulated (red) or down-regulated (blue) under PEG or NaCl treatment in different cotton tissues. **A**: represents the differential expression transcripts related to auxin (IAA) signalling transduction pathway. **B**: represents the differential expression transcripts related to cytokinin (CK) signalling transduction pathway. **C**: represents the differential expression transcripts related to gibberellin (GA) signalling transduction pathway. **D**: represents the differential expression transcripts related to abscisic acid (ABA) signalling transduction pathway. **E**: represents the differential expression transcripts related to ethylene signalling transduction pathway. **F**: represents the differential expression transcripts related to brassinosteroid (BR) signalling transduction pathway. **G**: represents the differential expression transcripts related to jasmonate (JA) signalling transduction pathway. **H**: represents the differential expression transcripts related to salicylic acid (SA) signalling transduction pathway.

SA signal transduction pathway also showed differential expression patterns (Fig. 4H), including homologs for *NPR1*, *TGA* and *PR-1*: e.g., the majority of changed *TGA* homologs were down-regulated under NaCl treatment in three tissues; while under PEG treatment, the changed *TGA* homologs were down-regulated in leaf and stem but up-regulated in root tissue. Cotton *NPR1* transcripts were only up-regulated under NaCl treatment in roots. There were up- and down-regulated *PR-1* transcripts under NaCl treatment; however, under PEG treatment, these were up-regulated in both leaf and root and down-regulated in stem tissue. Normally, ethylene, JA and SA form a complex network related to disease resistance and are mainly involved in biotic stress. Our differential expression analysis results indicated crosstalk between abiotic and biotic stresses mediating stress hormones, e.g., ABA, ethylene, JA and SA.

## Comparative Transcriptome Analysis between Cotton and Arabidopsis Transcripts Related to Hormone Signaling Pathways Under Water Stresses

In order to reveal the similarity and uniqueness of cotton drought and salt responses compared to other plant species, we conducted comparative transcriptome analysis between cotton and Arabidopsis responding to different water stresses, mainly focusing on the transcripts related to selected hormone signalling pathways. Through homolog search with BLAST tool, we mapped the cotton transcripts to Arabidopsis genes (TAIR9 version in http://www.arabidopsis.org). There were 61,631 cotton transcripts matching Arabidopsis genes with e-value cutoff as $10^{-6}$ (listed in Table S2).

The Arabidopsis transcriptome data were from publicly available Arabidopsis AtGenExpress project (stress treatments data were downloaded from GEO, http://www.ncbi.nlm.nih.gov/geo/and TAIR, http://www.arabidopsis.org) and we used the data sets for both root and shoot tissues under 3 h salt and osmotic treatments. The differentially expressed probe sets were calculated and mapped to plant hormone signaling pathways with MapMan. The individual gene expression patterns for the selected hormone signaling pathways (ABA, JA, IAA, and ethylene) were compared between different cotton and Arabidopsis tissues under water stresses (Fig. S4).

In the ABA signal transduction pathway (Fig. S4A), the transcripts in cotton and Arabidopsis were shown with similar proportion trend in the tissues under water stresses. The majority of changed *PYR/PYL* genes were down-regulated in Arabidopsis and cotton root and shoot tissues under salt and osmotic stresses. As to *PP2Cs* and *ABFs*, majority of their members in cotton and Arabidopsis were up-regulated by salt and osmotic stresses. These results indicated that the transcripts in cotton shared similar ABA signalling pathway as those in Arabidopsis responding to water stresses, while there was slight difference in the members of SnRK2, the cotton homologous were down-regulated by NaCl in root tissue but the Arabidopsis genes were not responded to salt stress in root tissue.

Like the genes involved in ABA signal transduction pathway, most cotton transcripts related to JA, auxin, and ethylene signalling pathways showed similar expression patterns as those in Arabidopsis, such as *JAZ* and *MYC2* genes in JA signal pathway, *AUX1* and *SAUR* genes in auxin signal pathway, *ETR1* and *ERF1/2* genes in ethylene pathway, etc. However, there were some significant differences compared between cotton and Arabidopsis in response to water stresses. For example, in the JA signal transduction pathway (Fig. S4B), the Arabidopsis *JAR1* and *COI1* genes were not responded to salt and osmotic stresses, whereas their homologous in cotton were mainly up-regulated under NaCl treatment. In the auxin signal transduction pathway (Fig. S4C), many *ARF* transcripts were down-regulated in cotton under water stresses, but there is no *ARF* gene with differential expression under similar conditions. Especially in the ethylene signal transduction pathways (Fig. S4D), there were a lower proportion of gene members responding to water stresses in Arabidopsis than in cotton, including *CTR1*, *MPK6*, *EBF1/2* and *EIN3* genes.

## Discussion

### *De novo* Assembly of *G. arboreum* mRNA-seq Data Sets and Tissue Selectivity of Transcripts' Response to Water Stress During Cotton Seedling Stage

In this study, through paired-end massively parallel mRNA-seq and transcriptome *de novo* assembly, we assembled the unigenes of nine samples and successfully obtained 123,579 transcripts of *G. arboreum* (N50 = 1065 bp and length ≥200 bp) with about 300× sequence depth, ≥60% transcripts without gaps (Ns) and ≤5% transcripts with ≥20% gaps. Through BLAST against the known cotton ESTs and recently published draft genome of *G. raimondii* [68], there were 89,128 transcripts with hits, and the remaining 34,451 may be considered as novel transcripts. Due to the high proportion of short sequences, the true number of novel transcripts may be lower. Among 123,579 transcripts, there were 20,008 transcripts with GO annotation and 20,071 transcripts mapped to 117 pathways belonging to all five categories of KEGG. We obtained adequate sequence depth of coverage and acceptable assembling results and expression level detection. Our *de novo* assembly of mRNA-seq will improve genome annotation of *G. arboreum* and the specificity of the transcript signal will allow us to distinguish individual members of gene families.

During water stresses, there is osmotic adjustment in cotton leaves and roots, and the growing root tips act as dehydration sensors in soil [26]. In addition, the Arabidopsis roots and leaves were reported to display very different changes in water stress-regulated genes; also, dynamic changes occurred during 3- and 27 h of stress, with ≤5% of the changes shared by all three stresses during 3 h of acute stress response; however, by 27 h, the shared responses were reduced to ≤0.5% [40]. Thus in the present study, we considered the importance of spatial resolution and conducted tissue-specific stress transcriptome analysis [26,36,38,72]. We generated transcriptome mapping for three different tissues (roots, stems (with hypocotyls) and leaves) with 3 h of acute response to salt and osmotic treatments. We applied PCA to determine the dimensionality of the mRNA-seq data set and to identify meaningful underlying expression variables of the transcripts under salt and osmotic stress in different seedling tissues (Fig. 1E). The PCA revealed that gene expression differences among leaf, stem and root tissues were much greater than differences among the three treatment conditions (i.e. mock, PEG and salt).

In addition, the results showed that tissue identity mediated water stress. PCA showed the difference in the three treatments in root samples was much greater than those in leaf and stem tissues. Statistical analysis was used to identify the differentially expressed transcripts that responded to PEG and salt stresses. Among 123,579 transcripts, about 36,961 were identified as either up- or down-regulated under NaCl or PEG treatments in different tissues. Some transcripts were selected for validation by real-time RT-PCR, and most were matched between mRNA-seq and real-time RT-PCR analysis (Fig. 2), which suggested that our *de novo* assembly transcripts were reliable and repeatable. Furthermore, Venn diagram analysis showed highly overlapping responses to salt and osmotic stresses in individual cotton tissues (Fig. 3A). There was a very similar directional trend of gene differential expression between salt and osmotic stresses. Compared to those in leaf and root, there were much lower numbers of differentially expressed transcripts under water stress in stem tissue. In leaf, relatively larger numbers of transcripts were up-regulated under PEG treatment, while in root many more transcripts were down-regulated under NaCl treatment. Both PCA and Venn diagrams indicated the strong tissue selectivity of transcripts under salt and osmotic stress in cotton seedlings. We further conducted SOM and hierarchical cluster analysis for the 36,961 differential expressed transcripts. The cluster results showed a great deal of detailed tissue selectivity in each transcript's response to water stress. We selected several groups of clustered transcripts with preferential expression in leaf, stem and root, respectively for GO enrichment analysis, which may elucidate the possible mechanism involved. There were common enriched GO terms for the differentially expressed genes, e.g. response to stimulus and oxidoreductase activity. However, compared to stem and root, the leaf-specific group of genes showed some unique significantly enriched GO terms, including those involved in photosynthesis, light reaction, carbon fixation, glucose metabolic process and tetrapyrrole binding. Unique GO terms were shown by stem-specific genes in relation to plant-type cell wall biogenesis, and by root-specific genes for monooxygenase activity. The GO enrichment analysis for the water stress regulated tissue-specific genes suggested that there were different adaptation mechanisms and sequential effects in the different tissues responding to water stress, adjusting and balancing the whole plant for tolerance during high salinity and drought stress conditions.

## Transcription Factors and Hormone Signal Transduction Pathways Involved in Water Stress

Transcription is a dynamic process and transcription factors are essential for regulation of gene expression. In the present study, we identified a total of 4,002 transcripts (Table 2) using global transcription factor classification for the differentially expressed transcripts. About 49% of members responded to salt or osmotic stresses in different cotton tissues, including some key regulatory gene families involved in abiotic and biotic stresses, e.g. AP2-EREBP, WRKY, bHLH, MYB, bZIP, NAC, and HSF. Some developmental-related transcription factor genes were also up- or down-regulated during water stress in cotton seedlings, e.g. growth-regulating factor (GRF), regulating cell expansion in leaf

and cotyledon tissues. In addition, several transcription families related to hormone signal transduction pathways were also largely regulated by water stresses, e.g. ABI3VP1 (ABA signal pathway), Tify (mainly includes the orthologs of Arabidopsis JAZ, regulating the JA signal pathway), ARR-B (CK signal pathway), and EIL (ethylene signal pathway). We also found similar results in a previous study on the transcription response to salt stress in roots of *G. hirsutum* [51].

Plant hormones are essential for plants to adapt to water stress conditions [73]. The assembled cotton genes for those hormone signal pathways showed differential expression under osmotic and salt stresses in different cotton tissues (Fig. 4). As one of the main plant hormones, ABA plays major roles in seed and bud dormancy, as well as in responses to water stress. ABA mediates the water-stress signalling transduction pathway through core signalling components [74,75,76,77,78,79], including the core group of ABA receptor PYR/PYL/RCAR family proteins, the type 2C protein phosphatase (PP2C), and members of SNF1-related protein kinase 2 (SnRK2). During osmotic stress, the PYR/PYL/RCAR ABA-receptor-PP2C complexes control the SnRK2-AREB/ABF pathways. In cotton under salt and osmotic stresses, some ABA receptor PYR/RCAR family genes showed significantly differential expression patterns (Fig. 4D). For example, the cotton orthologs of Arabidopsis PYL4/RCAR1 were down-regulated in all seedling tissues during 17% PEG (osmotic stress) and 150mM NaCl (salt stress) treatments, respectively, while the orthologs for Arabidopsis PYL9/RCAR1 (the paralog of PYL7/RCAR2) were up-regulated under osmotic stress in cotton leaf and stem and under salt stress in root and stem tissues. The orthologs for PYR1/RCAR12 and PYL2/RCAR14 were down-regulated under osmotic stress in cotton leaf and root, and were also down-regulated in root tissue during salt stress. The homologs of PYL8/RCAR3 were down-regulated in root tissue during salt stress. In addition, the homologs for atPYL1 and atPYL11 were down-regulated in stem tissue under salt and osmotic stresses, separately. It was reported that PYL8/RCAR3 was strongly down-regulated, but PYL7/RCAR2 was up-regulated by salt and osmotic stresses [80]. Some members of PP2Cs, such as ABI1, ABI2, HAB1 and AHG3, are involved in ABA signalling through direct interaction with PYR/PYL/RCAR ABA receptors [76,79,81]. PP2Cs play vital roles in negatively regulating ABA response. Interestingly, our assembled *PP2C* transcripts were up-regulated under osmotic and salt stresses in all seedling tissues, similarly to the downstream *ABF* genes. The assembled cotton *SnRK2* family genes showed tissue selectivity during salt stress, e.g. the majority were up-regulated under osmotic and salt stresses in leaf and stem, but were down-regulated under salt stress in root tissue. Our results indicated a conserved crosstalk between water stress and the ABA signal transduction pathway. The comparative analysis of expression profiles responding to water stresses in the seedling tissues showed the conservation between cotton and Arabidopsis in the ABA signal transduction pathway (Fig. S4A).

Auxin controls various developmental processes, such as apical dominance, root initiation and stem elongation, etc. Auxin is transported polarly in plant shoots and roots. Our mRNA-seq based transcriptome analysis showed that a large number of genes involved in the auxin signal pathway were differentially expressed in response to water stress in cotton seedlings (Fig. 4A). The majority of changed cotton *AUX1* and *ARF* genes were down-regulated under osmotic and salt stress conditions in root, stem and leaf tissues. Some down-stream genes such as GH3 and SAUR genes showed both up- or down- regulation during water stress. Plants can adjust their RSA and direction of root growth to deal with high soil salinity and the underlying mechanism may be related to ABA-dependent repression of lateral root formation and auxin distribution in the roots during osmotic and high salt stresses [82].

In addition, many cotton JA signal transduction genes, such as members of the JAR1, JAZ, MYC2 families of genes, were also differently expressed under salt stress and osmotic stresses (Fig. 4G). In particular, a large number of JAR1 and JAZ genes were up-regulated during salt stress in root tissue, consistent with our previous report concerning salt response in JA genes of roots of upland cotton using microarray analysis [51]. The expression of the Arabidopsis JA signaling repressor JAZ1/TIFY10A was reported to be stimulated by auxin [83]. Our identified salt-induced JAZ genes may play key roles in shaping plant roots and mediating the crosstalk between auxin and JA signaling during salt stress. Some cotton *JAR1* and *JAZ* genes were also up-regulated under both salt and osmotic stresses in leaf tissue; possibly related to involvement of endogenous ABA in JA-induced stomatal closure [84]. There is crosstalk between JA and ethylene signal transduction pathways. The modulation of ethylene responses may affect plant salt-stress responses in Arabidopsis [85]. Our mRNA-seq results showed that some key regulatory genes of the ethylene signal transduction pathway (Fig. 4E), e.g. *ETR*, *CTR*, *EBF* and *ERF1/2*, were differentially expressed during osmotic and salt stresses in cotton seedlings, with the majority up-regulated. Similar results were also found for upland cotton under salt stress [51]. This suggests that there is crosstalk between ethylene signaling and water stress in cotton.

Besides ABA, auxin, JA, and ethylene-mediated signaling pathways in the response to water stress in cotton seedling stage, other hormones (e.g. CK, SA, GA, and BR) also play direct or indirect roles during water stress. CK is an antagonist to ABA and water stress results in decreased levels of CK, which is a positive regulator of auxin biosynthesis. Both CK and auxin promote stomatal opening, while ABA, JA, and SA induce stomatal closure [86]. GA and BR have many similar properties, and regulate some common biological processes, e.g. short primary roots [73]. Our mRNA-seq results indicated multiple hormone crosstalks in response to osmotic and salt stresses in different tissues of cotton seedlings. We also suggest that there are hormone crosstalks between abiotic and biotic stresses.

In summary, through mRNA-seq analysis for nine cotton samples using NaCl and PEG treatments in three cotton tissues, we tried to provide an overview of transcriptome profiling of *G. arboreum*. The whole transcriptome shotgun sequencing in *G. arboreum* will allow us to gain a broad picture of the genomic response to water stress and some interesting clues for further research. In addition, our *de novo* assembled cotton transcriptome with three tissues and three treatments will be beneficial to cotton whole genome annotation and reconstruction.

## Materials and Methods

### Plant Material and Growth Conditions

Cotton (*G. arboreum* L. cv. Shixiya) seeds were immersed in water for 1 d at 30°C and then placed for germination on sterilized soil in plates maintained under the following conditions: 28/25°C, 12/12 h of light/darkness, and relative humidity of 80%. After 3–4 d, properly germinated seeds were transferred to black plastic tanks filled with nutrient solution [51] and allowed to grow until they had produced 6–7 leaves. Seedlings showing normal growth were randomly divided into three groups, one group placed into tanks filled with a 150 mM solution of NaCl in water; another group placed into tanks filled with a 17% solution of PEG 6000 in water; and the remaining seedlings transferred to tanks filled with plain water to serve as mock. After exposing the seedlings to different solutions for 3 h, leaf, stem (including hypocotyl), and root tissues were harvested at the same time.

### Isolation of RNA and Real-time PCR

All the cotton tissue samples were homogenized in liquid nitrogen before isolation of RNA. Total RNA was isolated using a modified CTAB method and purified using Qiagen RNeasy columns (Qiagen, Hilden, Germany).

Reverse transcription was performed using an M-MLV kit (Invitrogen). The samples, 10 µl each containing 2 µg of total RNA and 20 pmol of random hexamers (Invitrogen), were maintained at 70°C for 10 min to denature the RNA and then chilled on ice for 2 min. The reaction buffer and M-MLV enzyme (20 µl of the mixture contained 500 µM dNTPs, 50 mM Tris-HCl (pH 8.3), 75 mM KCl, 3 mM MgCl2, 5 mM dithiothreitol, 200 units of M-MLV, and 20 pmol random hexamers) was added to the chilled samples and the samples maintained at 37°C for 1 h. The cDNA samples were diluted to 8 ng/µl for RT-PCR analysis.

For real-time RT-PCR, assays were performed in triplicate on 1 µl of each cDNA dilution using the SYBR Green Master Mix (PN 4309155, Applied Biosystems) with an ABI 7500 sequence detection system as prescribed in the manufacturer's protocol (Applied Biosystems). The gene-specific primers were designed using PRIMER3 (http://frodo.wi.mit.edu/primer3/input.htm). The amplification of 18S rRNA was used as an internal control to normalize all data (forward primer, 5'-CGGCTACCACATC-CAAGGAA-3'; reverse primer, 5'- TGTCAC-TACCTCCCCGTGTCA-3'). The gene-specific primers are listed in Table S1. The relative quantification method ($\Delta\Delta$CT) was used for quantitative evaluation of the variation between replicates.

### mRNA-seq Experiment and Transcriptome *de novo* Assembly

From each cotton tissue sample, 10 µg total RNA was collected for isolate poly(A) mRNA using beads with Oligo(dT). Then the mRNA was interrupted into short fragments by fragmentation buffer. The suitable fragments were selected for the PCR amplification as templates to prepare Illumina RNA-Seq library. Each library had an insert size around 200 bp and was sequenced using Illumina HiSeq™ 2000. The read lengths were 75 bp for root, and 90 bp for leaf and stem samples.

Sequencing-received raw image data was transformed by base calling into sequence data and stored in fastq format. After filtering low quality and dirty raw reads, transcriptome *de novo* assembly was carried out with a short reads assembling program (SOAPdenovo; [66]). The first step was to combine reads with certain length of overlap to form longer fragments, which are called contigs. Next, SOAPdenovo connectted the contigs using N to represent unknown sequences between each two contigs based on the paired-end reads, and then scaffolds were made. Paired-end reads were used again for gap filling of scaffolds to get sequences with least Ns and that could not be extended on either end. Such sequences are defined as unigenes. Finally, unigenes from each sample's assembly were used for further processes of sequence splicing and redundancy removing with sequence clustering software to acquire non-redundant unigenes (here called 'transcripts') that were as long as possible.

To assign the possible annotation of the transcripts, blastx alignment (e-value <0.00001) between transcripts and protein databases (e.g. nr, Swiss-Prot, KEGG and COG) was performed, and the best aligning results used to decide sequence direction of transcripts. ESTScan [67] was introduced to predict the coding regions and determine the sequence direction of transcripts without annotation. The Blast2GO program [69] was used to get GO annotation of transcripts. We also collected all cotton ESTs from NCBI and DFCI Cotton Gene Index (http://compbio.dfci.harvard.edu/tgi/plant.html) for transcripts sequence analysis.

To identify possible transcription factors in the transcripts, we used the sequence information in the PlnTFDB (http://plntfdb.bio.uni-potsdam.de/v3.0/) and PlantTFDB (http://planttfdb.cbi.pku.edu.cn) by BLAST (basic local alignment and search tool), as well as the annotation from Swiss-Prot and nr which the transcripts hit.

## Identify Differential Expression Transcripts and Functional Analysis

The expression of transcripts was calculated by RPKM method [70]; the formula is shown below:

$$RPKM = 10^6 C / (NL/10^3).$$

where C is the number of reads that uniquely aligned to the transcript, N is the total number of reads that uniquely aligned to all transcripts in the specific sample, and L is number of bases of the transcript. The $P$-value corresponds to differential transcript expression in two samples was determined from Audic's algorithm [71], and FDR method was applied to determine the threshold of $P$-values in multiple tests. We use "FDR $\leq 0.001$ and the absolute value of $\log_2$Ratio $\geq 1$" as the threshold to judge the significance of gene expression difference.

GO enrichment analysis was performed for functional categorization of differentially expressed transcripts using agriGO software [87] and the $P$-values corrected by applying the FDR correction to control falsely rejected hypotheses during GO analysis.

MapMan (http://gabi.rzpd.de/projects/MapMan) was used for key regulation group analysis. The pathway analysis for plant hormone signalling were conducted using KEGG (www.genome.jp/kegg/) and the corresponding MapMan pathways were created through the mapping files for BLAST hits of transcripts to Arabidopsis TAIR9 version (www.arabidopsis.org).

## Supporting Information

**Figure S1  Workflow for sample preparation and experiment pipeline of *de novo* mRNA-seq transcriptome.** A total of nine samples were collected for mRNA-seq: leaf, stem (including hypocotyls), and root samples of cotton seedling under mock, 17% PEG, and 150 mM NaCl conditions.

**Figure S2  Pair-wise scatter plots for the raw reads of transcripts across all samples.**

**Figure S3  SOM (self-organized mapping) cluster of the transcripts response to PEG or NaCl treatments in different cotton tissue samples.** An overview of $10 \times 10$ SOM cluster for 36,961 transcripts' response to PEG or NaCl treatments in leaf, stem, or root sample of cotton seedlings. For each cluster, the red line represents the expression pattern of the centroid, the order from left to right is: leaf-mock, leaf-PEG, leaf-NaCl, stem-mock, stem-PEG, stem-NaCl, root-mock, root-PEG, and root-NaCl. The cluster id and the number of transcripts in the cluster are also listed in the figures.

**Figure S4  Comparative analysis of cotton and Arabidopsis transcripts related to selected plant hormone signal transduction pathways and their response to water stresses across different tissue samples.** The colored bars represent the percentage of the transcripts in each bin whether up-regulated (red) or down-regulated (blue) under water stresses in different tissues. The charts above the pathway represent the cotton tissues and the charts below the pathway represent the Arabidopsis tissues. **A**: represents the differential expression transcripts related to abscisic acid (ABA) signalling transduction pathway. **B**: represents the differential expression transcripts related to jasmonate (JA) signalling transduction pathway. **C**: represents the differential expression transcripts related to auxin (IAA) signalling transduction pathway. **D**: represents the differential expression transcripts related to ethylene signalling transduction pathway.

**Table S1  Primer list of transcripts for real-time RT-PCR.**

**Table S2  In cotton tissues, 36,961 transcripts responded to PEG and NaCl treatments.** Including the transcript length, raw RPKM data, $\log_2$ ratio, FDR-value, change call, and additional annotation of each transcript

## Acknowledgments

We wish to thank Beijing Genomics Institute at Shenzhen (BGI Shenzhen) for help in sequencing, assembly, and analysis support. We also thank Qunlian Zhang for technical support.

## Author Contributions

Conceived and designed the experiments: WX ZS FL XZ. Performed the experiments: DY WX XZ Q. Wang Q. Wei CW. Analyzed the data: WX ZS XZ. Contributed reagents/materials/analysis tools: DY Q. Wang XZ WX CL CZ HY YL FL. Wrote the paper: WX ZS XZ.

## References

1. Sexton PD, Gerard CJ (1982) Emergence Force of Cotton Seedlings as Influenced by Salinity. Agron J 74: 699–702.
2. Bharambe PR, Varade SB (1980) Effect of plant water deficits on abscission of leaves and fruiting forms in CJ-73 (Gossypium arboreum). Journal of Maharashtra Agricultural Universities 5: 201–204.
3. Berkant ÖDEMİŞ RK (2009) Effects of Irrigation Water Quality on Evapotranspiration, Yield and Biomass of Cotton. Journal of Plant & Environmental Sciences: 16–20.
4. Akhtar J, Saqib ZA, Sarfraz M, Saleem I, Haq SA (2010) Evaluating salt tolerant cotton genotypes at different levels of NaCl stress in solution and soil culture. Pak J Bot 42: 2857–2866.
5. Berlin J, Quisenberry JE, Bailey F, Woodworth M, McMichael BL (1982) Effect of water stress on cotton leaves : I. An electron microscopic stereological study of the palisade cells. Plant Physiol 70: 238–243.
6. Timpa JD, Burke JJ, Quisenberry JE, Wendt CW (1986) Effects of water stress on the organic Acid and carbohydrate compositions of cotton plants. Plant Physiol 82: 724–728.
7. Chinnusamy V, Jagendorf A, and Zhu JK (2005) Understanding and Improving Salt Tolerance in Plants. CROP SCIENCE 45: 437–448.
8. Munns R, Tester M (2008) Mechanisms of salinity tolerance. Annu Rev Plant Biol 59: 651–681.
9. Zhu JK (2002) Salt and drought stress signal transduction in plants. Annu Rev Plant Biol 53: 247–273.
10. Hasegawa PM, Bressan RA, Zhu JK, Bohnert HJ (2000) Plant Cellular and Molecular Responses to High Salinity. Annu Rev Plant Physiol Plant Mol Biol 51: 463–499.
11. Tuteja N (2007) Mechanisms of high salinity tolerance in plants. Methods Enzymol 428: 419–438.

12. Shinozaki K (1999) [Plant response to drought and salt stress: overview]. Tanpakushitsu Kakusan Koso 44: 2186–2187.

13. Gómez-Cadenas A, Tadeo FR, Primo-Millo E, Talon M (1998) Involvement of abscisic acid and ethylene in the responses of citrus seedlings to salt shock. Physiologia Plantarum 103: 475–484.

14. Galvan-Ampudia SC, Testerink C (2011) Salt stresssignals shape the plant root. Current Opinion in Plant Biology 14: 1–7.

15. Ackerson RC (1981) Osmoregulation in Cotton in Response to Water Stress : II. LEAF CARBOHYDRATE STATUS IN RELATION TO OSMOTIC ADJUSTMENT. Plant Physiol 67: 489–493.

16. Boyer JS (1965) Effects of Osmotic Water Stress on Metabolic Rates of Cotton Plants with Open Stomata. Plant Physiol 40: 229–234.

17. Brugnoli E, Lauteri M (1991) Effects of Salinity on Stomatal Conductance, Photosynthetic Capacity, and Carbon Isotope Discrimination of Salt-Tolerant (Gossypium hirsutum L.) and Salt-Sensitive (Phaseolus vulgaris L.) C(3) Non-Halophytes. Plant Physiol 95: 628–635.

18. Zhu JK (2003) Regulation of ion homeostasis under salt stress. Curr Opin Plant Biol 6: 441–445.

19. Serrano R, Rodriguez-Navarro A (2001) Ion homeostasis during salt stress in plants. Curr Opin Cell Biol 13: 399–404.

20. Maas EV (1986) Salt tolerance of plants. Appl Agric Res 1: 12–26.

21. Zhang L, Li FG, Liu CL, Zhang CJ, Zhang XY (2009) Construction and analysis of cotton (Gossypium arboreum L.) drought-related cDNA library. BMC Research Notes 2: 120.

22. Sattar S, Hussnain T, Javid A (2010) Effect of NaCl salinity on cotton (Gossypium arboreum L.) grown on MS medium and in hydroponics cultures. The Journal of Animal & Plant Sciences 20: 87–89.

23. Barozai MY, Husnain T (2012) Identification of biotic and abiotic stress up-regulated ESTs in Gossypium arboreum. Mol Biol Rep. 39: 1011–1018.

24. Maqbool A, Abbas W, Rao AQ, Irfan M, Zahur M, et al. (2010) Gossypium arboreum GHSP26 enhances drought tolerance in Gossypium hirsutum. Biotechnol Prog 26: 21–25.

25. Zhang L, Li FG, Liu CL, Zhang CJ, Zhang XY (2009) Construction and analysis of cotton (Gossypium arboreum L.) drought-related cDNA library. BMC Res Notes 2: 120.

26. Oosterhuis DM, Wullschleger SD (1987) Osmotic Adjustment in Cotton (Gossypium hirsutum L.) Leaves and Roots in Response to Water Stress. Plant Physiol 84: 1154–1157.

27. Radin JW (1984) Stomatal responses to water stress and to abscisic Acid in phosphorus-deficient cotton plants. Plant Physiol 76: 392–394.

28. Gossett DR, Millhollon EP, Lucas MC (1994) Antioxidant Response to NaCl Stress in Salt-Tolerant and Salt-Sensitive Cultivars of Cotton. CROP SCIENCE 34: 706–714.

29. Xie Z, Duan L, Tian X, Wang B, Eneji AE, et al. (2008) Coronatine alleviates salinity stress in cotton by improving the antioxidative defense system and radical-scavenging activity. J Plant Physiol 165: 375–384.

30. Jordan WR, Brown KW, Thomas JC (1975) Leaf Age as a Determinant in Stomatal Control of Water Loss from Cotton during Water Stress. Plant Physiol 56: 595–599.

31. Da Silva JV, Naylor AW, Kramer PJ (1974) Some ultrastructural and enzymatic effects of water stress in cotton (gossypium hirsutum L.) leaves. Proc Natl Acad Sci U S A 71: 3243–3247.

32. Rodriguez-Uribe L HS, Stewart JM, Wilkins T, Lindemann W, Sengupta-Gopalan C, et al. (2011) Identification of salt responsive genes using comparative microarray analysis in Upland cotton (Gossypium hirsutum L.). Plant Sci 180: 461–469.

33. Lin H, Salus SS, Schumaker KS (1997) Salt Sensitivity and the Activities of the H+-ATPases in Cotton Seedlings. CROP SCIENCE 37: 190–197.

34. Ackerson RC (1982) Synthesis and movement of abscisic Acid in water-stressed cotton leaves. Plant Physiol 69: 609–613.

35. Kawasaki S, Borchert C, Deyholos M, Wang H, Brazille S, et al. (2001) Gene expression profiles during the initial phase of salt stress in rice. Plant Cell 13: 889–905.

36. Dinneny JR, Long TA, Wang JY, Jung JW, Mace D, et al. (2008) Cell identity mediates the response of Arabidopsis roots to abiotic stress. Science 320: 942–945.

37. Deyholos MK (2010) Making the most of drought and salinity transcriptomics. Plant Cell Environ 33: 648–654.

38. Dinneny JR (2010) Analysis of the salt-stress response at cell-type resolution. Plant Cell Environ 33: 543–551.

39. Ueda A, Kathiresan A, Bennett J, Takabe T (2006) Comparative transcriptome analyses of barley and rice under salt stress. Theor Appl Genet 112: 1286–1294.

40. Kreps JA, Wu Y, Chang HS, Zhu T, Wang X, et al. (2002) Transcriptome changes for Arabidopsis in response to salt, osmotic, and cold stress. Plant Physiol 130: 2129–2141.

41. Luo ZB, Janz D, Jiang X, Gobel C, Wildhagen H, et al. (2009) Upgrading root physiology for stress tolerance by ectomycorrhizas: insights from metabolite and transcriptional profiling into reprogramming for stress anticipation. Plant Physiol 151: 1902–1917.

42. Buchanan CD, Lim S, Salzman RA, Kagiampakis I, Morishige DT, et al. (2005) Sorghum bicolor's transcriptome response to dehydration, high salinity and ABA. Plant Mol Biol 58: 699–720.

43. Oztur ZN, Talame V, Deyholos M, Michalowski CB, Galbraith DW, et al. (2002) Monitoring large-scale changes in transcript abundance in drought- and salt-stressed barley. Plant Mol Biol 48: 551–573.

44. Ma S, Gong Q, Bohnert HJ (2006) Dissecting salt stress pathways. J Exp Bot 57: 1097–1107.

45. Ma S, Gong Q, Bohnert HJ (2007) An Arabidopsis gene network based on the graphical Gaussian model. Genome Res 17: 1614–1625.

46. Jiang Y, Deyholos MK (2006) Comprehensive transcriptional profiling of NaCl-stressed Arabidopsis roots reveals novel classes of responsive genes. BMC Plant Biol 6: 25.

47. Kawaura K, Mochida K, Ogihara Y (2008) Genome-wide analysis for identification of salt-responsive genes in common wheat. Funct Integr Genomics 8: 277–286.

48. Chao DY, Luo YH, Shi M, Luo D, Lin HX (2005) Salt-responsive genes in rice revealed by cDNA microarray analysis. Cell Res 15: 796–810.

49. Taji T, Seki M, Satou M, Sakurai T, Kobayashi M, et al. (2004) Comparative genomics in salt tolerance between Arabidopsis and aRabidopsis-related halophyte salt cress using Arabidopsis microarray. Plant Physiol 135: 1697–1709.

50. Richardt S, Timmerhaus G, Lang D, Qudeimat E, Correa LG, et al. (2010) Microarray analysis of the moss Physcomitrella patens reveals evolutionarily conserved transcriptional regulation of salt stress and abscisic acid signalling. Plant Mol Biol 72: 27–45.

51. Yao D, Zhang X, Zhao X, Liu C, Wang C, et al. (2011) Transcriptome analysis reveals salt-stress-regulated biological processes and key pathways in roots of cotton (Gossypium hirsutum L.). Genomics 98: 47–55.

52. Nagalakshmi U, Wang Z, Waern K, Shou C, Raha D, et al. (2008) The transcriptional landscape of the yeast genome defined by RNA sequencing. Science 320: 1344–1349.

53. Sultan M, Schulz MH, Richard H, Magen A, Klingenhoff A, et al. (2008) A global view of gene activity and alternative splicing by deep sequencing of the human transcriptome. Science 321: 956–960.

54. Wilhelm BT, Marguerat S, Watt S, Schubert F, Wood V, et al. (2008) Dynamic repertoire of a eukaryotic transcriptome surveyed at single-nucleotide resolution. Nature 453: 1239–1243.

55. Lister R, O'Malley RC, Tonti-Filippini J, Gregory BD, Berry CC, et al. (2008) Highly integrated single-base resolution maps of the epigenome in Arabidopsis. Cell 133: 523–536.

56. Filichkin SA, Priest HD, Givan SA, Shen R, Bryant DW, et al. (2010) Genome-wide mapping of alternative splicing in Arabidopsis thaliana. Genome Res 20: 45–58.

57. Yang H, Lu P, Wang Y, Ma H (2011) The transcriptome landscape of Arabidopsis male meiocytes from high-throughput sequencing: the complexity and evolution of the meiotic process. Plant J 65: 503–516.

58. Gonzalez-Ballester D, Casero D, Cokus S, Pellegrini M, Merchant SS, et al. (2010) RNA-seq analysis of sulfur-deprived Chlamydomonas cells reveals aspects of acclimation critical for cell survival. Plant Cell 22: 2058–2084.

59. Lu T, Lu G, Fan D, Zhu C, Li W, et al. (2010) Function annotation of the rice transcriptome at single-nucleotide resolution by RNA-seq. Genome Res 20: 1238–1249.

60. Zhang G, Guo G, Hu X, Zhang Y, Li Q, et al. (2010) Deep RNA sequencing at single base-pair resolution reveals high complexity of the rice transcriptome. Genome Res 20: 646–654.

61. Zenoni S, Ferrarini A, Giacomelli E, Xumerle L, Fasoli M, et al. (2010) Characterization of transcriptional complexity during berry development in Vitis vinifera using RNA-Seq. Plant Physiol 152: 1787–1795.

62. Wilhelm BT, Marguerat S, Goodhead I, Bahler J (2010) Defining transcribed regions using RNA-seq. Nat Protoc 5: 255–266.

63. Graveley BR, Brooks AN, Carlson JW, Duff MO, Landolin JM, et al. (2011) The developmental transcriptome of Drosophila melanogaster. Nature 471: 473–479.

64. Severin AJ, Woody JL, Bolon YT, Joseph B, Diers BW, et al. (2010) RNA-Seq Atlas of Glycine max: a guide to the soybean transcriptome. BMC Plant Biol 10: 160.

65. Oshlack A, Robinson MD, Young MD (2010) From RNA-seq reads to differential expression results. Genome Biol 11: 220.

66. Li R, Zhu H, Ruan J, Qian W, Fang X, et al. (2009) De novo assembly of human genomes with massively parallel short read sequencing. Genome Res 20: 265–272.

67. Iseli C, Jongeneel CV, Bucher P (1999) ESTScan: a program for detecting, evaluating, and reconstructing potential coding regions in EST sequences. Proc Int Conf Intell Syst Mol Biol: 138–148.

68. Wang K, Wang Z, Li F, Ye W, Wang J, et al. (2012) The draft genome of a diploid cotton Gossypium raimondii. Nat Genet. 44: 1098–1103.

69. Conesa A, Gotz S, Garcia-Gomez JM, Terol J, Talon M, et al. (2005) Blast2GO: a universal tool for annotation, visualization and analysis in functional genomics research. Bioinformatics 21: 3674–3676.

70. Mortazavi A, Williams BA, McCue K, Schaeffer L, Wold B (2008) Mapping and quantifying mammalian transcriptomes by RNA-Seq. Nat Methods 5: 621–628.

71. Audic S, Claverie JM (1997) The significance of digital gene expression profiles. Genome Res 7: 986–995.

72. Wee CW, Dinneny JR (2010) Tools for high-spatial and temporal-resolution analysis of environmental responses in plants. Biotechnol Lett 32: 1361–1371.

73. Peleg Z, Blumwald E (2011) Hormone balance and abiotic stress tolerance in crop plants. Curr Opin Plant Biol. 14: 290–295.

74. Ma Y, Szostkiewicz I, Korte A, Moes D, Yang Y, et al. (2009) Regulators of PP2C phosphatase activity function as abscisic acid sensors. Science 324: 1064–1068.

75. Park SY, Fung P, Nishimura N, Jensen DR, Fujii H, et al. (2009) Abscisic acid inhibits type 2C protein phosphatases via the PYR/PYL family of START proteins. Science 324: 1068–1071.

76. Nishimura N, Sarkeshik A, Nito K, Park SY, Wang A, et al. (2009) PYR/PYL/RCAR family members are major in-vivo ABI1 protein phosphatase 2C-interacting proteins in Arabidopsis. Plant J. 61: 290–299.

77. Santiago J, Dupeux F, Round A, Antoni R, Park SY, et al. (2009) The abscisic acid receptor PYR1 in complex with abscisic acid. Nature 462: 665–668.

78. Fujii H, Chinnusamy V, Rodrigues A, Rubio S, Antoni R, et al. (2009) In vitro reconstitution of an abscisic acid signalling pathway. Nature 462: 660–664.

79. Cutler SR, Rodriguez PL, Finkelstein RR, Abrams SR (2010) Abscisic acid: emergence of a core signaling network. Annu Rev Plant Biol 61: 651–679.

80. Saavedra X, Modrego A, Rodriguez D, Gonzalez-Garcia MP, Sanz L, et al. (2010) The nuclear interactor PYL8/RCAR3 of Fagus sylvatica FsPP2C1 is a positive regulator of abscisic acid signaling in seeds and stress. Plant Physiol 152: 133–150.

81. Nishimura N, Sarkeshik A, Nito K, Park SY, Wang A, et al. (2010) PYR/PYL/RCAR family members are major in-vivo ABI1 protein phosphatase 2C-interacting proteins in Arabidopsis. Plant J 61: 290–299.

82. Galvan-Ampudia CS, Testerink C (2011) Salt stress signals shape the plant root. Curr Opin Plant Biol.

83. Grunewald W, Vanholme B, Pauwels L, Plovie E, Inze D, et al. (2009) Expression of the Arabidopsis jasmonate signalling repressor JAZ1/TIFY10A is stimulated by auxin. EMBO Rep 10: 923–928.

84. Hossain MA, Munemasa S, Uraji M, Nakamura Y, Mori IC, et al. (2011) Involvement of endogenous abscisic Acid in methyl jasmonate-induced stomatal closure in Arabidopsis. Plant Physiol 156: 430–438.

85. Cao WH, Liu J, He XJ, Mu RL, Zhou HL, et al. (2007) Modulation of ethylene responses affects plant salt-stress responses. Plant Physiol 143: 707–719.

86. Acharya BR, Assmann SM (2009) Hormone interactions in stomatal function. Plant Mol Biol 69: 451–462.

87. Du Z, Zhou X, Ling Y, Zhang Z, Su Z (2010) agriGO: a GO analysis toolkit for the agricultural community. Nucleic Acids Res 38: W64–70.

# PERMISSIONS

All chapters in this book were first published in PLOS ONE, by The Public Library of Science; hereby published with permission under the Creative Commons Attribution License or equivalent. Every chapter published in this book has been scrutinized by our experts. Their significance has been extensively debated. The topics covered herein carry significant findings which will fuel the growth of the discipline. They may even be implemented as practical applications or may be referred to as a beginning point for another development.

The contributors of this book come from diverse backgrounds, making this book a truly international effort. This book will bring forth new frontiers with its revolutionizing research information and detailed analysis of the nascent developments around the world.

We would like to thank all the contributing authors for lending their expertise to make the book truly unique. They have played a crucial role in the development of this book. Without their invaluable contributions this book wouldn't have been possible. They have made vital efforts to compile up to date information on the varied aspects of this subject to make this book a valuable addition to the collection of many professionals and students.

This book was conceptualized with the vision of imparting up-to-date information and advanced data in this field. To ensure the same, a matchless editorial board was set up. Every individual on the board went through rigorous rounds of assessment to prove their worth. After which they invested a large part of their time researching and compiling the most relevant data for our readers.

The editorial board has been involved in producing this book since its inception. They have spent rigorous hours researching and exploring the diverse topics which have resulted in the successful publishing of this book. They have passed on their knowledge of decades through this book. To expedite this challenging task, the publisher supported the team at every step. A small team of assistant editors was also appointed to further simplify the editing procedure and attain best results for the readers.

Apart from the editorial board, the designing team has also invested a significant amount of their time in understanding the subject and creating the most relevant covers. They scrutinized every image to scout for the most suitable representation of the subject and create an appropriate cover for the book.

The publishing team has been an ardent support to the editorial, designing and production team. Their endless efforts to recruit the best for this project, has resulted in the accomplishment of this book. They are a veteran in the field of academics and their pool of knowledge is as vast as their experience in printing. Their expertise and guidance has proved useful at every step. Their uncompromising quality standards have made this book an exceptional effort. Their encouragement from time to time has been an inspiration for everyone.

The publisher and the editorial board hope that this book will prove to be a valuable piece of knowledge for researchers, students, practitioners and scholars across the globe.

# LIST OF CONTRIBUTORS

**Wei Liu, Wei Li, Qiuling He, Jinhong Chen and Shuijin Zhu**
Department of Agronomy, Zhejiang University, Hangzhou, Zhejiang, China

**Muhammad Khan Daud**
Department of Biotechnology and Genetic Engineering, Kohat University of Science and Technology, Kohat, Pakistan

**Ji Chen, Fengjuan Lv, Jingran Liu, Yina Ma, Youhua Wang, Binglin Chen, Yali Meng and Zhiguo Zhou**
Key Laboratory of Crop Physiology & Ecology, Ministry of Agriculture, Nanjing Agricultural University, Nanjing, Jiangsu Province, PR China

**Derrick M. Oosterhuis**
Department of Crop, Soil, and Environmental Sciences, University of Arkansas, Fayetteville, Arkansas, United States of America

**Wen-Qin Bai, Yue-Hua Xiao, Juan Zhao, Shui-Qing Song, Lin Hu, Jian-Yan Zeng, Xian-Bi Li, Lei Hou, Ming Luo, De-Mou Li and Yan Pei**
Biotechnology Research Center, Southwest University, Beibei, Chongqing, China

**Kelsie LaSharr**
School of Natural Resources & the Environment, The University of Arizona, Tucson, Arizona, United States of America

**Laura López-Hoffman and Ruscena Wiederholt**
School of Natural Resources & the Environment, The University of Arizona, Tucson, Arizona, United States of America

Udall Center for Studies in Public Policy, The University of Arizona, Tucson, Arizona, United States of America

**Chris Sansone**
Bayer CropScience, Research Triangle Park, North Carolina, United States of America,

**Kenneth J. Bagstad, Jay E. Diffendorfer and Darius Semmens**
United States Geological Survey, Geosciences and Environmental Change Science Center, Denver, Colorado, United States of America

**Paul Cryan**
United States Geological Survey, Fort Collins Science Center, Fort Collins, Colorado, United States of America

**Joshua Goldstein**
Department of Human Dimensions of Natural Resources, Colorado State University, Fort Collins, Colorado, United States of America

**John Loomis**
Department of Agricultural and Resource Economics, Colorado State University, Fort Collins, Colorado, United States of America

**Gary McCracken**
Department of Ecology and Evolutionary Biology, University of Tennessee, Knoxville, Tennessee, United States of America

**Rodrigo A. Medellín**
Instituto de Ecología, Universidad Nacional Autónoma de México, Distrito Federal, México

**Amy Russell**
Department of Biology, Grand Valley State University, Allendale, Michigan, United States of America

**Marina Naoumkina, Gregory Thyssen, David D. Fang and Christopher Florane**
Cotton Fiber Bioscience Research Unit, USDA-ARS, Southern Regional Research Center, New Orleans, Louisiana, United States of America

**Doug J. Hinchliffe**
Cotton Chemistry & Utilization Research Unit, USDA-ARS, Southern Regional Research Center, New Orleans, Louisiana, United States of America

**Kathleen M. Yeater**
USDA-ARS-Southern Plains Area, College Station, Texas, United States of America

**Justin T. Page and Joshua A. Udall**
Plant and Wildlife Science Department, Brigham Young University, Provo, Utah, United States of America

**Matthew H. Meisner**
Department of Evolution and Ecology, University of California Davis, Davis, California, United States of America
Department of Statistics, University of California Davis, Davis, California, United States of America

**Jay A. Rosenheim**
Department of Entomology and Nematology, University of California Davis, Davis, California, United States of America

**Prue Talbot**
Department of Cell Biology and Neuroscience, University of California Riverside, Riverside, California, United States of America

**Vasundhra Bahl**
Department of Cell Biology and Neuroscience, University of California Riverside, Riverside, California, United States of America
Environmental Toxicology Graduate Program, University of California Riverside, Riverside, California, United States of America

**Peyton Jacob III and Christopher Havel**
Department of Clinical Pharmacology, University of California San Francisco, San Francisco, California, United States of America

**Suzaynn F. Schick**
Department of Medicine, Division of Occupational and Environmental Medicine, University of California San Francisco, San Francisco, California, United States of America

**Yina Ma, Youhua Wang, Jingran Liu, Fengjuan Lv, Ji Chen and Zhiguo Zhou**
Key Laboratory of Crop Growth Regulation, Ministry of Agriculture, Nanjing Agricultural University, Nanjing, Jiangsu Province, PR China

**Francisco S. Ramalho, Jéssica K. S. Pachú, Aline C. S. Lira, José B. Malaquias and José Francisco S. Fernandes**
Unidade de Controle Biológico, Embrapa Algodão, Campina Grande, Paraíba, Brazil

**C. Zanuncio**
Departamento de Entomologia, Universidade Federal de Viçosa, Minas Gerais, Brazi

**Peng Han**
Hubei Key Laboratory of Insect Resources Application and Sustainable Pest Control, Plant Science & Technology College, Huazhong Agricultural University, Wuhan, China

French National Institute for Agricultural Research (INRA), Sophia-Antipolis, France

**Chang-ying Niu**
Hubei Key Laboratory of Insect Resources Application and Sustainable Pest Control, Plant Science & Technology College, Huazhong Agricultural University, Wuhan, China

**Nicolas Desneux**
French National Institute for Agricultural Research (INRA), Sophia-Antipolis, France

**Yi-Wei Fang, Ling-Yun Liu, Hua-Li Zhang, De-Feng Jiang and Dong Chu**
Key Lab of Integrated Crop Pest Management of Shandong Province, College of Agronomy and Plant Protection, Qingdao Agricultural University, Qingdao, China

**Wilhelm Klümper and Matin Qaim**
Department of Agricultural Economics and Rural Development, Georg-August-University of Goettingen, Goettingen, Germany

**Diana Castillo Lopez, Keyan Zhu-Salzman and Gregory A. Sword**
Department of Entomology, Texas A&M University, College Station, Texas, United States of America

**Maria Julissa Ek-Ramos**
Department of Immunology and Microbiology, Autonomous University of Nuevo Leon, San Nicolás de los Garza, Nuevo Leon, Mexico

**Yunlei Zhao, Hongmei Wang and Wei Chen and Yunhai Li**
State Key Laboratory of Cotton Biology, Institute of Cotton Research of Chinese Academy of Agricultural Sciences (CAAS), Anyang, People's Republic of China

**Yizhou Chen, Daniel R. Bogema, Idris M. Barchia and Grant A. Herron**
Elizabeth Macarthur Agricultural Institute, NSW Department of Primary Industries, Menangle, New South Wales, Australia

**Xingju Zhang, Yanchao Yuan, Ze Wei, Xian Guo, Yuping Guo, Suqing Zhang, Xianliang Song and Xuezhen Sun**
State Key Laboratory of Crop Biology/Agronomy College, Shandong Agricultural University, Taian, Shandong, China

**Junsheng Zhao**
Cotton Research Center, Shandong Academy of Agricultural Sciences, Jinan, Shandong, China

**Guihua Zhang**
Heze Academy of Agricultual Sciences, Heze, Shandong, China

**Xiaoyan Wang and Qifeng Ma**
College of Agronomy, Northwest A&F University, Yangling, Shaanxi, People's Republic of China

State Key Laboratory of Cotton Biology, Institute of Cotton Research of CAAS, Anyang, Henan, People's Republic of China

**Shuli Fan, Meizhen Song, Chaoyou Pang, Hengling Wei, Jiwen Yu and Shuxun Yu**
State Key Laboratory of Cotton Biology, Institute of Cotton Research of CAAS, Anyang, Henan, People's Republic of China

**Nadja Sparding, Gert M. Nicolaisen, Steen B. Giese, Jón Elmlund, Nina R. Steenhard**
Centre for Biosecurity and Biopreparedness, Statens Serum Institut, Copenhagen, Denmark

**Hans-Christian Slotved**
Centre for Biosecurity and Biopreparedness, Statens Serum Institut, Copenhagen, Denmark
Department of Microbiology and Infection Control, Statens Serum Institut, Copenhagen, Denmark

**Hongsheng Wu**
State Key Laboratory of Biocontrol, School of Life Sciences, Sun Yat-sen University, Guangzhou, China
Department of Crop Protection, Faculty of Bioscience Engineering, Ghent University, Ghent, Belgium

**Yuhong Zhang, Ping Liu, Jiaqin Xie, Yunyu He, Congshuang Deng and Hong Pang**
State Key Laboratory of Biocontrol, School of Life Sciences, Sun Yat-sen University, Guangzhou, China

**Patrick De Clercq**
Department of Crop Protection, Faculty of Bioscience Engineering, Ghent University, Ghent, Belgium

**Yinhua Jia, Junling Sun, Zhaoe Pan, Xiwen Wang, Shoupu He, Zhongli Zhou, Baoyin Pang, Liru Wang and Xiongming Du**
Institute of Cotton Research of Chinese Academy of Agricultural Sciences (ICR, CAAS), State Key Laboratory of Cotton Biology, Key Laboratory of Cotton Genetic Improvement, Ministry of Agriculture, Anyang, China

**Xiwei Sun and Jun Zhu**
Key Laboratory of Crop Germplasm Resource of Zhejiang Province, Zhejiang University, Hangzhou, China

**Songhua Xiao and Jianguang Liu**
Institute of industrial Crops, Jiangsu Academy of Agricultural Sciences, Nanjing, China

**Jun Ma and Weijun Shi**
Economic Crop Research Institute, Xinjiang Academy of Agricultural Science, Urumqi1, China

**Jinfeng Han**
Department of Agronomy, Henan Agricultural University, Zhengzhou, Henan, China

**Peng Yang**
Department of Agronomy, Henan Agricultural University, Zhengzhou, Henan, China
Department of Rural Development, Shanxi Agricultural University, Taigu, Shanxi, China

**Jinling Huang**
Department of Agronomy, Shanxi Agricultural University, Taigu, Shanxi, China

**Jungang Wang, Jing Wang, Jun Peng and Tingting Zhou**
College of Agriculture, Shihezi University, Shihezi, Xinjiang, People's Republic of China

**Hailan Jiang and Xiaoxia Deng**
College of Agriculture, Shihezi University, Shihezi, Xinjiang, People's Republic of China
Kuerle Agricultural Technology Extension Center, Kuerle, Xinjiang, People's Republic of China

**Xueyan Zhang, Qianhua Wang, Chuanliang Liu, Chaojun Zhang and Fuguang Li**
State Key Laboratory of Cotton Biology, Institute of Cotton Research, Chinese Academy of Agriculture Sciences (CAAS), Anyang, Henan, China

**Dongxia Yao, Wenying Xu, Qiang Wei, Chunchao Wang, Hong Yan, Yi Ling and Zhen Su**
State Key Laboratory of Plant Physiology and Biochemistry, College of Biological Sciences, China Agricultural University, Beijing, China

# Index

## A

Alabama Argillacea, 78-79, 87
Allotetraploid Cotton, 40, 43, 45, 48, 50, 133-134, 155
Antioxidant Enzyme Activities, 198, 200
Aphid Population, 88-89, 91, 93-94, 96
Aphis Gossypii Glover, 88, 91, 95, 113, 120, 136-137, 147, 172, 175-176
Aqueous Extractions, 61-62, 64
Arabidopsis Thaliana, 11, 31, 134, 156-159, 164-165, 196, 206, 220

## B

Bat Pest-control Services, 33-34
Beauveria Bassiana, 113, 119-120
Bemisia Tabaci Laboratory Population, 98
Bioactive Gibberellins (GAS), 26
Black Nightshade, 198-200, 202, 204

## C

Calcium-dependent, 1, 10-11
Cellulose Synthesis, 12-13, 17, 20, 23-25, 29, 32, 69, 73-77
Chromosomal Locations, 2
Cotton Aphid Reproduction, 113-119
Cotton Cultivars, 24, 69-70, 78-82, 84, 86-88, 94, 122, 128-129, 134, 148, 173, 175, 177-179, 181
Cotton Fiber Secondary Wall, 12, 14
Cotton Leafworm, 78, 80, 86-87
Crop Rotational Histories, 52-54, 56-57
Cryptolaemus Montrouzieri, 167-168, 175
Cytoplasmic Male Sterility, 185, 196-197

## D

Defoliating Isolate, 148
Determination of Endogenous Ga Contents, 27
Different Environments, 132, 177, 179
Dispersal Behavior, 78

## E

Ecoinformatics, 52-53, 56, 58-59
Elongating Fibers, 40-41
Empoasca Biguttula Shiraki, 89, 91
Entomopathogenic Fungal, 113, 119-120
Enzymatic Analyses, 14, 71
Epistasis, 177-184
Expression Analysis, 1, 8-11, 27, 40, 43, 49-50, 192, 196, 208, 216

## F

Fiber Development, 10, 12-13, 15, 17, 20, 25-26, 28-30, 40-41, 48-50, 69-70, 72-74, 76
Field Microclimate Measurement, 13
Fitness Differences, 97
Fluctuations, 33-34, 36, 160
Fruiting Positions, 69-76

## G

Gaeumannomyces Graminis, 52
Genetic Structure, 86-87, 121, 130, 134
Genetically Modified Crops, 96, 106, 112, 174
Genome-wide Survey, 1
Genotyping Assay, 136
Germplasm Population, 121
Gibberellin, 1, 26, 32, 156, 165, 212, 216
Gossypium Arboretum, 40
Gossypium Hirsutum L., 12-13, 25, 27, 32, 40, 50, 53, 69, 76-77, 121-122, 128-129, 133-135, 148, 154-156, 163, 165, 177, 184, 205-206, 220
Gossypium Raimondii, 1-2, 156, 165, 220
Greenhouse Conditions, 149, 151, 153

## H

Hormone Signal Transduction Pathways, 216-217
Hypocotyls, 26-29, 31, 77, 156, 162, 164-165, 217, 219

## J

Jasmonic Acid, 10, 149, 155, 158, 191, 196-197

## L

Late Planting, 12-13, 76
Li2 Mutation, 40-43, 45-48
Linkage Disequilibrium, 121-122, 124, 126, 131, 134-135, 177

## M

Map Construction, 150
Meta-analysis, 35, 38-39, 106-110, 112, 134, 184
Microscopic Measurement, 27
Molecular Mapping, 50, 134, 148, 154-155

## N

Neonate Larvae Feeding, 79

## P

Phenotypic Evaluation, 122
Phylogenetic Trees, 2

Physiological And Biochemical Response, 199-200
Physiological Factors, 69
Physiological Processes, 1
Physiological Responses, 166, 198-199
Pirimicarb Resistance, 136-137, 140, 145-147
Pooled Cotton Aphid, 136
Prometryn, 198-205
Protein Kinase, 1-2, 4, 10-11, 182, 184, 217
Purpureocillium Lilacinum, 113

**Q**
Quantitative Trait Loci (QTLS), 121

**R**
Relative Kinship, 124, 127
Resistance Management, 78-79, 85, 87, 119

**S**
Salicylic Acid Treatment, 158
Sampling and Processing, 13, 70
Seedling Stage, 93, 123, 149-153, 206-207, 216, 218
Sequence Comparative Analysis, 157-158
Soluble Sugar Contents, 28
Subgenome Expression, 40

Sucrose Cleavage, 69
Sucrose Content, 13, 15-16, 18-20, 23, 69-70, 72-74
Sucrose Metabolism, 13, 17, 20, 23-25, 69-70, 74-77
Sucrose Synthase, 12, 21-22, 24-32, 69, 71, 76-77
Sucrose Synthase Activity Assays, 28

**T**
Technological Substitutes, 33-34, 36
Thirdhand Cigarette Smoke, 60
Tobacco Specific Nitrosamines (TSNAS), 60
Transcriptional Regulators, 40, 42, 46
Transcriptome Sequencing, 4, 10, 185-189, 196
Transgenic Bt Cotton, 36, 88, 96, 171, 175

**U**
Upland Cotton Gene, 156

**V**
Vector Construction, 27, 158
Vector Construction And Genetic Transformation, 158
Verticillium Wilt, 121-125, 127-128, 132-134, 148, 151-155

**W**
Water Stress, 87, 96, 205-208, 212, 217-220
Whitefly Bemisia Tabaci, 89, 97, 104